Fundamentals of Engineering Electromagnetics

Fundamentals of Engineering Electromagnetics

DAVID K. CHENG

CENTENNIAL PROFESSOR EMERITUS, SYRACUSE UNIVERSITY

ADDISON-WESLEY PUBLISHING COMPANY

Reading, Massachusetts • Menlo Park, California • New York
Don Mills, Ontario • Wokingham, England • Amsterdam • Bonn
Sydney • Singapore • Tokyo • Madrid • San Juan • Milan • Paris

This book is in the
Addison-Wesley Series in Electrical Engineering

Sponsoring Editor: Eileen Bernadette Moran
Production Supervisor: Helen Wythe
Production Coordinator: Amy Willcutt
Technical Art Supervisor: Joseph K. Vetere
Illustration: Tech-Graphics
Composition: Doyle Graphics Limited
Art Coordinator: Alena Konecny
Interior Designer: Sally Bindari, Designworks, Inc.
Cover Designer: Peter Blaiwas
Manufacturing Manager: Roy Logan

Library of Congress Cataloging-in-Publication Data

Cheng, David K. (David Keun)
Fundamentals of engineering electromagnetics/David K. Cheng.
p. cm.
Includes bibliographical references and indexes.
ISBN 0-201-56611-7
1. Electric engineering. 2. Electromagnetism. I. Title.
TK153.C442 1993
621.3—dc20 92-23490
CIP

Reprinted with corrections July, 1993

3 4 5 6 7 8 9 10-DO-959493

Preface

This book is designed for use as an undergraduate text on engineering electromagnetics. Electromagnetics is one of the most fundamental subjects in an electrical engineering curriculum. Knowledge of the laws governing electric and magnetic fields is essential to the understanding of the principle of operation of electric and magnetic instruments and machines, and mastery of the basic theory of electromagnetic waves is indispensible to explaining action-at-a-distance electromagnetic phenomena and systems.

Because most electromagnetic variables are functions of three-dimensional space coordinates as well as of time, the subject matter is inherently more involved than electric circuit theory, and an adequate coverage normally requires a sequence of two semester-courses, or three courses in a quarter system. However, some electrical engineering curricula do not schedule that much time for electromagnetics. The purpose of this book is to meet the demand for a textbook that not only presents the fundamentals of electromagnetism in a concise and logical manner, but also includes important engineering application topics such as electric motors, transmission lines, waveguides, antennas, antenna arrays, and radar systems.

I feel that one of the basic difficulties that students have in learning electromagnetics is their failure to grasp the concept of an electromagnetic model. The traditional inductive approach of starting with experimental laws and gradually synthesizing them into Maxwell's equations tends to be fragmented and incohesive; and the introduction of gradient, divergence, and curl operations appears to be ad hoc and arbitrary. On the other hand, the extreme of starting with the entire set of Maxwell's equations, which are of considerable complexity, as fundamental postulates is likely to cause consternation and resistance in students at the outset. The question of the necessity and sufficiency of these general equations is not addressed, and the concept of the electromagnetic model is left vague.

This book builds the electromagnetic model using an *axiomatic approach in steps*[†]—first for static electric fields, then for static magnetic fields, and finally for time-varying fields leading to Maxwell's equations. The mathematical basis for each step is Helmholtz's theorem, which states that a vector field is determined to within an additive constant if both its divergence and its curl are specified everywhere. A physical justification of this theorem may be based on the fact that the divergence of a vector field is a measure of the strength of its flow source and the curl of the field is a measure of the strength of its vortex source. When the strengths of both the flow source and the vortex source are specified, the vector field is determined.

For the development of the electrostatic model in free space, it is only necessary to define a single vector (namely, the electric field intensity **E**) by specifying its divergence and its curl as postulates. All other relations in electrostatics for free space, including Coulomb's law and Gauss's law, can be derived from the two rather simple postulates. Relations in material media can be developed through the concept of equivalent charge distributions of polarized dielectrics.

Similarly, for the magnetostatic model in free space it is necessary to define only a single magnetic flux density vector **B** by specifying its divergence and its curl as postulates; all other formulas can be derived from these two postulates. Relations in material media can be developed through the concept of equivalent current densities. Of course, the validity of the postulates lies in their ability to yield results that conform with experimental evidence.

For time-varying fields, the electric and magnetic field intensities are coupled. The curl **E** postulate for the electrostatic model must be modified to conform with Faraday's law. In addition, the curl **B** postulate for the magnetostatic model must also be modified in order to be consistent with the equation of continuity. We have, then, the four Maxwell's equations that constitute the electromagnetic model. I believe that this gradual development of the electromagnetic model based on Helmholtz's theorem is novel, systematic, pedagogically sound, and more easily accepted by students.

A short Chapter 1 of the book provides some motivation for the study of electromagnetism. It also introduces the source functions, the fundamental field quantities, and the three universal constants for free space in the electromagnetic model. Chapter 2 reviews the basics of vector algebra, vector calculus, and the relations of Cartesian, cylindrical, and spherical coordinate systems. Chapter 3 develops the governing laws and methods of solution of electrostatic problems. Chapter 4 is on steady electric current fields and resistance calculations. Chapter 5 deals with static magnetic fields. Chapter 6, on time-varying electromagnetic fields, starts with Faraday's law of

[†]D. K. Cheng, "An alternative approach for developing introductory electromagnetics," *IEEE Antennas and Propagation Society Newsletter*, pp. 11–13, Feb. 1983.

electromagnetic induction, and leads to Maxwell's equations and wave equations. The characteristics of plane electromagnetic waves are treated in Chapter 7. The theory and applications of transmission lines are studied in Chapter 8. Further engineering applications of electromagnetic fields and waves are discussed in Chapter 9 (waveguides and cavity resonators) and Chapter 10 (antennas, antenna arrays, and radar systems). Much of the material has been adapted and reduced from my larger book, *Field and Wave Electromagnetics*[†], but in this book I have incorporated a number of innovative pedagogical features.

Each chapter of this book starts with an overview section that provides qualitative guidance to the topics to be discussed in the chapter. Throughout the book worked-out examples follow important formulas and quantitative relations to illustrate methods for solving typical problems. Where appropriate, simple exercises with answers are included to test students' ability to handle related situations. At irregular intervals, a group of review questions are inserted after several related sections. These questions serve to provide an immediate feedback of the topics just discussed and to reinforce students' qualitative understanding of the material. Also, a number of pertinent remarks usually follow the review questions. These remarks contain some points of special importance that the students may have overlooked. When new definitions, concepts, or relations are introduced, short notes are added in the margins to emphasize their significance. At the end of each chapter there is a summary with bulleted items summarizing the main topics covered in the chapter. I hope that these pedagogical aids will prove to be useful in helping students learn electromagnetics and its applications.

Many dedicated people, besides the author, are involved in the publication of a book such as this one. I wish to acknowledge the interest and support of Senior Editor Eileen Bernadette Moran and Executive Editor Don Fowley since the inception of this project. I also wish to express my appreciation to Production Supervisor Helen Wythe for her friendly assistance in keeping the production on schedule, as well as to Roberta Lewis, Amy Willcutt, Laura Michaels, and Alena Konecny for their contributions. Jim and Rosa Sullivan of Tech-Graphics were responsible for the illustrations. To them I offer my appreciation for their fine work. Above all, I wish to thank my wife, Enid, for her patience, understanding and encouragement through all phases of my challenging task of completing this book.

D.K.C.

[†]D. K. Cheng, *Field and Wave Electromagnetics*, 2nd ed., Addison-Wesley, Reading, Mass., 1989.

An Introductory Note to the Student

This book is your guide as you embark on the journey of learning engineering electromagnetics. Two questions may immediately come to mind: What is electromagnetics and why is it important? A short answer to the first question is that electromagnetics is the study of the effects of electric charges at rest or in motion. It is important because electromagnetic theory is essential in explaining electromagnetic phenomena and in understanding the principle of operation and the characteristics of electric, magnetic, and electromagnetic engineering devices. Contemporary society relies heavily on electromagnetic devices and systems. Think, for example, of microwave ovens, cathode-ray oscilloscopes, radio, television, radar, satellite communication, automatic instrument-landing systems, and electromagnetic energy conversion (motors and generators).

The basic principles of electromagnetism have been known for over one hundred fifty years. To study a mature scientific subject in an organized and logical way it is necessary to establish a valid theoretical model, which usually consists of a few basic quantities and some fundamental postulates (hypotheses, or axioms). Other relations and consequences are then developed from these postulates. For instance, the study of classical mechanics is based on a theoretical model that defines the quantities mass, velocity, acceleration, force, momentum, and energy. The fundamental postulates of the model are Newton's laws of motion, conservation of momentum, and conservation of energy. These postulates cannot be derived from other theorems; but all other relations and formulas in non-relativistic mechanics (situations where the velocity of motion is negligible compared to the velocity of light) can be developed from these postulates.

Similarly, in our study of electromagnetics we need first to establish an electromagnetic model. Chapter 1 of this book defines the basic quantities of our electromagnetic model. The fundamental postulates are introduced in gradual steps as they are needed when we deal with static electric fields, static magnetic fields, and time-varying fields in separate chapters. Various theorems and other results are then derived from these postulates. Engineering applications of the principles and methods developed throughout the text are explored further in the last several chapters.

In order to express our postulates and derive useful results in succinct forms, we must have the proper mathematical tools. In electromagnetics we most often encounter vectors, quantities that have both a magnitude and a direction. Hence, we must be proficient in vector algebra and vector calculus. These are covered in Chapter 2 on Vector Analysis. We must not only acquire a facility in manipulating vectors, but also understand the physical meaning of the various operations involving vectors. A deficiency in vector analysis in the study of electromagnetism is similar to a deficiency in algebra and calculus in the study of physics. Proficiency in the use of mathematical tools is essential for obtaining fruitful results.

Most likely, you have already studied circuit theory. Circuit theory deals with lumped-parameter systems consisting of components characterized by lumped parameters as resistances, inductances, and capacitances. Voltages and currents are the main system variables. For d-c circuits the system variables are constants, and the governing equations are algebraic equations. The system variables in a-c circuits are time-dependent; they are scalar quantities and are independent of space coordinates. The governing equations are ordinary differential equations. On the other hand, most electromagnetic variables are functions of time as well as of space coordinates. Many are vectors. Even in static cases the governing equations are, in general, partial differential equations. But partial differential equations can be broken up into ordinary differential equations, which you have encountered in courses on physics and linear system analysis. In simple situations where symmetries exist, partial differential equations reduce to ordinary differential equations. The separation of time and space dependence is achieved by the use of phasors.

Because of the need for defining more quantities and using more mathematical manipulations in electromagnetics, you may initially get the impression that electromagnetic theory is abstract. In fact, electromagnetic theory is no more abstract than circuit theory in the sense that the validity of both can be verified by experimentally measured results. We simply have to do more work in order to develop a logical and complete theory that can explain a wider variety of phenomena. The challenge of electromagnetic theory is not in the abstractness of the subject matter but rather in the process of mastering the electromagnetic model and the associated rules of operation.

You will find that each chapter of this book starts with an OVERVIEW section that introduces the topics to be discussed in the chapter. As new definitions, concepts, or relations are introduced, short notes appear in the margins to draw your attention to them. At the end of some related sections, at irregular intervals, are REVIEW QUESTIONS, which serve to provide an immediate feedback of the topics just discussed and to reinforce your qualitative understanding of the material. You should be able to answer these questions with confidence. If not, you should go back to these sections and clear up your doubts. A REMARKS box usually follows the review questions. It contains some points of special importance that you may have overlooked, but that you should understand and remember. At the conclusion of each chapter there is a SUMMARY section that itemizes the more important results obtained in the chapter. Its function is to emphasize the significance of these results without repeating the mathematical formulas.

Throughout the book new terms and important statements are printed in ***boldface italic***, and the more important formulas are boxed. Worked-out examples are provided to illustrate methods for solving typical problems. Where appropriate, simple exercises with answers are included. You should do these exercises as they occur, so that you can see if you have mastered the basic quantitative skills just presented. The problems at the end of a chapter are used to extend what you have learned from the chapter and to test your ability in tackling new situations. Answers to odd-numbered problems, included at the end of the book, give you a self-check and reassurance on your progress.

The learning of electromagnetics is an intellectual journey; this book is your guide, but you must bring along your dedication and perseverance. As you explore the territory of engineering electromagnetics, we hope you will have a stimulating and rewarding experience.

The author.

Learning is not attained by chance;
it must be sought for with ardor and attended to with diligence.

—Abigail Adams
(in letter to John Quincy Adams, 1780)

Contents

CHAPTER 3 STATIC ELECTRIC FIELDS 72

CHAPTER 4 STEADY ELECTRIC CURRENTS 150

CHAPTER 5 STATIC MAGNETIC FIELDS 170

CHAPTER 6 TIME-VARYING FIELDS AND MAXWELL'S EQUATIONS 228

CHAPTER 7 PLANE ELECTROMAGNETIC WAVES 272

CHAPTER 8 TRANSMISSION LINES 336

CHAPTER 9 Waveguides and Cavity Resonators 386

CHAPTER 10 Antennas and Antenna Arrays 426

APPENDIXES A Symbols and Units

CHAPTER 1

1-1 OVERVIEW

Definition of electromagnetics

Electromagnetics is the study of the electric and magnetic phenomena caused by electric charges at rest or in motion. The existence of electric charges was discovered more than two and a half milleniums ago by a Greek astronomer and philosopher Thales of Miletus. He noted that an amber rod, after being rubbed with silk or wool, attracted straw and small bits of paper. He attributed this mysterious property to the amber rod. The Greek word for amber is *elektron*, from which was derived the words *electron*, *electronics*, *electricity*, and so on.

Two kinds of charges: positive and negative

Field: a spatial distribution of a quantity

From elementary physics we know that there are two kinds of charges: positive and negative. Both positive and negative charges are sources of an electric field. Moving charges produce a current, which gives rise to a magnetic field. Here we tentatively speak of electric field and magnetic field in a general way; more definitive meanings will be attached to these terms later. A ***field*** is a spatial distribution of a quantity, which may or may not be a function of time. A time-varying electric field is accompanied by a magnetic field, and vice versa. In other words, time-varying electric and magnetic fields are coupled, resulting in an electromagnetic field. Under certain conditions, time-dependent electromagnetic fields produce waves that radiate from the source.

Fields and waves help explain action at a distance.

The concept of fields and waves is essential in the explanation of action at a distance. For instance, we learned from elementary mechanics that masses attract each other. This is why objects fall toward the Earth's surface. But since there are no elastic strings connecting a free-falling object and the Earth, how do we

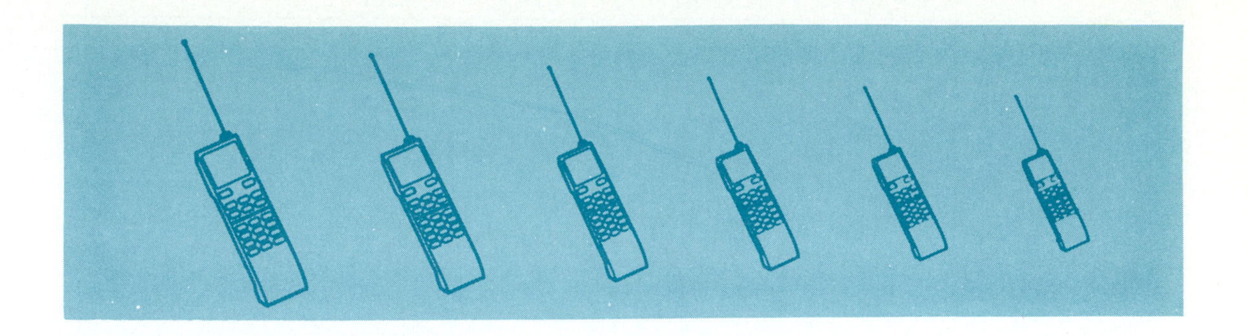

The Electromagnetic Model

explain this pheomenon? We explain this action-at-a-distance phenomenon by postulating the existence of a gravitational field. Similarly, the possibilities of satellite communication and of receiving signals from space probes millions of miles away can be explained only by postulating the existence of electric and magnetic fields and electromagnetic waves. In this book, *Fundamentals of Engineering Electromagnetics*, we study the fundamental laws of electromagnetism and some of their engineering applications.

Circuit theory cannot explain mobile phone communication.

A simple situation will illustrate the need for electromagnetic-field concepts. Figure 1-1 depicts a mobile telephone with an attached antenna. On transmit, a source at the base feeds the antenna with a message-carrying current at an appropriate carrier frequency. From a circuit-theory point of view, the source feeds into an open circuit because the upper tip of the antenna is not connected to anything physically; hence no current would flow, and nothing would happen. This viewpoint, of course, cannot explain why communication can be established between moving telephone units. Electromagnetic concepts must be used. We shall see in Chapter 10 that when the length of the antenna is an appreciable part of the carrier wavelength, a nonuniform current will flow along the open-ended antenna. This current radiates a time-varying electromagnetic field in space, which propagates as an electromagnetic wave and induces currents in other antennas at a distance. The message is then detected in the receiving unit.

Constructing a model

In this first chapter we begin the task of constructing an electromagnetic model, from which we shall develop the subject of engineering electromagnetics.

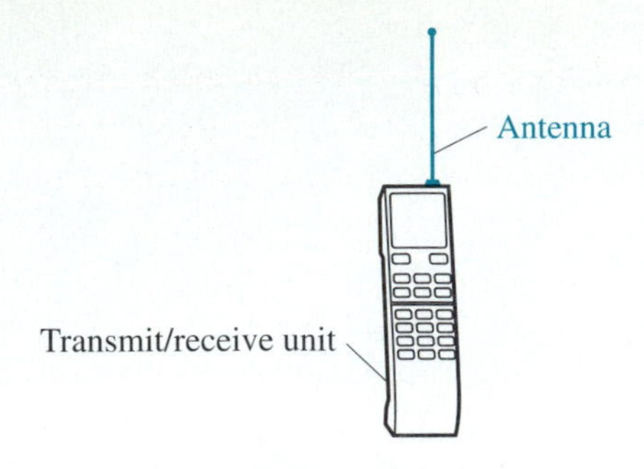

FIGURE 1-1 A mobile telephone.

1-2 THE ELECTROMAGNETIC MODEL

Inductive and deductive approaches

There are two approaches in the development of a scientific subject: the inductive approach and the deductive approach. Using the inductive approach, one follows the historical development of the subject, starting with the observations of some simple experiments and inferring from them laws and theorems. It is a process of reasoning from particular phenomena to general principles. The deductive approach, on the other hand, postulates a few fundamental relations for an idealized model. The postulated relations are axioms, from which particular laws and theorems can be derived. The validity of the model and the axioms is verified by their ability to predict consequences that check with experimental observations. In this book we prefer to use the deductive or axiomatic approach because it is more concise and enables the development of the subject of electromagnetics in an orderly way.

Three essential steps are involved in building a theory based on an idealized model.

Steps for developing a theory from an idealized model

STEP 1 Define some basic quantities germane to the subject of study.

STEP 2 Specify the rules of operation (the mathematics) of these quantities.

STEP 3 Postulate some fundamental relations. (These postulates or laws are usually based on numerous experimental observations acquired under controlled conditions and synthesized by brilliant minds.)

The circuit model

A familiar example is the circuit theory built on a *circuit model* of ideal sources and pure resistances, inductances, and capacitances. In this case the basic quantities are voltages (V), currents (I), resistances (R), inductances (L), and capacitances (C); the rules of operations are those of algebra, ordinary differential equations, and Laplace transformation; and the fundamental postulates are Kirchhoff's voltage and current laws. Many relations and formulas can be derived from this basically rather simple model, and the responses of very elaborate networks can be determined. The validity and value of the model have been amply demonstrated.

The three steps for developing an electromagnetic theory from an electromagnetic model

In a like manner, an electromagnetic theory can be built on a suitably chosen electromagnetic model. In this section we shall take the first step of defining the basic quantities of electromagnetics. The second step, the rules of operation, encompasses vector algebra, vector calculus, and partial differential equations. The fundamentals of vector algebra and vector calculus will be discussed in Chapter 2 (Vector Analysis), and the techniques for solving partial differential equations will be introduced when these equations arise later in the book. The third step, the fundamental postulates, will be presented in three substeps as we deal with static electric fields, static magnetic fields, and electromagnetic fields, respectively.

Basic quantities in the electromagnetic model: source quantities and field quantities

The quantities in our electromagnetic model can be divided roughly into two categories: source quantities and field quantities. The source of an electromagnetic field is invariably electric charges at rest or in motion. However, an electromagnetic field may cause a redistribution of charges, which will, in turn, change the field; hence the separation between the cause and the effect is not always so distinct.

Electric charges

We use the symbol q (sometimes Q) to denote ***electric charge***. Electric charge is a fundamental property of matter and exists only in positive or negative integral multiples of the charge on an electron, $-e$.

$$\boxed{e = 1.60 \times 10^{-19} \qquad \text{(C)},} \tag{1-1}$$

Unit of charge: coulomb (C)

where C is the abbreviation of the unit of charge, coulomb.[†] It is named after the French physicist Charles A. de Coulomb, who formulated Coulomb's law in 1785. (Coulomb's law will be discussed in Chapter 3.) A coulomb is a very large unit for electric charge; it takes $1/(1.60 \times 10^{-19})$ or 6.25 million trillion electrons to make up -1(C). In fact, two 1-(C) charges 1(m) apart will exert a force of approximately 1 million tons on each other. Some other physical constants for the electron are listed in Appendix B–2.

The principle of ***conservation of electric charge***, like the principle of conservation of energy, is a fundamental postulate or law of physics. It states that electric charge is conserved; that is, it can neither be created nor be destroyed. This is a law of nature and cannot be derived from other principles or relations.

Electric charges can move from one place to another and can be redistributed under the influence of an electromagnetic field; but the algebraic sum of the positive and negative charges in a closed (isolated) system remains

[†] The system of units will be discussed in Section 1–3.

Conservation of electric charge is a fundamental postulate of physics.

unchanged. ***The principle of conservation of electric charge must be satisfied at all times and under any circumstances.*** Any formulation or solution of an electromagnetic problem that violates the principle of conservation of electric charge *must be* incorrect.

Although, in a microscopic sense, electric charge either does or does not exist at a point in a discrete manner, these abrupt variations on an atomic scale are unimportant when we consider the electromagnetic effects of large aggregates of charges. In constructing a macroscopic or large-scale theory of electromagnetism we find that the use of smoothed-out average density functions yields very good results. (The same approach is used in mechanics where a smoothed-out mass density function is defined, in spite of the fact that mass is associated only with elementary particles in a discrete manner on an atomic scale.) We define a ***volume charge density***, ρ_v, as a source quantity as follows:

$$\rho_v = \lim_{\Delta v \to 0} \frac{\Delta q}{\Delta v} \qquad (\text{C/m}^3), \tag{1-2}$$

Volume, surface, and line charge densities— average densities in the macroscopic sense

where Δq is the amount of charge in a very small volume Δv. How small should Δv be? It should be small enough to represent an accurate variation of ρ_v but large enough to contain a very large number of discrete charges. For example, an elemental cube with sides as small as 1 micron (10^{-6} m or 1 μm) has a volume of 10^{-18}(m^3), which will still contain about 10^{11} (100 billion) atoms. A smoothed-out function of space coordinates, ρ_v, defined with such a small Δv is expected to yield accurate macroscopic results for nearly all practical purposes.

In some physical situations an amount of charge Δq may be identified with an element of surface Δs, or an element of line $\Delta \ell$. In such cases it will be more appropriate to define a ***surface charge density***, ρ_s, or a ***line charge density***, ρ_ℓ:

$$\rho_s = \lim_{\Delta s \to 0} \frac{\Delta q}{\Delta s} \qquad (\text{C/m}^2), \tag{1-3}$$

$$\rho_\ell = \lim_{\Delta \ell \to 0} \frac{\Delta q}{\Delta \ell} \qquad (\text{C/m}). \tag{1-4}$$

Charge densities are point functions.

Except for certain special situations, charge densities vary from point to point; hence ρ_v, ρ_s, and ρ_ℓ are, in general, *point functions* of space coordinates.

Current is the rate of change of charge with respect to time; that is,

$$I = \frac{dq}{dt} \qquad (\text{C/s or A}), \tag{1-5}$$

where I itself may be time-dependent. The unit of current is coulomb per second (C/s), which is the same as ampere (A). A current must flow through a

Current is not a point function, but current density is.

finite area (a conducting wire of a finite cross section, for instance); hence it is not a point function. In electromagnetics we define a vector point function ***current density*, J**, which measures the amount of current flowing through a unit area normal to the direction of current flow. The boldfaced **J** is a vector whose magnitude is the current per unit area (A/m^2) and those direction is the direction of current flow.

Four fundamental electromagnetic field quantities: E, B, D, H

There are four fundamental *vector* field quantities in electromagnetics: ***electric field intensity* E**, ***electric flux density*** (or ***electric displacement***) **D**, ***magnetic flux density* B**, and ***magnetic field intensity* H**. The definition and physical significance of these quantities will be explained fully when they are introduced later in the book. At this time we want only to establish the following. Electric field intensity **E** is the only vector needed in discussing electrostatics (effects of stationary electric charges) in free space; it is defined as the electric force on a unit test charge. Electric displacement vector **D** is useful in the study of electric field in material media, as we shall see in Chapter 3. Similarly, magnetic flux density **B** is the only vector needed in discussing magnetostatics (effects of steady electric currents) in free space and is related to the magnetic force acting on a charge moving with a given velocity. The magnetic field intensity vector **H** is useful in the study of magnetic field in material media. The definition and significance of **B** and **H** will be discussed in Chapter 5.

The four fundamental electromagnetic field quantities, together with their units, are tabulated in Table 1-1. In Table 1-1, V/m is volt per meter, and T stands for tesla or volt-second per square meter. When there is no time variation (as in static, steady, or stationary cases), the electric field quantities **E** and **D** and the magnetic field quantities **B** and **H** form two separate vector pairs. In time-dependent cases, however, electric and magnetic field quantities

TABLE **1-1** FUNDAMENTAL ELECTROMAGNETIC FIELD QUANTITIES

Symbols and Units for Field Quantities	Field Quantity	Symbol	Unit
Electric	Electric field intensity	**E**	V/m
	Electric flux density (Electric displacement)	**D**	C/m^2
Magnetic	Magnetic flux density	**B**	T
	Magnetic field intensity	**H**	A/m

are coupled; that is, time-varying **E** and **D** will give rise to **B** and **H**, and vice versa. All four quantities are point functions. Material (or medium) properties determine the relations between **E** and **D** and between **B** and **H**. These relations are called the ***constitutive relations*** of a medium and will be examined later.

The principal objective of studying electromagnetism is to understand the interaction between charges and currents at a distance based on the electromagnetic model. Fields and waves (time- and space-dependent fields) are basic conceptual quantities of this model. Fundamental postulates, to be enunciated in later chapters, will relate **E**, **D**, **B**, **H** and the source quantities; and derived relations will lead to the explanation and prediction of electromagnetic phenomena.

1-3 SI UNITS AND UNIVERSAL CONSTANTS

A measurement of any physical quantity must be expressed as a number followed by a unit. Thus we may talk about a length of three meters, a mass of two kilograms, and a time period of ten seconds. To be useful, a unit system should be based on some fundamental units of convenient (practical) sizes. In mechanics, all quantities can be expressed in terms of three basic units (for length, mass, and time). In electromagnetics a fourth basic unit (for current) is needed. The ***SI (International System of Units)*** is an ***MKSA system*** built from the four fundamental units listed in Table 1-2. All other units used in electromagnetics, including those appearing in Table 1-1, are derived units expressible in terms of *meters*, *kilograms*, *seconds*, and *amperes*. For example, the unit for charge, coulomb (C), is ampere-second (A · s); the unit for electric field intensity (V/m) is $kg \cdot m/A \cdot s^3$; and the unit for magnetic flux density, tesla (T), is $kg/A \cdot s^2$. More complete tables of the units for various quantities are given in Appendix A.

The SI, or MKSA, units

In our electromagnetic model there are three universal constants, in addition to the field quantities listed in Table 1-1. They relate to the

TABLE **1-2** FUNDAMENTAL SI UNITS

Quantity	Unit	Abbreviation
Length	meter	m
Mass	kilogram	kg
Time	second	s
Current	ampere	A

Three universal constants in electromagnetic model

properties of the free space (vacuum). They are as follows: ***velocity of electromagnetic wave*** (including light) in free space, c; ***permittivity*** of free space, ϵ_0; and ***permeability*** of free space, μ_0. Many experiments have been performed for precise measurement of the velocity of light, to many decimal places. For our purpose it is sufficient to remember that

$$c \cong 3 \times 10^8 \quad \text{(m/s).} \qquad \text{(in free space)} \tag{1-6}$$

The other two constants, ϵ_0 and μ_0, pertain to electric and magnetic phenomena, respectively: ϵ_0 is the proportionality constant between the electric flux density **D** and the electric field intensity **E** in free space, such that

$$\mathbf{D} = \epsilon_0 \mathbf{E}; \qquad \text{(in free space)} \tag{1-7}$$

μ_0 is the proportionality constant between the magnetic flux density **B** and the magnetic field intensity **H** in free space, such that

$$\mathbf{H} = \frac{1}{\mu_0}\mathbf{B}. \qquad \text{(in free space)} \tag{1-8}$$

The values of ϵ_0 and μ_0 are determined by the choice of the unit system, and they are not independent. In the ***SI system***, which is almost universally adopted for electromagnetics work, the permeability of free space is chosen to be

$$\mu_0 = 4\pi \times 10^{-7} \quad \text{(H/m),} \qquad \text{(in free space)} \tag{1-9}$$

where H/m stands for henry per meter. With the values of c and μ_0 fixed in Eqs. (1–6) and (1–9) the value of the permittivity of free space is then derived from the following relationships:

$$c = \frac{1}{\sqrt{\epsilon_0 \mu_0}} \quad \text{(m/s),} \qquad \text{(in free space)} \tag{1-10}$$

or

$$\begin{aligned}\epsilon_0 &= \frac{1}{c^2 \mu_0} \cong \frac{1}{36\pi} \times 10^{-9} \\ &\cong 8.854 \times 10^{-12} \quad \text{(F/m),}\end{aligned} \qquad \text{(in free space)} \tag{1-11}$$

TABLE **1-3** UNIVERSAL CONSTANTS IN SI UNITS

Universal Constants	Symbol	Value	Unit
Velocity of light in free space	c	3×10^8	m/s
Permeability of free space	μ_0	$4\pi \times 10^{-7}$	H/m
Permittivity of free space	ϵ_0	$\frac{1}{36\pi} \times 10^{-9}$	F/m

where F/m is the abbreviation for farad per meter. The three universal constants and their values are summarized in Table 1-3.

Now that we have defined the basic quantities and the universal constants of the electromagnetic model, we can develop the various subjects in electromagnetics. But, before we do that, we must be equipped with the appropriate mathematical tools. In the following chapter we discuss the basic rules of operation for vector algebra and vector calculus.

SUMMARY

This chapter laid the foundation for our study of engineering electromagnetism. We adopt a deductive or axiomatic approach and construct an electromagnetic model. Basic source quantities (charge, charge densities, current density) and field quantities (**E**, **D**, **B**, **H**) are defined, the unit system (SI) is specified, and the three universal constants for free space (μ_0, c, ε_0) are given. With this framework we can develop the various topics by introducing the fundamental postulates in subsequent chapters. We shall do this gradually, in steps. But first we need to be familiar with the mathematics that will be used to relate the different quantities. A secure knowledge of vector analysis is essential. Chapter 2 presents the required material on vector algebra and vector calculus.

REVIEW QUESTIONS

Q.1-1 What is electromagnetics?

Q.1-2 Describe two phenomena or situations, other than the mobile telephone depicted in Fig. 1-1, that cannot be adequately explained by circuit theory.

Q.1-3 What are the three essential steps in building an idealized model for the study of a scientific subject?

Q.1-4 What are the source quantities in the electromagnetic model?

Q.1-5 What is meant by a *point function*? Is charge density a point function? Is current a point function?

Q.1-6 What are the four fundamental SI units in electromagnetics?

Q.1-7 What are the four fundamental field quantities in the electromagnetic model? What are their units?

Q.1-8 What are the three universal constants in the electromagnetic model, and what are their relations?

CHAPTER 2

2-1 OVERVIEW

In our electromagnetic model some of the quantities (such as charge, current, and energy) are scalars; and some others (such as electric and magnetic field intensities) are vectors. Both scalars and vectors can be functions of time and position. At a given time and position, a ***scalar*** is completely specified by its magnitude (positive or negative, together with its unit). Thus we can specify, for instance, a charge of $-1(\mu C)$ at a certain location at $t=0$. The specification of a ***vector*** at a given location and time, on the other hand, requires both a magnitude and a direction. How do we specify the direction of a vector? In a three-dimensional space, three numbers are needed, and these numbers depend on the choice of a coordinate system.

Scalar

Vector

It is important to note that physical laws and theorems relating various scalar and vector quantities must hold irrespective of a coordinate system. *The general expressions of the laws of electromagnetism do not require the specification of a coordinate system.* A particular coordinate system is chosen only when a problem of a given geometry is to be analyzed. For example, if we are to determine the magnetic field at the center of a current-carrying wire loop, it is more convenient to use rectangular coordinates if the loop is rectangular, whereas polar coordinates will be more appropriate if the loop is circular in shape. The basic electromagnetic relation governing the solution of such a problem is the same for both geometries.

Coordinate-system independence

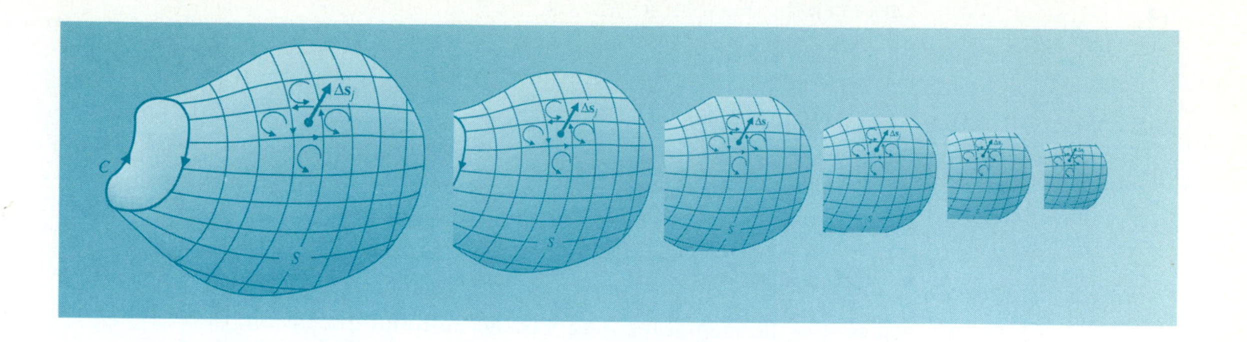

Vector Analysis

Because many electromagnetic quantities are vectors, we must be able to handle (add, subtract, and multiply) them with ease. In order to express specific results in a three-dimensional space, we must choose a suitable coordinate system. In this chapter we will discuss the three most common orthogonal coordinate systems: Cartesian, cylindrical, and spherical coordinates. We will see how to resolve a given vector into components in these coordinates and how to transform from one coordinate system to another.

The use of certain differential operators enables us to express the fundamental postulates and other formulas in electromagnetics in a succinct and general manner. We will discuss the significance of gradient, divergence, and curl operations and prove divergence and Stokes's theorems.

This chapter on vector analysis deals with three main topics:

1. Vector algebra—addition, subtraction, and multiplication of vectors.
2. Orthogonal coordinate systems—Cartesian, cylindrical, and spherical coordinates.
3. Vector calculus—differentiation and integration of vectors; gradient, divergence, and curl operations.

We also prove two important null identities involving repeated applications of differential operators.

2-2 VECTOR ADDITION AND SUBTRACTION

We know that a vector has a magnitude and a direction. A vector **A** can be written as

$$\mathbf{A} = \mathbf{a}_A A, \tag{2-1}$$

where A is the magnitude (and has the unit and dimension) of **A**:

$$A = |\mathbf{A}|, \tag{2-2}$$

which is a scalar. $\mathbf{a}_A$ is a dimensionless unit vector having a unity magnitude; it specifies the direction of **A**. We can find $\mathbf{a}_A$ from the vector **A** by dividing it by its magnitude.

Finding the unit vector from a vector

$$\mathbf{a}_A = \frac{\mathbf{A}}{|\mathbf{A}|} = \frac{\mathbf{A}}{A}. \tag{2-3}$$

The vector **A** can be represented graphically by a directed straight-line segment of a length $|\mathbf{A}| = A$ with its arrowhead pointing in the direction of $\mathbf{a}_A$, as shown in Fig. 2-1.

Two vectors are equal if they have the same magnitude and the same direction, even though they may be displaced in space. Since it is difficult to write boldfaced letters by hand, it is a common practice, in writing, to use an arrow or a bar over a letter ($\vec{A}$ or $\bar{A}$) or a wiggly line under a letter ($\underset{\sim}{A}$) to distinguish a vector from a scalar. *This distinguishing mark, once chosen, should never be omitted whenever and wherever vectors are written.*

Distinguishing marks for vectors

Two vectors **A** and **B**, which are not in the same direction nor in opposite directions, such as given in Fig. 2-2(a), determine a plane. Their sum is another vector **C** in the same plane. $\mathbf{C} = \mathbf{A} + \mathbf{B}$ can be obtained graphically in two ways:

1. By the parallelogram rule: The resultant **C** is the diagonal vector of the

FIGURE 2-1 Graphical representation of vector **A**.

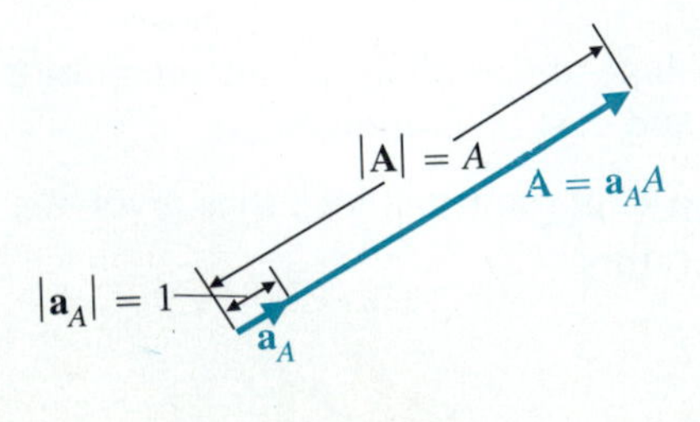

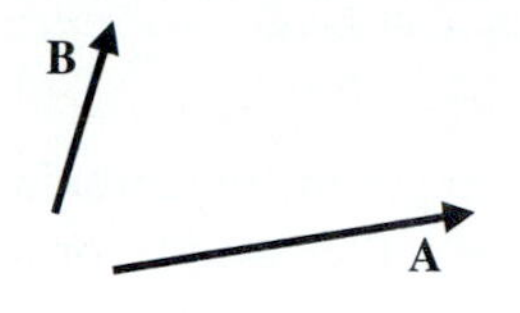

(a) Two vectors, **A** and **B**.

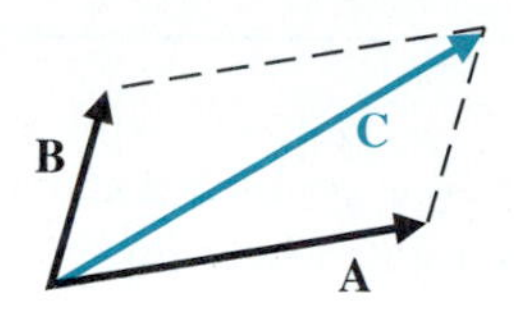

(b) Parallelogram rule.

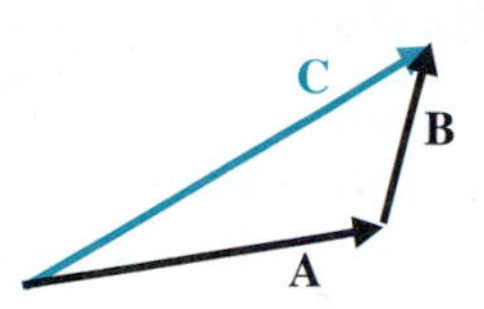

(c) Head-to-tail rule, **A** + **B**.

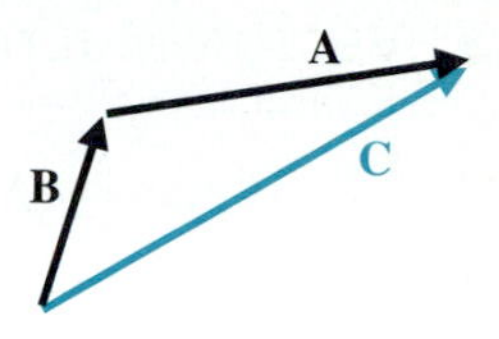

(d) Head-to-tail rule, **B** + **A**.

FIGURE 2-2 Vector addition, **C** = **A** + **B** = **B** + **A**.

parallelogram formed by **A** and **B** drawn from the same point, as shown in Fig. 2-2(b).

2. By the head-to-tail rule: The head of **A** connects to the tail of **B**. Their sum **C** is the vector drawn from the tail of **A** to the head of **B**; and vectors **A**, **B**, and **C** form a triangle, as shown in Fig. 2-2(c). **C** = **A** + **B** = **B** + **A**, as illustrated graphically in Fig. 2-2(d).

Vector subtraction can be defined in terms of vector addition in the following way:

$$\mathbf{A} - \mathbf{B} = \mathbf{A} + (-\mathbf{B}), \tag{2-4}$$

where $-\mathbf{B}$ is the negative of vector **B**. This is illustrated in Fig. 2-3.

NOTE: It is meaningless to add or subtract a scalar from a vector, or to add or subtract a vector from a scalar.

■ **EXERCISE 2.1** Three vectors **A**, **B**, and **C**, drawn in a head-to-tail fashion, form three sides of a triangle. What is **A** + **B** + **C**? What is **A** + **B** − **C**?

ANS. 0, $-2\mathbf{C}$.

FIGURE 2-3 Vector subtraction, **D** = **A** − **B** = **A** + (−**B**).

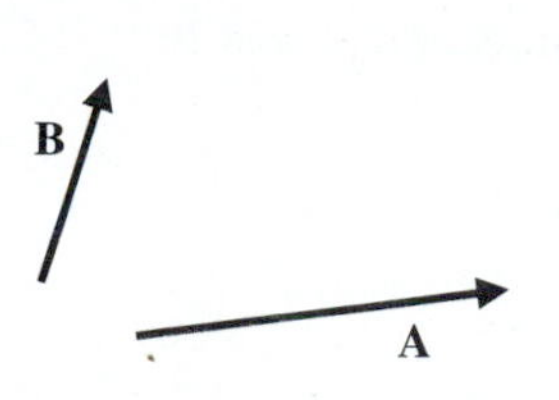

(a) Two vectors, **A** and **B**.

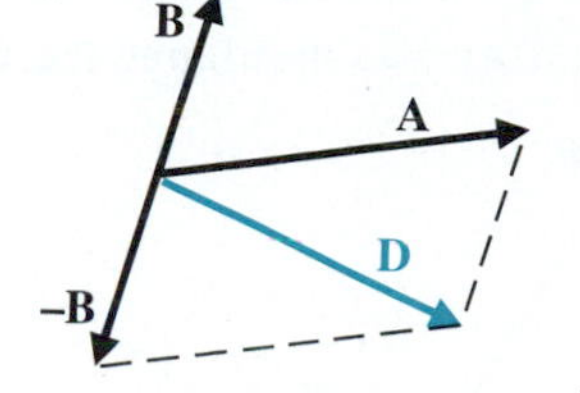

(b) Parallelogram rule.

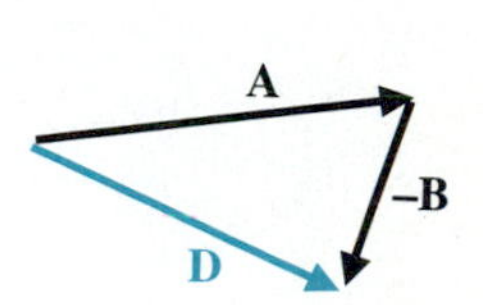

(c) Head-to-tail rule.

2-3 VECTOR MULTIPLICATION

Multiplication of a vector **A** by a positive scalar k changes the magnitude of **A** by k times without changing its direction (k can be either greater or less than 1).

$$k\mathbf{A} = \mathbf{a}_A(kA). \tag{2-5}$$

It is not sufficient to say "the multiplication of one vector by another" or "the product of two vectors" because there are two distinct and very different types of products of two vectors. They are (1) the scalar or dot product, and (2) the vector or cross product. These will be defined in the following subsections.

2-3.1 SCALAR OR DOT PRODUCT

The scalar or dot product of two vectors **A** and **B** is denoted by $\mathbf{A} \cdot \mathbf{B}$ ("**A** dot **B**"). The result of the dot product of two vectors is a scalar. It is equal to the product of the magnitudes of **A** and **B** and the cosine of the angle between them. Thus,

Definition of scalar or dot product of two vectors

$$\boxed{\mathbf{A} \cdot \mathbf{B} \triangleq AB \cos \theta_{AB}.} \tag{2-6}$$

In Eq. (2-6) the symbol $\triangleq$ signifies "equal by definition," and θ_{AB} is the *smaller* angle between **A** and **B** and is less than π radians (180°), as indicated in Fig. 2-4.

From the definition in Eq. (2-6) we see that the dot product of two vectors (1) is less than or equal to the product of their magnitudes; (2) can be either a positive or a negative quantity, depending on whether the angle between them is smaller or larger than $\pi/2$ radians (90°); (3) is equal to the product of the magnitude of one vector and the projection of the other vector

FIGURE 2-4 Illustrating the dot product of **A** and **B**.

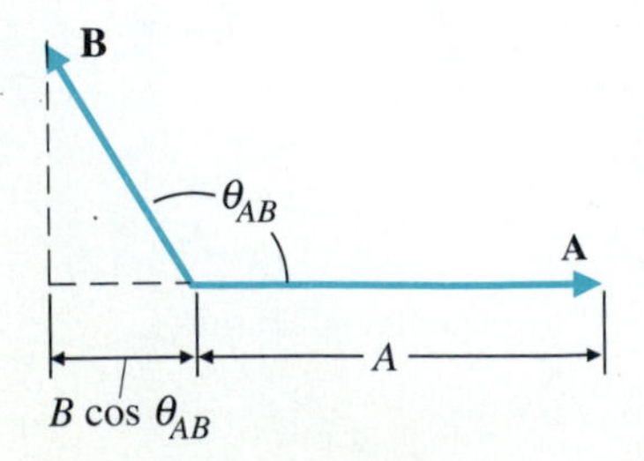

upon the first one; and (4) is zero when the vectors are perpendicular to each other.

From Eq. (2-6) we can see that

Dot product is commutative.

$$\mathbf{A}\cdot\mathbf{B} = \mathbf{B}\cdot\mathbf{A}. \tag{2-7}$$

Thus the order of the vectors in a dot product is not important. (The dot product is commutative.) Also,

$$\mathbf{A}\cdot\mathbf{A} = A^2 \tag{2-8}$$

or

Finding the magnitude of a vector

$$A = |\mathbf{A}|\, {}^{+}\!\sqrt{\mathbf{A}\cdot\mathbf{A}}. \tag{2-9}$$

Equation (2-9) enables us to find the magnitude of a vector when the expression of the vector is given in *any coordinate system.* We simply form the dot product of the vector with itself, $(\mathbf{A}\cdot\mathbf{A})$, and take the positive square root of the scalar result.

EXAMPLE 2-1

Use vectors to prove the law of cosines for a triangle.

SOLUTION

The law of cosines is a scalar relationship that expresses the length of a side of a triangle in terms of the lengths of the two other sides and the angle between them. For Fig. 2-5 the law of cosines states that

$$C = \sqrt{A^2 + B^2 - 2AB\cos\alpha}. \tag{2-10}$$

We prove this by considering the sides as vectors; that is,

$$\mathbf{C} = \mathbf{A} + \mathbf{B}.$$

FIGURE 2-5 Illustrating Example 2-1.

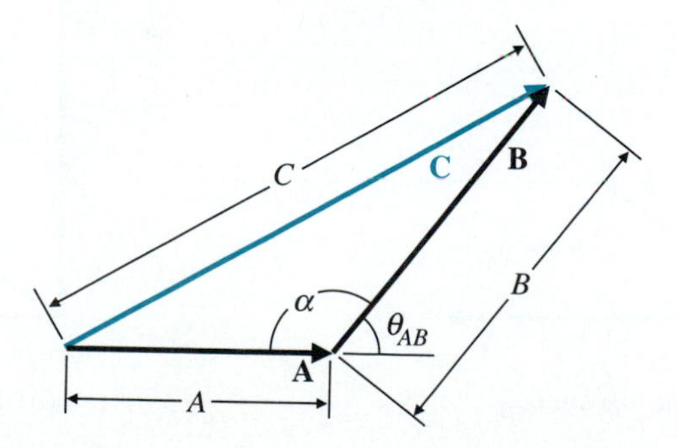

In order to find the magnitude of **C** we take the dot product of **C** with itself, as in Eq. (2-8).

$$\begin{aligned} C^2 = \mathbf{C}\cdot\mathbf{C} &= (\mathbf{A} + \mathbf{B})\cdot(\mathbf{A} + \mathbf{B}) \\ &= \mathbf{A}\cdot\mathbf{A} + \mathbf{B}\cdot\mathbf{B} + 2\mathbf{A}\cdot\mathbf{B} \\ &= A^2 + B^2 + 2AB\cos\theta_{AB}. \end{aligned}$$

Since θ_{AB} is, by definition, the *smaller* angle between **A** and **B** and is equal to $(180° - \alpha)$ we know that $\cos\theta_{AB} = \cos(180° - \alpha) = -\cos\alpha$. Therefore,

$$C^2 = A^2 + B^2 - 2AB\cos\alpha. \qquad (2\text{-}11)$$

The square root of both sides of Eq. (2-11) gives the law of cosines in Eq. (2-10). Note that no coordinate system needs to be specified in this problem.

2-3.2 VECTOR OR CROSS PRODUCT

A second kind of vector multiplication is the vector or cross product. Given two vectors **A** and **B**, the cross product, denoted by **A** × **B** ("**A** cross **B**"), is another vector defined by

Definition of vector or cross product of two vectors

$$\boxed{\mathbf{A} \times \mathbf{B} \triangleq \mathbf{a}_n AB\sin\theta_{AB},} \qquad (2\text{-}12)$$

where θ_{AB} is the *smaller* angle between the vectors **A** and **B** ($\leqslant \pi$), and $\mathbf{a}_n$ is a unit vector normal (perpendicular) to the plane containing **A** and **B**. The direction of $\mathbf{a}_n$ follows that of the thumb of a *right hand* when the fingers rotate from **A** to **B** through the angle θ_{AB} (the *right-hand rule*). This is illustrated in Fig. 2-6. From the figure we can see that $B\sin\theta_{AB}$ is the height of the parallelogram formed by the vectors **A** and **B**. We also recognize that the

FIGURE 2-6 Cross product of **A** and **B**, **A** × **B**.

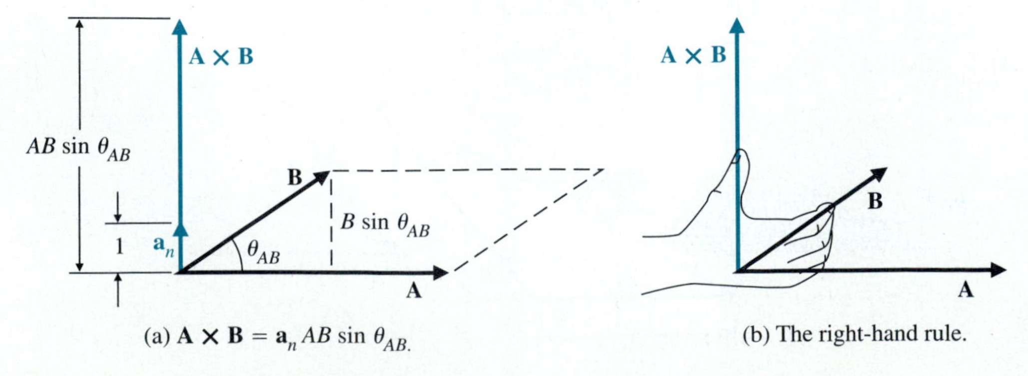

(a) $\mathbf{A} \times \mathbf{B} = \mathbf{a}_n AB\sin\theta_{AB}$.

(b) The right-hand rule.

quantity $AB \sin \theta_{AB}$, which is nonnegative (positive or 0), is numerically equal to the area of the parallelogram. Thus the cross product $\mathbf{A} \times \mathbf{B}$ results in another vector, whose direction $\mathbf{a}_n$ is obtained by the right-hand rule in rotating from **A** to **B**, and whose magnitude is equal to the area of the parallelogram found by **A** and **B**.

Using the definition in Eq. (2-12) and following the right-hand rule, we find that

Vector product is not commutative.

$$\mathbf{B} \times \mathbf{A} = -\mathbf{A} \times \mathbf{B}. \tag{2-13}$$

Hence the cross product is *not* commutative; and reversing the order of two vectors in a cross product changes the sign of the product.

2-3.3 PRODUCTS OF THREE VECTORS

There are two kinds of products of three vectors: (1) scalar triple product, and (2) vector triple product.

1. Scalar triple product. This is the dot product of one vector with the result of the cross product of two other vectors. A typical form of this is

   $\mathbf{A} \cdot (\mathbf{B} \times \mathbf{C})$,

   where **A**, **B**, and **C** are three arbitrary vectors, as illustrated in Fig. 2-7(a).

   According to Eq. (2-12), the cross product $\mathbf{B} \times \mathbf{C}$ has a magnitude $BC \sin \alpha$, which is equal to the area of the shaded parallelogram formed by sides **B** and **C**. The direction of $\mathbf{B} \times \mathbf{C}$ is $\mathbf{a}_n$, a unit normal vector perpendicular to the plane containing **B** and **C**, as shown. The given triple product is then

$$\mathbf{A} \cdot (\mathbf{B} \times \mathbf{C}) = (\mathbf{A} \cdot \mathbf{a}_n) BC \sin \alpha. \tag{2-14}$$

FIGURE 2-7 Illustrating scalar triple products.

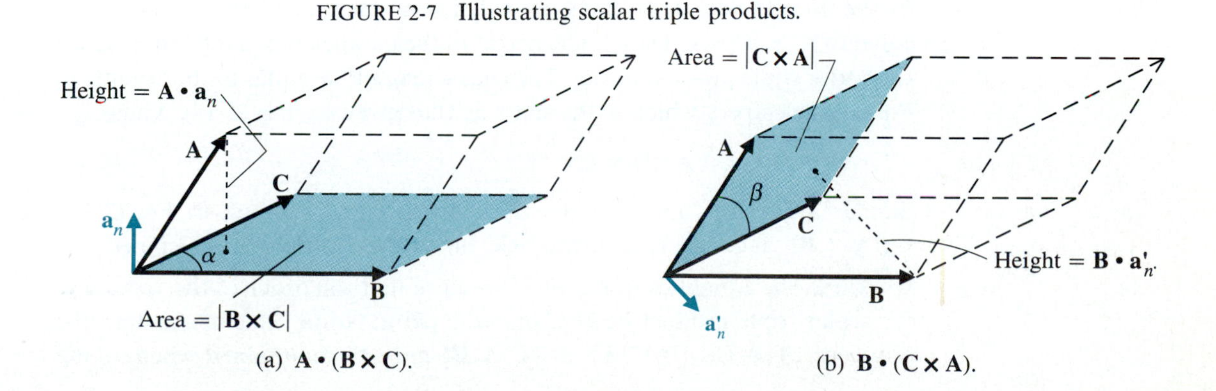

In Eq. (2-14), $(\mathbf{A} \cdot \mathbf{a}_n)$ is a scalar whose magnitude is the projection of **A** in the direction of the unit normal vector $\mathbf{a}_n$. Thus $(\mathbf{A} \cdot \mathbf{a}_n)$ is numerically equal to the height of the parallelepiped formed by the vectors **A**, **B**, and **C**, and the given scalar triple product is equal to the volume of the parallelepiped.

2. Vector triple product. This is the cross product of one vector with the result of the cross product of two other vectors. A typical form of this is

$$\mathbf{A} \times (\mathbf{B} \times \mathbf{C}).$$

This case is more complicated, and we will not attempt a general derivation here. However, it can be expanded quite simply when a coordinate system is given. (See Problem P.2-9.) We will discuss its use when the occasion arises in the future.

EXAMPLE 2-2

Given three vectors **A**, **B**, and **C**, prove the following relation for scalar triple products:

$$\boxed{\mathbf{A} \cdot (\mathbf{B} \times \mathbf{C}) = \mathbf{B} \cdot (\mathbf{C} \times \mathbf{A}) = \mathbf{C} \cdot (\mathbf{A} \times \mathbf{B}).} \tag{2-15}$$

SOLUTION

We have found that the first scalar triple product $\mathbf{A} \cdot (\mathbf{B} \times \mathbf{C})$ as expressed in Eq. (2-14) is equal to the volume of the parallelepiped formed by the three vectors **A**, **B**, and **C**. Let us now examine the second scalar triple product $\mathbf{B} \cdot (\mathbf{C} \times \mathbf{A})$. We have, from Fig. 2-7(b) and Eq. (2-12),

$$\mathbf{B} \cdot (\mathbf{C} \times \mathbf{A}) = (\mathbf{B} \cdot \mathbf{a}'_n) CA \sin \beta, \tag{2-16}$$

where $\mathbf{a}'_n$ and $CA \sin \beta$ represent, respectively, the direction and the magnitude of the cross product $\mathbf{C} \times \mathbf{A}$. Now visualize the parallelepiped formed by the three vectors **A**, **B**, and **C** as standing on the shaded base with an area equal to $|\mathbf{C} \times \mathbf{A}| = CA \sin \beta$. The height of the parallelepiped is $(\mathbf{B} \cdot \mathbf{a}'_n)$. Hence the scalar triple product in Eq. (2-16) has a magnitude equal to the volume of the parallelepiped, which is the same as that given by Eq. (2-14). Thus,

$$\mathbf{B} \cdot (\mathbf{C} \times \mathbf{A}) = \mathbf{A} \cdot (\mathbf{B} \times \mathbf{C}). \tag{2-17}$$

Similar arguments apply to the third scalar triple product in Eq. (2-15), $\mathbf{C} \cdot (\mathbf{A} \times \mathbf{B})$, since all three forms yield the volume of the parallelepiped.

CAUTION: The equalities in Eq. (2-15) require that the order of the vectors in the scalar triple product be kept in cyclic permutation. This means that the sequence $\{\mathbf{A}, \mathbf{B}, \mathbf{C}\}$, $\{\mathbf{B}, \mathbf{C}, \mathbf{A}\}$, or $\{\mathbf{C}, \mathbf{A}, \mathbf{B}\}$ must be maintained when taking the dot product of the first vector with the result of the cross product of the second and third vectors. $\mathbf{B}\cdot(\mathbf{A}\times\mathbf{C})$, which does not follow the cyclical sequence, is not the same as (but is the negative of) $\mathbf{B}\cdot(\mathbf{C}\times\mathbf{A})$ in Eq. (2-16).

REVIEW QUESTIONS

Q.2-1 Under what conditions can the dot product of two vectors be negative?

Q.2-2 Write down the results of $\mathbf{A}\cdot\mathbf{B}$ and $\mathbf{A}\times\mathbf{B}$ if (a) $\mathbf{A}\parallel\mathbf{B}$, and (b) $\mathbf{A}\perp\mathbf{B}$.

Q.2-3 Is $(\mathbf{A}\cdot\mathbf{B})\mathbf{C}$ equal to $\mathbf{A}(\mathbf{B}\cdot\mathbf{C})$? Explain.

Q.2-4 Given two vectors **A** and **B**, how do you find (a) the component of **A** in the direction of **B**, and (b) the component of **B** in the direction of **A**?

Q.2-5 Does $\mathbf{A}\cdot\mathbf{B}=\mathbf{A}\cdot\mathbf{C}$ imply $\mathbf{B}=\mathbf{C}$? Explain.

Q.2-6 Does $\mathbf{A}\times\mathbf{B}=\mathbf{A}\times\mathbf{C}$ imply $\mathbf{B}=\mathbf{C}$? Explain.

REMARKS

1. In writing a vector, *never* leave out the mark that distinguishes it from a scalar.
2. Do not add or subtract a vector from a scalar, or vice versa.
3. Division by a vector is not defined. Do not attempt to divide a quantity by a vector.
4. Two vectors are perpendicular to each other if their dot product is zero, and vice versa. ($\theta=\pi/2$, $\cos\theta=0$—Eq. 2-6.)
5. Two vectors are parallel to each other if their cross product is zero, and vice versa. ($\theta=0$, $\sin\theta=0$—Eq. 2-10.)

EXERCISE 2.2 Compare the values of the following scalar triple products of vectors: (a) $(\mathbf{A}\times\mathbf{C})\cdot\mathbf{B}$, (b) $\mathbf{A}\cdot(\mathbf{C}\times\mathbf{B})$, (c) $(\mathbf{A}\times\mathbf{B})\cdot\mathbf{C}$, and (d) $\mathbf{B}\cdot(\mathbf{a}_A\times\mathbf{A})$.

EXERCISE 2.3 Which of the following expressions do not make sense?
(a) $\mathbf{A}\times\mathbf{B}/|\mathbf{B}|$, (b) $\mathbf{C}\cdot\mathbf{D}/(\mathbf{A}\times\mathbf{B})$, (c) $\mathbf{AB}/\mathbf{CD}$, (d) $\mathbf{A}\times\mathbf{B}/(\mathbf{C}\cdot\mathbf{D})$, (e) $\mathbf{ABC}$, (f) $\mathbf{A}\times\mathbf{B}\times\mathbf{C}$.

2-4 ORTHOGONAL COORDINATE SYSTEMS

We have indicated before that although the laws of electromagnetism are invariant with coordinate system, solution of practical problems requires that the relations derived from these laws be expressed in a coordinate system appropriate to the geometry of the given problems. For example, to determine the electric field at a certain point in space, we at least need to describe the position of the source and the location of this point with respect to a coordinate system. In a three-dimensional space a point can be located as the intersection of three surfaces. Assume that the three families of surfaces are described by $u_1 =$ constant, $u_2 =$ constant, and $u_3 =$ constant, where the u's need not all be lengths, and some may be angles. (In the familiar Cartesian or rectangular coordinate system, u_1, u_2, and u_3 correspond to x, y, and z, respectively.) When these three surfaces are mutually perpendicular to one another, we have an ***orthogonal coordinate system***.

Orthogonal coordinate systems

Many orthogonal coordinate systems exist; but we shall be concerned only with the three that are most common and most useful:

1. Cartesian (or rectangular) coordinates.[†]
2. Cylindrical coordinates.
3. Spherical coordinates.

These will be discussed separately in the following subsections.

2-4.1 CARTESIAN COORDINATES

A point $P(x_1, y_1, z_1)$ in Cartesian coordinates is the intersection of *three planes* specified by $x = x_1$, $y = y_1$, and $z = z_1$, as shown in Fig. 2-8. We have

$$(u_1, u_2, u_3) = (x, y, z).$$

The three mutually perpendicular unit vectors, $\mathbf{a}_x$, $\mathbf{a}_y$, and $\mathbf{a}_z$, in the three coordinate directions are called the ***base vectors***. For a right-handed system we have the following cyclic properties:

$$\mathbf{a}_x \times \mathbf{a}_y = \mathbf{a}_z \tag{2-18a}$$

$$\mathbf{a}_y \times \mathbf{a}_z = \mathbf{a}_x \tag{2-18b}$$

$$\mathbf{a}_z \times \mathbf{a}_x = \mathbf{a}_y. \tag{2-18c}$$

The following relations follow directly.

$$\mathbf{a}_x \cdot \mathbf{a}_y = \mathbf{a}_y \cdot \mathbf{a}_z = \mathbf{a}_x \cdot \mathbf{a}_z = 0 \tag{2-19}$$

and

$$\mathbf{a}_x \cdot \mathbf{a}_x = \mathbf{a}_y \cdot \mathbf{a}_y = \mathbf{a}_z \cdot \mathbf{a}_z = 1. \tag{2-20}$$

The position vector to the point $P(x_1, y_1, z_1)$ is the vector drawn from

[†] The term "Cartesian coordinates" is preferred because the term "rectangular coordinates" is customarily associated with two-dimensional geometry. The adjective "Cartesian" is used in honor of French philosopher and mathematician Renatus Cartesius (Latinized form for René Descartes 1596–1650), who initiated analytic geometry.

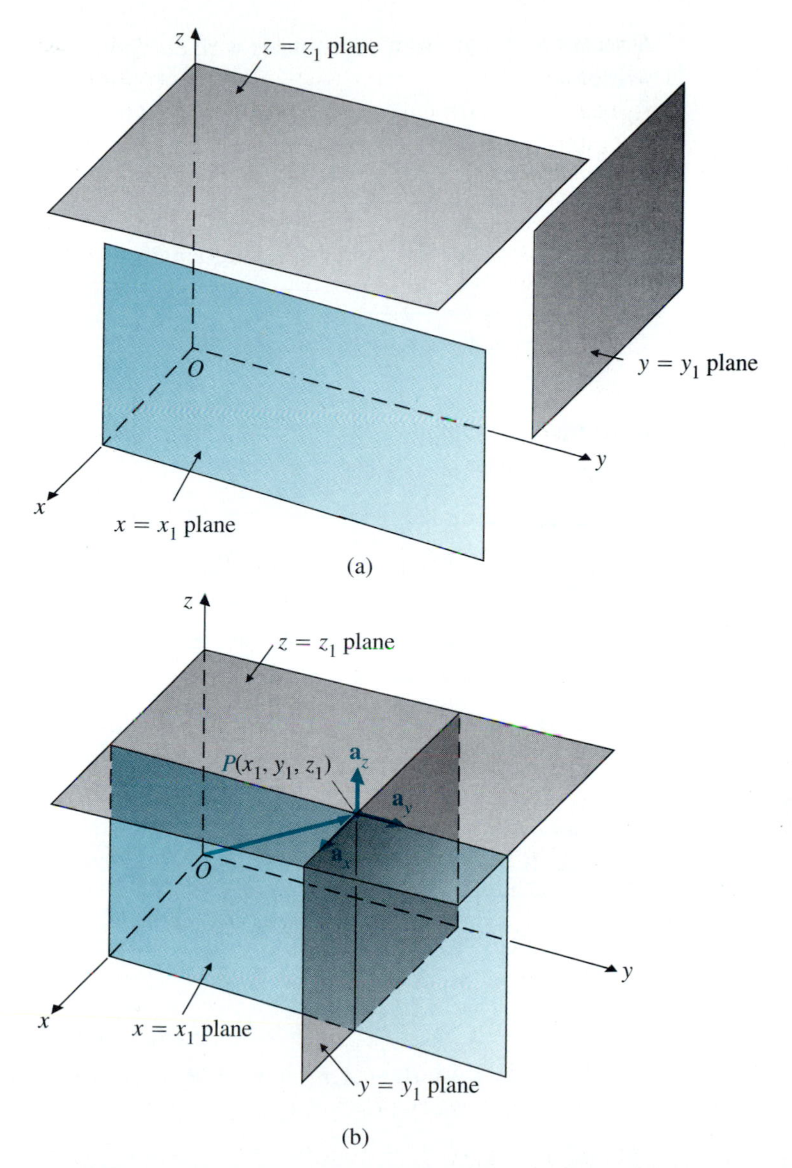

FIGURE 2-8 Cartesian coordinates. (a) Three mutually perpendicular planes. (b) Intersection of the three planes in (a) specifies the location of a point **P**.

the origin O to P; its components in the $\mathbf{a}_x$, $\mathbf{a}_y$, $\mathbf{a}_z$ directions are, respectively, x_1, y_1, and z_1†.

$$\overrightarrow{OP} = \mathbf{a}_x x_1 + \mathbf{a}_y y_1 + \mathbf{a}_z z_1. \tag{2-21}$$

†In writing vectors in this book we stick to the convention of always writing the direction (of a unit vector) first, followed by the magnitude.

A vector $\mathbf{A}$ in Cartesian coordinates with components A_x, A_y, and A_z can be written as

Vector A in Cartesian coordinates

$$\mathbf{A} = \mathbf{a}_x A_x + \mathbf{a}_y A_y + \mathbf{a}_z A_z. \tag{2-22}$$

The expression for a vector differential length is

Vector differential length in Cartesian coordinates

$$d\boldsymbol{\ell} = \mathbf{a}_x dx + \mathbf{a}_y dy + \mathbf{a}_z dz. \tag{2-23}$$

A differential volume is the product of the differential length changes in the three coordinate directions:

A differential volume in Cartesian coordinates

$$dv = dx\,dy\,dz. \tag{2-24}$$

The dot product of $\mathbf{A}$ in Eq. (2-22) and another vector $\mathbf{B} = \mathbf{a}_x B_x + \mathbf{a}_y B_y + \mathbf{a}_z B_z$ is

$$\mathbf{A} \cdot \mathbf{B} = (\mathbf{a}_x A_x + \mathbf{a}_y A_y + \mathbf{a}_z A_z) \cdot (\mathbf{a}_x B_x + \mathbf{a}_y B_y + \mathbf{a}_z B_z),$$

or

Scalar product of A and B in Cartesian coordinates

$$\mathbf{A} \cdot \mathbf{B} = A_x B_x + A_y B_y + A_z B_z, \tag{2-25}$$

in view of Eqs. (2-19) and (2-20).

The cross product of $\mathbf{A}$ and $\mathbf{B}$ is

$$\begin{aligned}\mathbf{A} \times \mathbf{B} &= (\mathbf{a}_x A_x + \mathbf{a}_y A_y + \mathbf{a}_z A_z) \times (\mathbf{a}_x B_x + \mathbf{a}_y B_y + \mathbf{a}_z B_z) \\ &= \mathbf{a}_x(A_y B_z - A_z B_y) + \mathbf{a}_y(A_z B_x - A_x B_z) + \mathbf{a}_z(A_x B_y - A_y B_x),\end{aligned} \tag{2-26}$$

in view of Eqs. (2-18a, b, and c). Equation (2-26) can be conveniently written in a determinant form for easy memory:

Vector product of A and B in Cartesian coordinates

$$\mathbf{A} \times \mathbf{B} = \begin{vmatrix} \mathbf{a}_x & \mathbf{a}_y & \mathbf{a}_z \\ A_x & A_y & A_z \\ B_x & B_y & B_z \end{vmatrix}. \tag{2-27}$$

EXAMPLE 2-3

Given a vector $\mathbf{A} = -\mathbf{a}_x + \mathbf{a}_y 2 - \mathbf{a}_z 2$ in Cartesian coordinates, find

a) its magnitude $A = |\mathbf{A}|$,

b) the expression of the unit vector $\mathbf{a}_A$ in the direction of **A**, and

c) the angle that **A** makes with the z-axis.

SOLUTION

a) We find A by using Eqs. (2-8) and (2-9) and noting Eqs. (2-19) and (2-20).

$$\begin{aligned}\mathbf{A}\cdot\mathbf{A} &= (-\mathbf{a}_x + \mathbf{a}_y 2 - \mathbf{a}_z 2)\cdot(-\mathbf{a}_x + \mathbf{a}_y 2 - \mathbf{a}_z 2)\\ &= (-1)(-1) + (2)(2) + (-2)(-2)\\ &= 1 + 4 + 4 = 9.\end{aligned}$$

Thus,

$$A = +\sqrt{\mathbf{A}\cdot\mathbf{A}} = +\sqrt{9} = 3.$$

b) The unit vector $\mathbf{a}_A$ is obtained by using Eq. (2-3). We have

$$\begin{aligned}\mathbf{a}_A = \frac{\mathbf{A}}{A} &= \frac{1}{3}(-\mathbf{a}_x + \mathbf{a}_y 2 - \mathbf{a}_z 2)\\ &= -\mathbf{a}_x\frac{1}{3} + \mathbf{a}_y\frac{2}{3} - \mathbf{a}_z\frac{2}{3}.\end{aligned}$$

c) To find the angle θ_z that **A** makes with the $+z$ axis, we take the dot product of **A** with the unit vector $\mathbf{a}_z$. From Eq. (2-6) we have

$$\mathbf{A}\cdot\mathbf{a}_z = A\cos\theta_z,$$

$$(-\mathbf{a}_x + \mathbf{a}_y 2 - \mathbf{a}_z 2)\cdot\mathbf{a}_z = -2 = 3\cos\theta_z,$$

from which we obtain

$$\theta_z = \cos^{-1}\left(\frac{-2}{3}\right) = 180^\circ - 48.2^\circ = 131.8^\circ.$$

QUESTION: Why is this answer not -48.2° or 228.2° $(180^\circ + 48.2^\circ)$?

EXAMPLE 2-4

Given $\mathbf{A} = \mathbf{a}_x 5 - \mathbf{a}_y 2 + \mathbf{a}_z$ and $\mathbf{B} = -\mathbf{a}_x 3 + \mathbf{a}_z 4$, find

a) $\mathbf{A}\cdot\mathbf{B}$,

b) $\mathbf{A}\times\mathbf{B}$, and

c) θ_{AB}.

SOLUTION

a) From Eq. (2-25) we find

$$\mathbf{A}\cdot\mathbf{B} = (5)(-3) + (-2)(0) + (1)(4) = -11.$$

b) From Eq. (2-27) we have

$$\mathbf{A}\times\mathbf{B} = \begin{vmatrix} \mathbf{a}_x & \mathbf{a}_y & \mathbf{a}_z \\ 5 & -2 & 1 \\ -3 & 0 & 4 \end{vmatrix} = -\mathbf{a}_x 8 - \mathbf{a}_y 23 - \mathbf{a}_z 6.$$

c) We can find θ_{AB}, the angle between vectors **A** and **B**, from the definition of $\mathbf{A}\cdot\mathbf{B}$ in Eq. (2-6). The magnitudes A of **A** and B of **B** are:

$$A = |\mathbf{A}| = +\sqrt{5^2 + (-2)^2 + 1^2} = +\sqrt{30}$$

and

$$B = |\mathbf{B}| = +\sqrt{(-3)^2 + 4^2} = 5.$$

From Eq. (2-6),

$$\cos\theta_{AB} = \frac{\mathbf{A}\cdot\mathbf{B}}{AB} = \frac{-11}{5\sqrt{30}} = -0.402.$$

Hence,

$$\theta_{AB} = \cos^{-1}(-0.402) = 180^\circ - 66.3^\circ = 113.7^\circ.$$

EXAMPLE 2-5

a) Write the expression of the vector going from point $P_1(1, 3, 2)$ to point $P_2(3, -2, 4)$ in Cartesian coordinates.

b) Determine the length of the line $\overline{P_1P_2}$.

c) Find the perpendicular distance from the origin to this line.

SOLUTION

a) From Fig. 2-9 we see that

$$\begin{aligned}\overrightarrow{P_1P_2} &= \overrightarrow{OP_2} - \overrightarrow{OP_1} \\ &= (\mathbf{a}_x 3 - \mathbf{a}_y 2 + \mathbf{a}_z 4) - (\mathbf{a}_x + \mathbf{a}_y 3 + \mathbf{a}_z 2) \\ &= \mathbf{a}_x 2 - \mathbf{a}_y 5 + \mathbf{a}_z 2.\end{aligned}$$

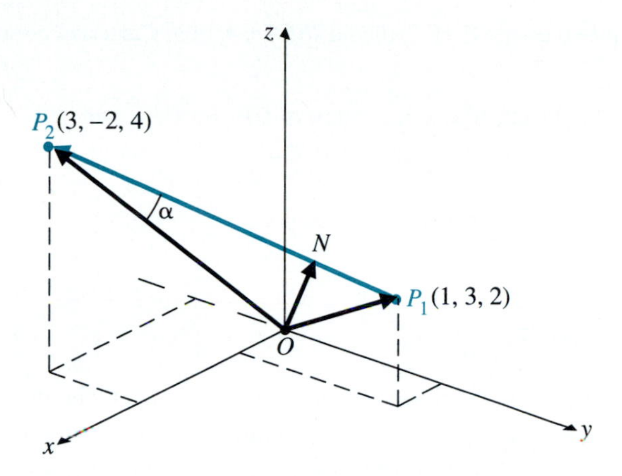

FIGURE 2-9 Illustrating Example 2-5.

b) The length of the line $\overline{P_1P_2}$ is

$$\begin{aligned}\overline{P_1P_2} &= |\overrightarrow{P_1P_2}| \\ &= \sqrt{2^2 + (-5)^2 + 2^2} \\ &= \sqrt{33}.\end{aligned}$$

c) The perpendicular (shortest) distance from the origin O to the line is $|\overrightarrow{ON}|$, which equals $|\overrightarrow{OP_2}|\sin\alpha = |\overrightarrow{OP_2} \times \mathbf{a}_{P_1P_2}|$. Thus,

$$\begin{aligned}|\overrightarrow{ON}| &= \frac{|\overrightarrow{OP_2} \times \overrightarrow{P_1P_2}|}{|\overrightarrow{P_1P_2}|} \\ &= \frac{|(\mathbf{a}_x3 - \mathbf{a}_y2 + \mathbf{a}_z4) \times (\mathbf{a}_x2 - \mathbf{a}_y5 + \mathbf{a}_z2)}{\sqrt{33}} \\ &= \frac{|\mathbf{a}_x16 + \mathbf{a}_y2 - \mathbf{a}_z11|}{\sqrt{33}} = \frac{\sqrt{381}}{\sqrt{33}} = 3.40.\end{aligned}$$

NOTE: Units have been omitted in this example for simplicity.

■ **EXERCISE 2.4** Given a vector $\mathbf{B} = \mathbf{a}_x2 - \mathbf{a}_y6 + \mathbf{a}_z3$, find

a) the magnitude of **B**,

b) the expression for $\mathbf{a}_B$,

c) the angles that **B** makes with the x, y, and z axes.

ANS. (a) 7, (b) $\mathbf{a}_B = \mathbf{a}_x0.296 - \mathbf{a}_y0.857 + \mathbf{a}_z0.429$, (c) 73.4°, 149.0°, 64.6°.

EXERCISE 2.5 Given two points $P_1(1, 2, 0)$ and $P_2(-3, 4, 0)$ in Cartesian coordinates with origin O, find

a) the length of the projection of $\overrightarrow{OP_2}$ on $\overrightarrow{OP_1}$, and

b) the area of the triangle OP_1P_2.

ANS. (a) 2.236, (b) 5.

2-4.2 CYLINDRICAL COORDINATES

In cylindrical coordinates a point $P(r_1, \phi_1, z_1)$ is the intersection of a circular cylindrical surface $r = r_1$, a half-plane with the z-axis as an edge and making an angle $\phi = \phi_1$ with the xy-plane, and a plane parallel to the xy-plane at $z = z_1$. We have

$$(u_1, u_2, u_3) = (r, \phi, z).$$

As indicated in Fig. 2-10, r is the radial distance measured from the z-axis, and angle ϕ is measured from the positive x-axis. The base vector $\mathbf{a}_\phi$ is tangential to the cylindrical surface. The directions of both $\mathbf{a}_r$ and $\mathbf{a}_\phi$ change with the location of the point P. The following right-hand relations hold for $\mathbf{a}_r$, $\mathbf{a}_\phi$, and $\mathbf{a}_z$:

$$\mathbf{a}_r \times \mathbf{a}_\phi = \mathbf{a}_z, \tag{2-28a}$$

$$\mathbf{a}_\phi \times \mathbf{a}_z = \mathbf{a}_r, \tag{2-28b}$$

$$\mathbf{a}_z \times \mathbf{a}_r = \mathbf{a}_\phi. \tag{2-28c}$$

Metric coefficient

Two of the three coordinates, r and z (u_1 and u_3), are themselves lengths. But, $\phi(u_2)$ is an angle, requiring a multiplying coefficient (a *metric coefficient*) r to convert a differential angle change $d\phi$ to a differential length change. This is illustrated in Fig. 2-11.

The metric coefficients for dr and dz are unity. Denoting the three metric coefficients in the three coordinate directions $\mathbf{a}_r$, $\mathbf{a}_\phi$, and $\mathbf{a}_z$ by h_1, h_2, and h_3, respectively, we have for cylindrical coordinates $h_1 = 1$, $h_2 = r$, $h_3 = 1$. These are listed in Table 2-1. The metric coefficients in Cartesian coordinates in all three coordinate directions are unity ($h_1 = h_2 = h_3 = 1$), because all three coordinates (x, y, z) are lengths themselves.

The general expression for a vector differential length in cylindrical coordinates is the vector sum of the differential length changes in the three coordinate directions.

Vector differential length in cylindrical coordinates

$$d\boldsymbol{\ell} = \mathbf{a}_r dr + \mathbf{a}_\phi r\, d\phi + \mathbf{a}_z\, dz. \tag{2-29}$$

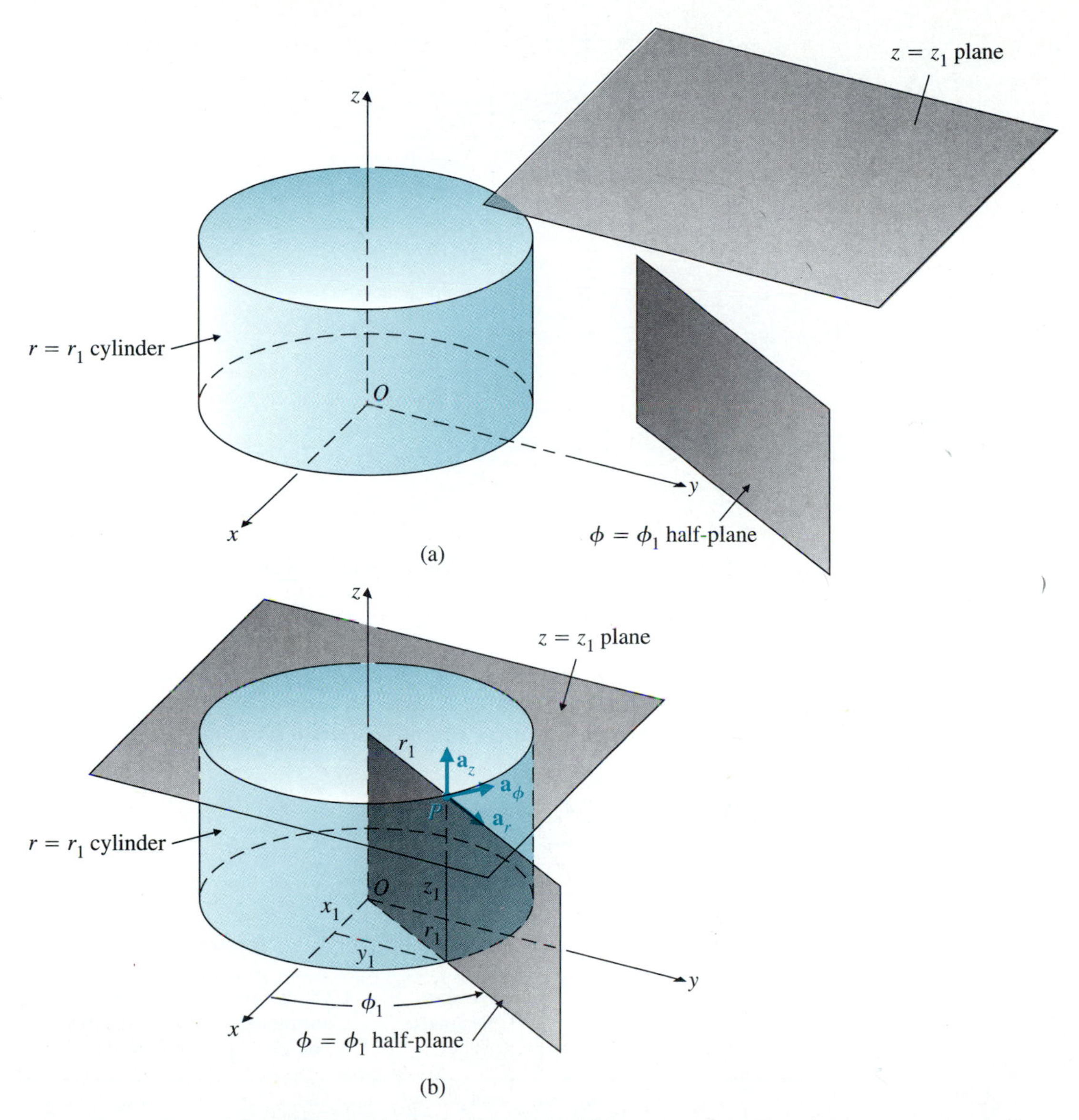

FIGURE 2-10 Cylindrical coordinates. (a) A circular cylindrical surface, a half-plane with the z-axis as an edge, and a plane perpendicular to the z-axis. (b) Intersection of the cylindrical surface and the two planes in (a) specifies the location of a point P.

A differential volume is the product of the differential length changes in the three coordinate directions. In cylindrical coordinates it is

Differential volume in cylindrical coordinates

$$dv = r\,dr\,d\phi\,dz. \tag{2-30}$$

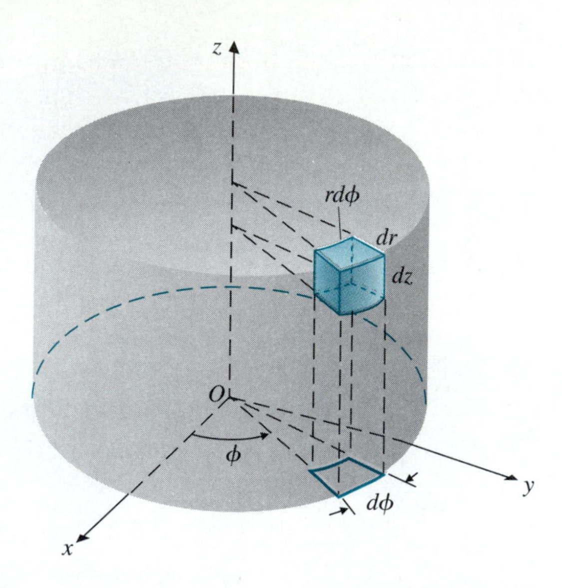

FIGURE 2-11 A differential volume element in cylindrical coordinates.

Cylindrical coordinates are important for problems with long line charges or currents, and in places where cylindrical or circular boundaries exist.

A vector in cylindrical coordinates is written as

Vector A in cylindrical coordinates

$$\mathbf{A} = \mathbf{a}_r A_r + \mathbf{a}_\phi A_\phi + \mathbf{a}_z A_z. \tag{2-31}$$

TABLE **2-1** THREE BASIC ORTHOGONAL COORDINATE SYSTEMS

		Cartesian Coordinates (x, y, z)	Cylindrical Coordinates (r, ϕ, z)	Spherical Coordinates (R, θ, ϕ)
Base Vectors	$\mathbf{a}_{u_1}$	$\mathbf{a}_x$	$\mathbf{a}_r$	$\mathbf{a}_R$
	$\mathbf{a}_{u_2}$	$\mathbf{a}_y$	$\mathbf{a}_\phi$	$\mathbf{a}_\theta$
	$\mathbf{a}_{u_3}$	$\mathbf{a}_z$	$\mathbf{a}_z$	$\mathbf{a}_\phi$
Metric Coefficients	h_1	1	1	1
	h_2	1	r	R
	h_3	1	1	$R \sin \theta$
Differential Volume	dv	$dx\,dy\,dz$	$r\,dr\,d\phi\,dz$	$R^2 \sin \theta\,dR\,d\theta\,d\phi$

Vectors given in cylindrical coordinates can be transformed and expressed in Cartesian coordinates, and vice versa. Suppose we want to express $\mathbf{A} = \mathbf{a}_r A_r + \mathbf{a}_\phi A_\phi + \mathbf{a}_z A_z$ in Cartesian coordinates; that is, we want to write $\mathbf{A}$ as $\mathbf{a}_x A_x + \mathbf{a}_y A_y + \mathbf{a}_z A_z$ and determine A_x, A_y, and A_z. First of all, we note that A_z, the z-component of $\mathbf{A}$, is not changed by the transformation from cylindrical to Cartesian coordinates. To find A_x, we equate the dot products of both expressions of $\mathbf{A}$ with $\mathbf{a}_x$. Thus

$$\begin{aligned} A_x &= \mathbf{A} \cdot \mathbf{a}_x \\ &= A_r \mathbf{a}_r \cdot \mathbf{a}_x + A_\phi \mathbf{a}_\phi \cdot \mathbf{a}_x. \end{aligned} \tag{2-32}$$

The term containing A_z disappears here because $\mathbf{a}_z \cdot \mathbf{a}_x = 0$. Referring to Fig. 2-12, which shows the relative positions of the base vectors $\mathbf{a}_x$, $\mathbf{a}_y$, $\mathbf{a}_r$, and $\mathbf{a}_\phi$ in the xy-plane, we see that

$$\mathbf{a}_r \cdot \mathbf{a}_x = \cos\phi \tag{2-33}$$

and

$$\mathbf{a}_\phi \cdot \mathbf{a}_x = \cos\left(\frac{\pi}{2} + \phi\right) = -\sin\phi. \tag{2-34}$$

Substituting Eqs. (2-33) and (2-34) into Eq. (2-32), we obtain

$$A_x = A_r \cos\phi - A_\phi \sin\phi. \tag{2-35}$$

Similarly, to find A_y, we take the dot products of both expressions of $\mathbf{A}$ with $\mathbf{a}_y$:

$$\begin{aligned} A_y &= \mathbf{A} \cdot \mathbf{a}_y \\ &= A_r \mathbf{a}_r \cdot \mathbf{a}_y + A_\phi \mathbf{a}_\phi \cdot \mathbf{a}_y. \end{aligned}$$

From Fig. 2-12 we find that

$$\mathbf{a}_r \cdot \mathbf{a}_y = \cos\left(\frac{\pi}{2} - \phi\right) = \sin\phi \tag{2-36}$$

FIGURE 2-12 Relations among $\mathbf{a}_x$, $\mathbf{a}_y$, $\mathbf{a}_r$, and $\mathbf{a}_\phi$.

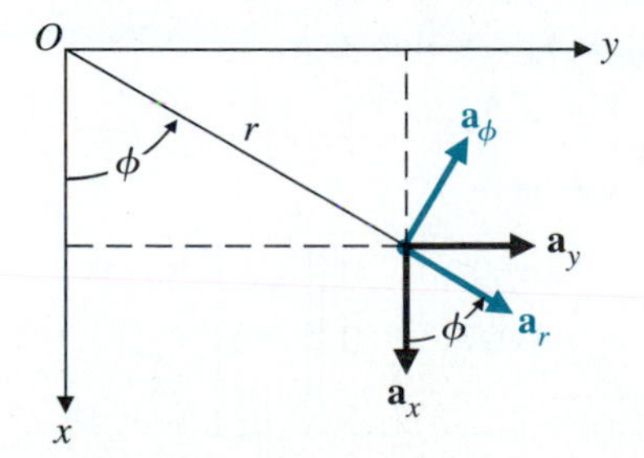

and

$$\mathbf{a}_\phi \cdot \mathbf{a}_y = \cos\phi. \qquad (2\text{-}37)$$

It follows that

$$A_y = A_r \sin\phi + A_\phi \cos\phi. \qquad (2\text{-}38)$$

It is convenient to write the relations between the components of a vector in Cartesian and cylindrical coordinates in a matrix form:

Transformation of vector components in cylindrical coordinates to Cartesian coordinates

$$\begin{bmatrix} A_x \\ A_y \\ A_z \end{bmatrix} = \begin{bmatrix} \cos\phi & -\sin\phi & 0 \\ \sin\phi & \cos\phi & 0 \\ 0 & 0 & 1 \end{bmatrix} \begin{bmatrix} A_r \\ A_\phi \\ A_z \end{bmatrix}. \qquad (2\text{-}39)$$

From Fig. 2-12 we see that the coordinates of a point in cylindrical coordinates (r, ϕ, z) can be transformed into those in Cartesian coordinates (x, y, z) as follows:

Transformation of the location of a point in cylindrical coordinates to Cartesian coordinates

$$x = r\cos\phi, \qquad (2\text{-}40a)$$

$$y = r\sin\phi, \qquad (2\text{-}40b)$$

$$z = z. \qquad (2\text{-}40c)$$

EXAMPLE 2-6

Assuming a vector field expressed in cylindrical coordinates to be $\mathbf{A} = \mathbf{a}_r(3\cos\phi) - \mathbf{a}_\phi 2r + \mathbf{a}_z z$,

a) what is the field at the point $P(4, 60^\circ, 5)$?

b) Express the field $\mathbf{A}_P$ at P in Cartesian coordinates.

c) Express the location of the point P in Cartesian coordinates.

SOLUTION

a) At point $P(r = 4, \phi = 60^\circ, z = 5)$ the field is

$$\mathbf{A}_P = \mathbf{a}_r(3\cos 60^\circ) - \mathbf{a}_\phi(2 \times 4) + \mathbf{a}_z 5$$
$$= \mathbf{a}_r(3/2) - \mathbf{a}_\phi 8 + \mathbf{a}_z 5.$$

b) Using Eq. (2-39), we have

$$\begin{bmatrix} A_x \\ A_y \\ A_z \end{bmatrix} = \begin{bmatrix} \cos 60^\circ & -\sin 60^\circ & 0 \\ \sin 60^\circ & \cos 60^\circ & 0 \\ 0 & 0 & 1 \end{bmatrix} \begin{bmatrix} 3/2 \\ -8 \\ 5 \end{bmatrix}$$

$$= \begin{bmatrix} 1/2 & -\sqrt{3}/2 & 0 \\ \sqrt{3}/2 & 1/2 & 0 \\ 0 & 0 & 1 \end{bmatrix} \begin{bmatrix} 3/2 \\ -8 \\ 5 \end{bmatrix} = \begin{bmatrix} 7.68 \\ -2.70 \\ 5 \end{bmatrix}.$$

Thus,

$$\mathbf{A}_P = \mathbf{a}_x 7.68 - \mathbf{a}_y 2.70 + \mathbf{a}_z 5.$$

c) Using Eqs. (2-40a, b, and c), we obtain the Cartesian coordinates of the point P as $(4\cos 60°, 4\sin 60°, 5)$, or $(2, 2\sqrt{3}, 5)$.

■ **EXERCISE 2.6**

Express the position vector $\overrightarrow{OQ}$ from the origin O to the point $Q(3, 4, 5)$ in cylindrical coordinates.

ANS. $\mathbf{a}_r 5 + \mathbf{a}_z 5$.

■ **EXERCISE 2.7**

The cylindrical coordinates of two points P_1 and P_2 are: $P_1(4, 60°, 1)$ and $P_2(3, 180°, -1)$. Determine the distance between these two points.

ANS. $\sqrt{41}$.

2-4.3 SPHERICAL COORDINATES

A point $P(R_1, \theta_1, \phi_1)$ in spherical coordinates is specified as the intersection of the following three surfaces: a spherical surface centered at the origin with a radius $R = R_1$; a right circular cone with its apex at the origin, its axis coinciding with the $+z$-axis and having a half-angle $\theta = \theta_1$; and a half-plane with the z-axis as an edge and making an angle $\phi = \phi_1$ with the xz-plane. We have

$$(u_1, u_2, u_3) = (R, \theta, \phi).$$

$\mathbf{a}_R$ and $\mathbf{a}_r$ are very different.

The three intersecting surfaces are shown in Fig. 2-13. Note that the *base vector* $\mathbf{a}_R$ *at P is radial from the origin and is quite different from* $\mathbf{a}_r$ *in cylindrical coordinates, the latter being perpendicular to the z-axis.* The base vector $\mathbf{a}_\theta$ lies in the $\phi = \phi_1$ plane and is tangential to the spherical surface, whereas the base vector $\mathbf{a}_\phi$ is the same as that in cylindrical coordinates. These are illustrated in Fig. 2-11. For a right-handed system we have

$$\mathbf{a}_R \times \mathbf{a}_\theta = \mathbf{a}_\phi, \qquad (2\text{-}41\text{a})$$

$$\mathbf{a}_\theta \times \mathbf{a}_\phi = \mathbf{a}_R, \qquad (2\text{-}41\text{b})$$

$$\mathbf{a}_\phi \times \mathbf{a}_R = \mathbf{a}_\theta. \qquad (2\text{-}41\text{c})$$

Spherical coordinates are important for problems involving point sources and regions with spherical boundaries. When an observer is very far from a

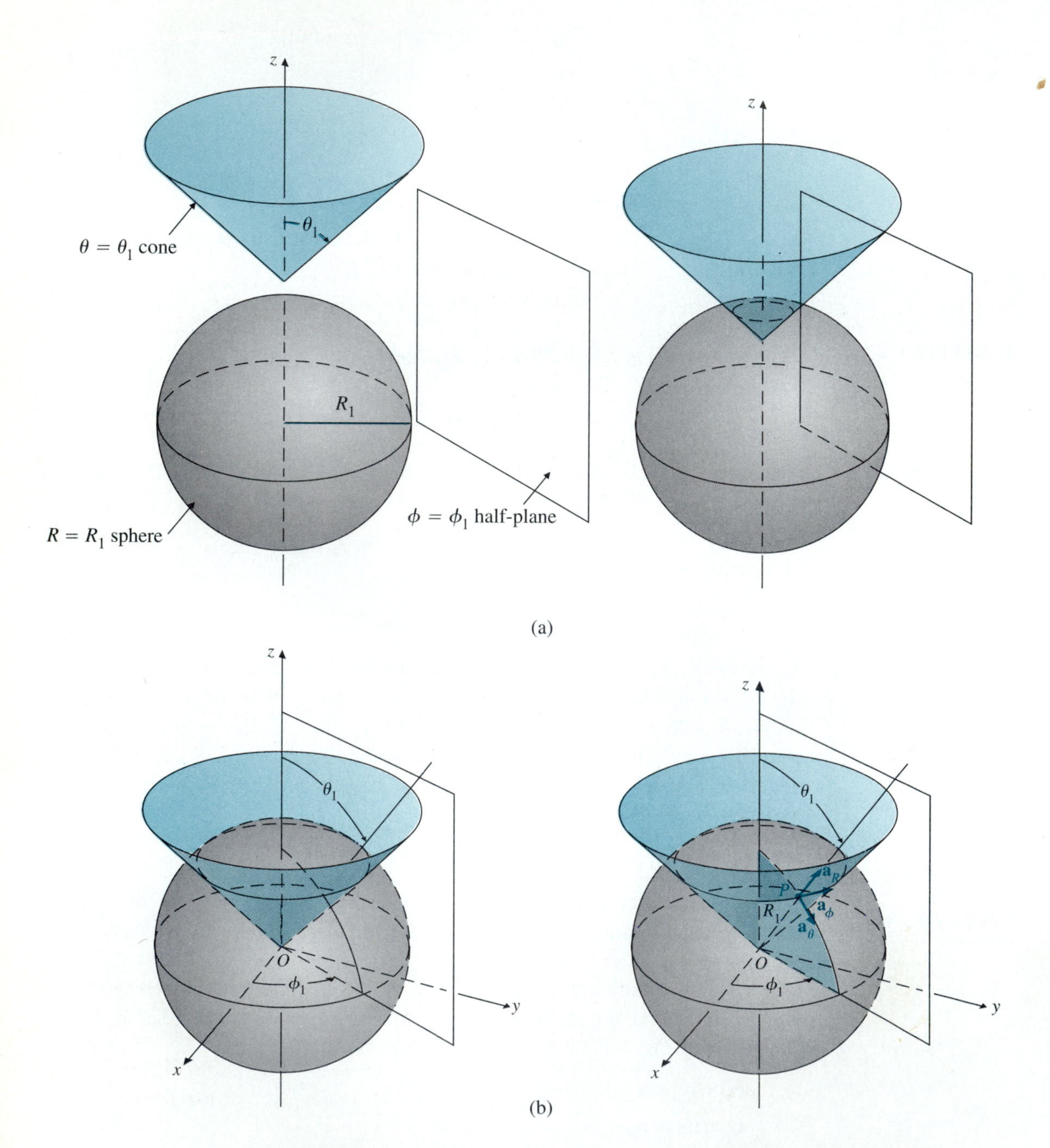

FIGURE 2-13 (a) A spherical surface, a right circular cone, and a half-plane containing the z-axis. (b) Intersection of the sphere, the cone, and the half-plane in (a) specifies the point P.

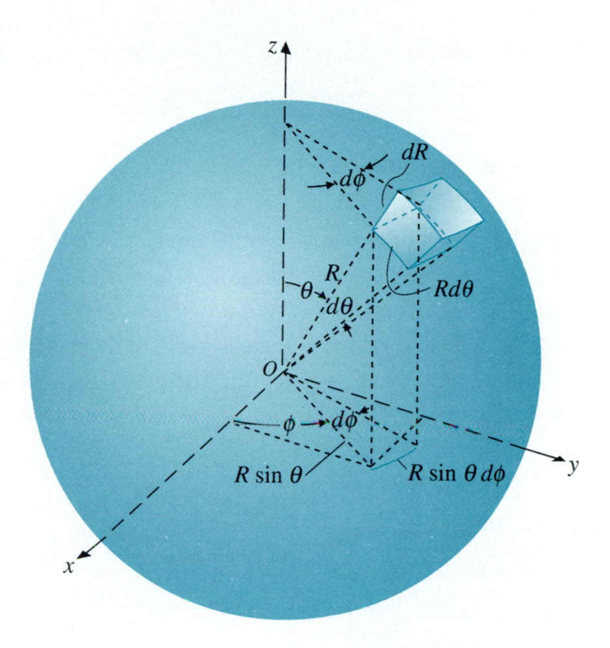

FIGURE 2-14 A differential volume element in spherical coordinates.

source region of a finite extent, the source could be considered approximately as a point. It could be chosen as the origin of a spherical coordinate system so that suitable simplifying approximations could be made. This is the reason that spherical coordinates are used in solving antenna problems in the far field.

A vector in spherical coordinates is written as

Vector A in spherical coordinates

$$\mathbf{A} = \mathbf{a}_R A_R + \mathbf{a}_\theta A_\theta + \mathbf{a}_\phi A_\phi. \tag{2-42}$$

In spherical coordinates, only R is a length. The other two coordinates, θ and ϕ are angles. Referring to Fig. 2-14, in which a typical differential volume element is shown, we see that metric coefficients $h_2 = R$ and $h_3 = R \sin\theta$ are required to convert $d\theta$ and $d\phi$, respectively, into differential lengths $(R)d\theta$ and $(R\sin\theta)d\phi$. The general expression for a vector differential length is

Vector differential length in spherical coordinates

$$d\boldsymbol{\ell} = \mathbf{a}_R\, dR + \mathbf{a}_\theta R\, d\theta + \mathbf{a}_\phi R \sin\theta\, d\phi. \tag{2-43}$$

A differential volume is the product of differential length changes in the three coordinate directions:

A differential volume in spherical coordinates

$$dv = R^2 \sin\theta \, dR \, d\theta \, d\phi. \tag{2-44}$$

The base vectors, metric coefficients, and expressions for differential volume for the three basic orthogonal coordinate systems are shown in Table 2-1.

Figure 2-15 shows the interrelationship of the space variables (x, y, z), (r, ϕ, z), and (R, θ, ϕ) that specify the location of a point P. The following equations transform the coordinate variables in spherical coordinates to those in Cartesian coordinates.

Transformation of the location of a point in spherical coordinates to Cartesian coordinates

$$x = R \sin\theta \cos\phi, \tag{2-45a}$$

$$y = R \sin\theta \sin\phi, \tag{2-45b}$$

$$z = R \cos\theta. \tag{2-45c}$$

■ **EXERCISE 2.8** Transform Cartesian coordinates $(4, -6, 12)$ into spherical coordinates.

ANS. $(14, 31°, 303.7°)$.

FIGURE 2-15 Showing interrelationship of space variables (x, y, z), (r, ϕ, z), and (R, θ, ϕ).

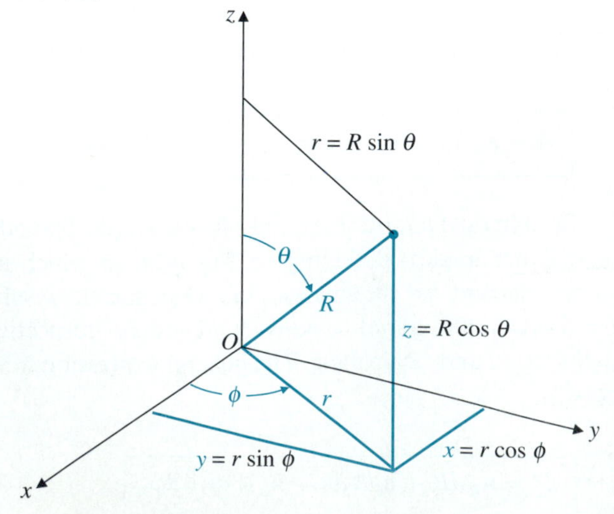

EXAMPLE 2-7

Express the unit vector $\mathbf{a}_z$ in spherical coordinates.

SOLUTION

First of all, we must not be tempted by Eq. (2-45c) to write $\mathbf{a}_z$ as $\mathbf{a}_R R\cos\theta$ or $\mathbf{a}_R \cos\theta$ because both the direction ($\mathbf{a}_z \neq \mathbf{a}_R$) and the magnitude ($1 \neq R\cos\theta$ or $\cos\theta$ for all θ) would be incorrect. Since the base vectors for spherical coordinates are $\mathbf{a}_R$, $\mathbf{a}_\theta$ and $\mathbf{a}_\phi$, let us proceed by finding the components of $\mathbf{a}_z$ in these directions. From Figs. 2-13 and 2-14 we have

$$\mathbf{a}_z \cdot \mathbf{a}_R = \cos\theta, \tag{2-46a}$$

$$\mathbf{a}_z \cdot \mathbf{a}_\theta = -\sin\theta, \tag{2-46b}$$

$$\mathbf{a}_z \cdot \mathbf{a}_\phi = 0. \tag{2-46c}$$

Thus,

$$\mathbf{a}_z = \mathbf{a}_R \cos\theta - \mathbf{a}_\theta \sin\theta. \tag{2-47}$$

EXAMPLE 2-8

Assuming that a cloud of electrons confined in a region between two spheres of radii 2 and 5 (cm) has a charge density of

$$\frac{-3 \times 10^{-8}}{R^4}\cos^2\phi \qquad (\mathrm{C/m^3}),$$

find the total charge contained in the region.

SOLUTION

We have

$$\rho_v = -\frac{3 \times 10^{-8}}{R^4}\cos^2\phi,$$

$$Q = \int \rho_v \, dv.$$

The given conditions of the problem obviously point to the use of spherical coordinates. Using the expression for dv in Eq. (2-44), we perform a triple integration:

$$Q = \int_0^{2\pi}\int_0^{\pi}\int_{0.02}^{0.05} \rho_v R^2 \sin\theta \, dR \, d\theta \, d\phi.$$

Two things are of importance here. First, since ρ_v is given in units of coulombs per cubic meter, the limits of integration for R must be converted to meters. Second, the full range of integration for θ is from 0 to π radians, *not* from 0 to 2π radians. A little reflection will convince us that a half-circle (not a full-circle) rotated about the z-axis through 2π radians (ϕ from 0 to 2π) generates a sphere. We have

$$
\begin{aligned}
Q &= -3 \times 10^{-8} \int_0^{2\pi} \int_0^{\pi} \int_{0.02}^{0.05} \frac{1}{R^2} \cos^2 \phi \sin \theta \, dR \, d\theta \, d\phi \\
&= -3 \times 10^{-8} \int_0^{2\pi} \int_0^{\pi} \left(-\frac{1}{0.05} + \frac{1}{0.02} \right) \sin \theta \, d\theta \cos^2 \phi \, d\phi \\
&= -0.9 \times 10^{-6} \int_0^{2\pi} (-\cos \theta) \Big|_0^{\pi} \cos^2 \phi \, d\phi \\
&= -1.8 \times 10^{-6} \left(\frac{\phi}{2} + \frac{\sin 2\phi}{4} \right) \Big|_0^{2\pi} = -1.8\pi \qquad (\mu\text{C}).
\end{aligned}
$$

■ **EXERCISE 2.9** Derive the formula for the surface of a sphere with a radius R_0 by integrating the differential surface area in spherical coordinates.

ANS. $4\pi R_0^2$.

REVIEW QUESTIONS

Q.2-7 What makes a coordinate system (a) orthogonal? and (b) right-handed?

Q.2-8 What are metric coefficients?

Q.2-9 Write $d\ell$ and dv (a) in Cartesian coordinates, (b) in cylindrical coordinates, and (c) in spherical coordinates.

Q.2-10 Given two points $P_1(1, 2, 3)$ and $P_2(-1, 0, 2)$ in Cartesian coordinates, write the expressions of the vectors $\overrightarrow{P_1P_2}$ and $\overrightarrow{P_2P_1}$.

Q.2-11 What are the expressions for $\mathbf{A} \cdot \mathbf{B}$ and $\mathbf{A} \times \mathbf{B}$ in Cartesian coordinates?

REMARKS

1. Proper metric coefficients must be used in converting angle-changes to length-changes.
2. Do not confuse cylindrical distance, r, measured from the z-axis with spherical distance, R, measured from the origin.
3. Cross products of base vectors in each coordinate system follow the right-hand rule in cyclic order.

2-5 GRADIENT OF A SCALAR FIELD

In electromagnetics we often deal with quantities that depend on both time and position. Since three coordinate variables are involved in a three-dimensional space, we expect to encounter scalar and vector fields that are functions of four variables: (t, u_1, u_2, u_3). In general, the fields may change as any one of the four variables changes. We now address the method for describing the space rate of change of a scalar field at a given time. Partial derivatives with respect to the three space-coordinate variables are involved, and, since the rate of change may be different in different directions, a vector is needed to define the space rate of change of a scalar field at a given point and at a given time.

Let us consider a scalar function of space coordinates $V(u_1, u_2, u_3)$, which may represent, say, the temperature distribution in a building, the altitude of a mountainous terrain, or the electric potential in a region. The magnitude of V, in general, depends on the position of the point in space, but it may be constant along certain lines or surfaces. Figure 2-16 shows two surfaces on which the magnitude of V is constant and has the values V_1 and $V_1 + dV$, respectively, where dV indicates a small change in V. We should note that constant-V surfaces need not coincide with any of the surfaces that define a particular coordinate system. Point P_1 is on surface V_1; P_2 is the corresponding point on surface $V_1 + dV$ along the normal vector $d\mathbf{n}$; and P_3 is a point close to P_2 along another vector $d\boldsymbol{\ell} \neq d\mathbf{n}$. For the same change dV in V, the space rate of change, $dV/d\ell$, is obviously greatest along $d\mathbf{n}$ because

FIGURE 2-16 Concerning gradient of a scalar.

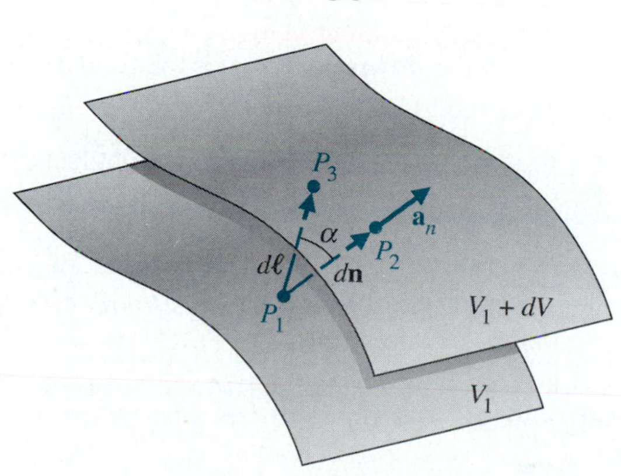

dn is the shortest distance between the two surfaces.† Since the magnitude of $dV/d\ell$ depends on the direction of $d\ell$, $dV/d\ell$ is a directional derivative. ***We define the vector that represents both the magnitude and the direction of the maximum space rate of increase of a scalar as the gradient of that scalar.*** We write

The gradient of a scalar field: physical definition

The gradient of a scalar field: mathematical definition

$$\mathbf{grad}\, V \triangleq \mathbf{a}_n \frac{dV}{dn}. \tag{2-48}$$

For brevity it is customary to employ the operator *del*, represented by the symbol ∇,‡ and write ∇V in place of **grad** V. Thus,

$$\nabla V \triangleq \mathbf{a}_n \frac{dV}{dn}. \tag{2-49}$$

We have assumed that dV is positive (an increase in V); if dV is negative (a decrease in V from P_1 to P_2), ∇V will be negative in the $\mathbf{a}_n$ direction.

The directional derivative along $d\ell$ is

$$\begin{aligned} \frac{dV}{d\ell} &= \frac{dV}{dn}\frac{dn}{d\ell} = \frac{dV}{dn}\cos\alpha \\ &= \frac{dV}{dn}\mathbf{a}_n \cdot \mathbf{a}_\ell = (\nabla V)\cdot \mathbf{a}_\ell. \end{aligned} \tag{2-50}$$

Equation (2-50) states that the space rate of increase of V in the $\mathbf{a}_\ell$ direction is equal to the projection (the component) of the gradient of V in that direction. We can also write Eq. (2-50) as

Space rate of increase of V in terms of ∇V

$$dV = (\nabla V)\cdot d\boldsymbol{\ell}, \tag{2-51}$$

where $d\boldsymbol{\ell} = \mathbf{a}_\ell\, d\ell$. Now, dV in Eq. (2-51) is the total differential of V as a result of a change in position (from P_1 to P_3 in Fig. 2-16); it can be expressed as:

$$dV = \frac{\partial V}{\partial \ell_1} d\ell_1 + \frac{\partial V}{\partial \ell_2} d\ell_2 + \frac{\partial V}{\partial \ell_3} d\ell_3, \tag{2-52}$$

where $d\ell_1$, $d\ell_2$, and $d\ell_3$ are the components of the vector differential displacement $d\boldsymbol{\ell}$ in a chosen coordinate system. In Cartesian coordinates,

†In a more formal treatment, changes ΔV and $\Delta\ell$ would be used, and the ratio $\Delta V/\Delta\ell$ would become the derivative $dV/d\ell$ as $\Delta\ell$ approaches zero. We avoid this formality in favor of simplicity.

‡∇ is sometimes also called the *nabla* operator.

$(u_1, u_2, u_3) = (x, y, z)$, and $d\ell_1, d\ell_2$, and $d\ell_3$ are respectively, dx, dy, and dz (see Eq. 2-23). We can write dV in Eq. (2-52) as the dot product of two vectors, as follows:

$$\begin{aligned} dV &= \left(\mathbf{a}_x \frac{\partial V}{\partial x} + \mathbf{a}_y \frac{\partial V}{dy} + \mathbf{a}_z \frac{\partial V}{\partial z}\right) \cdot (\mathbf{a}_x\, dx + \mathbf{a}_y\, dy + \mathbf{a}_z\, dz) \\ &= \left(\mathbf{a}_x \frac{\partial V}{\partial x} + \mathbf{a}_y \frac{\partial V}{\partial y} + \mathbf{a}_z \frac{\partial V}{\partial z}\right) \cdot d\boldsymbol{\ell}. \end{aligned} \tag{2-53}$$

Comparing Eq. (2-53) with Eq. (2-51), we obtain

∇V in Cartesian coordinates

$$\nabla V = \mathbf{a}_x \frac{\partial V}{\partial x} + \mathbf{a}_y \frac{\partial V}{\partial y} + \mathbf{a}_z \frac{\partial V}{\partial z}, \tag{2-54}$$

or

$$\nabla V = \left(\mathbf{a}_x \frac{\partial}{\partial x} + \mathbf{a}_y \frac{\partial}{\partial y} + \mathbf{a}_z \frac{\partial}{\partial z}\right) V. \tag{2-55}$$

In view of Eq. (2-55), it is convenient to consider ∇ in Cartesian coordinates as a vector differential operator.

$$\nabla \equiv \mathbf{a}_x \frac{\partial}{\partial x} + \mathbf{a}_y \frac{\partial}{\partial y} + \mathbf{a}_z \frac{\partial}{\partial z}. \tag{2-56}$$

In general orthogonal coordinates (u_1, u_2, u_3) with metric coefficients (h_1, h_2, h_3) we can define ∇ as

$$\nabla \equiv \left(\mathbf{a}_{u_1} \frac{\partial}{h_1\, \partial u_1} + \mathbf{a}_{u_2} \frac{\partial}{h_2\, \partial u_2} + \mathbf{a}_{u_3} \frac{\partial}{h_3\, \partial u_3}\right). \tag{2-57}$$

The expressions for ∇V in cylindrical and spherical coordinates are given on the inside of the back cover of the book.

EXAMPLE 2-9

The electrostatic field intensity $\mathbf{E}$ is derivable as the negative gradient of a scalar electric potential V; that is, $\mathbf{E} = -\nabla V$. Determine $\mathbf{E}$ at the point (1, 1, 0) if

a) $V = V_0 e^{-x} \sin \dfrac{\pi y}{4}$,

b) $V = E_0 R \cos \theta$.

SOLUTION

a) We use Eq. (2-54) to evaluate $\mathbf{E} = -\nabla V$ in Cartesian coordinates.

$$\mathbf{E} = -\left[\mathbf{a}_x \frac{\partial}{\partial x} + \mathbf{a}_y \frac{\partial}{\partial y} + \mathbf{a}_z \frac{\partial}{\partial z}\right] V_0 e^{-x} \sin \frac{\pi y}{4}$$

$$= \left(\mathbf{a}_x \sin \frac{\pi y}{4} - \mathbf{a}_y \frac{\pi}{4} \cos \frac{\pi y}{4}\right) V_0 e^{-x}.$$

Thus, $\mathbf{E}(1, 1, 0) = \left(\mathbf{a}_x - \mathbf{a}_y \frac{\pi}{4}\right) \frac{V_0}{\sqrt{2}} = \mathbf{a}_E E,$

where

$$E = V_0 \sqrt{\frac{1}{2}\left(1 + \frac{\pi^2}{16}\right)},$$

$$\mathbf{a}_E = \frac{1}{\sqrt{1 + (\pi^2/16)}} \left(\mathbf{a}_x - \mathbf{a}_y \frac{\pi}{4}\right).$$

b) Here V is given as a function of the spherical coordinate θ. For spherical coordinates, we have $(u_1, u_2, u_3) = (R, \theta, \phi)$ and $(h_1, h_2, h_3) = (1, R, R \sin \theta)$—see Table 2-1. We have, from Eq. (2-57),

$$\mathbf{E} = -\nabla V = -\left[\mathbf{a}_R \frac{\partial}{\partial R} + \mathbf{a}_\theta \frac{\partial}{R \partial \theta} + \mathbf{a}_\phi \frac{\partial}{R \sin \theta \, \partial \phi}\right] E_0 R \cos \theta$$

$$= -(\mathbf{a}_R \cos \theta - \mathbf{a}_\theta \sin \theta) E_0.$$

In view of Eqs. (2-47), the results above converts very simply to $\mathbf{E} = -\mathbf{a}_z E_0$ in Cartesian coordinates. This makes sense, since a careful examination of the given V reveals that $E_0 R \cos \theta$ is, in fact, equal to $E_0 z$. In Cartesian coordinates,

$$\mathbf{E} = -\nabla V = -\mathbf{a}_z \frac{\partial}{\partial z}(E_0 z) = -\mathbf{a}_z E_0.$$

EXERCISE 2.10 Assuming $V = xy - 2yz$, find, at point $P(2, 3, 6)$,

a) the direction and the magnitude of the maximum increase of V, and

b) the space rate of decrease of V in the direction toward the origin.

ANS. (a) $\mathbf{a}_x 3 - \mathbf{a}_y 10 - \mathbf{a}_z 6$, (b) $-60/7$.

2-6 DIVERGENCE OF A VECTOR FIELD

In the preceding section we considered the spatial derivatives of a scalar field, which led to the definition of the gradient. We now turn our attention to the spatial derivatives of a vector field. This will lead to the definitions of the ***divergence*** and the ***curl*** of a vector. We discuss the meaning of divergence in this section and that of curl in Section 2-8. Both are very important in the study of electromagnetism.

In the study of vector fields it is convenient to represent field variations graphically by directed field lines, which are called ***flux lines***. They are directed lines or curves that indicate at each point the direction of the vector field, as illustrated in Fig. 2-17. The magnitude of the field at a point is depicted either by the density or by the length of the directed lines in the vicinity of the point. Figure 2-17(a) shows that the field in region A is stronger than that in region B because there is a higher density of equal-length directed lines in region A. In Fig. 2-17(b), the decreasing arrow lengths away from the point q indicate a radial field that is strongest in the region closest to q. Figure 2-17(c) depicts a uniform field.

The vector field strength in Fig. 2-17(a) is measured by the number of flux lines passing through a unit surface normal to the vector. The flux of a vector field is analogous to the flow of an incompressible fluid such as water. For a volume with an enclosed surface there will be an excess of outward or inward flow through the surface only when the volume contains a source or a sink, respectively. That is, a net positive divergence indicates the presence of a source of fluid inside the volume, and a net negative divergence indicates the

FIGURE 2-17 Flux lines of vector fields.

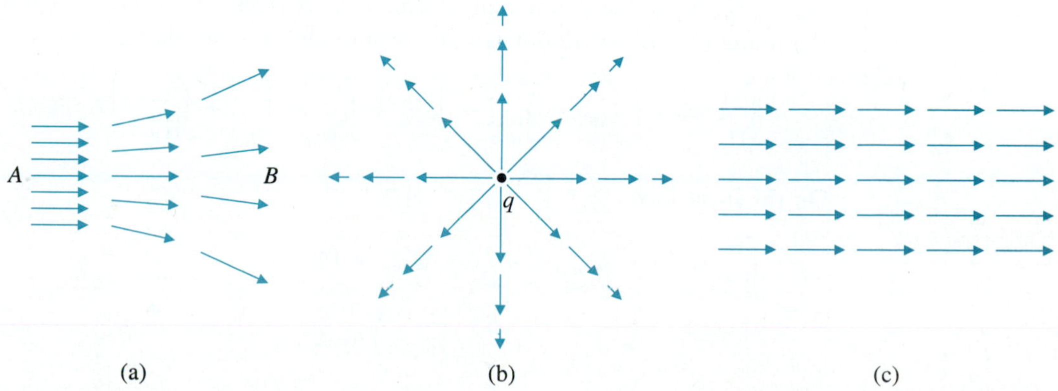

A B

(a)

q

(b)

(c)

presence of a sink. The net outward flow of the fluid per unit volume is therefore a measure of the strength of the enclosed source. In the uniform field shown in Fig. 2-17(c) there is an equal amount of inward and outward flux going through any closed volume containing no sources or sinks, resulting in a zero divergence.

The divergence of a vector field A: physical definition

We define the divergence of a vector field **A** *at a point, abbreviated div* **A**, *as the net outward flux of* **A** *per unit volume as the volume about the point tends to zero:*

The divergence of A: mathematical definition

$$\operatorname{div}\mathbf{A} \triangleq \lim_{\Delta v \to 0} \frac{\oint_S \mathbf{A} \cdot d\mathbf{s}}{\Delta v}. \tag{2-58}$$

The numerator in Eq. (2-58) is a surface integral. It is actually a double integral over two dimensions, but it is written with a single integral sign for simplicity. The small circle on the integral sign indicates that the integral is to be carried out over the *entire* surface S *enclosing a volume.* In the integrand, the vector differential surface element $d\mathbf{s} = \mathbf{a}_n\, ds$ has a magnitude ds and a direction denoted by the unit normal vector $\mathbf{a}_n$ pointing *outward* from the enclosed volume. The enclosed surface integral represents the net outward flux of the vector field **A**. Equation (2-58) is the general definition of div **A**, which is a *scalar quantity* whose magnitude may vary from point to point as **A** itself varies. This definition holds for any coordinate system; the expression for div **A**, like that for **A**, will, of course, depend on the choice of the coordinate system. We shall now derive the expression for div **A** in Cartesian coordinates.

Consider a differential volume of sides Δx, Δy, and Δz centered about a point $P(x_0, y_0, z_0)$ in the field of a vector **A**, as shown in Fig. 2-18. In Cartesian coordinates, $\mathbf{A} = \mathbf{a}_x A_x + \mathbf{a}_y A_y + \mathbf{a}_y A_z$. We wish to find div **A** at the point (x_0, y_0, z_0). Since the differential volume has six faces, the surface integral in the numerator of Eq. (2-58) can be decomposed into six parts:

$$\oint_S \mathbf{A} \cdot d\mathbf{s} = \left[\int_{\substack{\text{front}\\\text{face}}} + \int_{\substack{\text{back}\\\text{face}}} + \int_{\substack{\text{right}\\\text{face}}} + \int_{\substack{\text{left}\\\text{face}}} + \int_{\substack{\text{top}\\\text{face}}} + \int_{\substack{\text{bottom}\\\text{face}}}\right] \mathbf{A} \cdot d\mathbf{s}. \tag{2-59}$$

On the front face,

$$\begin{aligned} \int_{\substack{\text{front}\\\text{face}}} \mathbf{A} \cdot d\mathbf{s} &= \mathbf{A}_{\substack{\text{front}\\\text{face}}} \cdot \Delta\mathbf{s}_{\substack{\text{front}\\\text{face}}} = \mathbf{A}_{\substack{\text{front}\\\text{face}}} \cdot \mathbf{a}_x(\Delta y\, \Delta z) \\ &= A_x\left(x_0 + \frac{\Delta x}{2}, y_0, z_0\right) \Delta y\, \Delta z. \end{aligned} \tag{2-60}$$

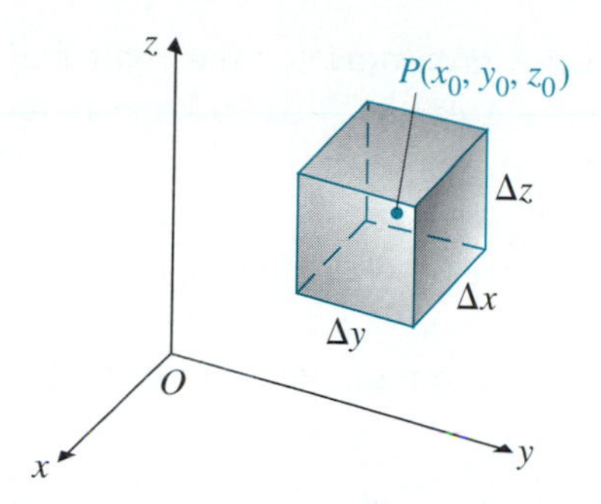

FIGURE 2-18 A differential volume in Cartesian coordinates.

The quantity $A_x(x_0+(\Delta x/2),\ y_0,\ z_0)$ can be expanded as a Taylor series about its value at (x_0, y_0, z_0), as follows:

$$A_x\left(x_0+\frac{\Delta x}{2},\ y_0,\ z_0\right)=A_x(x_0,\ y_0,\ z_0)+\frac{\Delta x}{2}\frac{\partial A_x}{\partial x}\bigg|_{(x_0,\,y_0,\,z_0)} + \text{higher-order terms}, \tag{2-61}$$

where the higher-order terms (H.O.T.) contain the factors $(\Delta x/2)^2$, $(\Delta x/2)^3$, etc. Similarly, on the back face,

$$\begin{aligned}\int_{\substack{\text{back}\\\text{face}}}\mathbf{A}\cdot d\mathbf{s}&=\mathbf{A}_{\substack{\text{back}\\\text{face}}}\cdot\Delta\mathbf{s}_{\substack{\text{back}\\\text{face}}}=\mathbf{A}_{\substack{\text{back}\\\text{face}}}\cdot(-\mathbf{a}_x\,\Delta y\,\Delta z)\\&=-A_x\left(x_0-\frac{\Delta x}{2},\ y_0,\ z_0\right)\Delta y\,\Delta z.\end{aligned} \tag{2-62}$$

The Taylor-series expansion of $A_x\left(x_0-\dfrac{\Delta x}{2},\ y_0,\ z_0\right)$ is

$$A_x\left(x_0-\frac{\Delta x}{2},\ y_0,\ z_0\right)=A_x(x_0, y_0, z_0)-\frac{\Delta x}{2}\frac{\partial A_x}{\partial x}\bigg|_{(x_0,\,y_0,\,z_0)}+\text{H.O.T.} \tag{2-63}$$

Substituting Eq. (2-61) in Eq. (2-60) and Eq. (2-63) in Eq. (2-62) and adding the contributions, we have

$$\left[\int_{\substack{\text{front}\\\text{face}}}+\int_{\substack{\text{back}\\\text{face}}}\right]\mathbf{A}\cdot d\mathbf{s}=\left(\frac{\partial A_x}{\partial x}+\text{H.O.T.}\right)\bigg|_{(x_0,\,y_0,\,z_0)}\Delta x\,\Delta y\,\Delta z. \tag{2-64}$$

Here a Δx has been factored out from the H.O.T. in Eqs. (2-61) and (2-63), but all terms of the H.O.T. in Eq. (2-64) still contain powers of Δx.

Following the same procedure for the right and left faces, where the coordinate changes are $+\Delta y/2$ and $-\Delta y/2$, respectively, and $\Delta s = \Delta x\,\Delta z$, we find

$$\left[\int_{\substack{\text{right}\\ \text{face}}} + \int_{\substack{\text{left}\\ \text{face}}}\right]\mathbf{A}\cdot d\mathbf{s} = \left(\frac{\partial A_y}{\partial y} + \text{H.O.T.}\right)\Bigg|_{(x_0,\,y_0,\,z_0)} \Delta x\,\Delta y\,\Delta z. \tag{2-65}$$

Here the higher-order terms contain the factors Δy, $(\Delta y)^2$, etc. For the top and bottom faces we have

$$\left[\int_{\substack{\text{top}\\ \text{face}}} + \int_{\substack{\text{bottom}\\ \text{face}}}\right]\mathbf{A}\cdot d\mathbf{s} = \left(\frac{\partial A_z}{\partial z} + \text{H.O.T.}\right)\Bigg|_{(x_0,\,y_0,\,z_0)} \Delta x\,\Delta y\,\Delta z, \tag{2-66}$$

where the higher-order terms contain the factors Δz, $(\Delta z)^2$, etc. Now the results from Eqs. (2-64), (2-65), and (2-66) are combined in Eq. (2-59) to obtain

$$\oint_S \mathbf{A}\cdot d\mathbf{s} = \left(\frac{\partial A_x}{\partial x} + \frac{\partial A_y}{\partial y} + \frac{\partial A_z}{\partial z}\right)\Bigg|_{(x_0,\,y_0,\,z_0)} \Delta x\,\Delta y\,\Delta z$$
$$+\text{higher-order terms in } \Delta x,\ \Delta y,\ \Delta z. \tag{2-67}$$

Since $\Delta v = \Delta x\,\Delta y\,\Delta z$, substitution of Eq. (2-67) in Eq. (2-58) yields the expression of div **A** in Cartesian coordinates:

$\nabla\cdot\mathbf{A}$ in Cartesian coordinates

$$\text{div}\,\mathbf{A} = \frac{\partial A_x}{\partial x} + \frac{\partial A_y}{\partial y} + \frac{\partial A_z}{\partial z}. \tag{2-68}$$

The higher-order terms vanish as the differential volume $\Delta x\,\Delta y\,\Delta z$ approaches zero. The value of div **A**, in general, depends on the position of the point at which it is evaluated. We have dropped the notation (x_0, y_0, z_0) in Eq. (2-68) because it applies to any point at which **A** and its partial derivatives are defined.

With the vector differential operator del, ∇, defined in Eq. (2-56) we can write Eq. (2-68) alternatively as $\nabla\cdot\mathbf{A}$ (read "del dot **A**"); that is,

$$\nabla\cdot\mathbf{A} \equiv \text{div}\,\mathbf{A}. \tag{2-69}$$

In general orthogonal curvilinear coordinates (u_1, u_2, u_3), Eq. (2-58) will lead to

$\nabla\cdot\mathbf{A}$ in general orthogonal coordinate system

$$\nabla\cdot\mathbf{A} = \frac{1}{h_1h_2h_3}\left[\frac{\partial}{\partial u_1}(h_2h_3A_1) + \frac{\partial}{\partial u_2}(h_1h_3A_2) + \frac{\partial}{\partial u_3}(h_1h_2A_3)\right]. \tag{2-70}$$

The expressions for $\nabla \cdot \mathbf{A}$ in cylindrical and spherical coordinates are given on the inside of the back cover of the book.

EXAMPLE 2-10

Find the divergence of the position vector to an arbitrary point.

SOLUTION

We will find the solution in Cartesian as well as in spherical coordinates.

a) *Cartesian coordinates.* The expression for the position vector to an arbitrary point (x, y, z) is

$$\overrightarrow{OP} = \mathbf{A} = \mathbf{a}_x x + \mathbf{a}_y y + \mathbf{a}_z z. \tag{2-71}$$

Using Eq. (2-68), we have

$$\nabla \cdot (\overrightarrow{OP}) = \nabla \cdot \mathbf{A} = \frac{\partial x}{\partial x} + \frac{\partial y}{\partial y} + \frac{\partial z}{\partial z} = 3.$$

b) *Spherical coordinates.* Here the position vector is simply

$$\overrightarrow{OP} = \mathbf{A} = \mathbf{a}_R R. \tag{2-72}$$

Its divergence in spherical coordinates (R, θ, ϕ) can be obtained from Eq. (2-70) by using Table 2-1 as follows:

$\nabla \cdot \mathbf{A}$ in spherical coordinates

$$\boxed{\nabla \cdot \mathbf{A} = \frac{1}{R^2}\frac{\partial}{\partial R}(R^2 A_R) + \frac{1}{R\sin\theta}\frac{\partial}{\partial\theta}(A_\theta \sin\theta) + \frac{1}{R\sin\theta}\frac{\partial A_\phi}{\partial\phi}.} \tag{2-73}$$

Substituting Eq. (2-72) in Eq. (2-73), we also obtain $\nabla \cdot (\overrightarrow{OP}) = 3$, as expected.

■ **EXERCISE 2.11** Solve Example 2-10 in cylindrical coordinates.

EXAMPLE 2-11

The magnetic flux density $\mathbf{B}$ outside a very long current-carrying wire is circumferential and is inversely proportional to the distance to the axis of the wire. Find $\nabla \cdot \mathbf{B}$.

SOLUTION

Let the long wire be coincident with the z-axis in a cylindrical coordinate system. The problem states that

$$\mathbf{B} = \mathbf{a}_\phi \frac{k}{r},$$

where k is a constant. The divergence of a vector field in cylindrical coordinates (r, ϕ, z) can be found from Eq. (2-70) by using Table 2-1.

$\nabla \cdot \mathbf{A}$ in cylindrical coordinates

$$\nabla \cdot \mathbf{B} = \frac{1}{r}\frac{\partial}{\partial r}(rB_r) + \frac{1}{r}\frac{\partial B_\phi}{\partial \phi} + \frac{\partial B_z}{\partial z}. \tag{2-74}$$

Now $B_\phi = k/r$, and $B_r = B_z = 0$. Equation (2-74) gives

$$\nabla \cdot \mathbf{B} = 0.$$

Solenoidal field

We have here a vector that is not a constant, but whose divergence is zero. A divergenceless field is called a ***solenoidal field***. We will see in Chapter 5 that magnetic field is solenoidal.

2-7 DIVERGENCE THEOREM

Divergence theorem

In the preceding section we defined the divergence of a vector field as the net outward flux per unit volume. We may expect intuitively that ***the volume integral of the divergence of a vector field equals the total outward flux of the vector through the surface that bounds the volume***; that is,

$$\int_V \nabla \cdot \mathbf{A}\, dv = \oint_S \mathbf{A} \cdot d\mathbf{s}. \tag{2-75}$$

This identity, which will be proved in the following paragraph, is called the ***divergence theorem.***† It applies to any volume V that is bounded by surface S. The direction of $d\mathbf{s}$ is always that of the *outward normal*, perpendicular to the surface ds and directed away from the volume.

For a very small differential volume element Δv_j bounded by a surface s_j, the definition of $\nabla \cdot \mathbf{A}$ in Eq. (2-58) gives directly

$$(\nabla \cdot \mathbf{A})_j \Delta v_j = \oint_{s_j} \mathbf{A} \cdot d\mathbf{s}. \tag{2-76}$$

†It is also known as ***Gauss's theorem***.

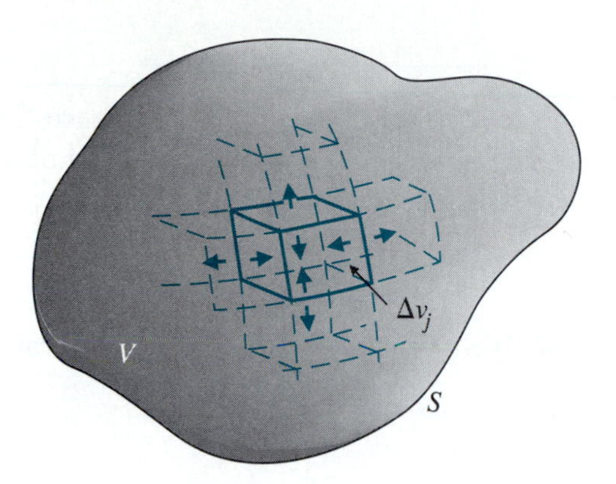

FIGURE 2-19 Subdivided volume for proof of divergence theorem.

In case of an arbitrary volume V, we can subdivide it into many, say, N, small differential volumes, of which Δv_j is typical. This is depicted in Fig. 2-19. Let us now combine the contributions of all these differential volumes to both sides of Eq. (2-76). We have

$$\lim_{\Delta v_j \to 0} \left[\sum_{j=1}^{N} (\nabla \cdot \mathbf{A})_j \Delta v_j \right] = \lim_{\Delta v_j \to 0} \left[\sum_{j=1}^{N} \oint_{s_j} \mathbf{A} \cdot d\mathbf{s} \right]. \tag{2-77}$$

The left side of Eq. (2-77) is, by definition, the volume integral of $\nabla \cdot \mathbf{A}$:

$$\lim_{\Delta v_j \to 0} \left[\sum_{j=1}^{N} (\nabla \cdot \mathbf{A})_j \Delta v_j \right] = \int_V (\nabla \cdot \mathbf{A})\, dv. \tag{2-78}$$

The surface integrals on the right side of Eq. (2-77) are summed over all the faces of all the differential volume elements. The contributions from the internal surfaces of adjacent elements will, however, cancel each other, because at a common internal surface the outward normals of the adjacent elements point in opposite directions. Hence the net contribution of the right side of Eq. (2-77) is due only to that of the external surface S bounding the volume V; that is,

$$\lim_{\Delta v_j \to 0} \left[\sum_{j=1}^{N} \oint_{s_j} \mathbf{A} \cdot d\mathbf{s} \right] = \oint_S \mathbf{A} \cdot d\mathbf{s}. \tag{2-79}$$

The substitution of Eqs. (2-78) and (2-79) in Eq. (2-77) yields the divergence theorem in Eq. (2-75).

The divergence theorem is an important identity in vector analysis. *It converts a volume integral of the divergence of a vector to a closed surface integral of the vector, and vice versa.* We use it frequently in establishing other theorems and relations in electromagnetics. We emphasize that, *although a single integral sign is used on both sides of Eq. (2-75) for simplicity, the volume and surface integrals represent triple and double integrations, respectively.*

EXAMPLE 2-12

Given $\mathbf{A} = \mathbf{a}_x x^2 + \mathbf{a}_y xy + \mathbf{a}_z yz$, verify the divergence theorem over a cube one unit on each side. The cube is situated in the first octant of the Cartesian coordinate system with one corner at the origin.

SOLUTION

Refer to Fig. 2-20. We first evaluate the surface integral over the six faces.

1. Front face: $x = 1$, $d\mathbf{s} = \mathbf{a}_x\, dy\, dz$;

$$\int_{\substack{\text{front}\\\text{face}}} \mathbf{A}\cdot d\mathbf{s} = \int_0^1 \int_0^1 dy\, dz = 1.$$

2. Back face: $x = 0$, $d\mathbf{s} = -\mathbf{a}_x\, dy\, dz$;

$$\int_{\substack{\text{back}\\\text{face}}} \mathbf{A}\cdot d\mathbf{s} = 0.$$

3. Left face: $y = 0$, $d\mathbf{s} = -\mathbf{a}_y\, dx\, dz$;

$$\int_{\substack{\text{left}\\\text{face}}} \mathbf{A}\cdot d\mathbf{s} = 0.$$

4. Right face: $y = 1$, $d\mathbf{s} = \mathbf{a}_y\, dx\, dz$;

$$\int_{\substack{\text{right}\\\text{face}}} \mathbf{A}\cdot d\mathbf{s} = \int_0^1 \int_0^1 x\, dx\, dz = \tfrac{1}{2}.$$

5. Top face: $z = 1$, $d\mathbf{s} = \mathbf{a}_z\, dx\, dy$;

$$\int_{\substack{\text{top}\\\text{face}}} \mathbf{A}\cdot d\mathbf{s} = \int_0^1 \int_0^1 y\, dx\, dy = \tfrac{1}{2}.$$

6. Bottom face: $z = 0$, $d\mathbf{s} = -\mathbf{a}_z\, dx\, dy$;

$$\int_{\substack{\text{bottom}\\\text{face}}} \mathbf{A}\cdot d\mathbf{s} = 0.$$

FIGURE 2-20 A unit cube (Example 2-12).

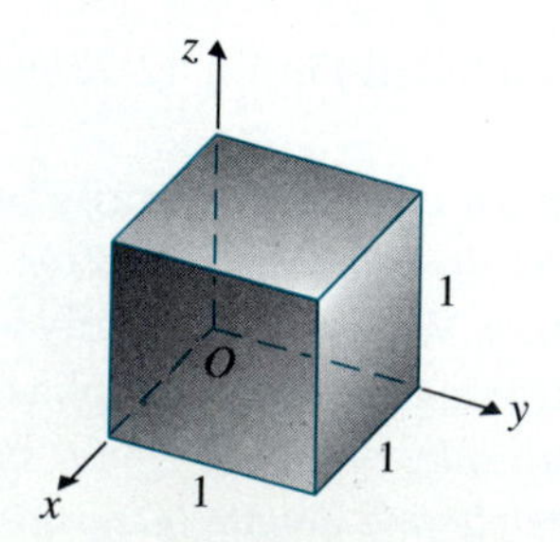

Adding the above six values, we have

$$\oint_S \mathbf{A} \cdot d\mathbf{s} = 1 + 0 + 0 + \tfrac{1}{2} + \tfrac{1}{2} + 0 = 2. \tag{2-80}$$

Now the divergence of **A** is

$$\nabla \cdot \mathbf{A} = \frac{\partial}{\partial x}(x^2) + \frac{\partial}{\partial y}(xy) + \frac{\partial}{\partial z}(yz) = 3x + y.$$

Hence,

$$\int_V \nabla \cdot \mathbf{A}\, dv = \int_0^1 \int_0^1 \int_0^1 (3x + y)\, dx\, dy\, dz = 2, \tag{2-81}$$

which is the same as the result of the closed surface integral in (2-80). The divergence theorem is therefore verified.

EXAMPLE 2-13

Given $\mathbf{F} = \mathbf{a}_R kR$, determine whether the divergence theorem holds for the shell region enclosed by spherical surfaces at $R = R_1$ and $R = R_2 (R_2 > R_1)$ centered at the origin, as shown in Fig. 2-21.

SOLUTION

Here the specified region has two surfaces, at $R = R_1$ and $R = R_2$.

At the outer surface: $R = R_2$, $d\mathbf{s} = \mathbf{a}_R R_2^2 \sin\theta\, d\theta\, d\phi$;

$$\int_{\substack{\text{outer}\\ \text{surface}}} \mathbf{F} \cdot d\mathbf{s} = \int_0^{2\pi} \int_0^{\pi} (kR_2) R_2^2 \sin\theta\, d\theta\, d\phi = 4\pi k R_2^3.$$

FIGURE 2-21 A spherical shell region (Example 2-13).

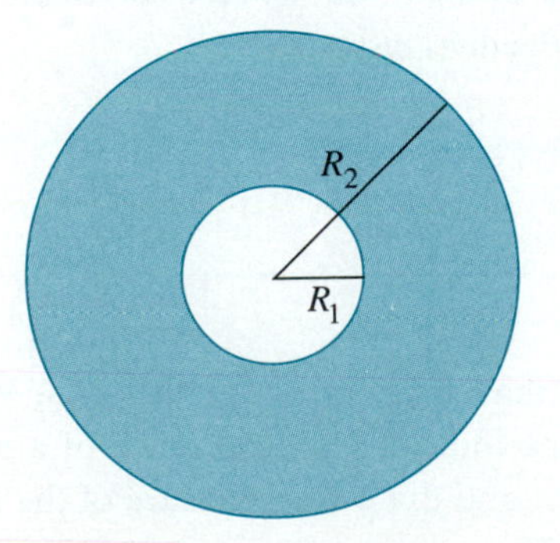

At the inner surface: $R = R_1$, $d\mathbf{s} = -\mathbf{a}_R R_1^2 \sin\theta \, d\theta \, d\phi$;

$$\int_{\substack{\text{inner}\\\text{surface}}} \mathbf{F} \cdot d\mathbf{s} = -\int_0^{2\pi} \int_0^{\pi} (kR_1) R_1^2 \sin\theta \, d\theta \, d\phi = -4\pi k R_1^3.$$

Actually, since the integrand is independent of θ or ϕ in both cases, the integral of a constant over a spherical surface is simply the constant multiplied by the area of the surface ($4\pi R_2^2$ for the outer surface and $4\pi R_1^2$ for the inner surface), and no integration is necessary. Adding the two results, we have

$$\oint_S \mathbf{F} \cdot d\mathbf{s} = 4\pi k (R_2^3 - R_1^3). \tag{2-82}$$

To find the volume integral, we first determine $\nabla \cdot \mathbf{F}$ for an $\mathbf{F}$ that has only an F_R component. From Eq. (2-73), we have

$$\nabla \cdot \mathbf{F} = \frac{1}{R^2} \frac{\partial}{\partial R} (R^2 F_R) = \frac{1}{R^2} \frac{\partial}{\partial R} (kR^3) = 3k.$$

Since $\nabla \cdot \mathbf{F}$ is a constant, its volume integral equals the product of the constant and the volume. The volume of the shell region between the two spherical surfaces with radii R_1 and R_2 is $4\pi(R_2^3 - R_1^3)/3$. Therefore,

$$\int_V \nabla \cdot \mathbf{F} \, dv = (\nabla \cdot \mathbf{F}) V = 4\pi k (R_2^3 - R_1^3), \tag{2-83}$$

which is the same as the result in Eq. (2-82).

This example shows that the divergence theorem holds even when the volume has holes inside—that is, even when the volume is enclosed by a multiply connected surface.

■ **EXERCISE 2.12** Given a vector field $\mathbf{A} = \mathbf{a}_r r + \mathbf{a}_z z$,

a) find the total outward flux over a circular cylinder around the z-axis with a radius 2 and a height 4 centered at origin.

b) Repeat (a) for the same cylinder with its base coinciding with the xy-plane.

c) Find $\nabla \cdot \mathbf{A}$ and verify the divergence theorem.

ANS. (a) 48π, (c) 3.

2-8 CURL OF A VECTOR FIELD

$\nabla \cdot \mathbf{A}$ is a measure of the strength of the flow source of A.

In Section 2-6 we stated that a net outward flux of a vector $\mathbf{A}$ through a surface bounding a volume indicates the presence of a source. This source may be called a *flow source*, and div $\mathbf{A}$ is a measure of the strength of the flow source. There is another kind of source, called ***vortex source***, which causes a circulation of a vector field around it. The ***net circulation*** (or simply ***circulation***) of a vector field around a *closed path* is defined as the scalar line integral of the vector over the path. We have

$$\text{Circulation of } \mathbf{A} \text{ around contour } C \triangleq \oint_C \mathbf{A} \cdot d\boldsymbol{\ell}. \tag{2-84}$$

Equation (2-87) is a mathematical definition. The physical meaning of circulation depends on what kind of field the vector **A** represents. If **A** is a force acting on an object, its circulation will be the work done by the force in moving the object once around the contour; if **A** represents an electric field intensity, then the circulation will be an electromotive force around the closed path. The familiar phenomenon of water whirling down a sink drain is an example of a *vortex sink* causing a circulation of fluid velocity. A circulation of **A** may exist even when $\operatorname{div} \mathbf{A} = 0$ (when there is no flow source).

EXAMPLE 2-14

Given a vector field $\mathbf{F} = \mathbf{a}_x xy - \mathbf{a}_y 2x$, find its circulation around the path $OABO$ shown in Fig. 2-22.

SOLUTION

Let us split the circulation integral into three parts.

$$\oint_{OABO} \mathbf{F} \cdot d\boldsymbol{\ell} = \int_O^A \mathbf{F} \cdot d\boldsymbol{\ell} + \int_A^B \mathbf{F} \cdot d\boldsymbol{\ell} + \int_B^O \mathbf{F} \cdot d\boldsymbol{\ell}.$$

Along path OA: $y = 0$, $\mathbf{F} = -\mathbf{a}_y 2x$, $d\boldsymbol{\ell} = \mathbf{a}_x dx$, $\mathbf{F} \cdot d\boldsymbol{\ell} = 0$.

$$\int_O^A \mathbf{F} \cdot d\boldsymbol{\ell} = 0.$$

FIGURE 2-22 Path for line integral (Examples 2-14 and 2-16).

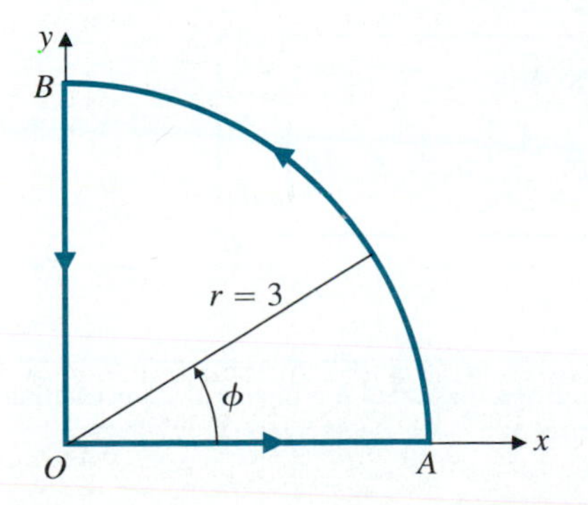

Along path BO: $x = 0$, $\mathbf{F} = \mathbf{0}$. $\int_B^O \mathbf{F} \cdot d\boldsymbol{\ell} = 0$.

Along path AB: $d\boldsymbol{\ell} = \mathbf{a}_x\,dx + \mathbf{a}_y\,dy$ (see Eq. 2-23).

$\mathbf{F} \cdot d\boldsymbol{\ell} = xy\,dx - 2x\,dy$.

The equation of the quarter-circle is $x^2 + y^2 = 9$ $(0 \leqslant x,\ y \leqslant 3)$. Therefore,

$$\begin{aligned}\int_A^B \mathbf{F} \cdot d\boldsymbol{\ell} &= \int_3^0 x\sqrt{9 - x^2}\,dx - 2\int_0^3 \sqrt{9 - y^2}\,dy \\ &= -\frac{1}{3}(9 - x^2)^{3/2}\Big|_3^0 - \left[y\sqrt{9 - y^2} + 9\sin^{-1}\frac{y}{3}\right]_0^3 \\ &= -9\left(1 + \frac{\pi}{2}\right).\end{aligned}$$

Hence,

$$\oint_{OABO} \mathbf{F} \cdot d\boldsymbol{\ell} = -9\left(1 + \frac{\pi}{2}\right).$$

EXERCISE 2.13 Find the clockwise circulation of the vector field $\mathbf{F}$ given in Example 2-14 around a square path in the xy-plane centered at the origin and having four units on each side $(-2 \leqslant x \leqslant 2$ and $-2 \leqslant y \leqslant 2)$.

ANS. 32.

Since circulation as defined in Eq. (2-84) is a line integral of a dot product, its value obviously depends on the orientation of the contour C relative to the vector $\mathbf{A}$. In order to define a point function, which is a measure of the strength of a vortex source, we must make C very small and orient it in such a way that the circulation is a maximum. We define[†]

Mathematical definition of a vector field A

$$\begin{aligned}\operatorname{curl}\mathbf{A} &\equiv \nabla \times \mathbf{A} \\ &\triangleq \lim_{\Delta s \to 0} \frac{1}{\Delta s}\left[\mathbf{a}_n \oint_C \mathbf{A} \cdot d\boldsymbol{\ell}\right]_{\max}.\end{aligned} \qquad \text{(2-85)}$$

[†] In books published in Europe, the curl of $\mathbf{A}$ is often called the rotation of $\mathbf{A}$ and written as rot $\mathbf{A}$. $\nabla \times \mathbf{A}$ is read as "del cross $\mathbf{A}$."

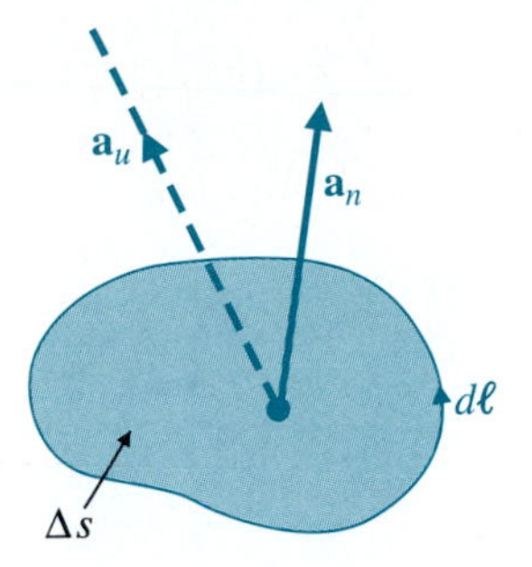

FIGURE 2-23 Relation between $\mathbf{a}_n$ and $d\boldsymbol{\ell}$ in defining curl.

Physical definition of $\nabla \times \mathbf{A}$, a measure of the strength of the vortex source of A

In words, Eq. (2-85) states that *the curl of a vector field* **A**, *denoted by curl* **A** *or* $\nabla \times \mathbf{A}$, *is a vector whose magnitude is the maximum net circulation of* **A** *per unit area as the area tends to zero and whose direction is the normal direction of the area when the area is oriented to make the net circulation maximum.* Because the normal to an area can point in two opposite directions, we adhere to the right-hand rule that when the fingers of the right hand follow the direction of $d\boldsymbol{\ell}$, the thumb points to the $\mathbf{a}_n$ direction. This is illustrated in Fig. 2-23. Curl **A** is a vector point function. Its component in any other direction $\mathbf{a}_u$ is $\mathbf{a}_u \cdot (\nabla \times \mathbf{A})$, which can be determined from the circulation per unit area normal to $\mathbf{a}_u$ as the area approaches zero.

$$\boxed{(\nabla \times \mathbf{A})_u = \mathbf{a}_u \cdot (\nabla \times \mathbf{A}) = \lim_{\Delta s_u \to 0} \frac{1}{\Delta s_u} \left(\oint_{C_u} \mathbf{A} \cdot d\boldsymbol{\ell} \right),} \tag{2-86}$$

where the direction of the line integration around the contour C_u bounding area Δs_u and the direction $\mathbf{a}_u$ follow the right-hand rule.

We now use Eq. (2-86) to find the three components of $\nabla \times \mathbf{A}$ in Cartesian coordinates. Refer to Fig. 2-24, in which a differential rectangular area parallel to the yz-plane and having sides Δy and Δz is drawn about a typical point $P(x_0, y_0, z_0)$. We have $\mathbf{a}_u = \mathbf{a}_x$ and $\Delta s_u = \Delta y\, \Delta z$, and the contour C_u consists of the four sides 1, 2, 3, and 4. Thus,

$$(\nabla \times \mathbf{A})_x = \lim_{\Delta y \Delta z \to 0} \frac{1}{\Delta y\, \Delta z} \left(\oint_{\substack{\text{sides}\\ 1,2,3,4}} \mathbf{A} \cdot d\boldsymbol{\ell} \right). \tag{2-87}$$

In Cartesian coordinates, $\mathbf{A} = \mathbf{a}_x A_x + \mathbf{a}_y A_y + \mathbf{a}_z A_z$. The contributions of the four sides to the line integral are as follows.

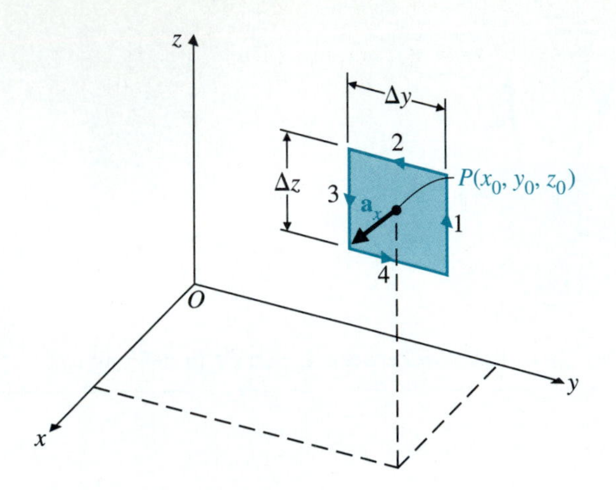

FIGURE 2-24 Determining $(\nabla \times \mathbf{A})_x$.

Side 1: $d\boldsymbol{\ell} = \mathbf{a}_z \Delta z,\ \mathbf{A}\cdot d\boldsymbol{\ell} = A_z\left(x_0, y_0 + \dfrac{\Delta y}{2}, z_0\right)\Delta z,$

where $A_z\left(x_0, y_0 + \dfrac{\Delta y}{2}, z_0\right)$ can be expanded as a Taylor series:

$$A_z\left(x_0, y_0 + \frac{\Delta y}{2}, z_0\right) = A_z(x_0, y_0, z_0) + \frac{\Delta y}{2}\frac{\partial A_z}{\partial y}\bigg|_{(x_0, y_0, z_0)} + \text{H.O.T.}, \tag{2-88}$$

where H.O.T. (higher-order terms) contain the factors $(\Delta y)^2$, $(\Delta y)^3$, etc.

Thus,

$$\int_{\text{side 1}} \mathbf{A}\cdot d\boldsymbol{\ell} = \left\{A_z(x_0, y_0, z_0) + \frac{\Delta y}{2}\frac{\partial A_z}{\partial y}\bigg|_{(x_0, y_0, z_0)} + \text{H.O.T.}\right\}\Delta z. \tag{2-89}$$

Side 3: $d\boldsymbol{\ell} = -\mathbf{a}_z \Delta z,\ \mathbf{A}\cdot d\boldsymbol{\ell} = A_z\left(x_0, y_0 - \dfrac{\Delta y}{2}, z_0\right)\Delta z,$

where

$$A_z\left(x_0, y_0 - \frac{\Delta y}{2}, z_0\right) = A_z(x_0, y_0, z_0) - \frac{\Delta y}{2}\frac{\partial A_z}{\partial y}\bigg|_{(x_0, y_0, z_0)} + \text{H.O.T.}; \tag{2-90}$$

$$\int_{\text{side } 3} \mathbf{A} \cdot d\boldsymbol{\ell} = \left\{ A_z(x_0, y_0, z_0) - \frac{\Delta y}{2} \frac{\partial A_z}{\partial y}\bigg|_{(x_0, y_0, z_0)} + \text{H.O.T.} \right\} (-\Delta z). \tag{2-91}$$

Combining Eqs. (2-89) and (2-91), we have

$$\int_{\substack{\text{sides} \\ 1 \text{ and } 3}} \mathbf{A} \cdot d\boldsymbol{\ell} = \left(\frac{\partial A_z}{\partial y} + \text{H.O.T.} \right)\bigg|_{(x_0, y_0, z_0)} \Delta y \, \Delta z. \tag{2-92}$$

The H.O.T. in Eq. (2-92) still contain powers of Δy. Similarly, it may be shown that

$$\int_{\substack{\text{sides} \\ 2 \text{ and } 4}} \mathbf{A} \cdot d\boldsymbol{\ell} = \left(-\frac{\partial A_y}{\partial z} + \text{H.O.T.} \right)\bigg|_{(x_0, y_0, z_0)} \Delta y \, \Delta z. \tag{2-93}$$

Substituting Eqs. (2-92) and (2-93) in Eq. (2-87) and noting that the higher-order terms tend to zero as Δx and $\Delta y \to 0$, we obtain the x-component of $\nabla \times \mathbf{A}$:

$$(\nabla \times \mathbf{A})_x = \frac{\partial A_z}{\partial y} - \frac{\partial A_y}{\partial z}. \tag{2-94}$$

A close examination of Eq. (2-94) will reveal a cyclic order in x, y, and z and enable us to write down the y- and z-components of $\nabla \times \mathbf{A}$. The entire expression for the curl of $\mathbf{A}$ in Cartesian coordinates is

Expression of $\nabla \times \mathbf{A}$ in Cartesian coordinates

$$\nabla \times \mathbf{A} = \mathbf{a}_x \left(\frac{\partial A_z}{\partial y} - \frac{\partial A_y}{\partial z} \right) + \mathbf{a}_y \left(\frac{\partial A_x}{\partial z} - \frac{\partial A_z}{\partial x} \right) + \mathbf{a}_z \left(\frac{\partial A_y}{\partial x} - \frac{\partial A_x}{\partial y} \right). \tag{2-95}$$

a scalar. Equation (2-95) can be remembered rather easily by arranging it in a determinantal form in the manner of the cross product exhibited in Eq. (2-27).

Another form of $\nabla \times \mathbf{A}$ in Cartesian coordinates

$$\nabla \times \mathbf{A} = \begin{vmatrix} \mathbf{a}_x & \mathbf{a}_y & \mathbf{a}_z \\ \dfrac{\partial}{\partial x} & \dfrac{\partial}{\partial y} & \dfrac{\partial}{\partial z} \\ A_x & A_y & A_z \end{vmatrix}. \tag{2-96}$$

The derivation of $\nabla \times \mathbf{A}$ in other coordinate systems follows the same procedure but is more involved. The expression for $\nabla \times \mathbf{A}$ in general orthogonal curvilinear coordinates (u_1, u_2, u_3) is given below:

Expression of $\nabla \times \mathbf{A}$ in general orthogonal coordinate system

$$\nabla \times \mathbf{A} = \frac{1}{h_1 h_2 h_3} \begin{vmatrix} \mathbf{a}_{u_1} h_1 & \mathbf{a}_{u_2} h_2 & \mathbf{a}_{u_3} h_3 \\ \dfrac{\partial}{\partial u_1} & \dfrac{\partial}{\partial u_2} & \dfrac{\partial}{\partial u_3} \\ h_1 A_1 & h_2 A_2 & h_3 A_3 \end{vmatrix}. \tag{2-97}$$

The expressions of $\nabla \times \mathbf{A}$ in cylindrical and spherical coordinates can be easily obtained from Eq. (2-97) by using the appropriate u_1, u_2, and u_3 and their metric coefficients h_1, h_2, and h_3 listed in Table 2-1. These expressions are given on the inside of the back cover.

EXAMPLE 2-15

Show that $\nabla \times \mathbf{A} = 0$ if

a) $\mathbf{A} = \mathbf{a}_\phi (k/r)$ in cylindrical coordinates, where k is a constant, or

b) $\mathbf{A} = \mathbf{a}_R f(R)$ in spherical coordinates, where $f(R)$ is any function of the radial distance R.

SOLUTION

a) In cylindrical coordinates the following apply: $(u_1, u_2, u_3) = (r, \phi, z)$; $h_1 = 1$, $h_2 = r$, and $h_3 = 1$. We have, from Eq. (2-97),

Expression of $\nabla \times \mathbf{A}$ in cylindrical coordinates

$$\nabla \times \mathbf{A} = \frac{1}{r} \begin{vmatrix} \mathbf{a}_r & \mathbf{a}_\phi r & \mathbf{a}_z \\ \dfrac{\partial}{\partial r} & \dfrac{\partial}{\partial \phi} & \dfrac{\partial}{\partial z} \\ A_r & rA_\phi & A_z \end{vmatrix}, \tag{2-98}$$

which yields, for the given **A**,

$$\nabla \times \mathbf{A} = \frac{1}{r} \begin{vmatrix} \mathbf{a}_r & \mathbf{a}_\phi r & \mathbf{a}_z \\ \dfrac{\partial}{\partial r} & \dfrac{\partial}{\partial \phi} & \dfrac{\partial}{\partial z} \\ 0 & k & 0 \end{vmatrix} = 0.$$

b) In spherical coordinates the following apply: $(u_1, u_2, u_3) = (R, \theta, \phi)$; $h_1 = 1$, $h_2 = R$, and $h_3 = R \sin\theta$. Hence,

Expression of $\nabla \times \mathbf{A}$ in spherical coordinates

$$\nabla \times \mathbf{A} = \frac{1}{R^2 \sin\theta} \begin{vmatrix} \mathbf{a}_R & \mathbf{a}_\theta R & \mathbf{a}_\phi R\sin\theta \\ \frac{\partial}{\partial R} & \frac{\partial}{\partial \theta} & \frac{\partial}{\partial \phi} \\ A_R & RA_\theta & (R\sin\theta)A_\phi \end{vmatrix}, \tag{2-99}$$

and, for the given $\mathbf{A}$,

$$\nabla \times \mathbf{A} = \frac{1}{R^2 \sin\theta} \begin{vmatrix} \mathbf{a}_R & \mathbf{a}_\theta R & \mathbf{a}_\phi R\sin\theta \\ \frac{\partial}{\partial R} & \frac{\partial}{\partial \theta} & \frac{\partial}{\partial \phi} \\ f(R) & 0 & 0 \end{vmatrix} = 0.$$

Definition of an irrotational or a conservative field

A curl-free vector field is called an ***irrotational*** or a ***conservative field***. Hence the two types of field given in this example are both conservative. We will see in the next chapter that an electrostatic field is conservative.

2-9 STOKES'S THEOREM

For a very small differential area Δs_j bounded by a contour c_j, the definition of $\nabla \times \mathbf{A}$ in Eq. (2-86) leads to

$$(\nabla \times \mathbf{A})_j \cdot (\Delta \mathbf{s}_j) = \oint_{c_j} \mathbf{A} \cdot d\boldsymbol{\ell}. \tag{2-100}$$

In obtaining Eq. (2-100), we have taken the dot product of both sides of Eq. (2-85) with $\mathbf{a}_n \Delta s_j$ or $\Delta \mathbf{s}_j$. For an arbitrary surface S, we can subdivide it into many, say N, small differential areas. Figure 2-25 shows such a scheme with Δs_j as a typical differential element. The left side of Eq. (2-100) is the flux of

FIGURE 2-25 Subdivided area for proof of Stokes's theorem.

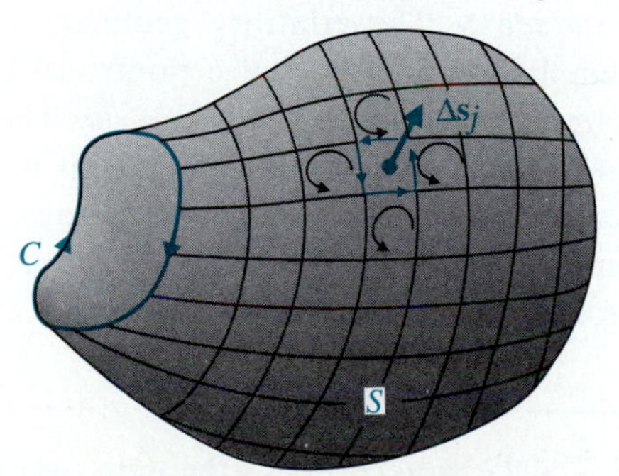

the vector $\nabla \times \mathbf{A}$ through the area Δs_j. Adding the contributions of all the differential areas to the flux, we have

$$\lim_{\Delta s_j \to 0} \sum_{j=1}^{N} (\nabla \times \mathbf{A})_j \cdot (\Delta \mathbf{s}_j) = \int_S (\nabla \times \mathbf{A}) \cdot d\mathbf{s}. \tag{2-101}$$

Now we sum up the line integrals around the contours of all the differential elements represented by the right side of Eq. (2-100). Since the common part of the contours of two adjacent elements is traversed in opposite directions by two contours, the net contribution of all the common parts in the interior to the total line integral is zero, and only the contribution from the external contour C bounding the entire area S remains after the summation:

$$\lim_{\Delta s_j \to 0} \sum_{j=1}^{N} \left(\oint_{C_j} \mathbf{A} \cdot d\boldsymbol{\ell} \right) = \oint_C \mathbf{A} \cdot d\boldsymbol{\ell}. \tag{2-102}$$

Combining Eqs. (2-101) and (2-102), we obtain ***Stokes's theorem***:

$$\boxed{\int_S (\nabla \times \mathbf{A}) \cdot d\mathbf{s} = \oint_C \mathbf{A} \cdot d\boldsymbol{\ell},} \tag{2-103}$$

Stokes's theorem

which states that ***the surface integral of the curl of a vector field over an open surface is equal to the closed line integral of the vector along the contour bounding the surface.***

Stokes's theorem converts a surface integral of the curl of a vector to a line integral of the vector, and vice versa. Like the divergence theorem, Stokes's theorem is an important identity in vector analysis, and we will use it frequently in establishing other theorems and relations in electromagnetics.

If the surface integral of $\nabla \times \mathbf{A}$ is carried over a closed surface, there will be no surface-bounding external contour, and Eq. (2-103) tells us that

$$\oint_S (\nabla \times \mathbf{A}) \cdot d\mathbf{s} = 0 \tag{2-104}$$

for any closed surface S. The arbitrary geometry in Fig. 2-25 is chosen deliberately to emphasize the fact that a nontrivial application of Stokes's theorem always implies *an open surface with a rim.* The simplest open surface would be a two-dimensional plane or disk with its circumference as the contour. We remind ourselves here that the *relative directions of $d\boldsymbol{\ell}$ and $d\mathbf{s}$* (its direction denoted by $\mathbf{a}_n$) *follow the right-hand rule*; that is, if the fingers of the right hand follow the direction of $d\boldsymbol{\ell}$, the thumb points in the direction of $\mathbf{a}_n$.

EXAMPLE 2-16

Given $\mathbf{F} = \mathbf{a}_x xy - \mathbf{a}_y 2x$, verify Stokes's theorem over a quarter-circular disk with a radius 3 in the first quadrant, as was shown in Fig. 2-22.

SOLUTION

We use Eq. (2-96) to find $\nabla \times \mathbf{F}$ in Cartesian coordinates.

$$\nabla \times \mathbf{F} = \begin{vmatrix} \mathbf{a}_x & \mathbf{a}_y & \mathbf{a}_z \\ \dfrac{\partial}{\partial x} & \dfrac{\partial}{\partial y} & \dfrac{\partial}{\partial z} \\ xy & -2x & 0 \end{vmatrix} = -\mathbf{a}_z(2 + x).$$

For the given geometry and the designated direction of $d\boldsymbol{\ell}$, $d\mathbf{s} = \mathbf{a}_n\, ds = \mathbf{a}_z\, dx\, dy$. We have

$$\begin{aligned}\int_S (\nabla \times \mathbf{F}) \cdot d\mathbf{s} &= \int_0^3 \int_0^{\sqrt{9-y^2}} (\nabla \times \mathbf{F}) \cdot (\mathbf{a}_z\, dx\, dy) \\ &= \int_0^3 \left[\int_0^{\sqrt{9-y^2}} -(2 + x)\, dx \right] dy \\ &= -\int_0^3 [2\sqrt{9 - y^2} + \tfrac{1}{2}(9 - y^2)]\, dy \\ &= -\left[y\sqrt{9 - y^2} + 9 \sin^{-1}\frac{y}{3} + \frac{9}{2}y - \frac{y^3}{6} \right]_0^3 \\ &= -9\left(1 + \frac{\pi}{2}\right).\end{aligned}$$

It is *important* to use the proper limits for the two variables of integration. We can interchange the order of integration as

$$\int_S (\nabla \times \mathbf{F}) \cdot d\mathbf{s} = \int_0^3 \left[\int_0^{\sqrt{9-x^2}} -(2 + x)\, dy \right] dx$$

and get the same result. But it would be quite wrong if the 0 to 3 range were used as the range of integration for both x and y. (Do you know why?)

The line integral of $\mathbf{F}$ around the quarter-circular disk along the path $OABO$, $\int \mathbf{F} \cdot d\boldsymbol{\ell}$, is the circulation found in Example 2-14, which is equal to the surface integral of $\nabla \times \mathbf{F}$ obtained above. Thus Stokes's theorem is verified.

■ **EXERCISE 2.14** Given $\mathbf{F} = \mathbf{a}_r \sin\phi + \mathbf{a}_\phi 3\cos\phi$ and the quarter-circular region shown in Fig. 2-22,

a) determine $\oint_{OABO} \mathbf{F} \cdot d\boldsymbol{\ell}$, and
b) find $\nabla \times \mathbf{F}$, and verify Stokes's theorem.

ANS. (a) 6, (b) $\mathbf{a}_z\left(\dfrac{2}{r}\cos\phi\right)$.

2-10 TWO NULL IDENTITIES

Two identities involving repeated del operations are of considerable importance in the study of electromagnetism, especially when we introduce potential functions. We shall discuss them separately below.

2-10.1 IDENTITY I

$$\boxed{\nabla \times (\nabla V) \equiv 0} \tag{2-105}$$

An important null identity

In words, ***the curl of the gradient of any scalar field is identically zero***. (The existence of V and its first derivatives everywhere is implied here.)

Equation (2-105) can be proved readily in Cartesian coordinates by using Eq. (2-56) for ∇ and performing the indicated operations. In general, if we take the surface integral of $\nabla \times (\nabla V)$ over any surface, the result is equal to the line integral of ∇V around the closed path bounding the surface, as asserted by Stokes's theorem:

$$\int_S [\nabla \times (\nabla V)] \cdot d\mathbf{s} = \oint_C (\nabla V) \cdot d\boldsymbol{\ell}. \tag{2-106}$$

However, from Eq. (2-51),

$$\oint_C (\nabla V) \cdot d\boldsymbol{\ell} = \oint_C dV = 0. \tag{2-107}$$

The combination of Eqs. (2-106) and (2-107) states that the surface integral of $\nabla \times (\nabla V)$ over *any* surface is zero. The integrand itself must therefore vanish, which leads to the identity in Eq. (2-105). Since a coordinate system is not specified in the derivation, the identity is a general one and is invariant with the choices of coordinate systems.

A converse statement of Identity I can be made as follows: ***If a vector field is curl-free, then it can be expressed as the gradient of a scalar field.*** Let a vector field be $\mathbf{E}$. Then, if $\nabla \times \mathbf{E} = 0$, we can define a scalar field V such that

$$\mathbf{E} = -\nabla V. \tag{2-108}$$

The negative sign here is unimportant as far as Identity I is concerned. (It is included in Eq. (2-108) because this relation conforms with a basic relation between *electric field intensity* $\mathbf{E}$ and *electric scalar potential* V in electrostatics, which we will take up in the next chapter. At this stage it is immaterial what $\mathbf{E}$ and V represent.) We know from Section 2-8 that a curl-free vector field is a conservative field; hence ***an irrotational (a conservative) vector field can always be expressed as the gradient of a scalar field.***

EXERCISE 2.15 Prove the identity in Eq. (2-105) in Cartesian coordinates.

2-10.2 IDENTITY II

$$\nabla \cdot (\nabla \times \mathbf{A}) \equiv 0 \tag{2-109}$$

Another important null identity

In words, ***the divergence of the curl of any vector field is identically zero.***

We can prove this identity without regard to a coordinate system by taking the volume integral of $\nabla \cdot (\nabla \times \mathbf{A})$ on the left side. Applying the divergence theorem, we have

$$\int_V \nabla \cdot (\nabla \times \mathbf{A})\, dv = \oint_S (\nabla \times \mathbf{A}) \cdot d\mathbf{s}. \tag{2-110}$$

Let us choose, for example, the arbitrary volume V enclosed by a surface S in Fig. 2-26. The closed surface S can be split into two open surfaces, S_1 and S_2, connected by a common boundary that has been drawn twice as C_1 and C_2. We then apply Stokes's theorem to surface S_1 bounded by C_1, and surface S_2 bounded by C_2, and we write the right side of Eq. (2-110) as

$$\oint_S (\nabla \times \mathbf{A}) \cdot d\mathbf{s} = \int_{S_1} (\nabla \times \mathbf{A}) \cdot \mathbf{a}_{n1}\, ds + \int_{S_2} (\nabla \times \mathbf{A}) \cdot \mathbf{a}_{n2}\, ds$$
$$= \oint_{C_1} \mathbf{A} \cdot d\boldsymbol{\ell} + \oint_{C_2} \mathbf{A} \cdot d\boldsymbol{\ell}. \tag{2-111}$$

The normals $\mathbf{a}_{n1}$ and $\mathbf{a}_{n2}$ to surfaces S_1 and S_2 are *outward* normals, and their relations with the path directions of C_1 and C_2 follow the right-hand rule. Since the contours C_1 and C_2 are, in fact, one and the same common boundary between S_1 and S_2, the two line integrals on the right side of Eq.

FIGURE 2-26 An arbitrary volume V enclosed by surface S.

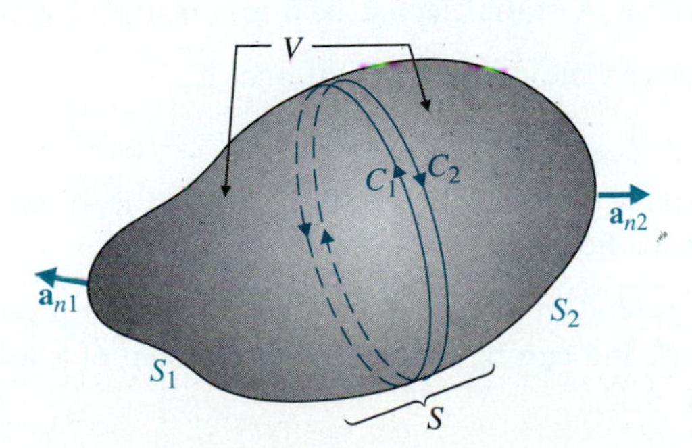

(2-111) traverse the same path in opposite directions. Their sum is therefore zero, and the volume integral of $\nabla \cdot (\nabla \times \mathbf{A})$ on the left side of Eq. (2-110) vanishes. Because this is true for any arbitrary volume, the integrand itself must be zero, as indicated by the identity in Eq. (2-109).

A converse statement of Identity II is as follows: ***If a vector field is divergenceless, then it is solenoidal and can be expressed as the curl of another vector field.*** Let a vector field be **B**. This converse statement asserts that if $\nabla \cdot \mathbf{B} = 0$, we can define a vector field **A** such that

$$\mathbf{B} = \nabla \times \mathbf{A}. \tag{2-112}$$

EXERCISE 2.16 Prove the identity in Eq. (2-109) in Cartesian coordinates.

2-11 FIELD CLASSIFICATION AND HELMHOLTZ'S THEOREM

Divergenceless field↔Solenoidal field

Curl-free field↔Irrotational (conservative) field

In previous sections we mentioned that *a divergenceless field is solenoidal* and *a curl-free field is irrotational* (*conservative*). We may classify vector fields in accordance with their being solenoidal and/or irrotational. A vector field **F** is

1. Solenoidal and irrotational if

   $\nabla \cdot \mathbf{F} = 0$ and $\nabla \times \mathbf{F} = 0.$

   EXAMPLE: A static electric field in a charge-free region.

2. Solenoidal but not irrotational if

   $\nabla \cdot \mathbf{F} = 0$ and $\nabla \times \mathbf{F} \neq 0.$

   EXAMPLE: A steady magnetic field in a current-carrying conductor.

3. Irrotational but not solenoidal if

   $\nabla \times \mathbf{F} = 0$ and $\nabla \cdot \mathbf{F} \neq 0.$

   EXAMPLE: A static electric field in a charged region.

4. Neither solenoidal nor irrotational if

   $\nabla \cdot \mathbf{F} \neq 0$ and $\nabla \times \mathbf{F} \neq 0.$

   EXAMPLE: An electric field in a charged medium with a time-varying magnetic field.

The most general vector field then has both a nonzero divergence and a nonzero curl, and can be considered as the sum of a solenoidal field and an irrotational field.

Helmholtz's theorem

Helmholtz's Theorem: ***A vector field is determined if both its divergence and its curl are specified everywhere.***[†]

Helmholtz's theorem can be proved as a mathematical theorem in a general way.[‡] For our purposes, we remind ourselves (see Section 2-8) that the divergence of a vector is a measure of the strength of the flow source and that the curl of a vector is a measure of the strength of the vortex source. When the strengths of both the flow source and the vortex source are specified, we expect that the vector field will be determined.

Axiomatic development of electromagnetism in steps

In following chapters we will rely on Helmholtz's theorem as a basic element in the axiomatic development of electromagnetism. For each topic of study (static electric fields, static magnetic fields, and time-varying electromagnetic fields) we will state the fundamental postulates (specify the divergence and the curl) of the basic field vector(s) needed in the electromagnetic model. Other theorems and relations will then be developed from the fundamental postulates.

■ **EXERCISE 2.17** Determine whether the following vector fields are irrotational, solenoidal, both, or neither:

a) $\mathbf{A} = \mathbf{a}_x xy - \mathbf{a}_y y^2 + \mathbf{a}_z xz$,

b) $\mathbf{B} = r(\mathbf{a}_r \sin \phi + \mathbf{a}_\phi 2 \cos \varphi)$,

c) $\mathbf{C} = \mathbf{a}_x x - \mathbf{a}_y 2y + \mathbf{a}_z z$,

d) $\mathbf{D} = \mathbf{a}_R k/R$.

ANS. (a) neither, (b) solenoidal, (c) both, (d) irrotational.

REVIEW QUESTIONS

Q.2-12 What is the difference between a scalar quantity and a scalar field? Between a vector quantity and a vector field?

Q.2-13 What is the physical definition of the gradient of a scalar field?

Q.2-14 Express the space rate of change of a scalar in a given direction in terms of its gradient.

Q.2-15 What is the expression for the del operator, ∇, in Cartesian coordinates?

Q.2-16 What is the physical definition of the divergence of a vector field?

Q.2-17 State the divergence theorem in words.

[†] More precisely, we need to require that in an unbounded region both the divergence and the curl of the vector field vanish at infinity. If the vector field is confined within a region bounded by a surface, then it is determined if its divergence and curl throughout the region, as well as the normal component of the vector over the bounding surface, are given.

[‡] See, for instance, G. Arfken, *Mathematical Methods for Physicists.* Section 1.15, Academic Press, New York, 1966.

Q.2-18 What is the physical definition of the curl of a vector field?

Q.2-19 State Stokes's theorem in words.

Q.2-20 What is the difference between an irrotational field and a solenoidal field?

Q.2-21 State Helmholtz's theorem in words.

REMARKS

1. The basic rules of vector algebra (vector addition, subtraction, dot and cross products) are independent of coordinate system.
2. The gradient of a scalar field is a vector point function.
3. The divergence of a vector field is a scalar point function.
4. The curl of a vector field is a vector point function.
5. Do not forget to draw a small circle on the integration sign ($\oint$) in writing a closed line integral or a surface integral over the entire surface enclosing a region.
6. The two null identities listed in Eqs. (2-105) and (2-109) and their implications are the bases for defining potential functions in later chapters. Learn these identities thoroughly.

SUMMARY

Vector analysis is an essential mathematical tool in electromagnetics. It provides a concise means for representing and expressing the relations of various quantities in the electromagnetic model. In this chapter we

- reviewed the basic rules of vector addition and subtraction, and of products of vectors,
- explained the properties of Cartesian, cylindrical, and spherical coordinate systems,
- introduced the differential del (∇) operator, and defined the gradient of a scalar field, and the divergence and the curl of a vector field,
- presented the divergence theorem that transforms the volume integral of the divergence of a vector field to a closed surface integral of the vector field, and vice versa,
- presented the Stokes's theorem that transforms the surface integral of the curl of a vector field to a closed line integral of the vector field, and vice versa,
- introduced two important null identities in vector field, and

- discussed the classification of vectors, and introduced Helmholtz's theorem, which will be used as a basic element in the axiomatic development of the various topics in electromagnetics.

PROBLEMS

P.2-1 A rhombus is an equilateral parallelogram. Denote two neighboring sides of a rhombus by vectors **A** and **B**.

a) Verify that the two diagonals are $\mathbf{A} + \mathbf{B}$ and $\mathbf{A} - \mathbf{B}$.

b) Prove that the diagonals are perpendicular to each other.

P.2-2 If the three sides of an arbitrary triangle are denoted by vectors **A**, **B**, and **C** in a clockwise or counterclockwise direction, then the equation $\mathbf{A} + \mathbf{B} + \mathbf{C} = 0$ holds. Prove the law of sines.

HINT: Cross multiply the equation separately by **A** and by **B**, and examine the magnitude relations of the products.

P.2-3 Given three vectors **A**, **B**, and **C** as follows:

$$\mathbf{A} = \mathbf{a}_x 6 + \mathbf{a}_y 2 - \mathbf{a}_z 3,$$
$$\mathbf{B} = \mathbf{a}_x 4 - \mathbf{a}_y 6 + \mathbf{a}_z 12,$$
$$\mathbf{C} = \mathbf{a}_x 5 - \mathbf{a}_z 2,$$

find

a) $\mathbf{a}_B$,

b) $|\mathbf{B} - \mathbf{A}|$,

c) the component of **A** in the direction of **B**,

d) $\mathbf{B} \cdot \mathbf{A}$,

e) the component of **B** in the direction of **A**,

f) θ_{AB},

g) $\mathbf{A} \times \mathbf{C}$, and

h) $\mathbf{A} \cdot (\mathbf{B} \times \mathbf{C})$ and $(\mathbf{A} \times \mathbf{B}) \cdot \mathbf{C}$.

P.2-4 Let unit vectors $\mathbf{a}_A$ and $\mathbf{a}_B$ denote the directions of vectors **A** and **B** in the xy-plane that make angles α and β, respectively, with the x-axis.

a) Obtain a formula for the expansion of the cosine of the difference of two angles, $\cos(\alpha - \beta)$, by taking the scalar product $\mathbf{a}_A \cdot \mathbf{a}_B$.

b) Obtain a formula for $\sin(\alpha - \beta)$ by taking the vector product $\mathbf{a}_B \times \mathbf{a}_A$.

P.2-5 The three corners of a right triangle are at $P_1(1, 0, 2)$, $P_2(-3, 1, 5)$, and $P_3(3, -4, 6)$.

a) Determine which corner is a right angle.

b) Find the area of the triangle.

P.2-6 Given two points $P_1(-2, 0, 3)$ and $P_2(0, 4, -1)$, find

a) the length of the line joining P_1 and P_2, and

b) the perpendicular distance from the point $P_3(3, 1, 3)$ to the line.

P.2-7 Given vector $\mathbf{A} = \mathbf{a}_x 5 - \mathbf{a}_y 2 + \mathbf{a}_z$, find the expression of

a) a unit vector $\mathbf{a}_B$ such that $\mathbf{a}_B \parallel \mathbf{A}$, and

b) a unit vector $\mathbf{a}_C$ in the xy-plane such that $\mathbf{a}_C \perp \mathbf{A}$.

P.2-8 Decompose vector $\mathbf{A} = \mathbf{a}_x 2 - \mathbf{a}_y 5 + \mathbf{a}_z 3$ into two components, $\mathbf{A}_1$ and $\mathbf{A}_2$, that are, respectively, perpendicular and parallel to another vector $\mathbf{B} = -\mathbf{a}_x + \mathbf{a}_y 4$.

P.2-9 Equation (2-15) in Example 2-2 describes the *scalar triple products* of three vectors **A**, **B**, and **C**. There is another important type of product of three vectors. It is a ***vector triple product***, $\mathbf{A} \times (\mathbf{B} \times \mathbf{C})$. Prove the following relation by expansion in Cartesian coordinates:

$$\mathbf{A} \times (\mathbf{B} \times \mathbf{C}) = \mathbf{B}(\mathbf{A} \cdot \mathbf{C}) - \mathbf{C}(\mathbf{A} \cdot \mathbf{B}). \tag{2-113}$$

Equation (2-113) is known as the ***"BAC-CAB" rule***.

P.2-10 Find the component of the vector $\mathbf{A} = \mathbf{a}_x z - \mathbf{a}_z x$ at the point $P_1(-1, 0, -2)$ that is directed toward the point $P_2(\sqrt{3}, 150°, 1)$.

P.2-11 The position of a point in cylindrical coordinates is given by $(3, 4\pi/3, -4)$. Specify the location of the point

a) in Cartesian coordinates, and

b) in spherical coordinates.

P.2-12 Find the results of the following products of unit vectors:

a) $\mathbf{a}_\phi \cdot \mathbf{a}_x$,

b) $\mathbf{a}_R \cdot \mathbf{a}_y$,

c) $\mathbf{a}_z \cdot \mathbf{a}_R$,

d) $\mathbf{a}_\phi \times \mathbf{a}_x$,

e) $\mathbf{a}_r \times \mathbf{a}_R$,

f) $\mathbf{a}_\theta \times \mathbf{a}_z$.

P.2-13 Express the r-component, A_r, of a vector **A**

a) in terms of A_x and A_y in Cartesian coordinates, and

b) in terms of A_R and A_θ in spherical coordinates.

P.2-14 Express the θ-component, E_θ, of a vector **E** at (R_1, θ_1, ϕ_1)

a) in terms of E_x, E_y, and E_z in Cartesian coordinates, and

b) in terms of E_r and E_z in cylindrical coordinates.

P.2-15 Given a vector field in spherical coordinates $\mathbf{F} = \mathbf{a}_R(12/R^2)$,

a) find $\mathbf{F}$ and F_y at the point $P(-2, -4, 4)$, and

b) find the angle that $\mathbf{F}$ makes with the vector $\mathbf{A} = \mathbf{a}_x 2 - \mathbf{a}_y 3 - \mathbf{a}_z 6$ at P.

P.2-16 Given a vector field $\mathbf{F} = \mathbf{a}_x y + \mathbf{a}_y x$, evaluate the integral $\int \mathbf{F} \cdot d\boldsymbol{\ell}$ from $P_1(2, 1, -1)$ to $P_2(8, 2, -1)$

a) along the straight line joining the two points, and

b) along the parabola $x = 2y^2$.

Is this $\mathbf{F}$ a conservative field? Explain.

P.2-17 Denote the position vector to a point $P(x, y, z)$ by $\mathbf{R}$. Determine $\nabla(1/R)$

a) in Cartesian coordinates, and

b) in spherical coordinates.

P.2-18 Given a scalar field $V = 2xy - yz + xz$,

a) find the vector representing the direction and the magnitude of the maximum rate of increase of V at point $P(2, -1, 0)$, and

b) find the rate of increase of V at point P in the direction toward the point $Q(0, 2, 6)$.

P.2-19 In a curvilinear coordinate system the differentiation of a base vector may lead to a new vector in a different direction.

a) Determine $\partial \mathbf{a}_r/\partial \phi$ and $\partial \mathbf{a}_\phi / d\phi$ in cylindrical coordinates.

b) Use the results in (a) to find the formula for $\nabla \cdot \mathbf{A}$ in cylindrical coordinates by using Eqs. (2-57) and (2-31).

P.2-20 Find the divergence of the following radial fields:

a) $f_1(\mathbf{R}) = \mathbf{a}_R R^n$,

b) $f_2(\mathbf{R}) = \mathbf{a}_R k/R^2$, where k is a constant.

P.2-21 Given a vector field $\mathbf{F} = \mathbf{a}_x xy + \mathbf{a}_y yz + \mathbf{a}_z zx$,

a) compute the total outward flux from the surface of a unit cube in the first octant with one corner at the origin, and

b) find $\nabla \cdot \mathbf{F}$ and verify the divergence theorem.

P.2-22 For a vector function $\mathbf{A} = \mathbf{a}_r r^2 + \mathbf{a}_z 2z$, verify the divergence theorem for the circular cylindrical region enclosed by $r = 5$, $z = 0$, and $z = 4$.

P.2-23 For a vector function $\mathbf{A} = \mathbf{a}_z z$,

a) find $\oint \mathbf{A} \cdot d\mathbf{s}$ over the surface of a hemispherical region that is the top half of a sphere of radius 3 centered at the origin with its flat base coinciding with the xy-plane,

b) find $\nabla \cdot \mathbf{A}$, and

c) verify the divergence theorem.

P.2-24 A vector field $\mathbf{D} = \mathbf{a}_R(\cos^2 \phi)/R^3$ exists in the region between two spherical shells defined by $R = 2$ and $R = 3$. Evaluate

a) $\oint \mathbf{D} \cdot d\mathbf{s}$, and

b) $\int \nabla \cdot \mathbf{D}\, dv$.

P.2-25 For a scalar function f and a vector function $\mathbf{A}$, prove

$$\boxed{\nabla \cdot (f\mathbf{A}) = f\nabla \cdot \mathbf{A} + \mathbf{A} \cdot \nabla f} \qquad (2\text{-}114)$$

in Cartesian coordinates.

P.2-26 Assume a vector field $\mathbf{A} = \mathbf{a}_x(2x^2 + y^2) + \mathbf{a}_y(xy - y^2)$.

a) Find $\oint \mathbf{A} \cdot d\boldsymbol{\ell}$ around the triangular contour shown in Fig. 2-27.

b) Find $\oint (\nabla \times \mathbf{A}) \cdot d\mathbf{s}$ over the triangular area.

c) Can $\mathbf{A}$ be expressed as the gradient of a scalar? Explain.

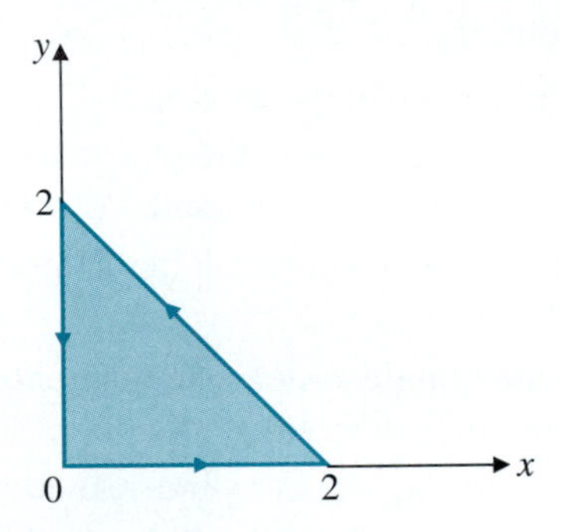

FIGURE 2-27 Graph for Problem P.2-26.

P.2-27 Assume a vector function $\mathbf{F} = \mathbf{a}_r 5r \sin\phi + \mathbf{a}_\phi r^2 \cos\phi$.

a) Evaluate $\oint \mathbf{F} \cdot d\boldsymbol{\ell}$ around the contour $ABCDA$ in the direction as indicated in Fig. 2-28.

b) Find $\nabla \times \mathbf{F}$.

c) Evaluate $\int (\nabla \times \mathbf{F}) \cdot d\mathbf{s}$ over the shaded area and compare the result with that obtained in Part (a).

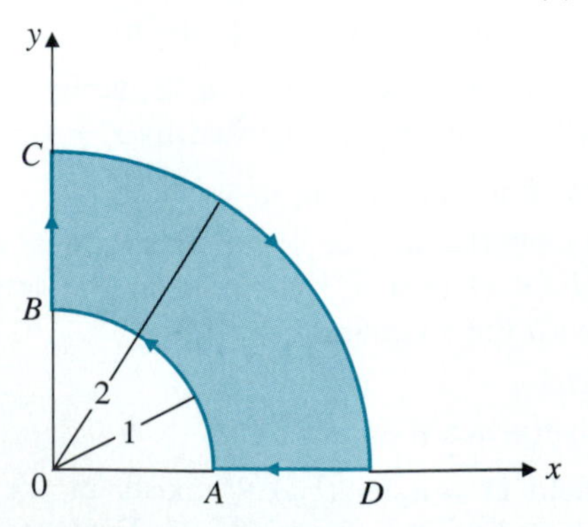

FIGURE 2-28 Graph for Problem P.2-27.

P.2-28 Given a vector function $\mathbf{A} = \mathbf{a}_\phi 3 \sin(\phi/2)$, verify Stokes's theorem over the surface of a hemispherical bowl of radius 4 and its circular rim.

P.2-29 For a scalar function f and a vector function $\mathbf{G}$, prove

$$\boxed{\nabla \times (f\mathbf{G}) = f(\nabla \times \mathbf{G}) + (\nabla f) \times \mathbf{G}} \tag{2-115}$$

in Cartesian coordinates.

P.2-30 Given a vector function

$$\mathbf{F} = \mathbf{a}_x(x + 3y - c_1 z) + \mathbf{a}_y(c_2 x + 5z) + \mathbf{a}_z(2x - c_3 y + c_4 z),$$

a) determine c_1, c_2, and c_3 if $\mathbf{F}$ is irrotational, and

b) determine c_4 if $\mathbf{F}$ is also solenoidal.

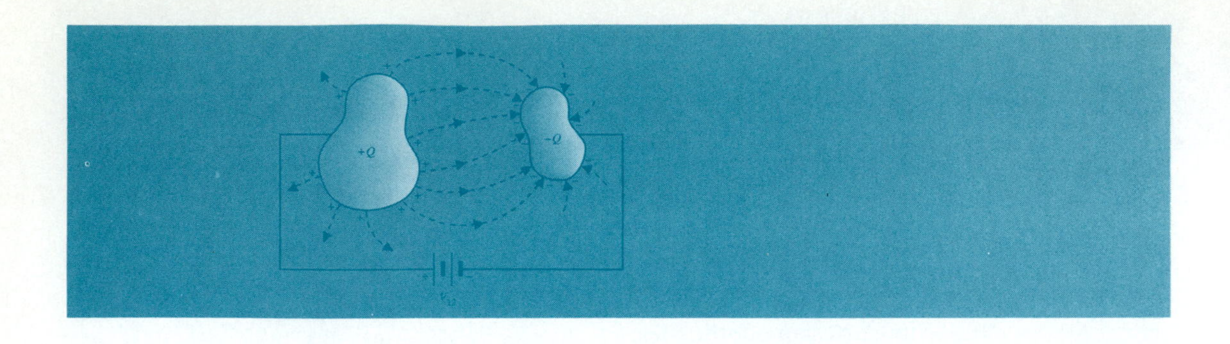

CHAPTER 3

3-1 OVERVIEW Static electric fields are caused by stationary electric charges. When we walk over a carpet in a dry room and touch a metal doorknob, we often draw a spark. This is because the static charges produced on our body as a result of rubbing shoe soles against the carpet tend to congregate on sharp points like fingertips and jump across the air to the doorknob. The potential difference generated may be several thousand volts, but no serious harm results, except for a minor shock, as the amount of charge involved is usually very small. Another example of static electricity is the phenomenon of a thin outer garment clinging to inner clothing of a different material due to opposite charges produced by relative motion and friction.

Examples of static electricity generation

Electrostatics is the study of the effects of electric charges at rest, and the electric fields do not change with time. Although it is the simplest situation in electromagnetics, its mastery is fundamental to the understanding of more complicated electromagnetic models. The explanation of many natural phenomena (such as lightning and corona) and the principles of some important industrial applications (such as oscilloscopes, ink-jet printers, xerography, capacitance keyboards and liquid crystal displays are based on electrostatics. A number of books on special applications of electrostatics have been published.[†]

[†]A. Klinkenberg and J. L. van der Minne, *Electrostatics in the Petroleum Industry*, Elsevier, Amsterdam, 1958. J. H. Dessauer and H. E. Clark, *Xerography and Related Processes*, Focal Press, London, 1965. A. D. Moore (Ed.), *Electrostatics and Its Applications*, John Wiley, New York, 1973. C. E. Jewett, *Electrostatics in the Electronics Environment*, John Wiley, New York, 1976. J. C. Crowley, *Fundamentals of Applied Electrostatics*, John Wiley, New York, 1986.

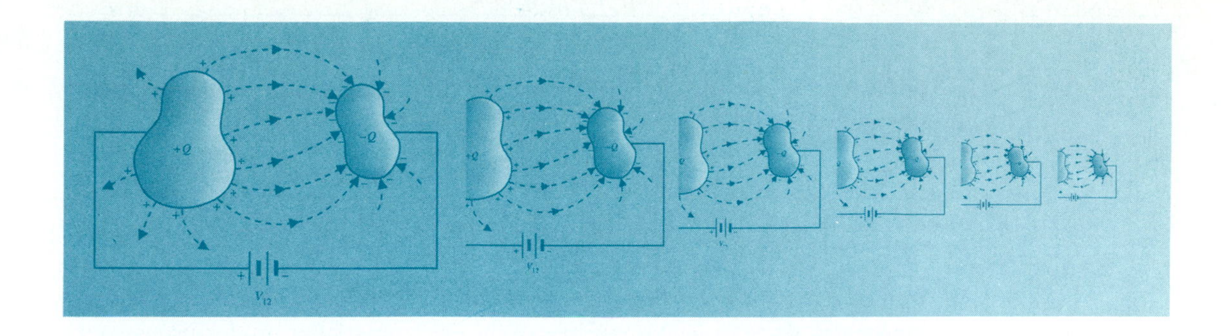

Static Electric Fields

Historically, quantitative relations in electrostatics began with the experiments of Charles Augustin de Coulomb, who formulated in 1785 what is now known as *Coulomb's law*. Later, Karl F. Gauss developed *Gauss's law*, and other scientists and engineers contributed various additional important results concerning stationary electric charges. The theory of static electric fields was gradually built up. This method of starting with experimental laws and synthesizing them in the form of Maxwell's equations is an *inductive approach*. It is an approach usually followed in an introductory physics course.

A deductive approach

Because the various results were obtained by uncoordinated individuals at different times, the inductive approach tends to appear fragmented and incohesive. In this book we prefer to use a *deductive approach*, which, as we have indicated in Section 1-2, is more concise and logical; it enables us to develop electromagnetics in an orderly way.

For the study of static electric fields in free space we define an electric field intensity vector by specifying its divergence and its curl. These constitute the fundamental postulates, from which we *derive* Coulomb's law and Gauss's law that together can be used to determine the electric field due to various charge distributions. The effects of conductors and dielectrics in electrostatic fields are then examined. Electrostatic potential is introduced, and the relations between electrostatic energy and forces explored. In situations where the exact charge distributions are not known everywhere but certain conditions at the boundaries must be satisfied, additional solution techniques are needed. We will discuss the procedure for solving simple Poisson's and Laplace's equations and explain the method of images.

3-2 FUNDAMENTAL POSTULATES OF ELECTROSTATICS IN FREE SPACE

For electrostatics in free space we need to consider only one of the four fundamental vector field quantities of the electromagnetic model discussed in Section 1-2, namely, the electric field intensity **E**. Furthermore, only the permittivity of free space, ϵ_0, of the three universal constants mentioned in Section 1-3 enters into our formulation.

Electric field intensity

Electric field intensity is defined as the force per unit charge that a very small stationary test charge experiences when it is placed in a region where an electric field exists. That is,

$$\mathbf{E} = \lim_{q \to 0} \frac{\mathbf{F}}{q} \qquad \text{(V/m)}. \tag{3-1}$$

SI unit for E is (V/m).

The electric field intensity **E** is, then, proportional to and in the direction of the force **F**. If **F** is measured in newtons (N) and charge q in coulombs (C), then **E** is in newtons per coulomb (N/C), which is the same as volts per meter (V/m). The test charge q, of course, cannot be zero in practice; as a matter of fact, it cannot be less than the charge on an electron. However, the finiteness of the test charge would not make the measured **E** differ appreciably from its calculated value if the test charge is small enough not to disturb the charge distribution of the source. An inverse relation of Eq. (3-1) gives the force **F** on a stationary charge q in an electric field **E**:

$$\mathbf{F} = q\mathbf{E} \qquad \text{(N)}. \tag{3-2}$$

The two fundamental postulates of electrostatics in free space specify the divergence and the curl of **E**. They are

Divergence of electrostatic E in free space

$$\nabla \cdot \mathbf{E} = \frac{\rho_v}{\epsilon_0} \qquad \text{(in free space)} \tag{3-3}$$

and

Curl of electrostatic E vanishes.

$$\nabla \times \mathbf{E} = 0. \tag{3-4}$$

In Eq. (3-3), ρ_v is the volume charge density of free charges (C/m^3), and ϵ_0 is the permittivity of free space, given in Eq. (1-11). Equation (3-4) asserts that ***static electric fields are irrotational***, whereas Eq. (3-3) implies that a static electric field is not solenoidal unless $\rho_v = 0$. These two postulates are concise, simple, and independent of any coordinate system; and they can be used to derive all other relations, laws, and theorems in electrostatics.

Equations (3-3) and (3-4) are point relations; that is, they hold at every point in space. They are referred to as the differential form of the postulates of electrostatics, since both divergence and curl operations involve spatial derivatives. In practical applications we are usually interested in the total field of an aggregate or a distribution of charges. This is more conveniently obtained by an integral form of Eq. (3-3). Taking the volume integral of both sides of Eq. (3-3) over an arbitrary volume V, we have

$$\int_V \nabla \cdot \mathbf{E}\, dv = \frac{1}{\epsilon_0} \int_V \rho_v\, dv. \tag{3-5}$$

In view of the divergence theorem in Eq. (2-75), Eq. (3-5) becomes

$$\boxed{\oint_S \mathbf{E} \cdot d\mathbf{s} = \frac{Q}{\epsilon_0},} \tag{3-6}$$

where Q is the total charge contained in volume V bounded by surface S. Equation (3-6) is a form of ***Gauss's law***—one of the most important relations in electrostatics. We will discuss it further in Section 3-4, along with illustrative examples.

An integral form can also be obtained for the curl relation in Eq. (3-4) by integrating $\nabla \times \mathbf{E}$ over an open surface and invoking Stokes's theorem as expressed in Eq. (2-103). We have

$$\boxed{\oint_C \mathbf{E} \cdot d\boldsymbol{\ell} = 0.} \qquad \text{(in free space)} \tag{3-7}$$

The line integral is performed over an arbitrary closed contour C. Equation (3-7) asserts that ***the scalar line integral of the static electric field intensity around any closed path vanishes***. The scalar product $\mathbf{E} \cdot d\boldsymbol{\ell}$ integrated over any path is the voltage along that path. Thus Eq. (3-7) is an expression of ***Kirchhoff's voltage law*** in circuit theory that ***the algebraic sum of voltage drops around any closed circuit is zero***.

Kirchhoff's voltage law

Equation (3-7) also implies that the scalar line integral of the irrotational $\mathbf{E}$ field from one point (say P_1) to any other point (say P_2) along any path is cancelled by that from P_2 to P_1 *along any other path*; that is, *the line integral of a static electric field depends only on the end points.* As we shall see in Section 3-5, the line integral of $\mathbf{E}$ from point P_1 to P_2 represents the work done *by* **E** in moving a unit charge from P_1 to P_2. Hence Eq. (3-7) says that the work done in moving a unit charge around a closed path in an electrostatic field is zero. It is a statement of conservation of work or energy in an electrostatic field. This is the reason allowing us to say that an irrotational field is a conservative field.†

The two fundamental postulates of electrostatics in free space are repeated below because they form the foundation upon which we build the structure of electrostatics.

Two fundamental postulates of electrostatics in free space

Postulates of Electrostatics in Free Space	
Differential Form	Integral Form
$\nabla \cdot \mathbf{E} = \dfrac{\rho_v}{\epsilon_0}$	$\oint_S \mathbf{E} \cdot d\mathbf{s} = \dfrac{Q}{\epsilon_0}$
$\nabla \times \mathbf{E} = 0$	$\oint_C \mathbf{E} \cdot d\boldsymbol{\ell} = 0$

We consider these postulates, like the principle of conservation of charge, to be representations of laws of nature. In the following section we shall derive Coulomb's law.

3-3 COULOMB'S LAW

A Gaussian surface is a hypothetical surface over which Gauss's law is applied.

We consider the simplest possible electrostatic problem of a single point charge, q, at rest in a boundless free space. In order to find the electric field intensity due to q, we draw a spherical surface of an arbitrary radius R centered at q—a hypothetical enclosed surface (a *Gaussian surface*) around the source, upon which Gauss's law is applied to determine the field. Since a point charge has no preferred directions, its electric field must be everywhere radial and has the same intensity at all points on the spherical surface. Applying Eq. (3-6) to Fig. 3-1(a), we have

$$\oint_S \mathbf{E} \cdot d\mathbf{s} = \oint_S (\mathbf{a}_R E_R) \cdot \mathbf{a}_R \, ds = \frac{q}{\epsilon_0},$$

or

$$E_R \oint_S ds = E_R(4\pi R^2) = \frac{q}{\epsilon_0}.$$

†We recall from mechanics that the gravitational field is a conservative field.

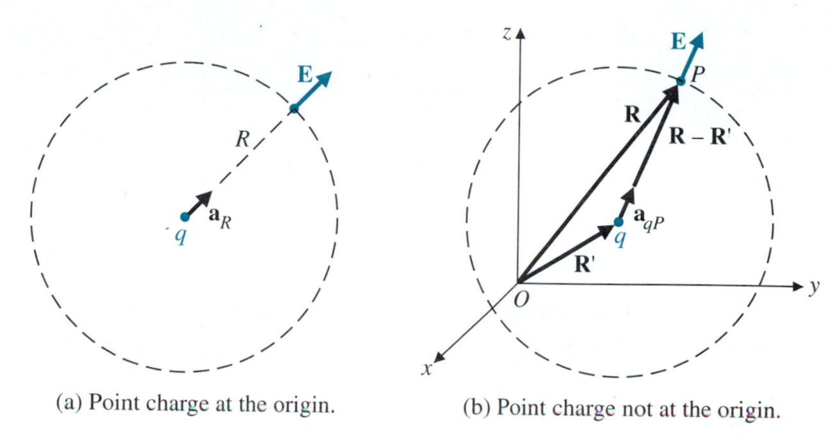

(a) Point charge at the origin. (b) Point charge not at the origin.

FIGURE 3-1 Electric field intensity due to a point charge.

Therefore,

Electric field intensity of an isolated point charge at the origin

$$\mathbf{E} = \mathbf{a}_R E_R = \mathbf{a}_R \frac{q}{4\pi\epsilon_0 R^2} \qquad \text{(V/m)}. \tag{3-8}$$

Equation (3-8) tells us that ***the electric field intensity of a positive point charge is in the outward radial direction and has a magnitude proportional to the charge and inversely proportional to the square of the distance from the charge.*** This is a very important basic formula in electrostatics. A flux-line graph for the electric field intensity of a positive point charge q will look like Fig. 2-17(b).

■ **EXERCISE 3.1** Verify that the $\mathbf{E}$ field in Eq. (3-8) satisfies Eq. (3-4) and hence is conservative.

If the charge q is not located at the origin of a chosen coordinate system, suitable changes should be made to the unit vector $\mathbf{a}_R$ and the distance R to reflect the locations of the charge and of the point at which $\mathbf{E}$ is to be determined. Let the position vector of q be $\mathbf{R}'$ and that of a field point P be $\mathbf{R}$, as shown in Fig. 3-1(b). Then, from Eq. (3-8),

$$\mathbf{E}_P = \mathbf{a}_{qP} \frac{q}{4\pi\varepsilon_0 |\mathbf{R} - \mathbf{R}'|^2}, \tag{3-9}$$

where $\mathbf{a}_{qP}$ is the unit vector drawn from q to P. Since

$$\mathbf{a}_{qP} = \frac{\mathbf{R} - \mathbf{R}'}{|\mathbf{R} - \mathbf{R}'|}, \tag{3-10}$$

we have

Electric field intensity of an isolated point charge at an arbitrary location

$$\mathbf{E}_P = \frac{q(\mathbf{R}-\mathbf{R}')}{4\pi\epsilon_0|\mathbf{R}-\mathbf{R}'|^3} \quad \text{(V/m)}. \tag{3-11}$$

EXAMPLE 3-1

Determine the electric field intensity at $P(-0.2, 0, -2.3)$ due to a point charge of $+5$ (nC) at $Q(0.2, 0.1, -2.5)$ in air. All dimensions are in meters.

SOLUTION

The position vector for the field point P

$$\mathbf{R} = \overrightarrow{OP} = -\mathbf{a}_x 0.2 - \mathbf{a}_z 2.3.$$

The position vector for the point charge Q is

$$\mathbf{R}' = \overrightarrow{OQ} = \mathbf{a}_x 0.2 + \mathbf{a}_y 0.1 - \mathbf{a}_z 2.5.$$

The difference is

$$\mathbf{R}-\mathbf{R}' = -\mathbf{a}_x 0.4 - \mathbf{a}_y 0.1 + \mathbf{a}_z 0.2,$$

which has a magnitude

$$|\mathbf{R}-\mathbf{R}'| = [(-0.4)^2 + (-0.1)^2 + (0.2)^2]^{1/2} = 0.458\,\text{(m)}.$$

Substituting in Eq. (3-11), we obtain

$$\begin{aligned}\mathbf{E}_P &= \left(\frac{1}{4\pi\epsilon_0}\right)\frac{Q(\mathbf{R}-\mathbf{R}')}{|\mathbf{R}-\mathbf{R}'|^3} \\ &= (9\times 10^9)\,\frac{5\times 10^{-9}}{0.458^3}\,(-\mathbf{a}_x 0.4 - \mathbf{a}_y 0.1 + \mathbf{a}_z 0.2) \\ &= 214.5(-\mathbf{a}_x 0.873 - \mathbf{a}_y 0.218 + \mathbf{a}_z 0.437) \quad \text{(V/m)}.\end{aligned}$$

The quantity within the parentheses is the unit vector $\mathbf{a}_{QP} = (\mathbf{R}-\mathbf{R}')/|\mathbf{R}-\mathbf{R}'|$, and $\mathbf{E}_P$ has a magnitude of 214.5 (V/m).

NOTE: The permittivity of air is essentially the same as that of the free space. The factor $1/(4\pi\epsilon_0)$ appears very frequently in electrostatics. From Eq. (1-11) we know that $\epsilon_0 = 1/(c^2\mu_0)$. But $\mu_0 = 4\pi \times 10^{-7}$ (H/m) in SI units; so

$$\frac{1}{4\pi\epsilon_0} = \frac{\mu_0 c^2}{4\pi} = 10^{-7} c^2 \qquad \text{(m/F)} \tag{3-12}$$

exactly. If we use the approximate value $c = 3\times 10^8$ (m/s), then $1/(4\pi\epsilon_0) = 9\times 10^9$ (m/F).

When a point charge q_2 is placed in the field of another point charge q_1, a force $\mathbf{F}_{12}$ is *experienced by* q_2 due to electric field intensity $\mathbf{E}_{12}$ of q_1 at q_2. Combining Eqs. (3-2) and (3-9), we have

$$\mathbf{F}_{12} = q_2\mathbf{E}_{12} = \mathbf{a}_{12}\frac{q_1 q_2}{4\pi\epsilon_0 R_{12}^2} \quad \text{(N)}. \tag{3-13}$$

Coulomb's law

Equation (3-13) is a mathematical form of ***Coulomb's law***. It states that ***the force between two point charges is proportional to the product of the charges and inversely proportional to the square of the distance of separation.*** We note from Eq. (3-13) that $\mathbf{F}_{12}$ is a force of repulsion when q_1 and q_2 are both positive or both negative (the direction of $\mathbf{a}_{12}$ is from q_1 to q_2, and the product q_1q_2 is positive), and a force of attraction when q_1 and q_2 are of opposite signs (the product q_1q_2 is negative).

■ **EXERCISE 3.2**

Given two point charges: $q_1 = 10\,(\mu\text{C})$ at $(2, 0, -4)$ and $q_2 = -60\,(\mu\text{C})$ at $(0, -1, -2)$, determine

a) the electric field intensity at q_1 due to q_2, and
b) the magnitude of the force experienced by q_1.

All dimensions are in meters.

ANS. (a) $-20(\mathbf{a}_x 2 + \mathbf{a}_y - \mathbf{a}_z 2)$ (kV/m), (b) 0.6 (N), attraction.

EXAMPLE 3-2

Electrostatic deflection system of a CRO

The electrostatic deflection system of a cathode-ray oscilloscope is depicted in Fig. 3-2. Electrons from a heated cathode are given an initial velocity $\mathbf{u}_0 = \mathbf{a}_z u_0$ by a positively charged anode (not shown). The electrons enter at $z = 0$ into a region of deflection plates where a uniform electric field $\mathbf{E}_d = -\mathbf{a}_y E_d$ is maintained over a width w. Ignoring gravitational effects, find the vertical deflection of the electrons on the fluorescent screen at $z = L$.

FIGURE 3-2 Electrostatic deflection system of a cathode-ray oscilloscope (Example 3-2).

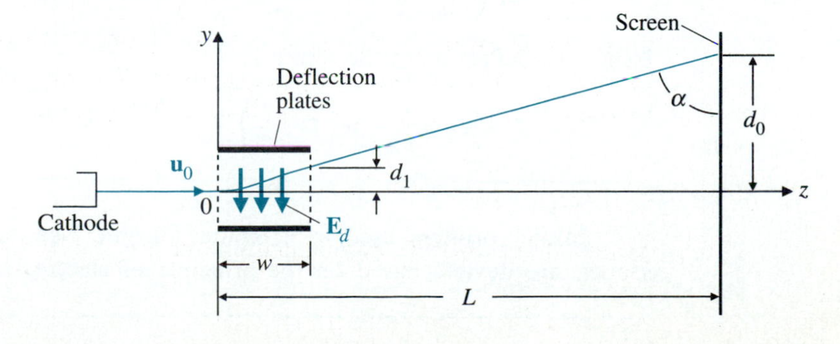

SOLUTION

Since there is no force in the z-direction in the $z > 0$ region, the horizontal velocity u_0 is maintained. The field $\mathbf{E}_d$ exerts a force on the electrons each carrying a charge $-e$, causing a deflection in the y-direction:

$$\mathbf{F} = (-e)\mathbf{E}_d = \mathbf{a}_y eE_d.$$

From Newton's second law of motion in the vertical direction we have

$$m\frac{du_y}{dt} = eE_d,$$

where m is the mass of an electron. Integrating both sides, we obtain

$$u_y = \frac{dy}{dt} = \frac{e}{m}E_d t,$$

where the constant of integration is set to zero because $u_y = 0$ at $t = 0$. Integrating again, we have

$$y = \frac{e}{2m}E_d t^2.$$

The constant of integration is again zero because $y = 0$ at $t = 0$. Note that the electrons have a parabolic trajectory between the deflection plates. At the exit from the deflection plates, $t = w/u_0$,

$$d_1 = \frac{eE_d}{2m}\left(\frac{w}{u_0}\right)^2$$

and

$$u_{y1} = u_y\left(\text{at } t = \frac{w}{u_0}\right) = \frac{eE_d}{m}\left(\frac{w}{u_0}\right).$$

When the electrons reach the screen, they have traveled a further horizontal distance of $(L-w)$, which takes them $(L-w)/u_0$ seconds. During that time there is an additional vertical deflection

$$d_2 = u_{y1}\left(\frac{L-w}{u_0}\right) = \frac{eE_d}{m}\frac{w(L-w)}{u_0^2}.$$

Hence the deflection at the screen is

$$d_0 = d_1 + d_2 = \frac{eE_d}{mu_0^2}w\left(L - \frac{w}{2}\right).$$

Ink-jet printers used in computer output, like cathode-ray oscilloscopes, are devices based on the principle of electrostatic deflection of a

Principle of operation of ink-jet printers

stream of charged particles. Minute droplets of ink are forced through a vibrating nozzle controlled by a piezo-electric transducer. Variable amounts of charges are imparted to the ink droplets, as determined by computer output. The charged ink droplets then pass through a pair of deflection plates where a uniform static electric field exists. The amount of droplet deflection depends on the charge it carries. As the print head moves in a horizontal direction, the ink droplets strike the print surface at various locations from the nozzle and thus form a printed image.

3-3.1 ELECTRIC FIELD DUE TO A SYSTEM OF DISCRETE CHARGES

Suppose an electrostatic field is created by a group of n discrete point charges located at different positions. Since electric field intensity is a linear function of (proportional to) $\mathbf{a}_R q/R^2$, the principle of superposition applies, and the total $\mathbf{E}$ field at a point is the *vector sum* of the fields caused by all the individual charges. Let the positions of the charges $q_1, q_2, \ldots, q_n$ (source points) be denoted by position vectors $\mathbf{R}'_1, \mathbf{R}'_2, \ldots, \mathbf{R}'_n$, and the position of the field point at which the electric intensity is to be calculated be denoted by $\mathbf{R}$.[†] We can write, from Eq. (3-11)

$$\mathbf{E} = \frac{1}{4\pi\epsilon_0}\left[\frac{q_1(\mathbf{R}-\mathbf{R}'_1)}{|\mathbf{R}-\mathbf{R}'_1|^3} + \frac{q_2(\mathbf{R}-\mathbf{R}'_2)}{|\mathbf{R}-\mathbf{R}'_2|^3} + \cdots + \frac{q_n(\mathbf{R}-\mathbf{R}'_n)}{|\mathbf{R}-\mathbf{R}'_n|^3}\right],$$

or

Electric field intensity of a system of discrete point charges

$$\boxed{\mathbf{E} = \frac{1}{4\pi\epsilon_0}\sum_{k=1}^{n}\frac{q_k(\mathbf{R}-\mathbf{R}'_k)}{|\mathbf{R}-\mathbf{R}'_k|^3} \qquad \text{(V/m)}.} \tag{3-14}$$

Although Eq. (3-14) is a succinct expression, it is somewhat inconvenient to use because we often need to add vectors of different magnitudes and directions. A simpler approach would be to find $\mathbf{E}$ from the electric potential. This will be discussed in Section 3-5.

3-3.2 ELECTRIC FIELD DUE TO A CONTINUOUS DISTRIBUTION OF CHARGE

The electric field caused by a continuous distribution of charge can be obtained by integrating (superposing) the contribution of an element of charge over the charge distribution. Refer to Fig. 3-3, where a volume charge

[†]When there is a need to distinguish the notation for the location of a source point from that of a field point, we follow the accepted convention of using primed coordinates for the former and unprimed coordinates for the latter.

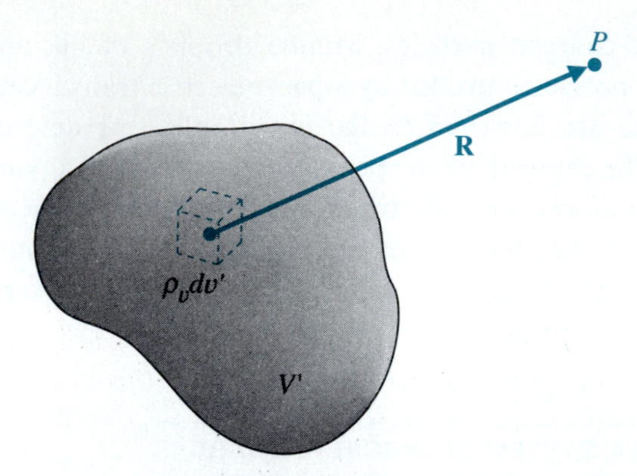

FIGURE 3-3 Electric field due to a continuous charge distribution.

distribution is shown. The volume charge density $\rho_v(\mathrm{C/m^3})$ is, in general, a function of the coordinates. Since a differential element of charge behaves like a point charge, the contribution of the charge $\rho_v\,dv'$ in a differential volume element dv' to the electric field intensity at the field point P is

$$d\mathbf{E} = \mathbf{a}_R \frac{\rho_v\,dv'}{4\pi\epsilon_0 R^2}. \tag{3-15}$$

We have

Electric field intensity of a volume distribution of charge

$$\boxed{\mathbf{E} = \frac{1}{4\pi\epsilon_0}\int_{V'} \mathbf{a}_R \frac{\rho_v}{R^2}\,dv' \qquad (\mathrm{V/m}).} \tag{3-16}$$

If the charge is distributed on a surface with a surface charge density $\rho_s(\mathrm{C/m^2})$, we write

Electric field intensity of a surface distribution of charge

$$\boxed{\mathbf{E} = \frac{1}{4\pi\epsilon_0}\int_{S'} \mathbf{a}_R \frac{\rho_s}{R^2}\,ds' \qquad (\mathrm{V/m}).} \tag{3-17}$$

For a line charge we have

Electric field intensity of a line charge

$$\boxed{\mathbf{E} = \frac{1}{4\pi\epsilon_0}\int_{L'} \mathbf{a}_R \frac{\rho_\ell}{R^2}\,d\ell' \qquad (\mathrm{V/m}),} \tag{3-18}$$

where $\rho_\ell(\mathrm{C/m})$ is the line charge density, and L' the line (not necessarily straight) along which the charge is distributed.

EXAMPLE 3-3

Determine the electric field intensity of an infinitely long, straight, line charge of a uniform density ρ_ℓ(C/m) in air.

SOLUTION

Let us assume that the line charge lies along the z'-axis as shown in Fig. 3-4. We are perfectly free to make this assumption because the field does not depend on how we designate the line. *Note the convention of using primed coordinates for source points and unprimed coordinates for field points.*

The problem asks us to find the electric field intensity at a point P, which is at a distance r from the line. Since the problem has a cylindrical symmetry (that is, the electric field is independent of the azimuth angle ϕ), it would be most convenient to work with cylindrical coordinates. We rewrite Eq. (3-18) as

$$\boxed{\mathbf{E} = \frac{1}{4\pi\epsilon_0}\int_{L'} \rho_\ell \frac{\mathbf{R}}{R^3}\, d\ell' \qquad \text{(V/m)}.} \tag{3-18a}$$

For the problem at hand, ρ_ℓ is constant, and a line element $d\ell' = dz'$ is chosen to be at an arbitrary distance z' from the origin. It is most important to

FIGURE 3-4 An infinitely long, straight, line charge.

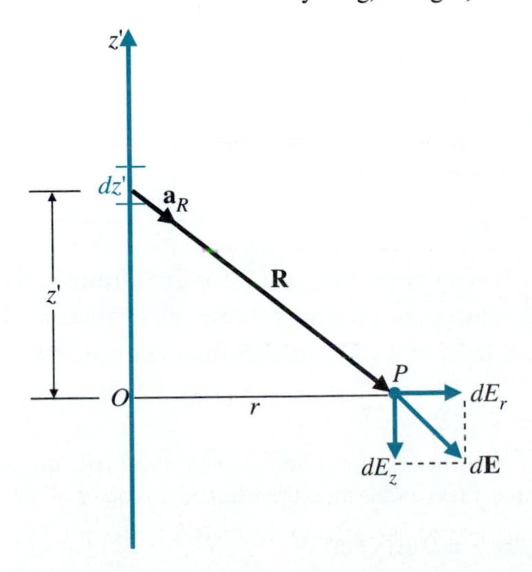

remember that $\mathbf{R}$ is the distance vector directed *from the source to the field point*, not the other way around. We have

$$\mathbf{R} = \mathbf{a}_r r - \mathbf{a}_z z'. \tag{3-19}$$

The electric field $d\mathbf{E}$ due to the differential line charge element $\rho_\ell \, d\ell' = \rho_\ell \, dz'$ is

$$\begin{aligned} d\mathbf{E} &= \frac{\rho_\ell \, dz'}{4\pi\epsilon_0} \frac{\mathbf{a}_r r - \mathbf{a}_z z'}{(r^2 + z'^2)^{3/2}} \\ &= \mathbf{a}_r \, dE_r + \mathbf{a}_z \, dE_z, \end{aligned} \tag{3-20}$$

where

$$dE_r = \frac{\rho_\ell r \, dz'}{4\pi\epsilon_0(r^2 + z'^2)^{3/2}} \tag{3-21}$$

and

$$dE_z = \frac{-\rho_\ell z' \, dz'}{4\pi\epsilon_0(r^2 + z'^2)^{3/2}}. \tag{3-22}$$

In Eq. (3-22) we have decomposed $d\mathbf{E}$ into its components in the $\mathbf{a}_r$ and $\mathbf{a}_z$ directions. For every $\rho_\ell \, dz'$ at $+z'$ there is a charge element $\rho_\ell \, dz'$ at $-z'$ that will produce a $d\mathbf{E}$ with components dE_r and $-dE_z$. Hence the $\mathbf{a}_z$ components will cancel in the integration process, and we only need to integrate the dE_r in Eq. (3-21):

$$\mathbf{E} = \mathbf{a}_r E_r = \mathbf{a}_r \frac{\rho_\ell r}{4\pi\epsilon_0} \int_{-\infty}^{\infty} \frac{dz'}{(r^2 + z'^2)^{3/2}},$$

or

Electric field intensity due to an infinite straight line charge of uniform density

$$\boxed{\mathbf{E} = \mathbf{a}_r \frac{\rho_\ell}{2\pi\epsilon_0 r} \quad \text{(V/m)}.} \tag{3-23}$$

Equation (3-23) is an important result for an infinite line charge. Of course, no physical line charge is infinitely long; nevertheless, Eq. (3-23) gives the approximate $\mathbf{E}$ field of a long straight line charge at a point close to the line charge.

■ **EXERCISE 3.3** Assuming that an infinitely long line charge of 50 (pC/m) parallel to the y-axis at $x = 2$ (m) and $z = 1$ (m), determine the electric intensity at the point $(-1, 5, -3)$.

ANS. $-0.18(\mathbf{a}_x 0.6 + \mathbf{a}_z 0.8)$ (V/m).

3-4 GAUSS'S LAW AND APPLICATIONS

Gauss's law follows directly from the divergence postulate of electrostatics, Eq. (3-3), by the application of the divergence theorem. It was derived as Eq. (3-6) and is repeated here because of its importance:

$$\oint_S \mathbf{E} \cdot d\mathbf{s} = \frac{Q}{\epsilon_0}. \tag{3-24}$$

Gauss's law

Gauss's law asserts that the total outward flux of the E-field over any closed surface in free space is equal to the total charge enclosed in the surface divided by ϵ_0. We note that the surface S can be *any hypothetical (mathematical) closed surface chosen for convenience*; it does not have to be, and usually is not, a physical surface.

Gauss's law is particularly useful in determining the **E**-field of charge distributions with some symmetry conditions, such that *the normal component of the electric field intensity is constant over an enclosed surface.* In such cases the surface integral on the left side of Eq. (3-24) would be very easy to evaluate, and Gauss's law would be a much more efficient way for finding the electric field intensity than Eqs. (3-16) through (3-18a).

Proper choice of Gaussian surface

On the other hand, when symmetry conditions do not exist, Gauss's law would not be of much help. The essence of applying Gauss's law lies first in the recognition of symmetry conditions and second in the suitable choice of a surface over which the normal component of **E** resulting from a given charge distribution is a constant. Such a surface is referred to as a ***Gaussian surface.*** This basic principle was used to obtain Eq. (3-8) for a point charge that possesses spherical symmetry; consequently, a proper Gaussian surface is the surface of a sphere centered at the point charge.

EXAMPLE 3-4

Use Gauss's law to determine the electric field intensity of an infinitely long, straight, line charge of a uniform density ρ_ℓ in air.

SOLUTION

This problem was solved in Example 3-3 by using Eq. (3-18). Since the line charge is infinitely long, the resultant **E** field must be radial and perpendicular to the line charge ($\mathbf{E} = \mathbf{a}_r E_r$), and a component of **E** along the line cannot exist. Taking advantage of cylindrical symmetry, we construct a cylindrical Gaussian surface of a radius r and an arbitrary length L with the line charge as its axis, as shown in Fig. 3-5. On this surface, E_r is constant, and

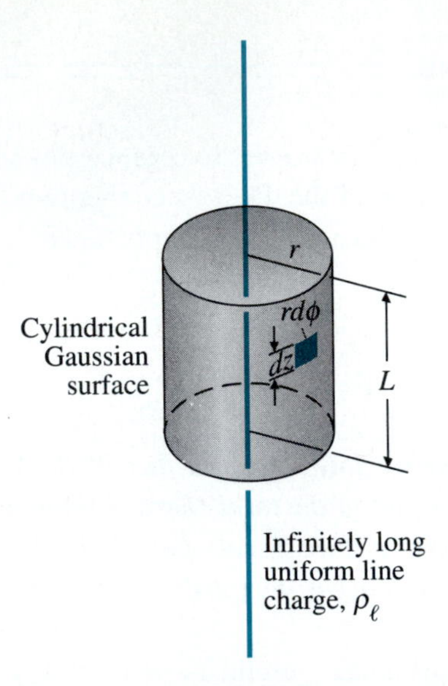

FIGURE 3-5 Applying Gauss's law to an infinitely long line charge (Example 3-4).

$d\mathbf{s} = \mathbf{a}_r r\,d\phi\,dz$. We have

$$\oint_S \mathbf{E}\cdot d\mathbf{s} = \int_0^L \int_0^{2\pi} E_r r\,d\phi\,dz = 2\pi r L E_r.$$

There is no contribution from the top or the bottom face of the cylinder because on the top face $d\mathbf{s} = \mathbf{a}_z r\,dr\,d\phi$ but $\mathbf{E}$ has no z-component there, making $\mathbf{E}\cdot d\mathbf{s} = 0$. Similarly for the bottom face. The total charge enclosed in the cylinder is $Q = \rho_\ell L$. Substitution into Eq. (3-24) gives us immediately

$$2\pi r L E_r = \frac{\rho_\ell L}{\epsilon_0},$$

or

$$\mathbf{E} = \mathbf{a}_r E_r = \mathbf{a}_r \frac{\rho_\ell}{2\pi\epsilon_0 r}.$$

This result is the same as that given in Eq. (3-23), but we arrived at it here in a much simpler way. Notice too that the length L of the cylindrical Gaussian surface does not appear in the final expression, so we could have chosen a cylinder of a unit length.

NOTE: The same cylindrical Gaussian surface does not work if the line charge is of a finite length. Do you know why?

EXAMPLE 3-5

Determine the electric field intensity of an infinite planar charge with a uniform surface charge density ρ_s.

SOLUTION

The $\mathbf{E}$ field caused by a charged sheet of an infinite extent is normal to the sheet. Equation (3-17) could be used to find $\mathbf{E}$, but this would involve a double integration between infinite limits of a general expression of $1/R^2$. Gauss's law can be used to much advantage here.

We choose as the Gaussian surface a rectangular box with top and bottom faces of an arbitrary area A equidistant from the planar charge, as shown in Fig. 3-6. The sides of the box are perpendicular to the charged sheet. If the charged sheet coincides with the xy-plane, then on the top face,

$$\mathbf{E} \cdot d\mathbf{s} = (\mathbf{a}_z E_z) \cdot (\mathbf{a}_z\, ds) = E_z\, ds.$$

On the bottom face,

$$\mathbf{E} \cdot d\mathbf{s} = (-\mathbf{a}_z E_z) \cdot (-\mathbf{a}_z\, ds) = E_z\, ds.$$

Since there is no contribution from the side faces, we have

$$\oint_S \mathbf{E} \cdot d\mathbf{s} = 2E_z \int_A ds = 2E_z A.$$

The total charge enclosed in the box is $Q = \rho_s A$. Therefore,

$$2E_z A = \frac{\rho_s A}{\epsilon_0},$$

FIGURE 3-6 Applying Gauss's law to an infinite planar charge (Example 3-5).

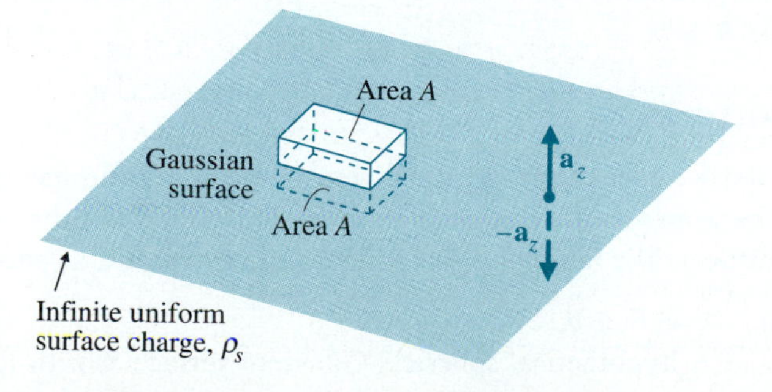

from which we obtain

$$\mathbf{E} = \mathbf{a}_z E_z = \mathbf{a}_z \frac{\rho_s}{2\epsilon_0}, \qquad z > 0, \tag{3-25a}$$

and

$$\mathbf{E} = -\mathbf{a}_z E_z = -\mathbf{a}_z \frac{\rho_s}{2\epsilon_0}, \qquad z < 0. \tag{3-25b}$$

The charged sheet does not always coincide with the xy-plane (so we do not always speak in terms of "above" and "below" the plane), but the **E** field always points *away* from the sheet if ρ_s is *positive.* The Gaussian surface we chose could have been a "pillbox" of any shape, not necessarily rectangular one.

NOTE: No suitable Gaussian surface can be chosen in this Example if the planar charge is not of infinite extent in both dimensions or is not flat. Can you explain why?

Comparing lighting schemes

The lighting scheme of an office or a classroom may consist of incandescent bulbs, long fluorescent tubes, or ceiling panel lights. These correspond roughly to point sources, line sources, and planar sources, respectively. From Eqs. (3-8), (3-23), and (3-25) we can estimate that light intensity will fall off rapidly—as the square of the distance from the source—in the case of incandescent bulbs, less rapidly—as the first power of the distance—for long fluorescent tubes, and not at all for ceiling panel lights.

EXAMPLE 3-6

Determine the **E** field caused by a spherical cloud of electrons with a volume charge density $\rho_v = -\rho_o$ for $0 \leqslant R \leqslant b$ (both ρ_o and b are positive) and $\rho_v = 0$ for $R > b$.

SOLUTION

First we recognize that the given source condition has spherical symmetry. The appropriate Gaussian surfaces must therefore be concentric spherical surfaces. We must find the **E** field in two regions, as shown in Fig. 3-7.

a) $0 \leqslant R \leqslant b$
A hypothetical spherical Gaussian surface S_i with $R < b$ is constructed

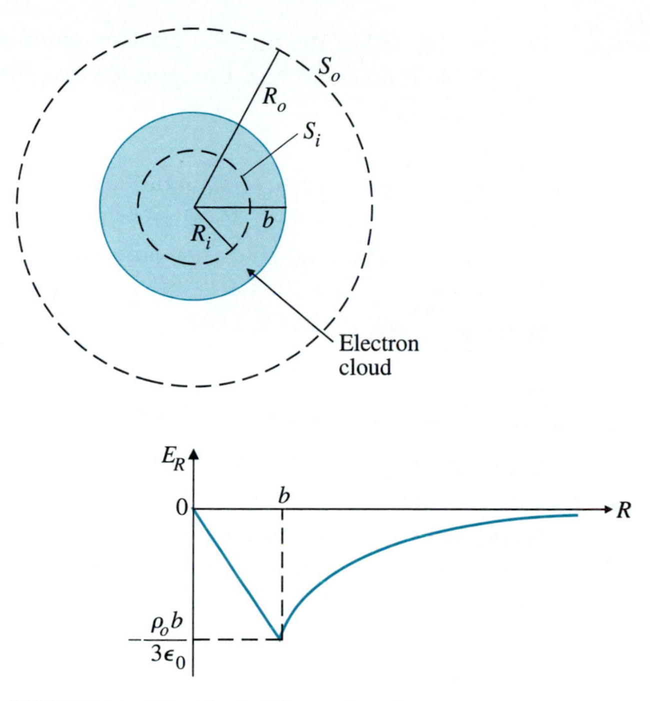

FIGURE 3-7 Electric field intensity of a spherical electron cloud (Example 3-6).

within the electron cloud. On this surface, $\mathbf{E}$ is radial and has a constant magnitude:

$$\mathbf{E} = \mathbf{a}_R E_R, \qquad d\mathbf{s} = \mathbf{a}_R\, ds.$$

The total outward E flux is

$$\oint_{S_i} \mathbf{E}\cdot d\mathbf{s} = E_R \int_{S_i} ds = E_R 4\pi R^2.$$

The total charge enclosed within the Gaussian surface is

$$Q = \int_V \rho_v\, dv$$

$$= -\rho_o \int_V dv = -\rho_o \frac{4\pi}{3} R^3.$$

Substitution into Eq. (3-6) yields

$$\mathbf{E} = -\mathbf{a}_R \frac{\rho_o}{3\epsilon_0} R, \qquad 0 \leqslant R \leqslant b.$$

We see that within the uniform electron cloud the **E** field is directed toward the center and has a magnitude proportional to the distance from the center.

b) $R \geqslant b$
For this case we construct a spherical Gaussian surface S_o with $R > b$ outside the electron cloud. We obtain the same expression for $\oint_{S_o} \mathbf{E} \cdot d\mathbf{s}$ as in case (a). The total charge enclosed is

$$Q = -\rho_o \frac{4\pi}{3} b^3.$$

Consequently,

$$\mathbf{E} = -\mathbf{a}_R \frac{\rho_o b^3}{3\epsilon_0 R^2}, \qquad R \geqslant b.$$

Notice that this relationship follows the inverse square law and could have been obtained directly from Eq. (3-8). We see that *outside* the charged cloud, the **E** field is exactly the same as if the total charge has been concentrated on a single point charge at the center. This result holds, in general, for any spherically symmetrical charged region even when ρ_v is a function of R.

EXERCISE 3.4 Given $\mathbf{E} = \mathbf{a}_r(20/r^2)$ (mV/m) in free space, find ρ_v at the point (3, −4, 1) (cm).

ANS. -1.42 (nC/m^3).

EXERCISE 3.5 A positive charge Q is uniformly distributed on a very thin spherical shell of radius b in air. Find **E** everywhere. Plot $|\mathbf{E}|$ versus R.

ANS. 0 for $0 < R < b$; $\mathbf{a}_R(Q/4\pi\epsilon_0 R^2)$ for $R > b$.

3-5 ELECTRIC POTENTIAL

Earlier, in connection with the null identity in Eq. (2-105), we noted that a curl-free vector field could always be expressed as the gradient of a scalar field. We can then define a scalar ***electric potential*** V from Eq. (3-4) such that

Electrostatic field intensity from electric potential

$$\boxed{\mathbf{E} = -\nabla V} \tag{3-26}$$

because scalar quantities are easier to handle than vector quantities. If we can determine V more easily, then **E** can be found by a gradient operation, which is a straightforward differentiation process. The reason for the inclusion of a negative sign in Eq. (3-26) will be explained presently.

Electric potential does have physical significance, and it is related to the work done in carrying a charge from one point to another. In Section 3-2 we defined the electric field intensity as the force acting on a unit test charge. Therefore in moving a unit charge from point P_1 to point P_2 in an electric field, work must be done *against the field* and is equal to

$$\frac{W}{q} = -\int_{P_1}^{P_2} \mathbf{E}\cdot d\boldsymbol{\ell} \qquad \text{(J/C or V).} \tag{3-27}$$

Many paths may be followed in going from P_1 to P_2. Two such paths are drawn in Fig. 3-8. Since the path between P_1 and P_2 is not specified in Eq. (3-27), the question naturally arises: How does the work depend on the path taken? A little thought will lead us to conclude that W/q in Eq. (3-27) should not depend on the path; if it did, one would be able to go from P_1 to P_2 along a path for which W is smaller and then to come back to P_1 along another path, achieving a net gain in work or energy. This result would be contrary to the principle of conservation of energy. We have already alluded to the path-independent nature of the scalar line integral of the irrotational (conservative) **E** field when we discussed Eq. (3-7).

Analogous to the concept of potential energy in mechanics, Eq. (3-27) represents the difference in electric potential energy of a unit charge between point P_2 and point P_1. If we denote the electric potential energy per unit charge by V (the **electric potential**) we have

Electrostatic potential difference between P_2 and P_1 equals the work done in moving a unit charge from P_1 to P_2

$$V_2 - V_1 = -\int_{P_1}^{P_2} \mathbf{E}\cdot d\boldsymbol{\ell} \qquad \text{(V).} \tag{3-28}$$

What we have defined in Eq. (3-28) is a ***potential difference (electrostatic voltage)*** between points P_2 and P_1. We cannot talk about the *absolute* potential of a point any more than we can talk about the absolute phase of a

FIGURE 3-8 Two paths leading from P_1 to P_2 in an electric field.

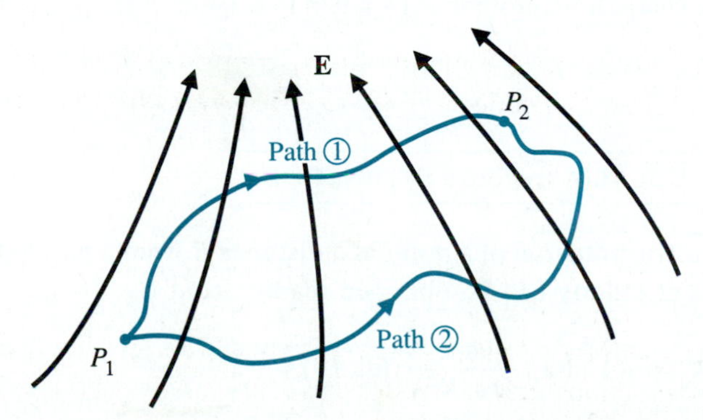

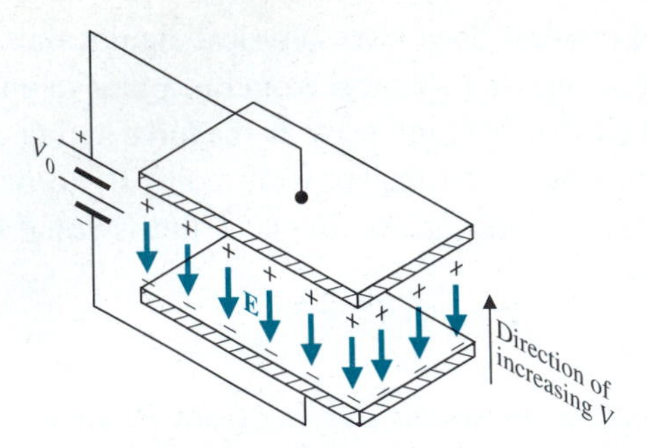

FIGURE 3-9 Relative directions of **E** and increasing V.

Choosing a reference zero-potential point

phasor or the absolute altitude of a geographical location. A reference zero-potential point, a reference zero phase (usually at $t = 0$), or a reference zero altitude (usually at sea level) must first be specified. In most (but not all) cases the zero-potential point is taken at infinity. When the reference zero-potential point is not at infinity (for example, when it is at "the ground"), it should be specifically stated.

We make two more important observations here about Eq. (3-28). First, the negative sign must be included in order to conform with the convention that in going *against* the **E** field the electric potential V *increases*. For instance, when a d-c battery of a voltage V_0 is connected between two parallel conducting plates, as in Fig. 3-9, positive and negative charges accumulate on the top and bottom plates, respectively. The **E** field is directed from positive to negative charges, while the potential increases in the *opposite* direction.

Electric field lines are perpendicular to equipotential lines and surfaces.

Second, we know from Section 2-5, when we defined the gradient of a scalar field, that the direction of ∇V is normal to the surfaces of constant V. Hence if we use ***directed field lines*** or ***flux lines*** to indicate the direction of the **E** field, they are everywhere perpendicular to ***equipotential lines*** and ***equipotential surfaces***.

EXERCISE 3.6 Determine the work done *by* the electric field $\mathbf{E} = \mathbf{a}_x x - \mathbf{a}_y 2y$ (V/m) in moving a unit positive charge from position $P_1(-2, 0, 0)$ to position $P_2(5, -1, 3)$. The distances are in (m).

ANS. 9.5 (J).

3-5.1 ELECTRIC POTENTIAL DUE TO A CHARGE DISTRIBUTION

The electric potential of a point at a distance R from a point charge q referred to that at infinity can be obtained readily from Eq. (3-28):

$$V = -\int_{\infty}^{R} \left(\mathbf{a}_R \frac{q}{4\pi\epsilon_0 R^2}\right) \cdot (\mathbf{a}_R \, dR),$$

which gives

Electrostatic potential of a point charge referred to infinity

$$V = \frac{q}{4\pi\epsilon_0 R} \quad \text{(V)}. \tag{3-29}$$

This is a scalar quantity and depends on, besides q, only the distance R. The potential difference between any two points P_2 and P_1 at distances R_2 and R_1, respectively, from q is

$$V_{21} = V_{P_2} - V_{P_1} = \frac{q}{4\pi\epsilon_0}\left(\frac{1}{R_2} - \frac{1}{R_1}\right). \tag{3-30}$$

The electric potential at $\mathbf{R}$ due to a system of n discrete charges $q_1, q_2, \ldots, q_n$ located at $\mathbf{R}'_1, \mathbf{R}'_2, \ldots, \mathbf{R}'_n$ is, by superposition, the sum of the potentials due to the individual charges:

$$V = \frac{1}{4\pi\epsilon_0} \sum_{k=1}^{n} \frac{q_k}{|\mathbf{R} - \mathbf{R}'_k|} \quad \text{(V)}. \tag{3-31}$$

Since this is a scalar sum, it is, in general, easier to determine $\mathbf{E}$ by taking the negative gradient of V than from the vector sum in Eq. (3-14) directly.

EXAMPLE 3-7

An electric dipole consisting of equal and opposite point charges $+q$ and $-q$ separated by a small distance d is shown in Fig. 3-10. Determine the potential V and the electric intensity $\mathbf{E}$ at an arbitrary point P at a distance $R \gg d$ from the dipole.

FIGURE 3-10 An electric dipole.

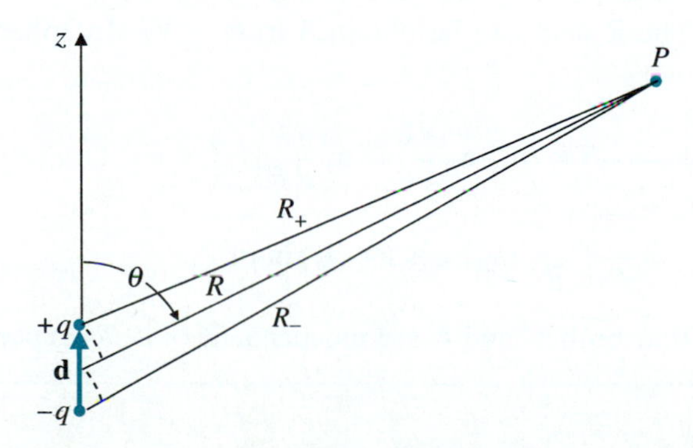

SOLUTION

Let the distances from the charges $+q$ and $-q$ to the field point P be R_+ and R_- respectively.

The potential at P can be written directly from Eq. (3-31).

$$V = \frac{q}{4\pi\epsilon_0}\left(\frac{1}{R_+} - \frac{1}{R_-}\right). \tag{3-32}$$

If $d \ll R$, we write

$$R_+ \cong \left(R - \frac{d}{2}\cos\theta\right) \tag{3-33}$$

and

$$R_- \cong \left(R + \frac{d}{2}\cos\theta\right). \tag{3-34}$$

Substituting Eqs. (3-33) and (3-34) in Eq. (3-32), we have

$$\begin{aligned} V &= \frac{q}{4\pi\epsilon_0}\left(\frac{1}{R - \frac{d}{2}\cos\theta} - \frac{1}{R + \frac{d}{2}\cos\theta}\right) \\ &= \frac{q}{4\pi\epsilon_0}\left(\frac{d\cos\theta}{R^2 - \frac{d^2}{4}\cos^2\theta}\right) \cong \frac{qd\cos\theta}{4\pi\epsilon_0 R^2}. \end{aligned} \tag{3-35}$$

Equation (3-35) can be written as

Finding electrostatic potential from electric dipole moment

$$\boxed{V = \frac{\mathbf{p}\cdot\mathbf{a}_R}{4\pi\epsilon_0 R^2} \qquad \text{(V)},} \tag{3-36}$$

where $\mathbf{p} = q\mathbf{d}$ is the ***electric dipole moment*** (SI unit: C·m). (The "approximate" sign ($\sim$) has been dropped for simplicity.)

The $\mathbf{E}$ field can be obtained from $-\nabla V$. In spherical coordinates we have

$$\begin{aligned} \mathbf{E} = -\nabla V &= -\mathbf{a}_R\frac{\partial V}{\partial R} - \mathbf{a}_\theta\frac{\partial V}{R\,\partial\theta} \\ &= \frac{p}{4\pi\epsilon_0 R^3}(\mathbf{a}_R 2\cos\theta + \mathbf{a}_\theta\sin\theta). \end{aligned} \tag{3-37}$$

Note that both V and $\mathbf{E}$ are independent of ϕ, as expected.

EXERCISE 3.7 An electric dipole at the origin has a dipole moment $\mathbf{a}_z 0.1$ (nC·m). Find V and $\mathbf{E}$ at (a) (0, 0, 5 (m)), and (b) (2(m), $\pi/3$, $\pi/8$).

ANS. (a) 36 (mV), $\mathbf{a}_z 14.4$ (mV/m); (b) 113 (mV), $\mathbf{a}_R 113 + \mathbf{a}_\theta 97.4$ (mV/m).

The electric potential due to a continuous distribution of charge confined in a given region is obtained by integrating the contribution of an element of charge over the charged region. We have, for a volume charge distribution,

Electric potential due to continuous charge distributions

$$V = \frac{1}{4\pi\epsilon_0} \int_{V'} \frac{\rho_v}{R} \, dv' \qquad \text{(V)}. \tag{3-38}$$

For surface charge distribution,

$$V = \frac{1}{4\pi\epsilon_0} \int_{S'} \frac{\rho_s}{R} \, ds' \qquad \text{(V)}; \tag{3-39}$$

and for a line charge,

$$V = \frac{1}{4\pi\epsilon_0} \int_{L'} \frac{\rho_\ell}{R} \, d\ell' \qquad \text{(V)}. \tag{3-40}$$

We note here again that the integrals in Eqs. (3-38) and (3-39) represent integrations in three and two dimensions, respectively.

EXAMPLE 3-8

Obtain a formula for the electric field intensity on the axis of a circular disk of radius b that carries a uniform surface charge density ρ_s.

SOLUTION

Although the disk has circular symmetry, we cannot visualize a surface around it over which the normal component of $\mathbf{E}$ has a constant magnitude; hence Gauss's law is not useful for the solution of this problem. We use Eq. (3-39). Working with the cylindrical coordinates indicated in Fig. 3-11, we have

$$ds' = r' \, dr' \, d\phi'$$

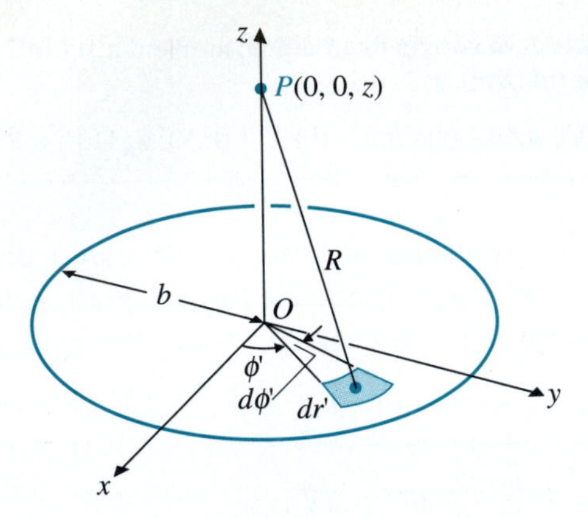

FIGURE 3-11 A uniformly charged disk (Example 3-8).

and

$$R = \sqrt{z^2 + r'^2}.$$

The electric potential at the point $P(0, 0, z)$ referring to the point at infinity is

$$\begin{aligned} V &= \frac{\rho_s}{4\pi\epsilon_0} \int_0^{2\pi} \int_0^b \frac{r'}{(z^2 + r'^2)^{1/2}} \, dr' \, d\phi' \\ &= \frac{\rho_s}{2\epsilon_0} [(z^2 + b^2)^{1/2} - |z|], \end{aligned} \tag{3-41}$$

where the absolute sign around z describes the fact that V is the same whether z is positive (a point above the disk) or z is negative (a point below the disk). Therefore,

$$\mathbf{E} = -\nabla V = -\mathbf{a}_z \frac{\partial V}{\partial z}$$

$$= \begin{cases} \mathbf{a}_z \dfrac{\rho_s}{2\epsilon_0} [1 - z(z^2 + b^2)^{-1/2}], & z > 0 \qquad \text{(3-42a)} \\ -\mathbf{a}_z \dfrac{\rho_s}{2\epsilon_0} [1 + z(z^2 + b^2)^{-1/2}]. & z < 0. \qquad \text{(3-42b)} \end{cases}$$

REVIEW QUESTIONS

Q.3-1 Write the differential form of the fundamental postulates of electrostatics in free space.

Q.3-2 Under what conditions will the electric field intensity be both solenoidal and irrotational?

Q.3-3 Write the integral form of the fundamental postulates of electrostatics in free space, and state their meaning in words.

Q.3-4 Explain why an irrotational field is also known as a conservative field.

Q.3-5 In what ways does the electric field intensity vary with distance for (a) a point charge? (b) an electric dipole?

Q.3-6 State *Coulomb's law.*

Q.3-7 State *Gauss's law.* Under what conditions is Gauss's law especially useful in determining the electric field intensity of a charge distribution?

Q.3-8 Describe the ways in which the electric field intensity of an infinitely long, straight line charge of uniform density varies with distance.

Q.3-9 If the electric potential at a point is zero, does it follow that the electrical field intensity is also zero at that point? Explain.

Q.3-10 If the electric field intensity at a point is zero, does it follow that the electric potential is also zero at that point? Explain.

REMARKS

1. In determining the electric field intensity, **E**, of a charge distribution, it is simplest to apply Gauss's law if a symmetrical Gaussian surface enclosing the charges can be found over which the normal component of the field is constant.
2. If a proper Gaussian surface cannot be found, then it is simpler to find V (a scalar) first, and then obtain **E** from $-\nabla V$.
3. Directed field lines (flux lines) are everywhere perpendicular to equipotential lines and equipotential surfaces.

3-6 MATERIAL MEDIA IN STATIC ELECTRIC FIELD

Conductors, semiconductors, and dielectrics

So far we have discussed only the electric field of stationary charge distributions in free space or air. We now examine the field behavior in material media. In general, we classify materials according to their electrical properties into three types: ***conductors***, ***semiconductors***, and ***insulators*** (or ***dielectrics***). In terms of the crude atomic model of an atom consisting of a positively charged nucleus with orbiting electrons, the electrons in the outermost shells of the atoms of ***conductors*** are very loosely held and migrate easily from one atom to another. Most metals belong to this group. The electrons in the atoms of ***insulators*** or dielectrics, however, are confined to their orbits; they cannot be liberated in normal circumstances, even by the application of an external electric field. The electrical properties of ***semiconductors*** fall between those of conductors and insulators in that they possess a relatively small number of freely movable charges.

In terms of the band theory of solids we find that there are allowed energy bands for electrons, each band consisting of many closely spaced, discrete energy states. Between these energy bands there may be forbidden regions or gaps where no electrons of the solid's atom can reside. Conductors have an upper energy band partially filled with electrons or an upper pair of overlapping bands that are partially filled so that the electrons in these bands can move from one to another with only a small change in energy. Insulators or dielectrics are materials with a completely filled upper band, so conduction could not normally occur because of the existence of a large energy gap to the next higher band. If the energy gap of the forbidden region is relatively small, small amounts of external energy may be sufficient to excite the electrons in the filled upper band to jump into the next band, causing conduction. Such materials are semiconductors. The macroscopic electrical property of a material medium is characterized by a constitutive parameter called ***conductivity***, which we will define in Chapter 4.

3-6.1 CONDUCTORS IN STATIC ELECTRIC FIELD

Assume for the present that some positive (or negative) charges are introduced in the interior of a good conductor. An electric field will be set up in the conductor, the field exerting a force on the charges and making them move away from one another. This movement will continue until *all* the charges reach the conductor surface and redistribute themselves in such a way that both the charge and the field inside vanish. Hence,

Inside a Conductor (Under Static Conditions)
$\rho_v = 0$ (3-43)
$\mathbf{E} = 0$ (3-44)

In the interior of a conductor under static conditions, both free charge and electric field intensity vanish.

When there are no free charges in the interior of a conductor ($\rho_v = 0$), $\mathbf{E}$ must be zero because, according to Gauss's law, the total outward electric flux through *any* closed surface constructed inside the conductor must vanish.

The charge distribution on the surface of a conductor depends on the shape of the surface. Obviously, the charges would not be in a state of equilibrium if there were a tangential component of the electric field intensity that produces a tangential force and moves the charges. Therefore, ***under static conditions the* E *field on a conductor surface is everywhere normal to the surface***. In other words, ***the surface of a conductor is an equipotential surface***

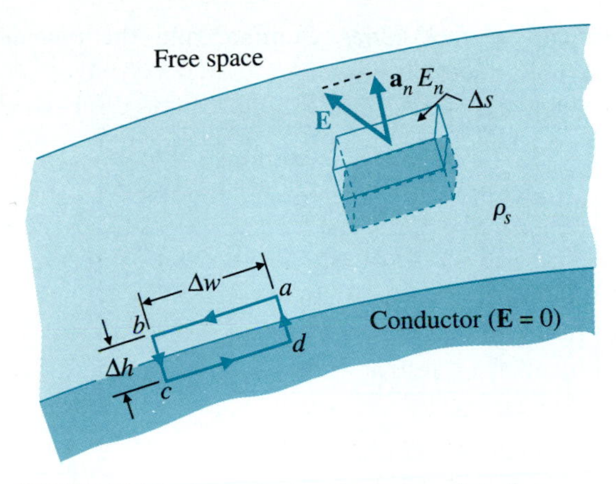

FIGURE 3-12 A conductor–free space interface.

under static conditions. As a matter of fact, since $\mathbf{E} = 0$ everywhere inside a conductor, the *whole* conductor has the same electrostatic potential.

Figure 3-12 shows an interface between a conductor and free space. Consider the contour *abcda*, which has width $ab = cd = \Delta w$ and height $bc = da = \Delta h$. Sides *ab* and *cd* are parallel to the interface. Applying Eq. (3-7), letting $\Delta h \to 0$, and noting that $\mathbf{E}$ in a conductor is zero, we obtain immediately

$$\oint_{abcda} \mathbf{E} \cdot d\boldsymbol{\ell} = E_t \Delta w = 0$$

or

$$E_t = 0, \tag{3-45}$$

which says that ***the tangential component of the* E *field on a conductor surface is zero under static conditions.*** In order to find E_n, the normal component of $\mathbf{E}$ at the surface of the conductor, we construct a Gaussian surface in the form of a thin pillbox with the top face in free space and the bottom face in the conductor where $\mathbf{E} = 0$. Using Eq. (3-6), we obtain

$$\oint_S \mathbf{E} \cdot d\mathbf{s} = E_n \Delta S = \frac{\rho_s \Delta S}{\epsilon_0},$$

or

$$E_n = \frac{\rho_s}{\epsilon_0}. \tag{3-46}$$

Hence, ***the normal component of the* E *field at a conductor-free space boundary is equal to the surface charge density on the conductor divided by the***

On the surface of a conductor under static conditions, the electric field is perpendicular to the surface, which is equipotential.

permittivity of free space. Summarizing the *boundary conditions* at the conductor surface, we have

Boundary Conditions at a Conductor–Free Space Interface
$E_t = 0$ (3-45)
$E_n = \dfrac{\rho_s}{\epsilon_0}$ (3-46)

EXAMPLE 3-9

A positive point charge Q is at the center of a spherical conducting shell of an inner radius R_i and an outer radius R_o. Determine $\mathbf{E}$ and V as functions of the radial distance R.

SOLUTION

The geometry of the problem is shown in Fig. 3-13(a). Since there is spherical symmetry, it is simplest to use Gauss's law to determine $\mathbf{E}$ and then find V by integration. There are three distinct regions: (a) $R > R_o$, (b) $R_i < R < R_o$, and (c) $R < R_i$. Suitable spherical Gaussian surfaces will be constructed in these regions. Symmetry requires that $\mathbf{E} = \mathbf{a}_R E_R$ in all three regions.

a) $R > R_o$ (Gaussian surface S_1):

$$\oint_S \mathbf{E}\cdot d\mathbf{s} = E_{R1} 4\pi R^2 = \frac{Q}{\epsilon_0},$$

or

$$E_{R1} = \frac{Q}{4\pi\epsilon_0 R^2}. \tag{3-47}$$

The $\mathbf{E}$ field is the same as that of a point charge Q without the presence of the shell; this relationship was given in Eq. (3-8). The potential referring to the point at infinity is

$$V_1 = -\int_\infty^R (E_{R1})\, dR = \frac{Q}{4\pi\epsilon_0 R}, \tag{3-48}$$

which is the same as that given in Eq. (3-29).

b) $R_i < R < R_o$ (Gaussian surface S_2): Because of Eq. (3-44), we know that

$$E_{R2} = 0. \tag{3-49}$$

Since $\rho_v = 0$ in the conducting shell and since the total charge enclosed

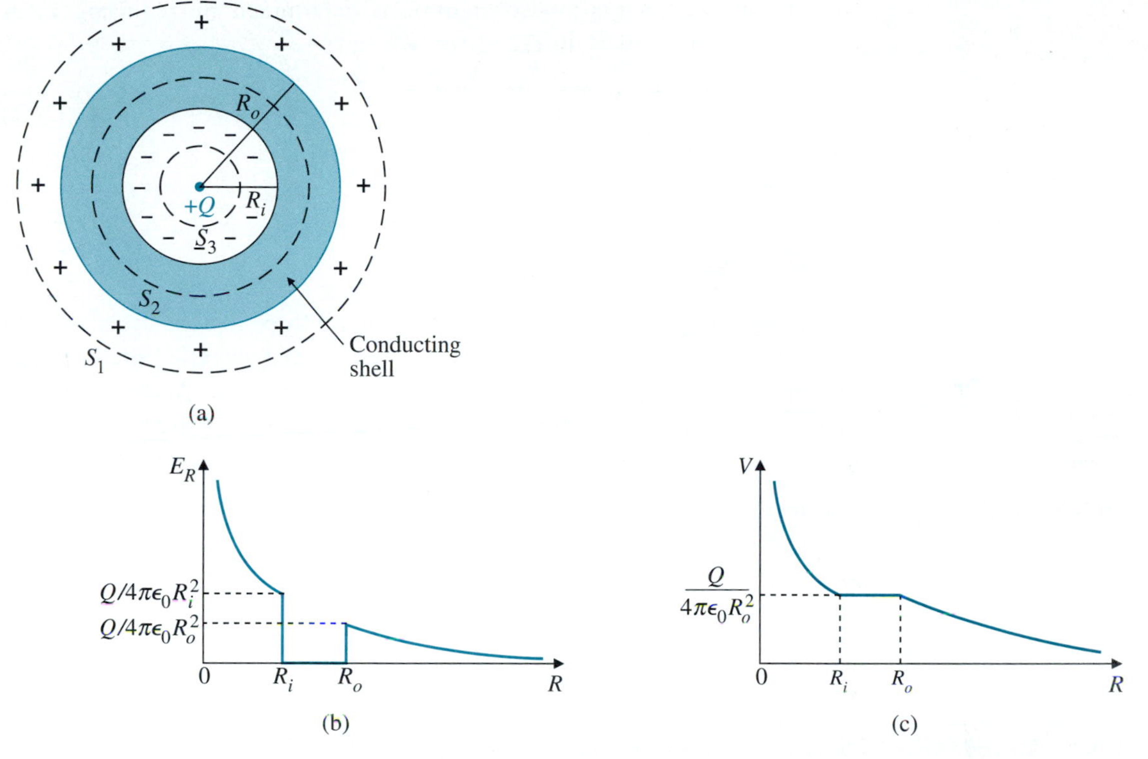

FIGURE 3-13 Electric field intensity and potential variations of a point charge $+Q$ at the center of a conducting shell (Example 3-9).

in surface S_2 must be zero, an amount of negative charge equal to $-Q$ must be induced on the inner shell surface at $R = R_i$. (This also means that an amount of positive charge equal to $+Q$ is induced on the outer shell surface at $R = R_o$.) The conducting shell is an equipotential body. Hence,

$$V_2 = V_1\bigg|_{R=R_o} = \frac{Q}{4\pi\epsilon_0 R_o}. \tag{3-50}$$

c) $R < R_i$ (Gaussian surface S_3): Application of Gauss's law yields the same formula for E_{R3} as E_{R1} in Eq. (3-47) for the first region:

$$E_{R3} = \frac{Q}{4\pi\epsilon_0 R^2}. \tag{3-51}$$

The potential in this region is

$$V_3 = -\int E_{R3}\,dR + K = \frac{Q}{4\pi\epsilon_0 R} + K,$$

where the integration constant K is determined by requiring V_3 at $R = R_i$ to equal V_2 in Eq. (3-50). We have

$$K = \frac{Q}{4\pi\epsilon_0}\left(\frac{1}{R_o} - \frac{1}{R_i}\right) \tag{3-52}$$

and

$$V_3 = \frac{Q}{4\pi\epsilon_0}\left(\frac{1}{R} + \frac{1}{R_o} - \frac{1}{R_i}\right). \tag{3-53}$$

The variations of E_R and V versus R in all three regions are plotted in Figs. 3-13(b) and 3-13(c). Note that whereas the electric intensity has discontinuous jumps, the potential remains continuous. A discontinuous jump in potential would mean an infinite electric field intensity.

Potential is continuous across boundaries.

EXERCISE 3.8

Assume that a very long copper tube with an outer radius 3 (cm) and inner radius 2 (cm) surrounds a line charge of 60 (pC/m) at its axis. Find

a) $\mathbf{E}$ at $r = 1$ (m), 2.5 (cm), and 1.5 (cm); and

b) the potential difference between the inner and outer tube surface.

ANS. (a) 1.08 (V/m), 0, 72 (V/m); (b) 0 (V).

3-6.2 DIELECTRICS IN STATIC ELECTRIC FIELD

All material media are composed of atoms with a positively charged nucleus surrounded by negatively charged electrons. Although the molecules of dielectrics are macroscopically neutral, the presence of an external electric field causes a force to be exerted on each charged particle and results in small displacements of positive and negative charges in opposite directions. These are ***bound charges***. The displacements, though small in comparison to atomic dimensions, nevertheless *polarize* a dielectric material and create electric dipoles. The situation is depicted in Fig. 3-14. Inasmuch as electric dipoles do have nonvanishing electric potential and electric field intensity (see Example 3-7), we expect that the ***induced electric dipoles*** will modify the electric field both inside and outside the dielectric material.

The molecules of some dielectrics possess permanent dipole moments, even in the absence of an external polarizing field. Such molecules usually consist of two or more dissimilar atoms and are called ***polar molecules***, in contrast to ***nonpolar molecules***, which do not have permanent dipole moments. An example is the water molecule H_2O, which consists of two hydrogen atoms and one oxygen atom. The atoms do not arrange themselves in a manner that makes the molecule have a zero dipole moment; that is, the hydrogen atoms do not lie exactly on diametrically opposite sides of the oxygen atom.

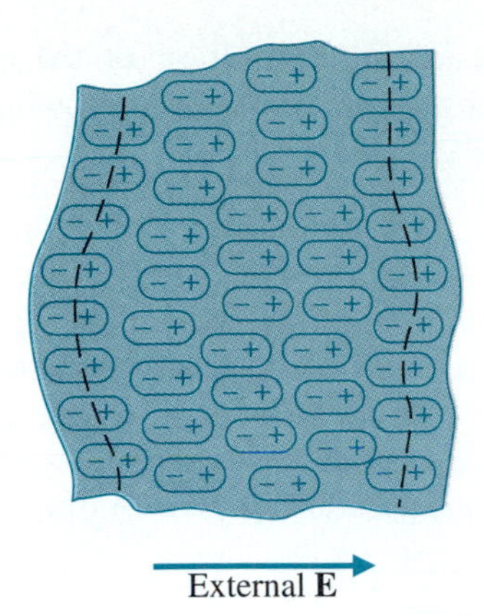

FIGURE 3-14 A cross section of a polarized dielectric medium.

The dipole moments of polar molecules are of the order of 10^{-30} (C·m). When there is no external field, the individual dipoles in a polar dielectric are randomly oriented, producing no net dipole moment macroscopically. An applied electric field will exert a torque on the individual dipoles and tend to align them with the field in a manner similar to that shown in Fig. 3-14.

To analyze the macroscopic effect of induced dipoles we define a ***polarization vector* P** as

Polarization vector is the volume density of electric dipole moment.

$$\mathbf{P} = \lim_{\Delta v \to 0} \frac{\sum_{k=1}^{n\Delta v} \mathbf{p}_k}{\Delta v} \qquad (\text{C/m}^2), \tag{3-54}$$

where n is the number of molecules per unit volume and the numerator represents the vector sum of the induced dipole moments contained in a very small volume Δv. The vector $\mathbf{P}$, a smoothed point function, is the *volume density of electric dipole moment.* The dipole moment $d\mathbf{p}$ of an elemental volume dv' is $d\mathbf{p} = \mathbf{P}\, dv'$, which produces an electrostatic potential (see Eq. 3-36)

$$dV = \frac{\mathbf{P} \cdot \mathbf{a}_R}{4\pi\epsilon_0 R^2}\, dv'. \tag{3-55}$$

Integrating over the volume V' of the dielectric, we obtain the potential due to the polarized dielectric.

Finding electrostatic potential from polarization vector

$$V = \frac{1}{4\pi\epsilon_0} \int_{V'} \frac{\mathbf{P} \cdot \mathbf{a}_R}{R^2}\, dv', \tag{3-56}$$

where R is the distance from the elemental volume dv' to a fixed field point.

A more useful physical interpretation of the effects of the induced electric dipoles can be had by observing the following surface and volume effects of the polarization vector **P**.[†]

1. *Equivalent polarization surface charge density,* ρ_{ps}.
   From Fig. 3-14 we see that the molecules effectively contribute to a distribution of positive surface charges on the right-hand boundary and a distribution of negative surface charges on the left-hand boundary. Since the surface charge density depends on the density of electric dipoles that protrude beyond the dashed lines at a surface, we can see that the equivalent ***polarization surface charge density*** is

Equivalent polarization surface charge density

$$\rho_{ps} = \mathbf{P}\cdot\mathbf{a}_n \qquad (\text{C/m}^2). \tag{3-57}$$

2. *Equivalent polarization volume charge density,* ρ_{pv}.
   For a surface S bounding a volume V, the net total charge flowing out of V as a result of polarization is obtained by integrating Eq. (3-57). The net charge *remaining* within the volume V is the *negative* of this integral:

$$\begin{aligned} Q &= -\oint_S \mathbf{P}\cdot\mathbf{a}_n\, ds \\ &= \int_V (-\nabla\cdot\mathbf{P})\, dv = \int_V \rho_{pv}\, dv, \end{aligned} \tag{3-58}$$

where we have applied divergence theorem to convert the closed surface integral to a volume integral. We can define the equivalent ***polarization volume charge density*** as

Equivalent polarization volume charge density

$$\rho_{pv} = -\nabla\cdot\mathbf{P} \qquad (\text{C/m}^3). \tag{3-59}$$

Therefore, where the divergence of **P** does not vanish, the polarized dielectric appears to be charged. However, since we started with an electrically neutral dielectric body, the total charge of the body after polarization must remain zero. This fact can be readily verified by noting that

$$\begin{aligned} \text{Total charge} &= \oint_S \rho_{ps}\, ds + \int_V \rho_{pv}\, dv \\ &= \oint_S \mathbf{P}\cdot\mathbf{a}_n\, ds - \int_V \nabla\cdot\mathbf{P}\, dv = 0, \end{aligned}$$

for a dielectric body of an arbitrary shape.

[†]A more formal derivation can be found in D. K. Cheng, *Field and Wave Electromagnetics*, Second Edition, Subsection 3-7.1, Addison-Wesley Publishing Co., Reading Mass., 1989.

Polarization charge densities ρ_{ps} and ρ_{pv} may be used to determine the potential and electric intensity fields due to a polarized dielectric:

$$V = \frac{1}{4\pi\epsilon_0} \oint_{S'} \frac{\rho_{ps}}{R} ds' + \frac{1}{4\pi\epsilon_0} \int_{V'} \frac{\rho_{pv}}{R} dv', \tag{3-60}$$

a relationship that is equivalent to Eq. (3-56). For electrostatic fields, $\mathbf{E} = -\nabla V$.

EXAMPLE 3-10

The polarization vector in a dielectric sphere of radius R_0 is $\mathbf{P} = \mathbf{a}_x P_0$. Determine

a) the equivalent polarization surface and volume charge densities, and

b) the total equivalent charge on the surface and inside of the sphere.

SOLUTION

a) The polarization surface charge density on the surface ($R = R_0$) of the sphere is

$$\rho_{ps} = \mathbf{P} \cdot \mathbf{a}_R = P_0(\mathbf{a}_x \cdot \mathbf{a}_R)$$
$$= P_0 \sin\theta \cos\phi.$$

The polarization volume charge density is

$$\rho_{pv} = -\nabla \cdot \mathbf{P} = -\nabla \cdot (\mathbf{a}_x P_0) = 0.$$

b) Total surface charge,

$$Q_s = \oint \rho_{ps} ds = \int_0^{\pi} \int_0^{2\pi} P_0 \sin\theta \cos\phi \, d\phi \, d\theta$$
$$= 0.$$

Total charge inside,

$$Q_v = \int \rho_{pv} dv = 0.$$

Thus, total charge on the sphere, $Q_s + Q_v = 0$, as expected.

3-7 ELECTRIC FLUX DENSITY AND DIELECTRIC CONSTANT

Because a polarized dielectric gives rise to an equivalent volume charge density ρ_{pv}, we expect the electric field intensity due to a given source distribution in a dielectric to be different from that in free space. In particular, the divergence postulated in Eq. (3-3) must be modified to include the effect of ρ_{pv}; that is,

$$\nabla \cdot \mathbf{E} = \frac{1}{\epsilon_0} (\rho_v + \rho_{pv}). \tag{3-60}$$

Using Eq. (3-59), we have

$$\nabla \cdot (\epsilon_0 \mathbf{E} + \mathbf{P}) = \rho_v. \tag{3-61}$$

We now define a new fundamental field quantity, the ***electric flux density***, or ***electric displacement***, **D**, such that

Definition of electric displacement D

$$\boxed{\mathbf{D} = \epsilon_0 \mathbf{E} + \mathbf{P} \qquad (\text{C/m}^2).} \tag{3-62}$$

The use of the vector **D** enables us to write a divergence relation between the electric field and the distribution of *free charges* in any medium without the necessity of dealing explicitly with the polarization vector **P** or the polarization charge density ρ_{pv}. Combining Eqs. (3-61) and (3-62) we obtain the new equation

$$\boxed{\nabla \cdot \mathbf{D} = \rho_v \qquad (\text{C/m}^3),} \tag{3-63}$$

where ρ_v is the volume density of *free charges*. Equations (3-63) and (3-4) are the two fundamental governing differential equations for electrostatics in any medium. Note that the permittivity of free space, ϵ_0, does not appear explicitly in these two equations.

The corresponding integral form of Eq. (3-63) is obtained by taking the volume integral of both sides. We have

$$\int_V \nabla \cdot \mathbf{D} \, dv = \int_V \rho_v \, dv, \tag{3-64}$$

or

Generalized Gauss's law—applicable to free space as well as to dielectric medium

$$\boxed{\oint_S \mathbf{D} \cdot d\mathbf{s} = Q \qquad (\text{C}).} \tag{3-65}$$

Equation (3-65), another form of ***Gauss's law***, states that ***the total outward flux of the electric displacement (or, simply, the total outward electric flux) over any closed surface is equal to the total free charge enclosed in the surface.***

When the dielectric properties of the medium are *linear* and *isotropic*, the polarization is directly proportional to the electric field intensity, and the proportionality constant is independent of the direction of the field. We write

Electric susceptibility

$$\mathbf{P} = \epsilon_0 \chi_e \mathbf{E}, \tag{3-66}$$

where χ_e is a dimensionless quantity called ***electric susceptibility***. A dielectric medium is linear if χ_e is independent of E and homogeneous if χ_e is independent of space coordinates. Substitution of Eq. (3-66) in Eq. (3-62) yields

Definition of a linear dielectric medium and a homogeneous dielectric medium

$$\boxed{\begin{aligned}\mathbf{D} &= \epsilon_0(1 + \chi_e)\mathbf{E} \\ &= \epsilon_0\epsilon_r\mathbf{E} = \epsilon\mathbf{E} \qquad (\text{C/m}^2),\end{aligned}} \tag{3-67}$$

where

$$\epsilon_r = 1 + \chi_e = \frac{\epsilon}{\epsilon_0} \tag{3-68}$$

Dielectric constant (relative permittivity)

is a dimensionless quantity known as the ***relative permittivity*** or the ***dielectric constant*** of the medium. The coefficient $\epsilon = \epsilon_0\epsilon_r$ is the ***absolute permittivity*** (often called simply ***permittivity***) of the medium and is measured in farads per meter (F/m). Air has a dielectric constant of 1.00059; hence its permittivity is usually taken as that of free space. The dielectric constants of some common materials are included in Table 3-1 and in Appendix B-3.

A simple medium

Note that ϵ_r can be a function of space coordinates. If ϵ_r is independent of position, the medium is said to be ***homogenous***. A linear, homogeneous, and isotropic medium is called a ***simple medium***. The relative permittivity of a simple medium is a constant. For ***anisotropic*** materials (such as crystals) the dielectric constant is different for different directions of the electric field, and **D** and **E** vectors have different directions.

TABLE **3-1** DIELECTRIC CONSTANTS AND DIELECTRIC STRENGTHS OF SOME COMMON MATERIALS

Material	Dielectric Constant	Dielectric Strength (V/m)
Air (atmospheric pressure)	1.0	3×10^6
Mineral oil	2.3	15×10^6
Paper	2–4	15×10^6
Polystyrene	2.6	20×10^6
Rubber	2.3–4.0	25×10^6
Glass	4–10	30×10^6
Mica	6.0	200×10^6

3-7.1 DIELECTRIC STRENGTH

We have explained that an electric field causes small displacements of the bound charges in a dielectric material, resulting in polarization. If the electric field is very strong, it will pull electrons completely out of the molecules. The electrons will accelerate under the influence of the electric field, collide violently with the molecular lattice structure, and cause permanent dislocations and damage in the material. Avalanche effect of ionization due to collisions may occur. The material will become conducting, and large currents may result. This phenomenon is called a ***dielectric breakdown***. The maximum electric field intensity that a dielectric material can withstand without breakdown is the ***dielectric strength*** of the material. The approximate dielectric strengths of some common substances are given in Table 3-1. The dielectric strength of a material must not be confused with its dielectric constant.

Dielectric strength

A convenient number to remember is that the dielectric strength of air at the atmospheric pressure is 3(kV/mm). When the electric field intensity exceeds this value, air breaks down. Massive ionization takes place, and sparking (corona discharge) follows. Charge tends to concentrate at sharp points. This is the principle upon which a lightning arrester with a sharp metal lightning rod on top of tall buildings works. When a cloud containing an abundance of electric charges approaches a tall building equipped with a lightning rod connected to the ground, charges of an opposite sign are attracted from the ground to the tip of the rod, where the electric field intensity is the strongest. As the electric field intensity exceeds the dielectric strength of the wet air, breakdown occurs, and the air near the tip is ionized and becomes conducting. The electric charges in the cloud are then discharged safely to the ground through the conducting path.

Dielectric strength of air is 3 (kV/mm)

Principle of a lightning arrester

The fact that the electric field intensity tends to be higher at a point near the surface of a charged conductor with a larger curvature is illustrated quantitatively in the following example.

Electric field intensity at a conductor surface is higher at points of larger curvature.

EXAMPLE 3-11

Consider two spherical conductors with radii b_1 and b_2 ($b_2 > b_1$) that are connected by a conducting wire. The distance of separation between the conductors is assumed to be very large in comparison to b_2 so that the charges on the spherical conductors may be considered as uniformly distributed. A total charge Q is deposited on the spheres. Find

a) the charges on the two spheres, and

b) the electric field intensities at the sphere surfaces.

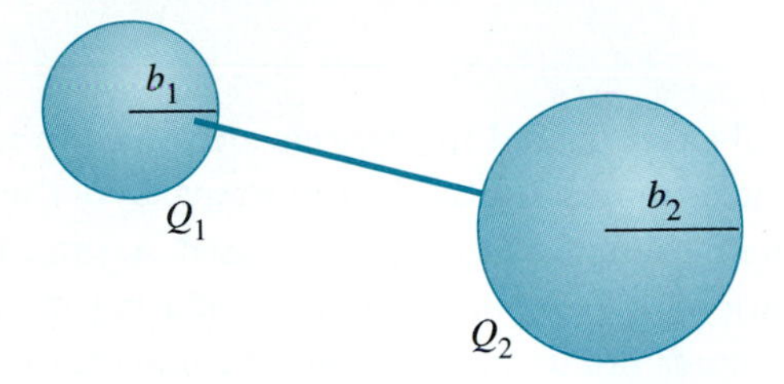

FIGURE 3-15 Two connected conducting spheres (Example 3-11).

SOLUTION

a) Refer to Fig. 3-15. Since the spherical conductors are at the same potential, we have

$$\frac{Q_1}{4\pi\epsilon_0 b_1} = \frac{Q_2}{4\pi\epsilon_0 b_2},$$

or

$$\frac{Q_1}{Q_2} = \frac{b_1}{b_2}.$$

Hence the charges on the spheres are directly proportional to their radii. But, since

$$Q_1 + Q_2 = Q,$$

we find that

$$Q_1 = \frac{b_1}{b_1 + b_2} Q \quad \text{and} \quad Q_2 = \frac{b_2}{b_1 + b_2} Q.$$

b) The electric field intensities at the surfaces of the two conducting spheres are

$$E_{1n} = \frac{Q_1}{4\pi\epsilon_0 b_1^2} \quad \text{and} \quad E_{2n} = \frac{Q_2}{4\pi\epsilon_0 b_2^2},$$

so

$$\frac{E_{1n}}{E_{2n}} = \left(\frac{b_2}{b_1}\right)^2 \frac{Q_1}{Q_2} = \frac{b_2}{b_1}. \tag{3-69}$$

The electric field intensities are therefore inversely proportional to the radii, being higher at the surface of the smaller sphere which has a larger curvature.

EXAMPLE 3-12

When a coaxial cable is used to carry electric power, the radius of the inner conductor is determined by the load current, and the overall size by the voltage and the type of insulating material used. Assume that the radius of the inner conductor is $r_i = 2\,(\text{mm})$ and the insulating material is polystyrene, determine the inner radius, r_o, of the outer conductor so that the cable is to work at a voltage rating of $10\,(\text{kV})$. In order to avoid breakdown due to voltage surges caused by lightning and other abnormal external conditions, the maximum electric field intensity in the insulating material is not to exceed 25% of its dielectric strength.

SOLUTION

From Table 3-1, we find the dielectric constant and dielectric strength of polystyrene to be 2.6 and 20×10^6 (V/m), respectively. The electric intensity due to a line charge ρ_ℓ is, from Eq. (3-23),

$$\mathbf{E} = \mathbf{a}_r E_r = \mathbf{a}_r \frac{\rho_\ell}{2\pi\epsilon_0\epsilon_r r}. \tag{3-70}$$

As the cable is to work at a potential difference of 10^4 (V) between the inner and outer conductors, we set

$$10^4 = -\int_{r_o}^{r_i} E_r\,dr = \frac{\rho_\ell}{2\pi\epsilon_0(2.6)} \ln\frac{r_o}{r_i},$$

or

$$\ln\frac{r_o}{r_i} = \left(\frac{5.2\pi\epsilon_0}{\rho_\ell}\right) \times 10^4. \tag{3-71}$$

In order to limit the maximum electric intensity to 25% of 20×10^6, we require, from Eq. (3-70),

$$\text{Max } E_r = 0.25 \times (20 \times 10^6) = \frac{\rho_\ell}{2\pi\epsilon_0(2.6)r_i},$$

or

$$\left(\frac{\rho_\ell}{5.2\pi\epsilon_0}\right) = (0.25 \times 20 \times 10^6)r_i = (5 \times 10^6) \times (2 \times 10^{-3})$$
$$= 10^4.$$

Substituting the above value into Eq. (3-71), we obtain $\ln(r_o/r_i) = 1$, or

$$\ln r_o = 1 + \ln r_i = 1 + \ln(2 \times 10^{-3})$$
$$= 1 - 6.215 = -5.215.$$

Hence,

$$r_o = 0.0054\,(\text{m}), \quad \text{or} \quad 5.4\,(\text{mm}).$$

■ EXERCISE 3.9 If the polystyrene of the coaxial cable in Example 3-12 is replaced by air, what would the maximum allowable working voltage of the cable be? (Maintain the restriction that the maximum field intensity is not to exceed 25% of the dielectric strength of the insulating material.)

ANS. 1.5 (kV).

■ EXERCISE 3.10 If it is desired that the working voltage of an air-filled coaxial cable having a radius $r_i = 2\,(\text{mm})$ for the inner conductor is to remain at 10 (kV) as in Example 3-9, what should r_o be?

ANS. 1.571 (m).

3-8 BOUNDARY CONDITIONS FOR ELECTROSTATIC FIELDS

Electromagnetic problems often involve media with different physical properties and require the knowledge of the relations of the field quantities at an interface between two media. For instance, we may wish to determine how the **E** and **D** vectors change in crossing an interface. We already know the boundary conditions that must be satisfied at a conductor–free space interface. These conditions have been given in Eqs. (3-45) and (3-46). We now consider an interface between two general media, shown in Fig. 3-16.

Let us construct a small path *abcda* with sides *ab* and *cd* in media 1 and 2, respectively, both being parallel to the interface and equal to Δw. Equation (3-7) is applied to this path. If we let sides $bc = da = \Delta h$ approach zero, their

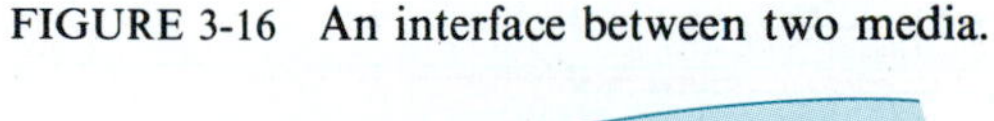
FIGURE 3-16 An interface between two media.

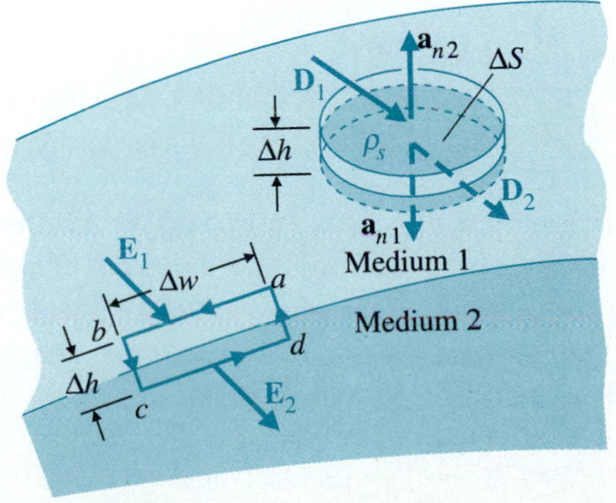

contributions to the line integral of **E** around the path can be neglected. We have

$$\oint_{abcda} \mathbf{E}\cdot d\boldsymbol{\ell} = \mathbf{E}_1\cdot\Delta\mathbf{w} + \mathbf{E}_2\cdot(-\Delta\mathbf{w}) = E_{1t}\Delta w - E_{2t}\Delta w = 0.$$

Therefore

Boundary condition for tangential component of E

$$E_{1t} = E_{2t} \qquad (\text{V/m}), \tag{3-72}$$

which states that ***the tangential component of an* E *field is continuous across an interface***. When media 1 and 2 are dielectrics with permittivities ϵ_1 and ϵ_2, respectively, we have

$$\frac{D_{1t}}{\epsilon_1} = \frac{D_{2t}}{\epsilon_2}. \tag{3-73}$$

In order to find a relation between the normal components of the fields at a boundary, we construct a small pillbox with its top face in medium 1 and bottom face in medium 2, as illustrated in Fig. 3-16. The faces have an area ΔS, and the height of the pillbox Δh is vanishingly small. Applying Gauss's law, Eq. (3-65) to the pillbox, we have

$$\begin{aligned}\oint_S \mathbf{D}\cdot d\mathbf{s} &= (\mathbf{D}_1\cdot\mathbf{a}_{n2} + \mathbf{D}_2\cdot\mathbf{a}_{n1})\Delta S\\ &= \mathbf{a}_{n2}\cdot(\mathbf{D}_1 - \mathbf{D}_2)\Delta S\\ &= \rho_s\Delta S,\end{aligned} \tag{3-74}$$

where we have used the relation $\mathbf{a}_{n2} = -\mathbf{a}_{n1}$. Unit vectors $\mathbf{a}_{n1}$ and $\mathbf{a}_{n2}$ are, respectively, *outward* unit normals from media 1 and 2. From Eq. (3-74) we obtain

$$\mathbf{a}_{n2}\cdot(\mathbf{D}_1 - \mathbf{D}_2) = \rho_s, \tag{3-75a}$$

or

$$D_{1n} - D_{2n} = \rho_s \qquad (\text{C/m}^2), \tag{3-75b}$$

where the reference unit normal is outward from medium 2.

Boundary condition for normal component of D

Equation (3-75b) states that ***the normal component of* D *field is discontinuous across an interface where a surface charge exists—the amount of discontinuity being equal to the surface charge density***. If medium 2 is a conductor, $\mathbf{D}_2 = 0$ and Eq. (3-75b) becomes

$$D_{1n} = \epsilon_1 E_{1n} = \rho_s, \tag{3-76}$$

which simplifies to Eq. (3-46) when medium 1 is free space.

When two dielectrics are in contact with *no free charges* at the interface, $\rho_s = 0$, we have

$$D_{1n} = D_{2n} \tag{3-77}$$

or

$$\epsilon_1 E_{1n} = \epsilon_2 E_{2n}. \tag{3-78}$$

Recapitulating, we find that the boundary conditions that must be satisfied for static electric fields are as follows:

Boundary conditions for electrostatic fields

Tangential components:	$E_{1t} = E_{2t}$	(3-79)
Normal components:	$\mathbf{a}_{n2} \cdot (\mathbf{D}_1 - \mathbf{D}_2) = \rho_s.$	(3-80)

■ **EXERCISE 3.11** State and explain the boundary conditions that must be satisfied by the electric potential at an interface between perfect dielectrics with dielectric constants ϵ_{r1} and ϵ_{r2}.

ANS. $\epsilon_{r1}\, \partial V_1/\partial n = \epsilon_{r2}\, \partial V_2/\partial n$, $V_1 = V_2$.

EXAMPLE 3-13

A lucite sheet ($\epsilon_r = 3.2$) is introduced perpendicularly in a uniform electric field $\mathbf{E}_o = \mathbf{a}_x E_o$ in free space. Determine $\mathbf{E}_i$, $\mathbf{D}_i$, and $\mathbf{P}_i$ inside the lucite.

SOLUTION

We assume that the introduction of the lucite sheet does not disturb the original uniform electric field $\mathbf{E}_o$. The situation is depicted in Fig. 3-17. Since

FIGURE 3-17 A lucite sheet in a uniform electric field (Example 3-13).

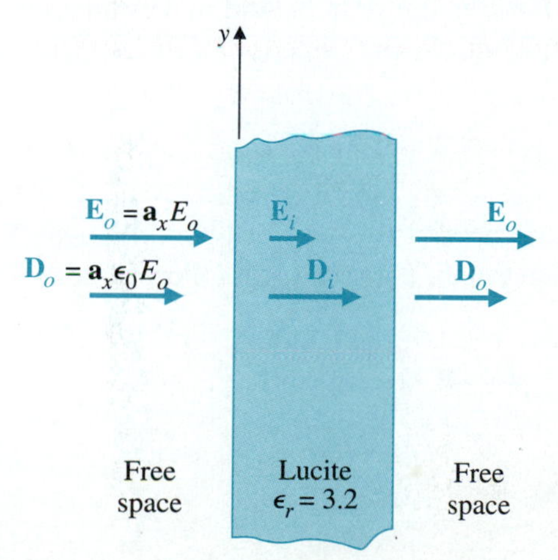

the interfaces are perpendicular to the electric field, only the normal field components need be considered. No free charges exist.

Boundary condition Eq. (3-77) at the left interface gives

$$\mathbf{D}_i = \mathbf{a}_x D_i = \mathbf{a}_x D_o,$$

or

$$\mathbf{D}_i = \mathbf{a}_x \epsilon_0 E_o.$$

There is no change in electric flux density across the interface. The electric field intensity inside the lucite sheet is

$$\mathbf{E}_i = \frac{1}{\epsilon}\mathbf{D}_i = \frac{1}{\epsilon_o \epsilon_r}\mathbf{D}_i = \mathbf{a}_x \frac{E_o}{3.2}.$$

Hence the effect of the lucite sheet is to reduce electric intensity. The polarization vector is zero outside the lucite sheet ($\mathbf{P}_o = 0$). Inside the sheet,

$$\mathbf{P}_i = \mathbf{D}_i - \epsilon_0 \mathbf{E}_i = \mathbf{a}_x \left(1 - \frac{1}{3.2}\right)\epsilon_0 E_o$$
$$= \mathbf{a}_x 0.6875 \epsilon_0 E_o \qquad (\text{C/m}^2).$$

Clearly, a similar application of the boundary condition Eq. (3-77) on the right interface will yield the original $\mathbf{E}_o$ and $\mathbf{D}_o$ in the free space on the right of the lucite sheet.

Does the solution of this problem change if the original electric field is not uniform; that is, if $\mathbf{E}_o = \mathbf{a}_x E(y)$?

EXAMPLE 3-14

Two dielectric media with permittivities ϵ_1 and ϵ_2 are separated by a charge-free boundary as shown in Fig. 3-18. The electric field intensity in medium 1 at the point P_1 has a magnitude E_1 and makes an angle α_1 with the normal. Determine the magnitude and direction of the electric field intensity at point P_2 in medium 2.

SOLUTION

Two equations are needed to solve for two unknowns E_{2t} and E_{2n}. After E_{2t} and E_{2n} have been found, E_2 and α_2 will follow directly. Using Eqs. (3-72) and (3-77), we have

$$E_2 \sin \alpha_2 = E_1 \sin \alpha_1 \tag{3-81}$$

and

$$\epsilon_2 E_2 \cos \alpha_2 = \epsilon_1 E_1 \cos \alpha_1. \tag{3-82}$$

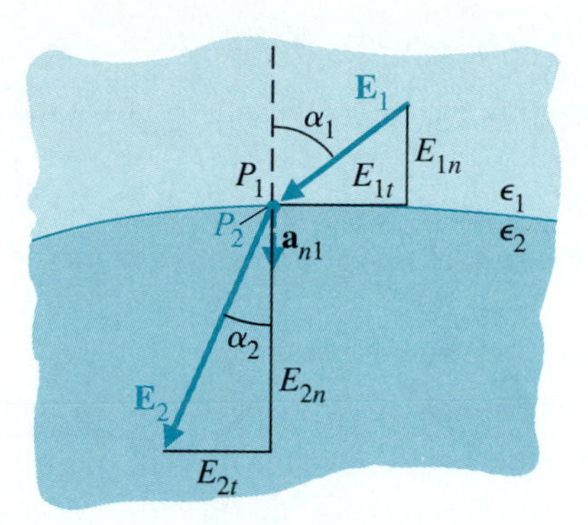

FIGURE 3-18 Boundary conditions at the interface between two dielectric media (Example 3-14).

Division of Eq. (3-81) by Eq. (3-82) gives

$$\boxed{\frac{\tan \alpha_2}{\tan \alpha_1} = \frac{\epsilon_2}{\epsilon_1}.} \tag{3-83}$$

The magnitude of $\mathbf{E}_2$ is

$$E_2 = \sqrt{E_{2t}^2 + E_{2n}^2} = \sqrt{(E_2 \sin \alpha_2)^2 + (E_2 \cos \alpha_2)^2}$$
$$= \left[(E_1 \sin \alpha_1)^2 + \left(\frac{\epsilon_1}{\epsilon_2} E_1 \cos \alpha_1\right)^2\right]^{1/2},$$

or

$$\boxed{E_2 = E_1 \left[\sin^2 \alpha_1 + \left(\frac{\epsilon_1}{\epsilon_2} \cos \alpha_1\right)^2\right]^{1/2}.} \tag{3-84}$$

By examining Fig. 3-18, can you tell whether ϵ_1 is larger or smaller than ϵ_2?

If medium 2 is a conductor, there can be no electric field in medium 2 under static conditions, and $\mathbf{E}_1$ at the boundary has only a normal component ($\alpha_1 = 0$). We have $\mathbf{E}_1 = \mathbf{a}_n E_{1n} = \mathbf{a}_n D_{1n}/\epsilon_1 = \mathbf{a}_n \rho_s/\epsilon_1$, where ρ_s is the surface charge density and $\mathbf{a}_n$ is the outward normal from the conductor surface.

EXERCISE 3.12 Assume that two homogeneous isotropic dielectric media with dielectric constants $\epsilon_{r1} = 3$ and $\epsilon_{r2} = 2$ are separated by the xy-plane. At a common point, $\mathbf{E}_1 = \mathbf{a}_x - \mathbf{a}_y 5 - \mathbf{a}_z 4$. Find $\mathbf{E}_2$, $\mathbf{D}_2$, α_1, and α_2.

ANS. $\mathbf{D}_2 = 2\epsilon_0 \mathbf{E}_2 = 2\epsilon_0(\mathbf{a}_x - \mathbf{a}_y 5 - \mathbf{a}_z 6)$, 51.9°, 40.4°.

REVIEW QUESTIONS

Q.3-11 Why are there no free charges in the interior of a good conductor under static conditions?

Q.3-12 Define *polarization vector*. What is its SI unit?

Q.3-13 What are *polarization charge densities*? What are the SI units for $\mathbf{P} \cdot \mathbf{a}_n$ and $\nabla \cdot \mathbf{P}$?

Q.3-14 What do we mean by a *simple medium*?

Q.3-15 Define *electric displacement vector*. What is its SI unit?

Q.3-16 Define *electric susceptibility*. What is its unit?

Q.3-17 What is the difference between the *dielectric constant* and the *dielectric strength* of a dielectric material?

Q.3-18 Explain the principle of operation of lightning arresters.

Q.3-19 What are the general boundary conditions for **E** and **D** at an interface between two different dielectric media with dielectric constants ϵ_{r1} and ϵ_{r2}?

Q.3-20 What are the boundary conditions for electrostatic fields at an interface between a conductor and a dielectric with permittivity ϵ?

Q.3-21 What is the boundary condition for electrostatic potential at an interface between two different dielectric media?

REMARKS

1. Under static conditions the **E** field inside a conductor is zero.
2. Under static conditions the surface of a conductor is an equipotential surface, and the **E** field there is everywhere normal to the surface.
3. At an interface between two different dielectric media the electric potential is continuous.
4. Do not confuse the dielectric constant of a medium, ϵ_r, with its permittivity, ϵ. The former is dimensionless; the SI unit for the latter is (F/m).

3-9 CAPACITANCES AND CAPACITORS

From Section 3-6 we understand that a conductor in a static electric field is an equipotential body and that charges deposited on a conductor will distribute themselves on its surface in such a way that the electric field inside vanishes. Suppose the potential due to a charge Q is V. Increasing the total charge by some factor k would merely increase the surface charge density ρ_s everywhere by the same factor without affecting the charge distribution, because the conductor remains an equipotential body in a static situation. We may conclude from Eq. (3-39) that the potential of an isolated conductor is directly proportional to the total charge on it. This may also be seen from

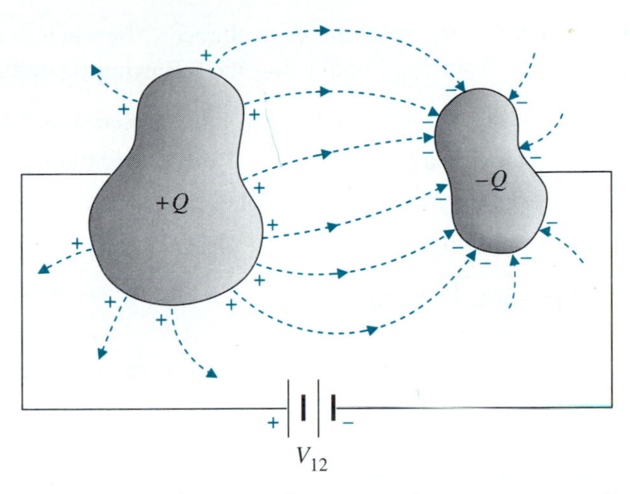

FIGURE 3-19 A two-conductor capacitor.

the fact that increasing V by a factor of k increases $\mathbf{E} = -\nabla V$ by a factor of k. But from Eq. (3-46). $\mathbf{E} = \mathbf{a}_n \rho_s/\epsilon_0$; it follows that ρ_s, and consequently the total charge Q will also increase by a factor of k. The ratio Q/V therefore remains unchanged. We write

$$\boxed{Q = CV,} \tag{3-85}$$

where the constant of proportionality C is called the **capacitance** of the isolated conducting body. Its SI unit is coulomb per volt, or farad (F).

Of considerable importance in practice is the **capacitor** (or **condenser**), which consists of two conductors separated by free space or a dielectric medium. The conductors may be of arbitrary shapes as in Fig. 3-19. When a d-c voltage source is connected between the conductors, a charge transfer occurs, resulting in a charge $+Q$ on one conductor and $-Q$ on the other. Several electric field lines originating from positive charges and terminating on negative charges are shown in Fig. 3-19. Note that the field lines are perpendicular to the conductor surfaces, which are equipotential surfaces. Equation (3-85) applies here if V is taken to mean the potential difference between the two conductors, V_{12}. That is,

Definition of capacitance

$$\boxed{C = \frac{Q}{V_{12}} \quad \text{(F).}} \tag{3-86}$$

The capacitance of a capacitor is a physical property of the two-conductor system. It depends on the geometry of the capacitor and on the permittivity of the medium. Capacitance C between two conductors can be determined from Eq. (3-86) using the following procedure:

Procedure for determining C

1. Choose an appropriate coordinate system for the given geometry.
2. Assume charges $+Q$ and $-Q$ on the conductors.
3. Find $\mathbf{E}$ from Q by Eq. (3-76), Gauss's law, or other relations.
4. Find V_{12} by evaluating

$$V_{12} = -\int_{2}^{1} \mathbf{E} \cdot d\boldsymbol{\ell}$$

from the conductor carrying $-Q$ to the other carrying $+Q$.

5. Find C by taking the ratio Q/V_{12}.

EXAMPLE 3-15

A parallel-plate capacitor consists of two parallel conducting plates of area S separated by a uniform distance d. The space between the plates is filled with a dielectric of a constant permittivity ϵ. Determine the capacitance.

SOLUTION

A cross section of the capacitor is shown in Fig. 3-20. The appropriate coordinate system to use here is the Cartesian coordinate system. Following the procedure outlined above, we put charges $+Q$ and $-Q$ on the upper and lower conducting plates, respectively. The charges are assumed to be uniformly distributed over the conducting plates with surface densities $+\rho_s$ and $-\rho_s$, where

$$\rho_s = \frac{Q}{S}.$$

From Eq. (3-76) we have

$$\mathbf{E} = -\mathbf{a}_y \frac{\rho_s}{\epsilon} = -\mathbf{a}_y \frac{Q}{\epsilon S},$$

FIGURE 3-20 Cross section of a parallel-plate capacitor (Example 3-15).

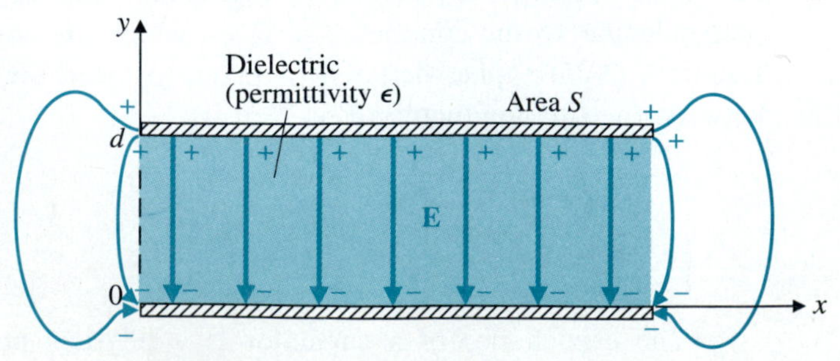

which is constant within the dielectric if the fringing of the electric field at the edges of the plates is neglected. Now

$$V_{12} = -\int_{y=0}^{y=d} \mathbf{E} \cdot d\boldsymbol{\ell} = -\int_0^d \left(-\mathbf{a}_y \frac{Q}{\epsilon S}\right) \cdot (\mathbf{a}_y \, dy) = \frac{Q}{\epsilon S} d.$$

Therefore, *for a parallel-plate capacitor,*

Capacitance of a parallel-plate capacitor

$$C = \frac{Q}{V_{12}} = \epsilon \frac{S}{d}, \tag{3-87}$$

which is independent of Q or V_{12}.

EXERCISE 3.13 Determine the capacitance of the parallel-plate capacitor in Fig. 3-20 by starting with an assumed potential difference V_{12} between the upper and lower plates, then finding Q and taking the ratio Q/V_{12}.

EXAMPLE 3-16

A cylindrical capacitor, shown in Fig. 3.21, consists of an inner conductor of radius a and an outer conductor whose inner radius is b. The space between the conductors is filled with a dielectric of permittivity ϵ, and the length of the capacitor is L. Determine the capacitance of this capacitor.

SOLUTION

We use cylindrical coordinates for this problem. First we assume charges $+Q$ and $-Q$ on the surface of the inner conductor and the inner surface of the outer conductor, respectively. The $\mathbf{E}$ field in the dielectric can be obtained by

FIGURE 3-21 A cylindrical capacitor (Examples 3-16 and 3-19).

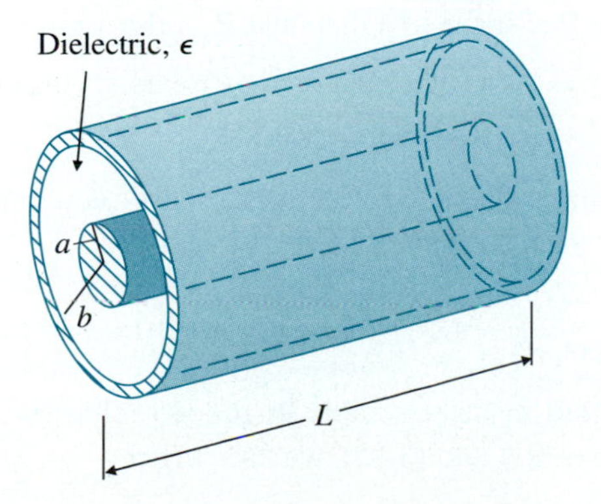

applying Gauss's law to a cylindrical Gaussian surface within the dielectric $a < r < b$. Noting that $\rho_\ell = Q/L$, we have, from Eq. (3-23),

$$\mathbf{E} = \mathbf{a}_r E_r = \mathbf{a}_r \frac{Q}{2\pi\epsilon L r}. \tag{3-88}$$

Again we neglect the fringing effect of the field near the edges of the conductors. The potential difference between the inner and outer conductors is

$$\begin{aligned} V_{ab} &= -\int_{r=b}^{r=a} \mathbf{E}\cdot d\boldsymbol{\ell} = -\int_b^a \left(\mathbf{a}_r \frac{Q}{2\pi\epsilon L r}\right)\cdot(\mathbf{a}_r\, dr) \\ &= \frac{Q}{2\pi\epsilon L}\ln\left(\frac{b}{a}\right). \end{aligned} \tag{3-89}$$

Therefore, *for a cylindrical capacitor,*

Capacitance of a cylindrical capacitor

$$\boxed{C = \frac{Q}{V_{ab}} = \frac{2\pi\epsilon L}{\ln\left(\frac{b}{a}\right)}.} \tag{3-90}$$

■ **EXERCISE 3.14** Assume the Earth to be a large conducting sphere (radius $= 6.37 \times 10^3$ km) surrounded by air. Find its capacitance referring to infinity.

ANS. 7.08×10^{-4} (F).

3-10 ELECTROSTATIC ENERGY AND FORCES

In Section 3-5 we indicated that electric potential at a point in an electric field is the work required to bring a unit positive charge from infinity (at reference zero-potential) to that point. To bring a charge Q_2 (slowly, so that kinetic energy and radiation effects may be neglected) from infinity *against* the field of a charge Q_1 in free space to a distance R_{12}, the amount of work required is

$$W_2 = Q_2 V_2 = Q_2 \frac{Q_1}{4\pi\epsilon_0 R_{12}}. \tag{3-91}$$

Because electrostatic fields are conservative, W_2 is independent of the path followed by Q_2. Another form of Eq. (3-91) is

$$W_2 = Q_1 \frac{Q_2}{4\pi\epsilon_0 R_{12}} = Q_1 V_1. \tag{3-92}$$

This work is stored in the assembly of the two charges as potential energy. Combining Eqs. (3-91) and (3-92), we can write

$$W_2 = \tfrac{1}{2}(Q_1 V_1 + Q_2 V_2). \tag{3-93}$$

Now suppose another charge Q_3 is brought from infinity to a point that is R_{13} from Q_1 and R_{23} from Q_2; an additional amount of work is required that equals

$$\Delta W = Q_3 V_3 = Q_3 \left(\frac{Q_1}{4\pi\epsilon_0 R_{13}} + \frac{Q_2}{4\pi\epsilon_0 R_{23}} \right). \tag{3-94}$$

The sum of ΔW in Eq. (3-94) and W_2 in Eq. (3-91) is the potential energy, W_3, stored in the assembly of the three charges Q_1, Q_2, and Q_3. That is,

$$W_3 = W_2 + \Delta W = \frac{1}{4\pi\epsilon_0} \left(\frac{Q_1 Q_2}{R_{12}} + \frac{Q_1 Q_3}{R_{13}} + \frac{Q_2 Q_3}{R_{23}} \right). \tag{3-95}$$

We can rewrite W_3 in the following form:

$$\begin{aligned} W_3 &= \frac{1}{2}\left[Q_1 \left(\frac{Q_2}{4\pi\epsilon_0 R_{12}} + \frac{Q_3}{4\pi\epsilon_0 R_{13}} \right) + Q_2 \left(\frac{Q_1}{4\pi\epsilon_0 R_{12}} + \frac{Q_3}{4\pi\epsilon_0 R_{23}} \right) \right. \\ &\qquad \left. + Q_3 \left(\frac{Q_1}{4\pi\epsilon_0 R_{13}} + \frac{Q_2}{4\pi\epsilon_0 R_{23}} \right) \right] \\ &= \tfrac{1}{2}(Q_1 V_1 + Q_2 V_2 + Q_3 V_3). \end{aligned} \tag{3-96}$$

In Eq. (3-96), V_1, the potential at the position of Q_1, is caused by charges Q_2 and Q_3; it is *different* from the V_1 in Eq. (3-92) in the two-charge case. Similarly V_2 and V_3 are the potentials at Q_2 and Q_3, respectively, in the three-charge assembly.

Extending this procedure of bringing in additional charges, we arrive at the following general expression for the potential energy of a group of N discrete point charges at rest. (The purpose of the subscript e on W_e is to denote that the energy is of an electric nature.) We have

Electric energy stored in a system of discrete point charges

$$\boxed{W_e = \frac{1}{2} \sum_{k=1}^{N} Q_k V_k \quad \text{(J)},} \tag{3-97}$$

where V_k, the electric potential at Q_k, is caused by all the other charges.

The SI unit for energy, ***joule*** (J), is too large a unit for work in physics of elementary particles, where energy is more conveniently measured in terms of a much smaller unit called ***electron-volt*** (eV). An electron-volt is the energy or work required to move an electron against a potential difference of one volt.

Relation between joule and electron-volt

$$1 \text{ (eV)} = (1.60 \times 10^{-19}) \times 1 = 1.60 \times 10^{-19} \quad \text{(J)}. \tag{3-98}$$

Energy in (eV) is essentially that in (J) per unit electronic charge.

EXERCISE 3.15 Convert the kinetic energy of 2 (TeV) of the proton beam of a very powerful high-energy particle accelerator into joules.

ANS. 3.20×10^{-7} (J).

■ **EXERCISE 3.16** Determine the amount of energy needed to arrange three point charges -1 (μC), 2 (μC), and 3 (μC) at the corners of an equilateral triangle of sides 10 (cm) in free space.

ANS. 0.09 (J).

EXAMPLE 3-17

Find the energy required to assemble a uniform sphere of charge of radius b and volume charge density ρ_v.

SOLUTION

Because of symmetry, it is simplest to assume that the sphere of charge is assembled by bringing up a succession of spherical layers of thickness dR. At a radius R shown in Fig. 3-22 the potential is

$$V_R = \frac{Q_R}{4\pi\epsilon_0 R},$$

where Q_R is the total charge contained in a sphere of radius R:

$$Q_R = \rho_v \tfrac{4}{3}\pi R^3.$$

The differential charge in a spherical layer of thickness dR is

$$dQ_R = \rho_v 4\pi R^2\, dR,$$

and the work or energy in bringing up dQ_R is

$$dW_e = V_R\, dQ_R = \frac{4\pi}{3\epsilon_0}\rho_v^2 R^4\, dR.$$

Hence the total work or energy required to assemble a uniform sphere of charge of radius b and charge density ρ_v is

$$W_e = \int dW_e = \frac{4\pi}{3\epsilon_0}\rho_v^2 \int_0^b R^4\, dR = \frac{4\pi\rho_v^2 b^5}{15\epsilon_0} \quad \text{(J)}. \tag{3-99}$$

In terms of the total charge

$$Q = \rho_v \frac{4\pi}{3} b^3,$$

we have

$$W_e = \frac{3Q^2}{20\pi\epsilon_0 b} \quad \text{(J)}. \tag{3-100}$$

Equation (3-100) shows that the energy is directly proportional to the square of the total charge. The sphere of charge in Fig. 3-22 could be a cloud of electrons, for instance.

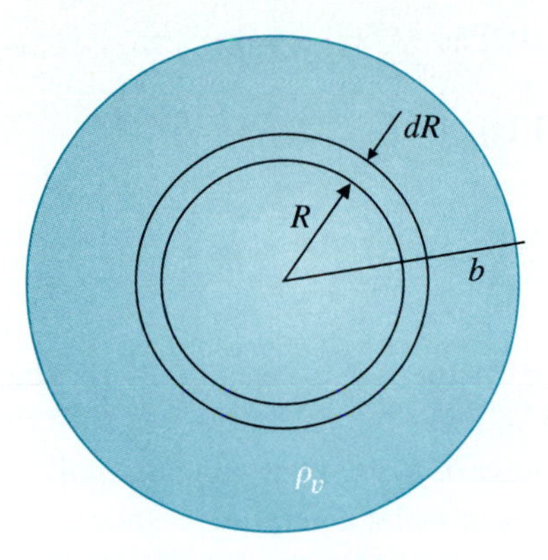

FIGURE 3-22 Assembling a uniform sphere of charge (Example 3-17).

For a continuous charge distribution of density ρ_v the formula for W_e in Eq. (3-97) for discrete charges must be modified. Without going through a separate proof, we replace Q_k by $\rho_v\, dv$ and the summation by an integration to obtain

Electric energy stored in a continuous distribution of charge

$$\boxed{W_e = \frac{1}{2}\int_{V'} \rho_v V\, dv \qquad \text{(J)}.} \tag{3-101}$$

In Eq. (3-101), V is the potential at the point where the volume charge density is ρ_v and V' is the volume of the region where ρ_v exists. Note that W_e in Eq. (3-101) includes the work (self-energy) required to assemble the distribution of macroscopic charges, because it is the energy of interaction of every infinitesimal charge element with all other infinitesimal charge elements.

3-10.1 ELECTROSTATIC ENERGY IN TERMS OF FIELD QUANTITIES

In Eq. (3-101) the expression of electrostatic energy of a charge distribution contains the source charge density ρ_v and the potential function V. We frequently find it more convenient to have an expression of W_e in terms of field quantities $\mathbf{E}$ and/or $\mathbf{D}$, without knowing ρ_v explicitly. To this end, we substitute $\nabla \cdot \mathbf{D}$ for ρ_v in Eq. (3-101):

$$W_e = \frac{1}{2}\int_{V'} (\nabla \cdot \mathbf{D}) V\, dv. \tag{3-102}$$

Now, using the vector identity (from Problem P.2-25), Eq. (2-114),

$$\nabla \cdot (V\mathbf{D}) = V\nabla \cdot \mathbf{D} + \mathbf{D} \cdot \nabla V, \tag{3-103}$$

we can write Eq. (3-102) as

$$W_e = \frac{1}{2}\int_{V'} \nabla\cdot(V\mathbf{D})\,dv - \frac{1}{2}\int_{V'} \mathbf{D}\cdot\nabla V\,dv$$
$$= \frac{1}{2}\oint_{S'} V\mathbf{D}\cdot\mathbf{a}_n\,ds + \frac{1}{2}\int_{V'} \mathbf{D}\cdot\mathbf{E}\,dv, \tag{3-104}$$

where the divergence theorem has been used to change the first volume integral into a closed surface integral and $\mathbf{E}$ has been substituted for $-\nabla V$ in the second volume integral. Since V' can be any volume that includes all the charges, we may choose it to be a very large sphere with radius R. As we let $R \to \infty$, electric potential V and the magnitude of electric displacement D fall off at least as fast as $1/R$ and $1/R^2$, respectively.† The area of the bounding surface S' increases as R^2. Hence the surface integral in Eq. (3-104) decreases at least as fast as $1/R$ and will vanish as $R \to \infty$. We are then left with only the second integral on the right side of Eq. (3-104):

Electric energy in terms of E and D

$$W_e = \frac{1}{2}\int_{V'} \mathbf{D}\cdot\mathbf{E}\,dv \qquad \text{(J)}. \tag{3-105}$$

Using the relation $\mathbf{D} = \epsilon\mathbf{E}$ for a linear and isotrophic medium, we can write W_e in terms of $\mathbf{E}$ alone.

Electric energy in terms of E and ϵ

$$W_e = \frac{1}{2}\int_{V'} \epsilon E^2\,dv \qquad \text{(J)}. \tag{3-106}$$

We can also define an ***electrostatic energy density*** w_e such that its volume integral equals the total electrostatic energy:

$$W_e = \int_{V'} w_e\,dv, \tag{3-107}$$

where

$$w_e = \tfrac{1}{2}\epsilon E^2 \qquad (\text{J/m}^3). \tag{3-108}$$

EXAMPLE 3-18

In Fig. 3-23 a parallel-plate capacitor of area S and separation d is charged to a voltage V. The permittivity of the dielectric is ϵ. Find the stored electrostatic energy.

† For point charges $V \propto 1/R$ and $D \propto 1/R^2$; for dipoles $V \propto 1/R^2$ and $D \propto 1/R^3$.

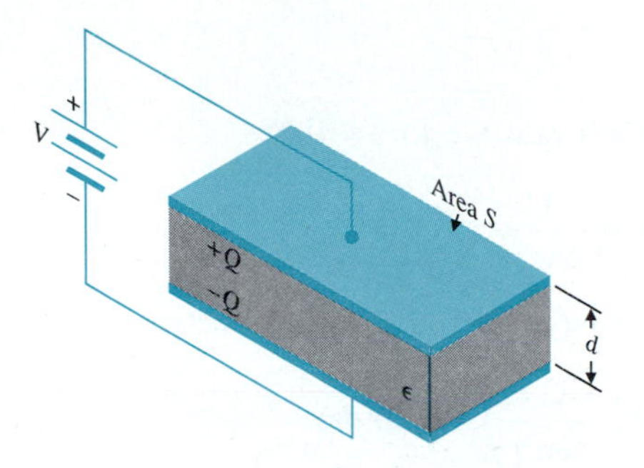

SOLUTION

With the d-c source (batteries) connected as shown, the upper and lower plates are charged positive and negative, respectively. If the fringing of the field at the edges is neglected, the electric field in the dielectric is uniform (over the plate) and constant (across the dielectric), and has a magnitude

$$E = \frac{V}{d}.$$

Using Eq. (3-106), we have

$$W_e = \frac{1}{2}\int_{V'} \epsilon\left(\frac{V}{d}\right)^2 dv = \frac{1}{2}\epsilon\left(\frac{V}{d}\right)^2 (Sd) = \frac{1}{2}\left(\epsilon\frac{S}{d}\right)V^2. \tag{3-109}$$

The quantity in the parentheses of the last expression, $\epsilon S/d$, is the capacitance of the parallel-plate capacitor (see Eq. 3-87). So,

Electric energy stored in a capacitor

$$\boxed{W_e = \tfrac{1}{2}CV^2 \qquad \text{(J).}} \tag{3-110}$$

The following example illustrates how Eq. (3-110) can be used in conjunction with Eq. (3-106) to determine capacitances.

EXAMPLE 3-19

Use energy formulas (3-106) and (3-110) to find the capacitance of a cylindrical capacitor having a length L, an inner conductor of radius a, an outer conductor of inner radius b, and a dielectric of permittivity ϵ, as shown in Fig. 3-21.

SOLUTION

By applying Gauss's law, we know that

$$\mathbf{E} = \mathbf{a}_r E_r = \mathbf{a}_r \frac{Q}{2\pi\epsilon L r}, \qquad a < r < b.$$

The electrostatic energy stored in the dielectric region is, from Eq. (3-106),

$$\begin{aligned} W_e &= \frac{1}{2}\int_a^b \epsilon\left(\frac{Q}{2\pi\epsilon L r}\right)^2 (L2\pi r\,dr) \\ &= \frac{Q^2}{4\pi\epsilon L}\int_a^b \frac{dr}{r} = \frac{Q^2}{4\pi\epsilon L}\ln\left(\frac{b}{a}\right). \end{aligned} \tag{3-111}$$

On the other hand, W_e can also be expressed in terms of capacitance C. From Eqs. (3-110) and (3-111) we have

$$W_e = \frac{C}{2}\left(\frac{Q}{C}\right)^2 = \frac{Q^2}{2C} = \frac{Q^2}{4\pi\epsilon L}\ln\left(\frac{b}{a}\right).$$

Solving for C, we get

$$C = \frac{2\pi\epsilon L}{\ln\left(\dfrac{b}{a}\right)},$$

which is the same as that given in Eq. (3-90).

EXERCISE 3.17 Two capacitors having capacitances 20 (μF) and 40 (μF) are connected in series across a 60-(V) battery. Calculate the energy stored in each capacitor.

ANS. 16 (mJ), 8 (mJ).

3-10.2 ELECTROSTATIC FORCES

Coulomb's law governs the force between two point charges. In a more complex system of charged bodies, using Coulomb's law to determine the force on one of the bodies that is caused by the charges on other bodies would be very tedious. This would be so even in the simple case of finding the force between the plates of a charged parallel-plate capacitor. We now discuss a method for calculating the force on an object in a charged system from the electrostatic energy of the system. This method is based on the ***principle of virtual displacement.***

We consider an *isolated system* of charged conducting, as well as dielectric, bodies separated from one another with no connection to the outside world. The charges on the bodies are constant. Imagine that the electric forces have displaced one of the bodies by a differential distance $d\boldsymbol{\ell}$ (a virtual displacement). The mechanical work done *by the system* would be

$$dW = \mathbf{F}_Q \cdot d\boldsymbol{\ell}, \tag{3-112}$$

where $\mathbf{F}_Q$ is the total electric force acting on the body under the condition of constant charges. Since we have an isolated system with no external supply of energy, this mechanical work must be done at the expense of the stored electrostatic energy; that is,

$$dW = -dW_e = \mathbf{F}_Q \cdot d\boldsymbol{\ell}. \tag{3-113}$$

Noting from Eq. (2-51) in Section 2-5 that the differential change of a scalar resulting from a position change $d\boldsymbol{\ell}$ is the dot product of the gradient of the scalar, and $d\boldsymbol{\ell}$, we write

$$dW_e = (\nabla W_e) \cdot d\boldsymbol{\ell}. \tag{3-114}$$

Since $d\boldsymbol{\ell}$ is arbitrary, comparison of Eqs. (3-113) and (3-114) leads to the following relation:

Determining electrostatic force on charged bodies by method of virtual displacement

$$\boxed{\mathbf{F}_Q = -\nabla W_e \qquad \text{(N)}.} \tag{3-115}$$

In a three-dimensional space, the vector equation (3-115) is actually three equations. For instance, in Cartesian coordinates the force in the x-direction is

$$(F_Q)_x = -\frac{\partial W_e}{\partial x}. \tag{3-116}$$

Similarly for the other directions.

EXAMPLE 3-20

Determine the force on the conducting plates of a charged parallel-plate capacitor, whose plates have an area S and are separated in air by a distance x.

SOLUTION

Assuming fixed charges $\pm Q$ on the plates, we write the stored electric energy as, from Eq. (3-110),

$$W_e = \tfrac{1}{2}CV^2 = \tfrac{1}{2}QV. \tag{3-117}$$

If fringing effect is neglected, there exists a constant electric field intensity $\mathbf{E}$ between the plates, where

$$\mathbf{E} = -\mathbf{a}_x \frac{\rho_s}{\epsilon_0} = -\mathbf{a}_x \frac{Q}{\epsilon_0 S}. \tag{3-118}$$

The potential difference V between the upper and lower plates is

$$V = -\int_{\substack{\text{Lower}\\ \text{plate}}}^{\substack{\text{Upper}\\ \text{plate}}} \mathbf{E}\cdot\mathbf{a}_x\,dx = \frac{Q}{\epsilon_0 S}x. \tag{3-119}$$

Substituting V from Eq. (3-119) in Eq. (3-117) and using Eq. (3-116), we obtain

$$(F_Q)_x = -\frac{Q}{2}\frac{\partial V}{\partial x} = -\frac{Q^2}{2\epsilon_0 S}. \tag{3-120}$$

The negative sign in Eq. (3-120) indicates that the force is opposite to the direction of increase of x.

REVIEW QUESTIONS

Q.3-22 Define *capacitance* and *capacitor*.

Q.3-23 Write the capacitance formula for a parallel-plate capacitor of area S whose plates are separated by a medium of dielectric constant ϵ_r and thickness d.

Q.3-24 What is the definition of an *electron-volt*? How does it compare with a joule?

Q.3-25 Write the expression for electrostatic energy in terms of **E**.

Q.3-26 Discuss the meaning and use of the *principle of virtual displacement*.

REMARKS

1. The capacitance of a capacitor is independent of the charge on the conductors or the potential difference between them.
2. The electrostatic energy stored in a system of discrete charges can be either positive or negative.
3. Energy formula Eq. (3-105) holds for a general medium, but Eq. (3-106) holds only for a linear and isotrophic medium.
4. It can be proved that the formula for stored electrostatic energy in Eq. (3-110) holds not only for parallel-plate capacitors but also for any two-conductor system.

3-11 SOLUTION OF ELECTROSTATIC BOUNDARY-VALUE PROBLEMS

Earlier in this chapter we dealt with techniques for determining the electric field intensity, electric potential, and electric flux density for given charge distributions. In many practical problems, however, the exact charge distribution is not known everywhere, and the formulas we have developed so far cannot be applied directly. For instance, if the charges at certain discrete points in space and the potentials of some conducting bodies are given, it is rather difficult to find the distribution of surface charges on the conducting bodies and/or the electric field intensity in space. In this section we discuss certain methods of solution for problems where the conditions at conductor/free space (or dielectric) boundaries are specified. These types of problems are called ***boundary-value problems***. First, we formulate the differential equations that govern the electric potential in an electrostatic situation.

3-11.1 POISSON'S AND LAPLACE'S EQUATIONS

In Section 3-7 we pointed out that Eqs. (3-63) and (3-4) are the two fundamental governing differential equations for electrostatics in any medium. These equations are repeated below for convenience.

$$\text{Eq. (3-63):} \qquad \nabla \cdot \mathbf{D} = \rho_v. \tag{3-121}$$

$$\text{Eq. (3-4):} \qquad \nabla \times \mathbf{E} = 0. \tag{3-122}$$

The irrotational nature of $\mathbf{E}$ indicated by Eq. (3-122) enables us to define a scalar electric potential V, as in Eq. (3-26).

$$\text{Eq. (3-26):} \qquad \mathbf{E} = -\nabla V. \tag{3-123}$$

In a linear and isotropic medium $\mathbf{D} = \epsilon \mathbf{E}$, and Eq. (3-121) becomes

$$\nabla \cdot \epsilon \mathbf{E} = \rho_v. \tag{3-124}$$

Substitution of Eq. (3-123) in Eq. (3-124) yields

$$\nabla \cdot (\epsilon \nabla V) = -\rho_v. \tag{3-125}$$

where ϵ can be a function of position. For a simple medium—that is, for a medium that is also homogeneous—ϵ is a constant and can then be taken out of the divergence operation. We have

Poisson's equation in operator form

$$\boxed{\nabla^2 V = -\frac{\rho_v}{\epsilon}.} \tag{3-126}$$

In Eq. (3-126) we have introduced a new operator, ∇^2 (del square), the ***Laplacian operator***, which stands for "the divergence of the gradient of," or $\nabla \cdot \nabla$. Equation (3-126) is known as ***Poisson's equation***. In Cartesian coordinates,

$$\nabla^2 V = \nabla \cdot \nabla V = \left(\mathbf{a}_x \frac{\partial}{\partial x} + \mathbf{a}_y \frac{\partial}{\partial y} + \mathbf{a}_z \frac{\partial}{\partial z}\right) \cdot \left(\mathbf{a}_x \frac{\partial V}{\partial x} + \mathbf{a}_y \frac{\partial V}{\partial y} + \mathbf{a}_z \frac{\partial V}{\partial z}\right),$$

and Eq. (3-126) becomes

Poisson's equation in Cartesian coordinates

$$\frac{\partial^2 V}{\partial x^2} + \frac{\partial^2 V}{\partial y^2} + \frac{\partial^2 V}{\partial z^2} = -\frac{\rho_v}{\epsilon} \quad (\text{V/m}^2). \tag{3-127}$$

Similarly, by using Eqs. (2-70) and (2-57) we can verify the following expressions for $\nabla^2 V$ in cylindrical and spherical coordinates.

Cylindrical coordinates:

$$\nabla^2 V = \frac{1}{r}\frac{\partial}{\partial r}\left(r\frac{\partial V}{\partial r}\right) + \frac{1}{r^2}\frac{\partial^2 V}{\partial \phi^2} + \frac{\partial^2 V}{\partial z^2}. \tag{3-128}$$

Spherical coordinates:

$$\nabla^2 V = \frac{1}{R^2}\frac{\partial}{\partial R}\left(R^2\frac{\partial V}{\partial R}\right) + \frac{1}{R^2 \sin\theta}\frac{\partial}{\partial \theta}\left(\sin\theta\frac{\partial V}{\partial \theta}\right) + \frac{1}{R^2 \sin^2\theta}\frac{\partial^2 V}{\partial \phi^2}. \tag{3-129}$$

The solution of Poisson's equation in three dimensions subject to prescribed boundary conditions is, in general, not an easy task.

At points in a simple medium where there is no free charge, $\rho_v = 0$ and Eq. (3-126) reduces to

Laplace's equation in operator form

$$\nabla^2 V = 0, \tag{3-130}$$

which is known as ***Laplace's equation***. Laplace's equation occupies a very important position in electromagnetics. In Cartesian coordinates we have

Laplace's equation in cartesian coordinates

$$\frac{\partial^2 V}{\partial x^2} + \frac{\partial^2 V}{\partial y^2} + \frac{\partial^2 V}{\partial z^2} = 0. \tag{3-131}$$

3-11.2 BOUNDARY-VALUE PROBLEMS IN CARTESIAN COORDINATES

The simplest situation is when V is a function of only one coordinate. We illustrate this situation with an example.

EXAMPLE 3-21

Two large parallel conducting plates are separated by a distance d and maintained at potentials 0 and V_0, as shown in Fig. 3-24. The region between the plates is filled with a continuous distribution of electrons having a volume density of charge $\rho_v = -\rho_0 y/d$. Assuming negligible fringing effect at the edges, determine

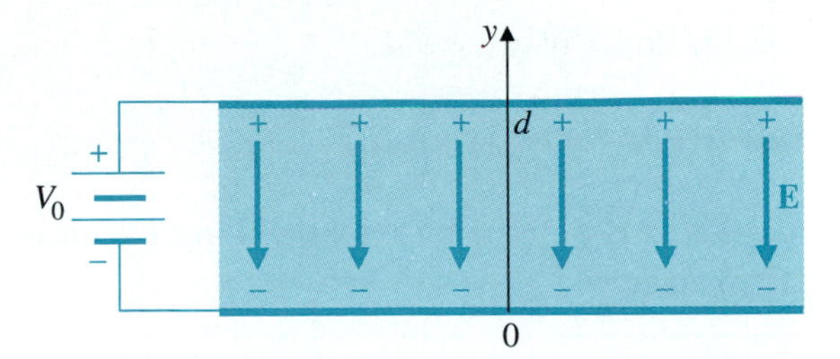

FIGURE 3-24 A parallel-plate capacitor (Example 3-21).

a) the potential at any point between the plates, and

b) the surface charge densities on the plates.

SOLUTION

a) Here the governing equation is Poisson's equation (3-127), which simplifies to

$$\frac{d^2V(y)}{dy^2} = \frac{\rho_0}{\epsilon_0 d} y. \tag{3-132}$$

Integrating Eq. (3-132) twice, we have

$$V(y) = \frac{\rho_0}{6\epsilon_0 d} y^3 + C_1 y + C_2. \tag{3-133}$$

The boundary conditions on the two conducting plates are:

At $y = 0$, $\quad V = 0 = C_2$.

At $y = d$, $\quad V = V_0 = \dfrac{\rho_0 d^2}{6\epsilon_0} + C_1 d, \qquad C_1 = \dfrac{V_0}{d} - \dfrac{\rho_0 d}{6\epsilon_0}$.

Substituting the above values of C_1 and C_2 in Eq. (3-133), we obtain the solution for Poisson's equation (3-132).

$$V(y) = \frac{\rho_0}{6\epsilon_0 d} y^3 + \left(\frac{V_0}{d} - \frac{\rho_0 d}{6\epsilon_0}\right) y. \tag{3-134}$$

The electric field intensity is $-\nabla V$.

$$\mathbf{E}(y) = -\mathbf{a}_y \frac{\partial V}{\partial y} = -\mathbf{a}_y \left[\frac{\rho_0}{2\epsilon_0 d} y^2 + \left(\frac{V_0}{d} - \frac{\rho_0 d}{6\epsilon_0}\right)\right]. \tag{3-135}$$

b) The surface charge densities on the conducting plates can be found from the boundary condition given in Eq. (3-46).
At the lower plate, $y = 0$.

$$\mathbf{a}_n = \mathbf{a}_y, \qquad \rho_{s\ell} = \epsilon_0 E_{y\ell} = -\frac{\epsilon_0 V_0}{d} + \frac{\rho_0 d}{6}.$$

At the upper plate, $y = d$.

$$\mathbf{a}_n = -\mathbf{a}_y, \qquad \rho_{su} = -\epsilon_0 E_{yu} = \frac{\epsilon_0 V_0}{d} + \frac{\rho_0 d}{3}.$$

In this case, $\rho_{su} \neq -\rho_{s\ell}$, and it no longer is meaningful to compute the capacitance.

■ **EXERCISE 3.18** Show that the following potential functions satisfy two-dimensional Laplace's equation:

a) $Ae^{-kx} \sin ky$, and

b) $A \sin\left(\frac{n\pi}{b} x\right) \cosh\left[\frac{n\pi}{b}(a - y)\right]$,

where A, k, a, and b are constants.

3-11.3 BOUNDARY-VALUE PROBLEMS IN CYLINDRICAL COORDINATES

Laplace's equation for scalar electric potential V in cylindrical coordinates is, from Eq. (3-128),

Laplace's equation in cylindrical coordinates

$$\frac{1}{r}\frac{\partial}{\partial r}\left(r\frac{\partial V}{\partial r}\right) + \frac{1}{r^2}\frac{\partial^2 V}{\partial \phi^2} + \frac{\partial^2 V}{\partial z^2} = 0. \tag{3-136}$$

A general solution of Eq. (3-136) is rather involved. In situations where there is cylindrical symmetry ($\partial^2 V/\partial \phi^2 = 0$) and the lengthwise dimension is very large in comparison to the radius ($\partial^2 V/\partial z^2 \cong 0$), Eq. (3-136) reduces to

$$\frac{d}{dr}\left(r\frac{dV}{dr}\right) = 0, \tag{3-137}$$

and V is a function of the radial dimension r only. Equation (3-137) can be integrated twice to yield

$$V(r) = C_1 \ln r + C_2, \tag{3-138}$$

where the integration constants C_1 and C_2 are to be determined by the boundary conditions of the problems.

■ **EXERCISE 3.19** The radius of the inner conductor of a long coaxial cable is a. The inner radius of the outer conductor is b. If the inner and outer conductors are kept at potentials V_0 and 0 respectively, determine the electric potential and the electric field intensity in the insulating material by solving Laplace's equation.

ANS. $V = V_0 \ln(b/r)/\ln(b/a)$, $\mathbf{E} = \mathbf{a}_r V_0/r \cdot \ln(b/a)$.

If the problem is such that electric potential changes only in the circumferential direction and not in r- and z-directions, Eq. (3-136) reduces to

$$\frac{1}{r^2}\frac{\partial^2 V}{\partial \phi^2} = 0. \tag{3-139}$$

We illustrate this case with the following example.

EXAMPLE 3-22

Two infinite insulated conducting planes maintained at potentials 0 and V_0 form a wedge-shaped configuration, as shown in Fig. 3-25. Determine the potential distributions for the regions:

a) $0 < \phi < \alpha$, and

b) $\alpha < \phi < 2\pi$.

SOLUTION

Here we have $\partial V/\partial r = 0$ and $\partial V/\partial z = 0$. Since the region at $r = 0$ is excluded, Eq. (3-139) becomes

$$\frac{d^2 V}{d\phi^2} = 0. \tag{3-140}$$

V depends only on ϕ, and can be obtained from Eq. (3-140) by integrating twice.

$$V(\phi) = K_1\phi + K_2. \tag{3-141}$$

The two integration constants K_1 and K_2 are to be determined from boundary conditions.

a) For $0 \leq \phi \leq \alpha$:

At $\alpha = 0$, $\quad V(0) = 0 = K_2$. (3-142a)

At $\phi = \alpha$, $\quad V(\alpha) = V_0 = K_1\alpha, \quad K_1 = V_0/\alpha$. (3-142b)

Therefore, from Eq. (3-141),

$$V(\phi) = \frac{V_0}{\alpha}\phi, \qquad 0 \leq \phi \leq \alpha. \tag{3-143}$$

FIGURE 3-25 Two infinite insulated conducting planes maintained at constant potentials (Example 3-22).

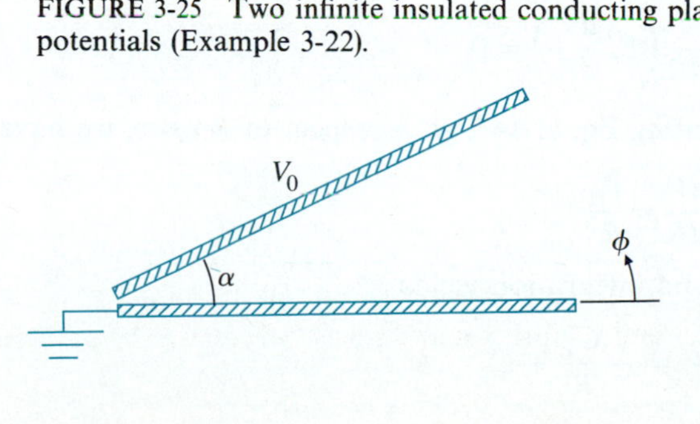

b) For $\alpha \leq \phi \leq 2\pi$:

At $\phi = \alpha$, $\quad V(\alpha) = \alpha K_1 + K_2.$ (3-144a)

At $\phi = 2\pi$, $\quad V(2\pi) = 2\pi K_1 + K_2.$ (3-144b)

Solving Eqs. (3-144a) and (3-144b), we find

$$K_1 = -\frac{V_0}{2\pi - \alpha} \tag{3-145a}$$

and

$$K_2 = \frac{2\pi V_0}{2\pi - \alpha}. \tag{3-145b}$$

Finally, from Eq. (3-141), we obtain

$$V(\phi) = \frac{V_0}{2\pi - \alpha}(2\pi - \phi), \qquad \alpha \leq \phi \leq 2\pi. \tag{3-146}$$

3-11.4 BOUNDARY-VALUE PROBLEMS IN SPHERICAL COORDINATES

Poisson's and Laplace's equations for scalar electric potential V in spherical coordinates can be obtained by using Eq. (3-129). In the following example we discuss a simplified, one-dimensional case.

EXAMPLE 3-23

The inner and outer radii of two concentric, thin, conducting, spherical shells are R_i and R_o respectively. The space between the shells is filled with an insulating material. The inner shell is maintained at a potential V_1 and the outer shell at V_2. Determine the potential distribution in the insulating material by solving Laplace's equation.

SOLUTION

Since the given situation as shown in Fig. 3-26 has spherical symmetry, the electric potential is independent of θ or ϕ. Substitution of the simplified Eq. (3-129) in Eq. (3-130) yields the following one-dimensional Laplace's equation.

$$\frac{d}{dR}\left(R^2 \frac{dV}{dR}\right) = 0. \tag{3-147}$$

Integrating Eq. (3-147) with respect to R once, we have

$$\frac{dV}{dR} = \frac{C_1}{R^2}. \tag{3-148}$$

A second integration yields

$$V = -\frac{C_1}{R} + C_2. \tag{3-149}$$

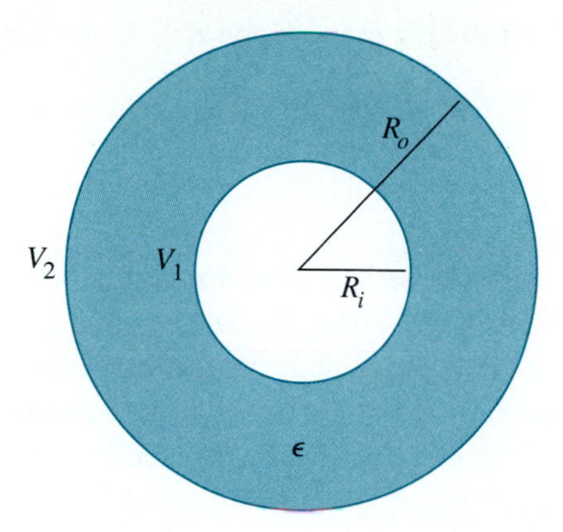

FIGURE 3-26 Two concentric conducting shells maintained at constant potentials (Example 3-23).

The two integration constants C_1 and C_2 are to be determined from the boundary conditions at the two conducting shells.

At $R = R_i$,

$$V_1 = -\frac{C_1}{R_i} + C_2. \tag{3-150a}$$

At $R = R_o$,

$$V_2 = -\frac{C_1}{R_o} + C_2. \tag{3-150b}$$

The solution of Eqs. (3-150a) and (3-150b) gives

$$C_1 = -\frac{R_o R_i (V_1 - V_2)}{R_o - R_i} \tag{3-151a}$$

and

$$C_2 = \frac{R_o V_2 - R_i V_1}{R_o - R_i}. \tag{3-151b}$$

Therefore, the potential distribution between the two shells is, from Eqs. (3-149), (3-151a), and (3-151b),

$$V(R) = \frac{1}{R_o - R_i}\left[\frac{R_i R_o}{R}(V_1 - V_2) + R_o V_2 - R_i V_1\right], \quad R_i \le R \le R_o. \tag{3-152}$$

We note from Eq. (3-152) that V is independent of the dielectric constant of the insulating material.

■ **EXERCISE 3.20** Find the potential distribution in the region $R \geq R_o$ in Example 3-23.

ANS. $V = R_o V_2 / R$.

3-11.5 METHOD OF IMAGES

Method of images simplifies solution of certain problems.

There is a class of electrostatic problems with boundary conditions that appear to be difficult to satisfy if the governing Poisson's or Laplace's equation is to be solved directly, but the conditions on the bounding surfaces in these problems can be set up by appropriate ***image charges***, and the potential distributions can then be determined in a straightforward manner. This method of replacing bounding surfaces by appropriate image charges instead of attempting a formal solution of a Poisson's or Laplace's equation is called the ***method of images***.

Uniqueness theorem

Before we discuss the method of images, it is important to know that ***a solution of Poisson's equation*** (of which Laplace's equation is a special case) ***that satisfies a given set of boundary conditions is a unique solution***. This statement is called the ***uniqueness theorem***. Because of the uniqueness theorem, a solution of an electrostatic problem satisfying that set of boundary conditions is *the only possible solution*, irrespective of the method by which the solution is obtained. A solution obtained even by intelligent guessing is the only correct solution. The uniqueness theorem can be proved formally,† but we will just accept its verity.

In the following we will discuss several important applications of the method of images.

A. *Point Charges Near Conducting Planes*

We illustrate this type of problem with an example.

EXAMPLE 3-24

A positive point charge Q is located at a distance d above a large grounded (zero-potential) conducting plane, as shown in Fig. 3-27(a). Find (a) the potential at an arbitrary point $P(x, y, z)$ in the $y > 0$ region, and (b) the induced charge distribution on the surface of the conducting plane.

SOLUTION

a) A formal procedure for solving this problem would require the solution of Poisson's equation in the $y > 0$ region subject to the boundary condition that $V = 0$ at $y = 0$ and at infinity. The direct

† See, for instance, D. K. Cheng, *Field and Wave Electromagnetics*, Second Edition, pp. 158–159, Addison-Wesley Publishing Co., Reading, Mass., 1989.

Image of a point charge near a large conducting plane

construction of such a solution is difficult. On the other hand, if we removed the conducting plane and replaced it by an image point charge $-Q$ at $y = -d$, as shown in Fig. 3-27(b), neither the situation in the $y > 0$ region nor the boundary conditions would be changed. The potential at a point $P(x, y, z)$ due to the two charges can be written down by inspection.

$$V(x, y, z) = \frac{Q}{4\pi\epsilon_0}\left(\frac{1}{R_+} - \frac{1}{R_-}\right)$$

$$= \frac{Q}{4\pi\varepsilon_0}\left[\frac{1}{\sqrt{x^2 + (y-d)^2 + z^2}} - \frac{1}{\sqrt{x^2 + (y+d)^2 + z^2}}\right], \quad y \geqslant 0; \tag{3-153a}$$

and

$$V(x, y, z) = 0, \quad y \leqslant 0. \tag{3-153b}$$

Since the given boundary conditions are satisfied, Eqs. (3-153a) and (3-153b) represent the correct solution by virtue of the uniqueness theorem.

b) In order to find the induced charge distribution on the conducting surface, we first determine the electric field intensity. From Eq. (3-153a) we have

$$\mathbf{E} = -\nabla V$$

$$= \frac{Q}{4\pi\epsilon_0}\left\{\frac{\mathbf{a}_x x + \mathbf{a}_y(y-d) + \mathbf{a}_z z}{[x^2 + (y-d)^2 + z^2]^{3/2}} - \frac{\mathbf{a}_x + \mathbf{a}_y(y+d) + \mathbf{a}_z z}{[x^2 + (y+d)^2 + z^2]^{3/2}}\right\},$$
$$y \geqslant 0. \tag{3-154}$$

FIGURE 3-27 Point charge and grounded plane conductor (Example 3-24).

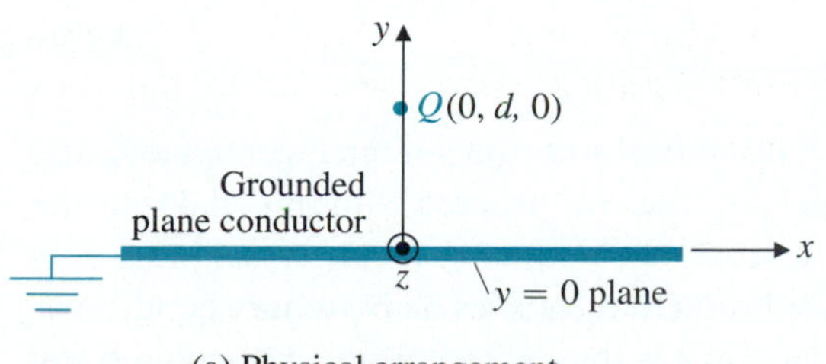

(a) Physical arrangement.

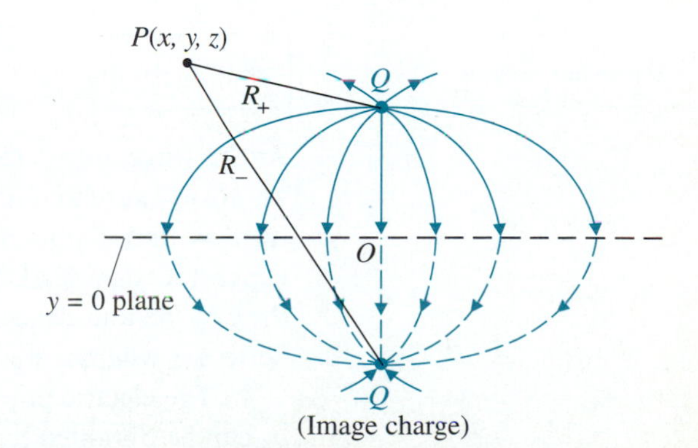

(b) Image charge and field lines.

The induced surface charge density is, from Eqs. (3-46) and (3-154),

$$\rho_s = \epsilon_0 E_y\Big|_{y=0} = -\frac{Qd}{2\pi(x^2 + d^2 + z^2)^{3/2}}. \tag{3-155}$$

It must be emphasized that the method of images can be used to determine the fields only in the region where the image charges are *not* located. Thus, in Example 3-24, the point charges $+Q$ and $-Q$ *cannot* be used to calculate V or $\mathbf{E}$ in the $y < 0$ region. As a matter of fact, both V and $\mathbf{E}$ vanish in the $y < 0$ region.

EXERCISE 3.21 Find the total amount of charge induced on the surface of the infinite conducting plane in Example 3-24.

ANS. $-Q$.

B. *Line Charge Near a Parallel Conducting Cylinder*

We now consider the problem of a line charge ρ_ℓ(C/m) located at a distance d from the axis of a parallel, conducting, circular cylinder of radius a. Both the line charge and the conducting cylinder are assumed to be infinitely long. Figure 3-28(a) shows a cross section of this arrangement. Before we approach this problem by the method of images, we note the following: (1) The image must be a parallel line charge inside the cylinder in order to make the cylindrical surface at $r = a$ an equipotential surface. Let us call this image line charge ρ_i. (2) Because of symmetry with respect to the line OP, the image line charge must lie somewhere along OP, say at point P_i, which is at a distance d_i from the axis (Fig. 3-28b). We need to determine the two unknowns, ρ_i and d_i.

As a first step, let us assume that

Image line charge

$$\boxed{\rho_i = -\rho_\ell.} \tag{3-156}$$

At this stage, Eq. (3-156) is just a trial solution (an intelligent guess), and we are not sure that it will hold true. We proceed with this trial solution until we find that it fails to satisfy the boundary conditions. However, if Eq. (3-156) does lead to a solution that satisfies all boundary conditions, then by the uniqueness theorem it is the only solution. Our next job will be to see whether we can determine d_i.

The electric potential at a distance r from a line charge of density ρ_ℓ can be obtained by integrating the electric field intensity $\mathbf{E}$ given in Eq. (3-23):

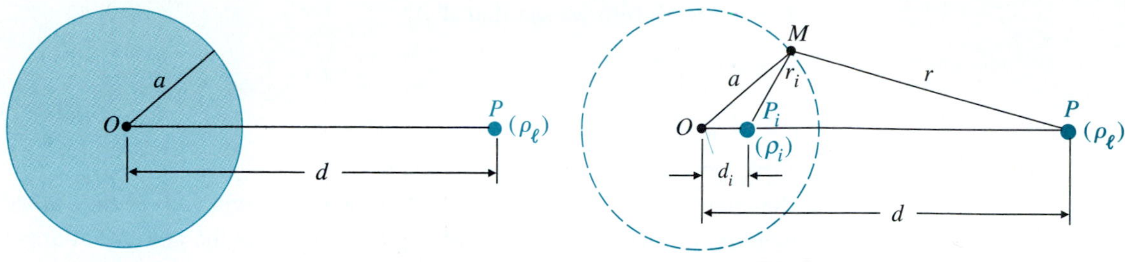

(a) Line charge and parallel conducting cylinder.

(b) Line charge and its image.

FIGURE 3-28 Cross section of line charge and its image in a parallel, conducting, circular cylinder.

$$V = -\int_{r_0}^{r} E_r\,dr = -\frac{\rho_\ell}{2\pi\epsilon_0}\int_{r_0}^{r}\frac{1}{r}\,dr = \frac{\rho_\ell}{2\pi\epsilon_0}\ln\frac{r_0}{r}. \tag{3-157}$$

Note that the reference point for zero potential, r_0, cannot be at infinity because setting $r_0 = \infty$ in Eq. (3-157) would make V infinite everywhere else. Let us leave r_0 unspecified for the time being. The potential at a point on or outside the cylindrical surface is obtained by adding the contributions of ρ_ℓ and ρ_i. In particular, at a point M on the cylindrical surface shown in Fig. 3-28(b) we have

$$V_M = \frac{\rho_\ell}{2\pi\epsilon_0}\ln\frac{r_0}{r} - \frac{\rho_\ell}{2\pi\epsilon_0}\ln\frac{r_0}{r_i} = \frac{\rho_\ell}{2\pi\epsilon_0}\ln\frac{r_i}{r}. \tag{3-158}$$

In Eq. (3-158) we have chosen, for simplicity, a point equidistant from ρ_ℓ and ρ_i as the reference point for zero potential so that the $\ln r_0$ terms cancel. Otherwise, a constant term should be included in the right side of Eq. (3-158), but it would not affect what follows. Equipotential surfaces are specified by

$$\frac{r_i}{r} = \text{Constant}. \tag{3-159}$$

If an equipotential surface is to coincide with the cylindrical surface ($\overline{OM} = a$), the point P_i must be located in such a way as to make triangles OMP_i and OPM similar. These two triangles already have one common angle, $\angle MOP_i$. Point P_i should be chosen to make $\angle OMP_i = \angle OPM$. We have

$$\frac{\overline{P_iM}}{\overline{PM}} = \frac{\overline{OP_i}}{\overline{OM}} = \frac{\overline{OM}}{\overline{OP}},$$

or

$$\frac{r_i}{r} = \frac{d_i}{a} = \frac{a}{d} = \text{Constant}. \tag{3-160}$$

From Eq. (3-160) we see that if

$$d_i = \frac{a^2}{d} \tag{3-161}$$

the image line charge $-\rho_\ell$, together with ρ_ℓ, will make the dashed cylindrical surface in Fig. 3-28(b) equipotential. As the point M changes its location on the dashed circle, both r_i and r will change; but their ratio remains a constant that equals a/d. Point P_i is called the ***inverse point*** of P with respect to a circle of radius a.

The image line charge $\rho_i = -\rho_\ell$ can then replace the cylindrical conducting surface, and V and $\mathbf{E}$ at any point *outside the surface* can be determined from the line charges ρ_ℓ and $-\rho_\ell$. By symmetry we find that the parallel cylindrical surface surrounding the original line charge ρ_ℓ with radius a and its axis at a distance d_i to the right of P is also an equipotential surface. This observation enables us to calculate the capacitance per unit length of an open-wire transmission line consisting of two parallel conductors of circular cross section.

EXAMPLE 3-25

Determine the capacitance per unit length between two long, parallel, circular conducting wires of radius a. The axes of the wires are separated by a distance D.

SOLUTION

Refer to the cross section of the two-wire transmission line shown in Fig. 3-29. The equipotential surfaces of the two wires can be considered to have been generated by a pair of line charges $+\rho_\ell$ and $-\rho_\ell$ separated by a distance $(D - 2d_i) = d - d_i$. The potential difference between the two wires is that between any two points on the respective wires. Let subscripts 1 and 2 denote the wires surrounding the equivalent line charges $+\rho_\ell$ and $-\rho_\ell$, respectively. We have, from Eqs. (3-158) and (3-160),

$$V_2 = \frac{\rho_\ell}{2\pi\epsilon_0} \ln \frac{a}{d}$$

and, similarly,

$$V_1 = -\frac{\rho_\ell}{2\pi\epsilon_0} \ln \frac{a}{d}.$$

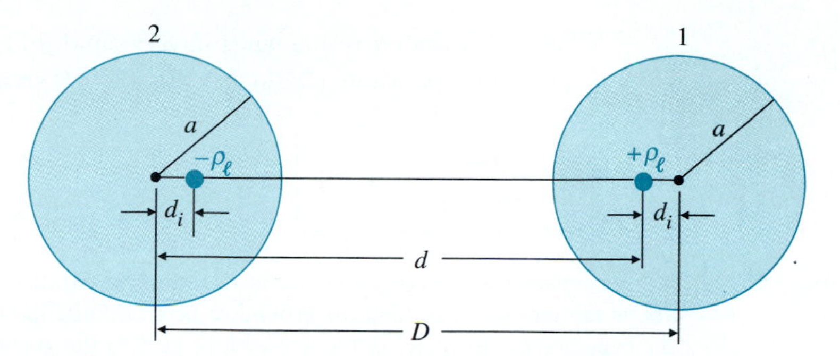

FIGURE 3-29 Cross section of two-wire transmission line and equivalent line charges (Example 3-25).

We note that V_1 is a positive quantity, whereas V_2 is negative because $a < d$. The capacitance per unit length is

$$C = \frac{\rho_\ell}{V_1 - V_2} = \frac{\pi\epsilon_0}{\ln(d/a)}, \tag{3-162}$$

where

$$d = D - d_i = D - \frac{a^2}{d},$$

from which we obtain†

$$d = \tfrac{1}{2}(D + \sqrt{D^2 - 4a^2}). \tag{3-163}$$

Using Eq. (3-163) in Eq. (3-162), we have

Capacitance per unit length of parallel wires

$$C = \frac{\pi\epsilon_0}{\ln[(D/2a) + \sqrt{(D/2a)^2 - 1}]} \quad \text{(F/m)}. \tag{3-164}$$

Since

$$\ln[x + \sqrt{x^2 - 1}] = \cosh^{-1}x \text{ for } x > 1,$$

Eq. (3-164) can be written alternatively as

$$C = \frac{\pi\epsilon_0}{\cosh^{-1}(D/2a)} \quad \text{(F/m)}. \tag{3-165}$$

† The other solution, $d = \frac{1}{2}(D - \sqrt{D^2 - 4a^2})$, is discarded because both D and d are usually much larger than a.

When the diameter of the wires is very small in comparison with the distance of separation, $(D/2a) \gg 1$, Eq. (3-164) simplifies to

$$C = \frac{\pi\epsilon_0}{\ln(D/a)} \quad \text{(F/m)}. \tag{3-166}$$

EXERCISE 3.22 A long power transmission line, 2 (cm) in radius, is parallel to and situated 10 (m) above the ground. Assuming the ground to be an infinite flat conducting plane, find the capacitance per meter of the line with respect to the ground.

ANS. 8.04 (pF/m).

REVIEW QUESTIONS

Q.3-27 Write Poisson's and Laplace's equations in vector notation for a simple medium.

Q.3-28 Write Poisson's and Laplace's equations in Cartesian coordinates for a simple medium.

Q.3-29 If $\nabla^2 U = 0$, why does it not follow that U is identically zero?

Q.3-30 A fixed voltage is connected across a parallel-plate capacitor.

- **a)** Does the electric field intensity in the space between the plates depend on the permittivity of the medium?
- **b)** Does the electric flux density depend on the permittivity of the medium? Explain.

Q.3-31 Assume that fixed charges $+Q$ and $-Q$ are deposited on the plates of an isolated parallel-plate capacitor.

- **a)** Does the electric field intensity in the space between the plates depend on the permittivity of the medium?
- **b)** Does the electric flux density depend on the permittivity of the medium? Explain.

Q.3-32 State in words the *uniqueness theorem of electrostatics.*

Q.3-33 What is the image of a spherical cloud of electrons with respect to an infinite conducting plane?

Q.3-34 What is the image of an infinitely long line charge of density ρ_ℓ with respect to a parallel conducting circular cylinder?

Q.3-35 Where is the zero-potential surface of the two-wire transmission line in Fig. 3-29?

REMARKS

1. Poisson's Eq. (3-126) and Laplace's Eq. (3-130) do not hold if the medium is nonlinear, inhomogeneous, or anisotropic.
2. The method of images can be used to determine the fields *only* in the region where the image charges are *not* located.

SUMMARY

This chapter deals with the static electric fields of charges that are at rest and do not change with time. After having defined the electric field intensity **E** as the force per unit charge, we

- presented the two fundamental postulates of electrostatics in free space that specify the divergence and the curl of **E**,
- derived Coulomb's law and Gauss's law, which enabled us to determine the electric field due to discrete and continuous charge distributions,
- introduced the concept of the scalar electric potential,
- considered the effect of material media on static electric field,
- discussed the macroscopic effect of induced dipoles by finding the equivalent polarization charge densities,
- defined electric flux density or electric displacement, **D**, and the dielectric constant,
- discussed the boundary conditions for static electric fields,
- defined capacitance and explained the procedure for its determination,
- found the formulas for stored electrostatic energy,
- used the principle of virtual displacement to calculate the force on an object in a charged system,
- introduced Poisson's and Laplace's equations and illustrated the method of solution for simple problems, and
- explained the method of images for solving electrostatic boundary-value problems.

PROBLEMS

P.3-1 The cathode-ray oscilloscope (CRO) shown in Fig. 3-2 is used to measure the voltage applied to the parallel deflection plates.

a) Assuming no breakdown in insulation, what is the maximum voltage that can be measured if the distance of separation between the plates is h?

b) What is the restriction on L if the diameter of the screen is D?

c) What can be done with a fixed geometry to double the CRO's maximum measurable voltage?

P.3-2 Three 2-(μC) point charges are located in air at the corners of an equilateral triangle that is 10 (cm) on each side. Find the magnitude and direction of the force experienced by each charge.

P.3-3 Two point charges, Q_1 and Q_2, are located at (0, 5, -1) and (0, -2, 6), respectively. Find the relation between Q_1 and Q_2 such that the total force on a test charge at the point $P(0, 2, 3)$ will have

a) no y-component, and

b) no z-component.

P.3-4 Three point charges $Q_1 = -9$ (μC), $Q_2 = 4$ (μC), and $Q_3 = -36$ (μC) are arranged on a straight line. The distance between Q_1 and Q_3 is 9 (cm). It is claimed that a location can be selected for Q_2 such that each charge will experience a zero force. Find this location.

P.3-5 In Example 3-8 determine the position of the point P on the z-axis beyond which the disk may be regarded as a point charge if the error in the calculation of $\mathbf{E}$ is not more than 1%.

P.3-6 A line charge of uniform charge density ρ_ℓ forms a circle of radius b that lies in the xy-plane in air with its center at the origin.

a) Find the electric field intensity $\mathbf{E}$ at the point $(0, 0, h)$.

b) At what value of h will $\mathbf{E}$ in part (a) be a maximum? What is this maximum?

c) Explain why $\mathbf{E}$ has a maximum at that location.

P.3-7 A line charge of uniform density ρ_ℓ forms a semicircle of radius b in the upper half xy-plane. Determine the magnitude and direction of the electric field intensity at the center of the semicircle.

P.3-8 A spherical distribution of charge $\rho = \rho_0[1 - (R^2/b^2)]$ exists in the region $0 \leqslant R \leqslant b$. This charge distribution is concentrically surrounded by a conducting shell with inner radius $R_i(>b)$ and outer radius R_o. Determine $\mathbf{E}$ everywhere.

P.3-9 Two infinitely long coaxial cylindrical surfaces, $r = a$ and $r = b$ $(b > a)$, carry surface charge densities ρ_{sa} and ρ_{sb}, respectively.

a) Determine $\mathbf{E}$ everywhere.

b) What must be the relation between a and b in order that $\mathbf{E}$ vanishes for $r > b$?

P.3-10 Determine the work done in carrying a $+5$(μC) charge from $P_1(1, 2, -4)$ to $P_2(-2, 8, -4)$ in the field $\mathbf{E} = \mathbf{a}_x y + \mathbf{a}_y x$

a) along the parabola $y = 2x^2$, and

b) along the straight line joining P_1 and P_2.

P.3-11 Repeat problem P.3-10 if the field is $\mathbf{E} = \mathbf{a}_x y - \mathbf{a}_y x$.

P.3-12 A finite line charge of length L carrying uniform line charge density ρ_ℓ is coincident with the x-axis.

a) Determine V in the plane bisecting the line charge.

b) Determine $\mathbf{E}$ from ρ_ℓ directly by applying Coulomb's law.

c) Check the answer in part (b) with $-\nabla V$.

P.3-13 The polarization in a dielectric cube of side L centered at the origin is given by $\mathbf{P} = \mathrm{P}_0(\mathbf{a}_x x + \mathbf{a}_y y + \mathbf{a}_z z)$.

a) Determine the surface and volume bound-charge densities.

b) Show that the total bound charge is zero.

P.3-14 The polarization vector in a dielectric sphere of radius b is $\mathbf{P} = \mathbf{a}_x \mathrm{P}_0$.

a) Determine the surface and volume charge densities.

b) Show that the total bound charge is zero.

P.3-15 The axis of a long dielectric tube of inner radius r_i and outer radius r_o coincides with the z-axis. A polarization vector $\mathbf{P} = \mathrm{P}_0(\mathbf{a}_x 3x + \mathbf{a}_y 4y)$ exists in the dielectric.

a) Determine the surface and volume charge densities.

b) Show that the total bound charge is zero.

P.3-16 A positive point charge Q is at the center of a spherical dielectric shell of an inner radius R_i and an outer radius R_o. The dielectric constant of the shell is ϵ_r. Determine **E**, V, **D**, and **P** as functions of the radial distance R.

P.3-17 Solve the following problems:

a) Find the breakdown voltage of a parallel-plate capacitor, assuming that the conducting plates are 50 (mm) apart and the medium between them is air.

b) Find the breakdown voltage if the entire space between the conducting plates is filled with plexiglass, which has a dielectric constant 3 and a dielectric strength 20 (kV/mm).

c) If a 10-(mm) thick plexiglass is inserted between the plates, what is the maximum voltage that can be applied to the plates without a breakdown?

P.3-18 Assume that the $z = 0$ plane separates two lossless dielectric regions with $\epsilon_{r1} = 2$ and $\epsilon_{r2} = 3$. If we know that $\mathbf{E}_1$ in region 1 is $\mathbf{a}_x 2y - \mathbf{a}_y 3x + \mathbf{a}_z(5 + z)$, what do we also know about $\mathbf{E}_2$ and $\mathbf{D}_2$ in region 2? Can we determine $\mathbf{E}_2$ and $\mathbf{D}_2$ at any point in region 2? Explain.

P.3-19 Dielectric lenses can be used to collimate electromagnetic fields. In Fig. 3-30 the left surface of the lens is that of a circular cylinder, and the right surface is a plane. If $\mathbf{E}_1$ at point $P(r_o, 45°, z)$ in region 1 is $\mathbf{a}_r 5 - \mathbf{a}_\phi 3$, what must be the dielectric constant of the lens in order that $\mathbf{E}_3$ in region 3 is parallel to the x-axis?

P.3-20 The space between a parallel-plate capacitor of area S is filled with a dielectric whose permittivity varies linearly from ϵ_1 at one plate ($y = 0$) to ϵ_2 at the other plate ($y = d$). Neglecting fringing effect, find the capacitance.

P.3-21 Assume that the outer conductor of the cylindrical capacitor in Example 3-16 is grounded and that the inner conductor is maintained at a potential V_0.

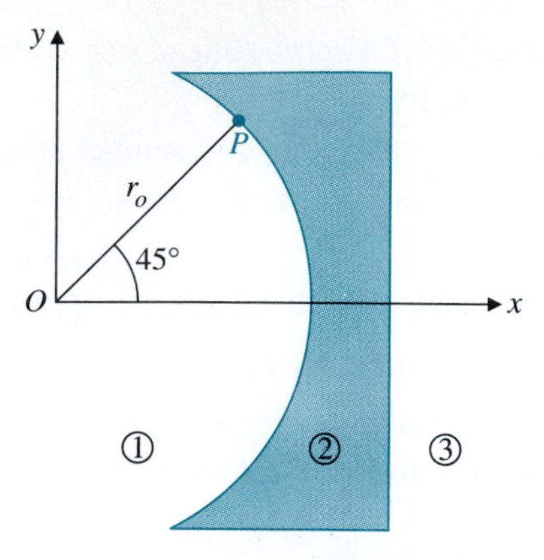

FIGURE 3-30 A dielectric lens (Problem P.3-19).

a) Find the electric field intensity, $\mathbf{E}(a)$, at the surface of the inner conductor.

b) With the inner radius, b, of the outer conductor fixed, find a so that $E(a)$ is minimized.

c) Find this minimum $E(a)$.

d) Determine the capacitance under the conditions of part (b).

P.3-22 The radius of the core and the inner radius of the outer conductor of a very long coaxial transmission line are r_i and r_o, respectively. The space between the conductors is filled with two coaxial layers of dielectrics. The dielectric constants of the dielectrics are ϵ_{r1} for $r_i < r < b$ and ϵ_{r2} for $b < r < r_o$. Determine its capacitance per unit length.

P.3-23 A spherical capacitor consists of an inner conducting sphere of radius R_i and an outer conductor with a spherical inner wall of radius R_o. The space in between is filled with a dielectric of permittivity ϵ. Determine the capacitance.

P.3-24 Three capacitors 1 (μF), 2 (μF), and 3 (μF) are connected as shown in Fig. 3-31 across a 120-volt source. Calculate the electric energy stored in each capacitor.

P.3-25 Calculate the energy expended in moving a point charge 500 (pC) from $P_1(2, \pi/3, -1)$ to $P_2(4, -\pi/2, -1)$ in an electric field $\mathbf{E} = \mathbf{a}_r 6r \sin \phi + \mathbf{a}_\phi 3r \cos \phi$ (V/m) by

a) first moving from $\phi = \pi/3$ to $-\pi/2$ at $r = 2$, then from $r = 2$ to 4 at $\phi = -\pi/2$, and

b) first moving from $r = 2$ to 4 at $\phi = \pi/3$, then from $\phi = \pi/3$ to $-\pi/2$ at $r = 4$.

P.3-26 For a coaxial capacitor of length L, use the method of virtual displacement to find the force between the inner conductor (radius a) and the

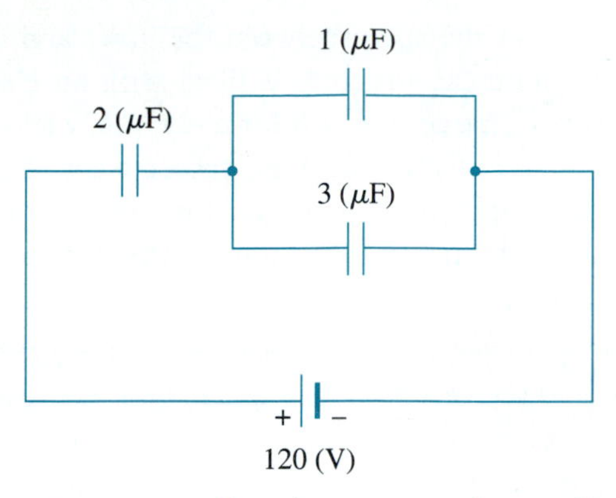

FIGURE 3-31 Capacitors across a battery (Problem P.3-24).

outer one (radius b) that carry charges $+Q$ and $-Q$, respectively. The permittivity of the insulating material is ϵ.

HINT: First assume the inner and outer radii to be a and $a + r$, respectively; then differentiate with respect to r.

P.3-27 A parallel-plate capacitor of width w, length L, and separation d has a solid dielectric slab of permittivity ϵ in the space between the plates. The capacitor is charged to a voltage V_0 by a battery, as indicated in Fig. 3-32. Assuming that the dielectric slab is withdrawn to the position shown and the switch is opened, determine the force acting on the slab.

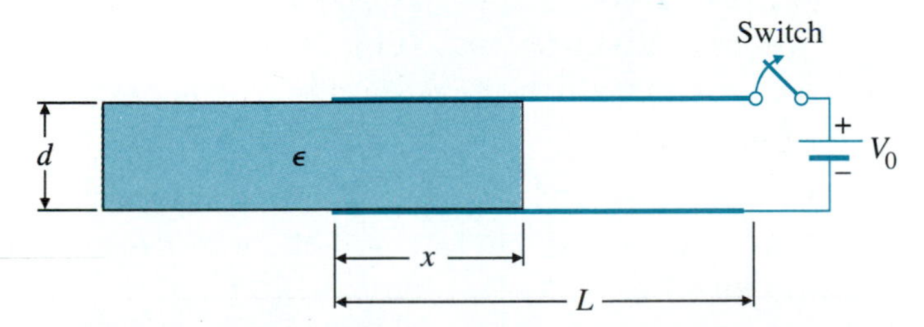

FIGURE 3-32 A partially filled parallel-plate capacitor (Problem P.3-27).

P.3-28 The upper and lower conducting plates of a large parallel-plate capacitor are separated by a distance d and maintained at potentials V_0 and 0, respectively. A dielectric slab of dielectric constant 6.0 and uniform thickness $0.8d$ is placed over the lower plate. Assuming negligible fringing effect, determine the following by solving Laplace's equation:

a) the potential and electric field distribution in the dielectric slab,

b) the potential and electric field distribution in the air space between the dielectric slab and the upper plate, and

c) the surface charge densities on the upper and lower plates.

P.3-29 Assume that the space between the inner and outer conductors of a long coaxial cylindrical structure is filled with an electron cloud having a volume density of charge $\rho_v = A/r$ for $a < r < b$, where a and b are, the radii of the inner and outer conductors, respectively. The inner conductor is maintained at a potential V_0, and the outer conductor is grounded. Determine the potential distribution in the region $a < r < b$ by solving Poisson's equation.

P.3-30 If the space between the inner and outer conductors of the coaxial structure in problem P.3-29 is free space, find the expression for $V(r)$ in the region $a \leqslant r \leqslant b$ by solving Laplace's equation. From $V(r)$, obtain the surface charge densities on the conductors and the capacitance per unit length of the structure. Compare your result with Eq. (3-90).

P.3-31 An infinite conducting cone of half-angle α is maintained at potential V_0 and insulated from a grounded conducting plane, as illustrated in Fig. 3-33. Determine

a) the potential distribution $V(\theta)$ in the region $\alpha < \theta < \pi/2$,

b) the electric field intensity in the region $\alpha < \theta < \pi/2$, and

c) the charge densities on the cone surface and on the grounded plane.

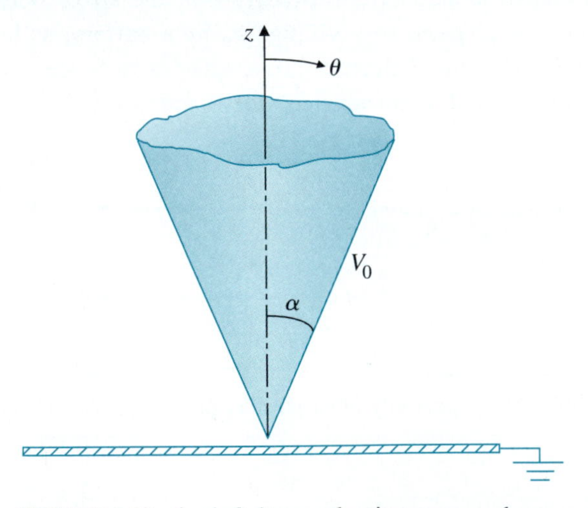

FIGURE 3-33 An infinite conducting cone and a grounded conducting plane (Problem P.3-31).

P.3-32 For a positive point charge Q located at a distance d from each of two grounded perpendicular conducting half-planes shown in Fig. 3-34, find the expressions for

a) the potential and the electric field intensity at an arbitrary point $P(x, y)$, and

b) the surface charge densities induced on the two half-planes.

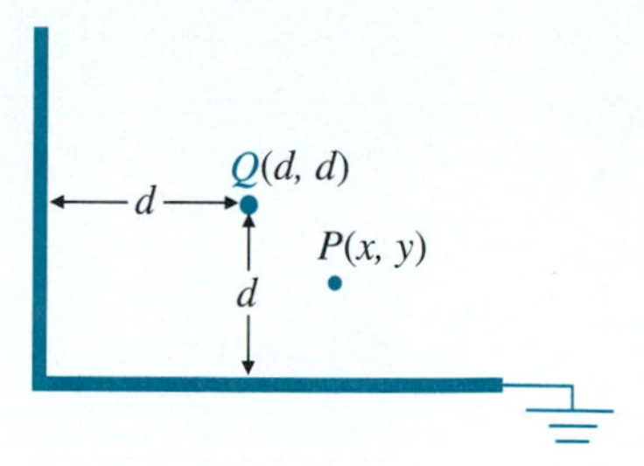

FIGURE 3-34. A point charge Q equidistant from two grounded perpendicular conducting half-planes (Problem P.3-32).

P.3-33 Determine the systems of image charges that will replace the conducting boundaries that are maintained at zero potential for

a) a point charge Q located between two large, grounded, parallel conducting planes as shown in Fig. 3-35(a), and

b) an infinite line charge ρ_ℓ located midway between two large, intersecting conducting planes forming a 60-degree angle, as shown in Fig. 3-35(b).

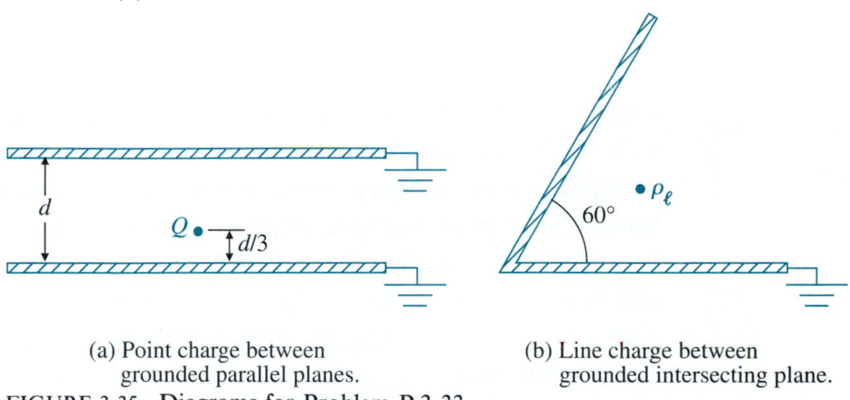

(a) Point charge between grounded parallel planes.

(b) Line charge between grounded intersecting plane.

FIGURE 3-35 Diagrams for Problem P.3-33.

P.3-34 An infinite line charge of 50 (nC/m) lies 3 (m) above the ground, which is at zero potential. Choosing the ground as the xy-plane and the line charge as parallel to the x-axis, find by the method of images the following:

a) **E** at (0, 4, 3), and

b) **E** and ρ_s at (0, 4, 0).

P.3-35 The axis of a long two-wire parallel transmission line are 2 (cm) apart. The wires have a radius 3 (mm) and are maintained at potentials +100 (V) and −100 (V). Find

a) the location of the equivalent line charges relative to the wire axes,

b) the equivalent line charge density of each wire, and

c) the electric field intensity at a point midway between the wires.

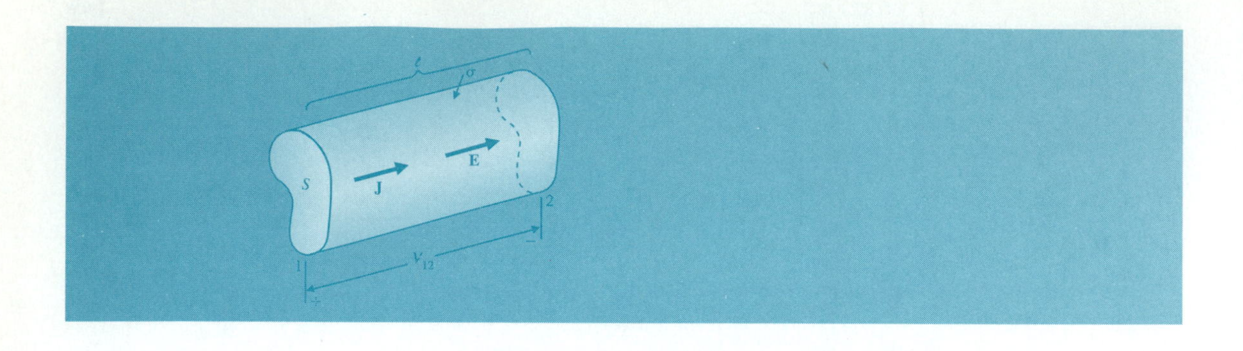

CHAPTER 4

Ohm's law

4-1 OVERVIEW In Chapter 3 we dealt with electrostatic problems, field problems associated with electric charges at rest. We will now consider charges in motion that constitute current flow. From d-c circuit theory you are familiar with problems of current flow in a conductive medium such as a metal wire. The governing relation in such cases is ***Ohm's law***, which states that the voltage across two terminals equals the product of the current and the resistance between the terminals. If the voltage is applied across a good insulator, little current will flow because of a very high resistance. How then do we explain the fact that current flows in a cathode-ray tube (such as that shown in Fig. 3-2), where the medium is a vacuum, an open circuit? Apparently Ohm's law does not apply in this case.

Two types of currents caused by motion of free charges

Convection electric currents are not governed by Ohm's law.

There are two types of electric current caused by the motion of free charges: *convection currents* and *conduction currents.* Convection currents are due to the motion of positively or negatively charged particles in a vacuum or rarefied gas. Familiar examples are electron beams in a cathode-ray tube and the violent motions of charged particles in a thunderstorm. Convection currents, the result of hydrodynamic motion involving a mass transport, are not governed by Ohm's law.

The mechanism of conduction currents is different from that of convection currents. In their normal state the atoms of a conductor occupy regular positions in a crystalline structure. The atoms consist of positively charged nuclei

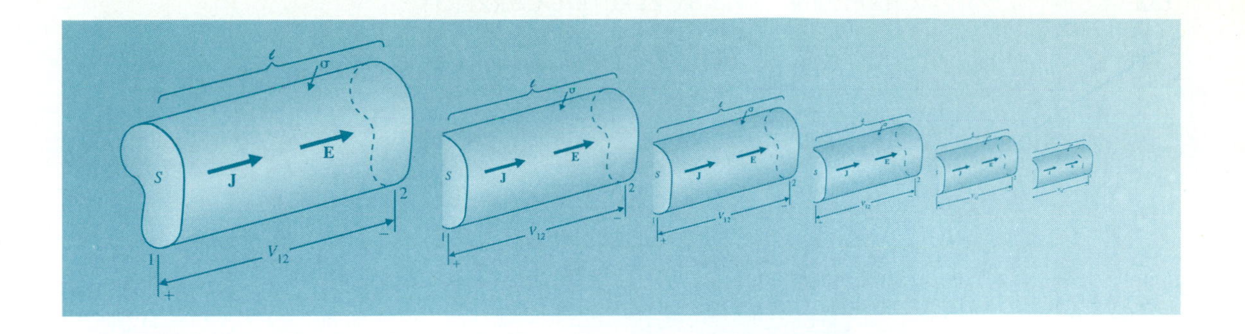

Steady Electric Currents

surrounded by electrons in a shell-like arrangement. The electrons in the inner shells are tightly bound to the nuclei and are not free to move away. The electrons in the outermost shells of a conductor atom do not completely fill the shells; they are valence or conduction electrons and are only very loosely bound to the nuclei. These latter electrons may wander from one atom to another in a random manner. The atoms, on the average, remain electrically neutral, and there is no net drift motion of electrons. When an external electric field is applied on a conductor, an organized motion of the conduction electrons will result, producing an electric current. The average drift velocity of the electrons is very low (on the order of 10^{-4} or 10^{-3} m/s) even for very good conductors because they collide with the atoms in the course of their motion, dissipating part of their kinetic energy as heat. This phenomenon manifests itself as a damping force, or resistance, to current flow. The relation between conduction current density and electric intensity gives us a point form of Ohm's law. Both types of currents will be discussed in this chapter.

Conduction electric currents are governed by Ohm's law.

4-2 Current Density and Ohm's Law

A) Convection current

Consider the steady motion of one kind of charge carriers, each of charge q (which is negative for electrons), across an element of surface Δs with a velocity $\mathbf{u}$, as shown in Fig. 4-1. If N is the number of charge

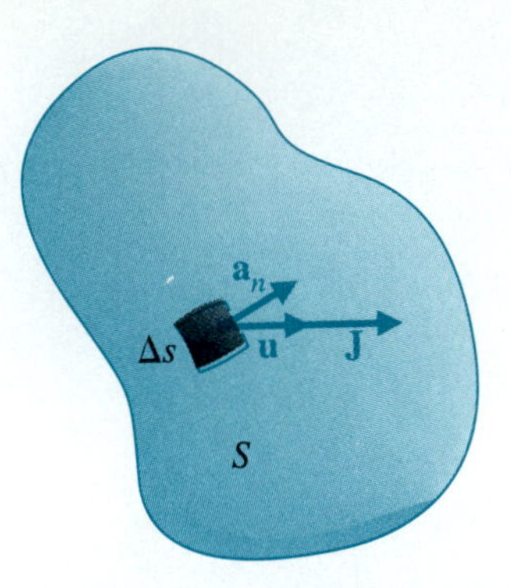

FIGURE 4-1 Conduction current due to drift motion of charge carriers across a surface.

carriers per unit volume, then in time Δt each charge carrier moves a distance $\mathbf{u}\Delta t$, and the amount of charge passing through the surface Δs is

$$\Delta Q = Nq\mathbf{u} \cdot \mathbf{a}_n \Delta s \Delta t \qquad \text{(C)}. \tag{4-1}$$

Since current is the time rate of change of charge, we have

$$\Delta I = \frac{\Delta Q}{\Delta t} = Nq\mathbf{u} \cdot \mathbf{a}_n \Delta s = Nq\mathbf{u} \cdot \Delta \mathbf{s} \qquad \text{(A)}. \tag{4-2}$$

Definition of current density

In Eq. (4-2) we have written $\Delta\mathbf{s} = \mathbf{a}_n \Delta s$ as a vector quantity. It is convenient to define a vector point function, ***volume current density***, or simply ***current density***, $\mathbf{J}$, in amperes per *square* meter,

$$\mathbf{J} = Nq\mathbf{u} \qquad (\text{A/m}^2), \tag{4-3}$$

so that Eq. (4-2) can be written as

$$\Delta I = \mathbf{J} \cdot \Delta \mathbf{s}. \tag{4-4}$$

The total current I flowing through an arbitrary surface S is then the flux of the $\mathbf{J}$ vector through S:

$$\boxed{I = \int_S \mathbf{J} \cdot d\mathbf{s} \qquad \text{(A)}.} \tag{4-5}$$

The product Nq is, in fact, free charge per unit volume, so we may rewrite Eq. (4-3) as

Relation between convection current density and velocity of charge carrier

$$\boxed{\mathbf{J} = \rho_v \mathbf{u} \qquad (\text{A/m}^2),} \tag{4-6}$$

which is the relation between the ***convection current density*** and the velocity of the charge carrier.

EXAMPLE 4-1

Assume a free charge density of -0.3 (nC/mm³) in a vacuum tube. For a current density of $-\mathbf{a}_z 2.4$ (A/mm²), find (a) the total current passing through a hemispherical cap specified by $R = 5$ (mm), $0 \leqslant \theta \leqslant \pi/2$, $0 \leqslant \phi \leqslant 2\pi$; and (b) the velocity of the free charges.

SOLUTION

Given:

$$\rho_v = -0.3 \text{ (nC/mm}^3\text{)},$$

$$\mathbf{J} = -\mathbf{a}_z 2.4 \text{ (A/mm}^2\text{)},$$

$$R = 5 \text{ (mm)}.$$

a)
$$I = \int \mathbf{J} \cdot d\mathbf{s} = -\int 2.4\,(\mathbf{a}_z \cdot \mathbf{a}_R)\,ds$$

$$= \int_0^{2\pi} \int_0^{\pi/2} -2.4\,(\cos\theta)(5^2 \sin\theta\, d\theta\, d\phi)$$

$$= -2\pi \int_0^{\pi/2} 60 \sin\theta\, d(\sin\theta)$$

$$= -120\pi \left(\frac{\sin^2\theta}{2}\right)_0^{\pi/2} = -60\pi = -188.5\,(\text{A}).$$

b)
$$\mathbf{u} = \frac{\mathbf{J}}{\rho_v} = \mathbf{a}_z \frac{-2.4}{-0.3 \times 10^{-9}} = \mathbf{a}_z 8 \times 10^9 \text{ (mm/s)}$$

$$= \mathbf{a}_z 8 \text{ (Mm/s)}.$$

B) Conduction current

In the case of conduction currents there may be more than one kind of charge carriers (electrons, holes, and ions) drifting with different velocities. Equation (4-3) should be generalized to read

$$\mathbf{J} = \sum_i N_i q_i \mathbf{u}_i \qquad (\text{A/m}^2). \tag{4-7}$$

As indicated in Section 4-1, conduction currents are the result of the drift motion of charge carriers under the influence of an applied electric field. The atoms remain neutral ($\rho_v = 0$). It can be justified analytically that, for most conducting materials, the average drift velocity is directly proportional to the electric field intensity. For metallic conductors we write

$$\mathbf{u}_e = -\mu_e \mathbf{E} \qquad (\text{m/s}), \tag{4-8}$$

where μ_e is the electron ***mobility*** measured in (m²/V·s). The electron mobility for copper is 3.2×10^{-3} (m²/V·s). It is 1.4×10^{-4} (m²/V·s) for aluminum and 5.2×10^{-3} (m^2/V·s) for silver. From Eqs. (4-3) and (4-8) we have

$$\mathbf{J} = -\rho_e \mu_e \mathbf{E}, \tag{4-9}$$

where $\rho_e = -Ne$ is the charge density of the drifting electrons and is a negative quantity. Equation (4-9) can be rewritten as

Point form of Ohm's law

$$\boxed{\mathbf{J} = \sigma \mathbf{E} \qquad (\text{A/m}^2),} \tag{4-10}$$

Definition of conductivity

where the proportionality constant, $\sigma = -\rho_e \mu_e$, is a macroscopic constitutive parameter of the medium called ***conductivity***. Equation (4-10) is the point form of ***Ohm's law***.

For semiconductors, conductivity depends on the concentration and mobility of both electrons and holes:

$$\sigma = -\rho_e \mu_e + \rho_h \mu_h, \tag{4-11}$$

where the subscript h denotes hole. In general, $\mu_e \neq \mu_h$. For germanium, typical values are $\mu_e = 0.38$, $\mu_h = 0.18$; for silicon, $\mu_e = 0.12$, $\mu_h = 0.03$ (m^2/V·s).

SI unit for conductivity

Equation (4-10) is a constitutive relation of a conducting medium. The unit for σ is ampere per volt-meter (A/V·m) or siemens per meter (S/m). Copper, the most commonly used conductor, has a conductivity 5.80×10^7 (S/m). On the other hand, the conductivity of germanium is around 2.2 (S/m), and that of silicon is 1.6×10^{-3} (S/m). The conductivity of semiconductors is highly dependent on (increases with) temperature. Hard rubber, a good insulator, has a conductivity of only 10^{-15} (S/m). Appendix B-4 lists the conductivities of some other frequently used materials. The reciprocal of conductivity is called ***resistivity***, in ohm-meters (Ω·m). We prefer to use conductivity; there is really no compelling need to use both conductivity and resistivity.

EXERCISE 4.1 For a current density of 7 (A/mm^2) in copper, find

a) the electric intensity, and
b) the electron drift velocity.

ANS. (a) 0.121 (V/m), (b) 3.57×10^{-4} (m/s).

We recall ***Ohm's law*** from circuit theory that the voltage V_{12} across a resistance R, in which a current I flows from point 1 to point 2, is equal to RI; that is,

$$V_{12} = RI. \tag{4-12}$$

Here R is usually a piece of conducting material of a given length; V_{12} is the voltage between two terminals 1 and 2; and I is the total current flowing from terminal 1 to terminal 2 through a finite cross section. Equation (4-12) is *not* a point relation.

We now use the point form of Ohm's law, Eq. (4-10), to derive the voltage–current relationship of a piece of homogeneous material of conductivity σ, length ℓ, and uniform cross section S, as shown in Fig. 4-2. Within the conducting material, $\mathbf{J} = \sigma\mathbf{E}$, where both $\mathbf{J}$ and $\mathbf{E}$ are in the direction of current flow. The potential difference or voltage between terminals 1 and 2 is

$$V_{12} = E\ell,$$

or

$$E = \frac{V_{12}}{\ell}. \tag{4-13}$$

The total current is

$$I = \int_S \mathbf{J}\cdot d\mathbf{s} = JS,$$

or

$$J = \frac{I}{S}. \tag{4-14}$$

Using Eqs. (4-13) and (4-14) in Eq. (4-10), we obtain

$$\frac{I}{S} = \sigma\frac{V_{12}}{\ell},$$

or

$$V_{12} = \left(\frac{\ell}{\sigma S}\right)I = RI, \tag{4-15}$$

FIGURE 4-2 Homogeneous conductor with a constant cross section.

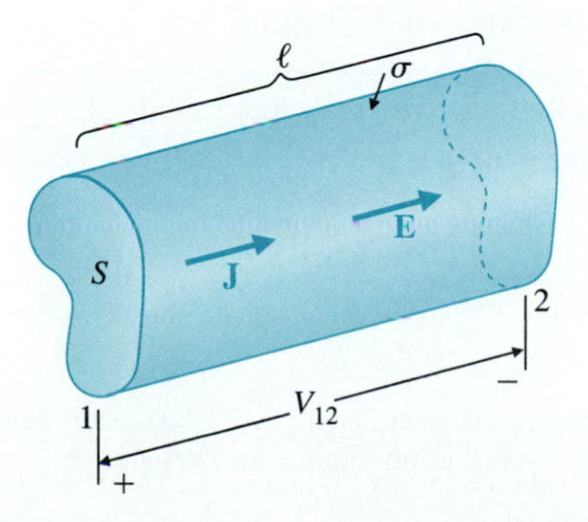

which is the same as Eq. (4-12). From Eq. (4-15) we have the formula for the ***resistance*** of a straight piece of homogeneous material of a uniform cross section for steady current (d.c.):

Resistance of a straight homogeneous material of uniform cross section

$$R = \frac{\ell}{\sigma S} \quad (\Omega). \tag{4-16}$$

EXAMPLE 4-2

a) Determine the d-c resistance of 1 (km) of copper wire having a 1-(mm) radius.

b) If an aluminium wire of the same length is to have the same resistance, what should its radius be?

SOLUTION

Since we are dealing with conductors of a uniform cross section, Eq. (4-16) applies.

a) For copper wire, $\sigma_{cu} = 5.80 \times 10^7$ (S/m):

$$\ell = 10^3\,(\text{m}), \qquad S_{cu} = \pi(10^{-3})^2 = 10^{-6}\pi \quad (\text{m}^2).$$

We have

$$R_{cu} = \frac{\ell}{\sigma_{cu} S_{cu}} = \frac{10^3}{5.80 \times 10^7 \times 10^{-6}\pi} = 5.49 \quad (\Omega).$$

b) For aluminum wire, $\sigma_{al} = 3.54 \times 10^7$ (S/m):

$$R_{al} = \frac{\ell}{\sigma_{al} S_{al}} = R_{cu},$$

$$S_{al} = \frac{\sigma_{cu}}{\sigma_{al}} S_{cu} = \frac{5.80}{3.54}(10^{-6}\pi) = \pi r^2,$$

$$r = 1.28 \times 10^{-3}\,(\text{m}), \text{ or } 1.28\,(\text{mm}).$$

Conductance, and its SI unit

The ***conductance***, G, or the reciprocal of resistance, is useful in combining resistances in parallel. The unit for conductance is (Ω^{-1}), or siemens (S).

$$G = \frac{1}{R} = \sigma \frac{S}{\ell} \quad (\text{S}). \tag{4-17}$$

■ **EXERCISE 4.2** Three resistors having resistances 1 (MΩ), 2 (MΩ), and 4 (MΩ) are connected in parallel. Calculate the overall conductance and resistance.

ANS. 1.75 (μS), 0.571 (MΩ).

REVIEW QUESTIONS

Q.4-1 Explain the difference between conduction current and convection current.

Q.4-2 What is the relation between convection current density and the velocity of charge carriers?

Q.4-3 Define *mobility* of the electron in a conductor. What is its SI unit?

Q.4-4 What is the point form for *Ohm's law*?

Q.4-5 Define *conductivity*. What is its SI unit?

Q.4-6 How does the resistance of a round conducting wire change if its radius is doubled?

REMARKS

1. Conduction currents are governed by Ohm's law; but convection currents are not.
2. Conductivity is not conductance, and resistivity is not resistance.
3. Resistance formula in Eq. (4-16) applies only to straight homogeneous material of a uniform cross section.

4-3 EQUATION OF CONTINUITY AND KIRCHHOFF'S CURRENT LAW

Principle of conservation of charges

The ***principle of conservation of charge*** is one of the fundamental postulates of physics. Electric charges may not be created or destroyed; all charges either at rest or in motion must be accounted for at all times. Consider an arbitrary volume V bounded by surface S. A net charge Q exists within this region. If a net current I flows across the surface *out* of this region, the charge in the volume must *decrease* at a rate that equals the current. Conversely, if a net current flows across the surface *into* the region, the charge in the volume must *increase* at a rate equal to the current. The current leaving the region is the total outward flux of the current density vector through the surface S. We have

$$I = \oint_S \mathbf{J} \cdot d\mathbf{s} = -\frac{dQ}{dt} = -\frac{d}{dt}\int_V \rho_v \, dv. \tag{4-18}$$

Divergence theorem, Eq. (2-69), may be invoked to convert the surface integral of $\mathbf{J}$ to the volume integral of $\nabla \cdot \mathbf{J}$. We obtain, for a stationary volume,

$$\int_V \nabla \cdot \mathbf{J} \, dv = -\int_V \frac{\partial \rho_v}{\partial t} \, dv. \tag{4-19}$$

In moving the time derivative of ρ_v inside the volume integral, it is necessary to use partial differentiation because ρ_v may be a function of time as well as of space coordinates. Since Eq. (4-19) must hold regardless of the choice of V, the integrands must be equal. Thus we have

Equation of continuity

$$\nabla \cdot \mathbf{J} = -\frac{\partial \rho_v}{\partial t} \qquad (\text{A/m}^3). \tag{4-20}$$

This point relationship derived from the principle of conservation of charge is called the ***equation of continuity***.

For steady currents, charge density does not vary with time, $\partial \rho_v / \partial t = 0$. Equation (4-20) becomes

$$\nabla \cdot \mathbf{J} = 0. \tag{4-21}$$

Steady electric current is solenoidal.

Thus, steady electric currents are divergenceless or solenoidal. Equation (4-21) is a point relationship and holds also at points where $\rho_v = 0$ (no flow source). It means that the field lines or streamlines of steady currents close upon themselves, unlike those of electrostatic field intensity that originate and end on charges. Over any enclosed surface, Eq. (4-21) leads to the following integral form:

$$\oint_S \mathbf{J} \cdot d\mathbf{s} = 0, \tag{4-22}$$

which can be written as

$$\sum_j I_j = 0 \qquad (\text{A}). \tag{4-23}$$

Kirchhoff's current law

Equation (4-23) is an expression of ***Kirchhoff's current law***. It states that ***the algebraic sum of all the currents flowing out of a junction in an electric circuit is zero***.

In Subsection 3-6.1 we stated that charges introduced in the interior of a conductor will move to the conductor surface and redistribute themselves in such a way as to make $\rho_v = 0$ and $\mathbf{E} = 0$ inside under equilibrium conditions. We are now in a position to prove this statement and to calculate the time it takes to reach an equilibrium. Combining Ohm's law, Eq. (4-10), with the equation of continuity and assuming a constant σ, we have

$$\sigma \nabla \cdot \mathbf{E} = -\frac{\partial \rho_v}{\partial t}. \tag{4-24}$$

In a simple medium, $\nabla \cdot \mathbf{E} = \rho_v / \epsilon$, and Eq. (4-24) becomes

$$\frac{\partial \rho_v}{\partial t} + \frac{\sigma}{\epsilon} \rho_v = 0. \tag{4-25}$$

The solution of Eq. (4-25) is

Volume density of charge decays exponentially with time.

$$\rho_v = \rho_0 e^{-(\sigma/\epsilon)t} \qquad (\text{C/m}^3), \tag{4-26}$$

where ρ_0 is the initial charge density at $t = 0$. Both ρ_v and ρ_0 can be functions of the space coordinates, and Eq. (4-26) says that the charge density at a given location will decrease with time exponentially. An initial charge density ρ_0 will decay to $1/e$ or 36.8% of its value in a time equal to

$$\tau = \frac{\epsilon}{\sigma} \qquad (\text{s}). \tag{4-27}$$

Definition of relaxation time

The time constant τ is called the ***relaxation time***. For a good conductor such as copper—$\sigma = 5.80 \times 10^7$ (S/m), $\epsilon \cong \epsilon_0 = 8.85 \times 10^{-12}$ (F/m)—τ equals 1.53×10^{-19} (s), a very short time indeed.

■ **EXERCISE 4.3**

Given that the dielectric constant and conductivity of rubber to be 3.0 and 10^{-15} (S/m) respectively, find

a) the relaxation time, and

b) the time required for a charge density to decay to 1% of its initial value.

ANS. (a) 7.38 hrs., (b) 1 day 10 hrs.

4-4 POWER DISSIPATION AND JOULE'S LAW

We indicated earlier that under the influence of an electric field, conduction electrons in a conductor undergo a drift motion macroscopically. Microscopically, these electrons collide with atoms on lattice sites. Energy is thus transmitted from the electric field to the atoms in thermal vibration. The work Δw done by an electric field $\mathbf{E}$ in moving a charge q a distance $\Delta \ell$ is $q\mathbf{E}\cdot(\Delta \ell)$, which corresponds to a power

$$p = \lim_{\Delta t \to 0} \frac{\Delta w}{\Delta t} = q\mathbf{E}\cdot\mathbf{u}, \tag{4-28}$$

where $\mathbf{u}$ is the drift velocity. The total power delivered to all the charge carriers in a volume dv is

$$dP = \sum_i p_i = \mathbf{E}\cdot\left(\sum_i N_i q_i \mathbf{u}_i\right) dv,$$

which, by virtue of Eq. (4-7), is

$$dP = \mathbf{E}\cdot\mathbf{J}\, dv,$$

or

$$\frac{dP}{dv} = \mathbf{E}\cdot\mathbf{J} \qquad (\text{W/m}^3). \tag{4-29}$$

Thus the point function $\mathbf{E}\cdot\mathbf{J}$ is a ***power density*** under steady-current conditions. For a given volume V the total electric power converted into heat is

Joule's law

$$P = \int_V \mathbf{E}\cdot\mathbf{J}\,dv \qquad \text{(W)}. \tag{4-30}$$

This is known as ***Joule's law***. (Note that the SI unit for P is watt, not joule, which is the unit for energy or work.) Equation (4-29) is the corresponding point relationship.

In a conductor of a constant cross section, $dv = ds\,d\ell$, with $d\ell$ measured in the direction $\mathbf{J}$. Equation (4-30) can be written as

$$P = \int_L E\,d\ell \int_S J\,ds = VI,$$

where I is the current in the conductor. Since $V = RI$, we have

$$P = I^2R \qquad \text{(W)}. \tag{4-31}$$

Equation (4-31) is, of course, the familiar expression for ohmic power, representing the heat dissipated in resistance R per unit time.

4-5 GOVERNING EQUATIONS FOR STEADY CURRENT DENSITY

As we see, the current density vector $\mathbf{J}$ is a basic quantity in the study of steady electric currents. According to Helmholtz's theorem the description of $\mathbf{J}$ requires the specification of its divergence and its curl. For steady currents, $\nabla\cdot\mathbf{J} = 0$, as in Eq. (4-21). The curl equation is obtained by combining Ohm's law ($\mathbf{J} = \sigma\mathbf{E}$) with $\nabla\times\mathbf{E} = 0$; that is, $\nabla\times(\mathbf{J}/\sigma) = 0$. The differential form and the corresponding integral form of the governing equations for steady current density are given below.

Governing Equations for Steady Current Density		
Differential Form	Integral Form	
$\nabla\cdot\mathbf{J} = 0$	$\oint_S \mathbf{J}\cdot d\mathbf{s} = 0$	(4-32)
$\nabla\times\left(\dfrac{\mathbf{J}}{\sigma}\right) = 0$	$\oint_C \dfrac{1}{\sigma}\mathbf{J}\cdot d\boldsymbol{\ell} = 0$	(4-33)

We recall from Section 3-8 that at an interface between two different media: (i) *a divergenceless field has a continuous normal component* (see Eq. 3-77); and (ii) *a curl-free field has a continuous tangential component* (see Eq. 3-72). The consequence of (i) and Eq. (4-32) is

Boundary condition for normal component of current density

$$J_{1n} = J_{2n} \qquad (\mathrm{A/m^2}). \tag{4-34}$$

At the boundary of two ohmic media with conductivities σ_1 and σ_2, the consequence of (ii) and Eq. (4-33) is

$$\frac{J_{1t}}{\sigma_1} = \frac{J_{2t}}{\sigma_2},$$

or

Boundary condition for tangential component of current density

$$\frac{J_{1t}}{J_{2t}} = \frac{\sigma_1}{\sigma_2}. \tag{4-35}$$

■ **EXERCISE 4.4** Two blocks of conducting material are in contact at the $z = 0$ plane. At a point P in the interface, the current density is $\mathbf{J}_1 = 10(\mathbf{a}_y 3 + \mathbf{a}_z 4)\,(\mathrm{A/m^2})$ in medium 1 (conductivity σ_1). Determine $\mathbf{J}_2$ at P in medium 2 if $\sigma_2 = 2\sigma_1$.

ANS. $20(\mathbf{a}_y 3 + \mathbf{a}_z 2)\ (\mathrm{A/m^2})$.

REVIEW QUESTIONS

Q.4-7 What is the physical significance of the *equation of continuity*?

Q.4-8 State *Kirchhoff's current law* in words.

Q.4-9 Define *relaxation time*. What is the order of magnitude of the relaxation time in copper?

Q.4-10 State Joule's law. Express the power dissipated in a volume (a) in terms of **E** and σ, and (b) in terms of **J** and σ.

Q.4-11 What are the boundary conditions of the normal and tangential components of steady current at the interface of two media with different conductivities?

REMARKS

1. Equations (4-24) and (4-26) hold only for simple media.
2. Steady current **J** is solenoidal.
3. Steady current density **J** is not conservative in inhomogeneous media.

4-6 RESISTANCE CALCULATIONS

In Section 3-9 we discussed the procedure for finding the capacitance between two conductors separated by a dielectric medium. These conductors may be of arbitrary shapes, as was shown in Fig. 3-19, which is reproduced here as Fig. 4-3. In terms of electric field quantities the basic formula for capacitance can be written as

$$C = \frac{Q}{V} = \frac{\oint_S \mathbf{D} \cdot d\mathbf{s}}{-\int_L \mathbf{E} \cdot d\boldsymbol{\ell}} = \frac{\oint_S \epsilon \mathbf{E} \cdot d\mathbf{s}}{-\int_L \mathbf{E} \cdot d\boldsymbol{\ell}}, \tag{4-36}$$

where the surface integral in the numerator is carried out over a surface enclosing the positive conductor and the line integral in the denominator is from the negative (lower-potential) conductor to the positive (higher-potential) conductor.

When the dielectric medium is lossy (having a small but nonzero conductivity), a current will flow from the positive to the negative conductor, and a current-density field will be established in the medium. Ohm's law, $\mathbf{J} = \sigma \mathbf{E}$, ensures that the streamlines for $\mathbf{J}$ and $\mathbf{E}$ will be the same in an isotropic medium. The resistance between the conductors is

$$R = \frac{V}{I} = \frac{-\int_L \mathbf{E} \cdot d\boldsymbol{\ell}}{\oint_S \mathbf{J} \cdot d\mathbf{s}} = \frac{-\int_L \mathbf{E} \cdot d\boldsymbol{\ell}}{\oint_S \sigma \mathbf{E} \cdot d\mathbf{s}}, \tag{4-37}$$

FIGURE 4-3 Two conductors in a lossy dielectric medium.

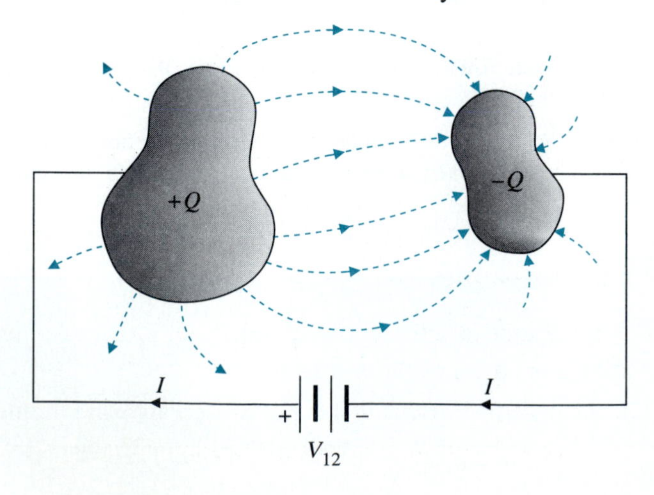

where the line and surface integrals are taken over the same L and S as those in Eq. (4-36). Comparison of Eqs. (4-36) and (4-37) shows the following interesting relationship:

Relation connecting *C* and *R* (or *G*) between two conductors

$$RC = \frac{C}{G} = \frac{\epsilon}{\sigma}. \tag{4-38}$$

Equation (4-38) holds if ϵ and σ of the medium have the same space dependence or if the medium is homogeneous (independent of space coordinates). In these cases, if the capacitance between two conductors is known, the resistance (or conductance) can be obtained directly from the ϵ/σ ratio without recomputation.

EXAMPLE 4-3

Find the leakage resistance per unit length

a) between the inner and outer conductors of a coaxial cable that has an inner conductor of radius a, an outer conductor of inner radius b, and a medium with conductivity σ, and

b) of a parallel-wire transmission line consisting of wires of radius a separated by a distance D in a medium with conductivity σ.

SOLUTION

a) The capacitance per unit length of a coaxial cable has been obtained as Eq. (3-90) in Example 3-16:

$$C_1 = \frac{2\pi\epsilon}{\ln(b/a)} \qquad \text{(F/m)}.$$

Hence the leakage resistance per unit length is, from Eq. (4-38),

$$R_1 = \frac{\epsilon}{\sigma}\left(\frac{1}{C_1}\right) = \frac{1}{2\pi\sigma}\ln\left(\frac{b}{a}\right) \qquad (\Omega\cdot\text{m}). \tag{4-39}$$

The conductance per unit length is $G_1 = 1/R_1$.

b) For the parallel-wire transmission line, Eq. (3-165) in Example 3-25 gives the capacitance per unit length:

$$C_1' = \frac{\pi\epsilon}{\cosh^{-1}\left(\dfrac{D}{2a}\right)} \qquad \text{(F/m)}.$$

Therefore if we use the relationship in Eq. (4-38), the leakage resistance per unit length is quickly found to be

$$R_1' = \frac{\epsilon}{\sigma}\left(\frac{1}{C_1'}\right) = \frac{1}{\pi\sigma}\cosh^{-1}\left(\frac{D}{2a}\right)$$

$$= \frac{1}{\pi\sigma}\ln\left[\frac{D}{2a} + \sqrt{\left(\frac{D}{2a}\right)^2 - 1}\right] \quad (\Omega\cdot\text{m}). \tag{4.40}$$

The conductance per unit length is $G_1' = 1/R_1'$.

In certain situations, electrostatic and steady current problems are not exactly analogous, even when the geometrical configurations are the same. This is because current flow can be confined strictly within a conductor (which has a *very large* σ in comparison to that of the surrounding medium), whereas electric flux usually cannot be contained within a dielectric slab of finite dimensions. The range of the dielectric constant of available materials is very limited (see Appendix B-3), and the flux-fringing around conductor edges makes the computation of capacitance less accurate.

The procedure for computing the resistance of a piece of conducting material between specified equipotential surfaces (or terminals) is as follows:

1. Choose an appropriate coordinate system for the given geometry.
2. Assume a potential difference V_0 between conductor terminals.
3. Find electric field intensity $\mathbf{E}$ within the conductor. (If the material is homogeneous, having a *constant* conductivity, the general method is to solve Laplace's equation $\nabla^2 V = 0$ for V in the chosen coordinate system, and then obtain $\mathbf{E} = -\nabla V$.)
4. Find the total current

   $$I = \int_S \mathbf{J}\cdot d\mathbf{s} = \int_S \sigma\mathbf{E}\cdot d\mathbf{s},$$

   where S is the cross-sectional area over which I flows.
5. Find resistance R by taking the ratio V_0/I.

EXAMPLE 4-4

A conducting material of uniform thickness h and conductivity σ has the shape of a quarter of a flat circular washer, with inner radius a and outer radius b, as shown in Fig. 4-4. Determine the resistance between the end faces.

SOLUTION

Obviously, the appropriate coordinate system to use for this problem is the cylindrical coordinate system. Following the foregoing procedure, we first

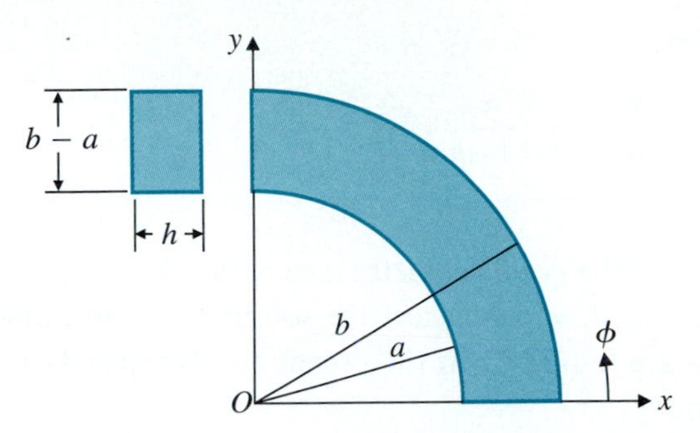

FIGURE 4-4 A quarter of a flat conducting circular washer (Example 4-4).

assume a potential difference V_0 between the end faces, say $V = 0$ on the end face at $y = 0\ (\phi = 0)$ and $V = V_0$ on the end face at $x = 0\ (\phi = \pi/2)$. We are to solve Laplace's equation in V subject to the following boundary conditions:

$$V = 0 \quad \text{at} \quad \phi = 0, \tag{4-41a}$$

$$V = V_0 \quad \text{at} \quad \phi = \pi/2. \tag{4-41b}$$

Since potential V is a function of ϕ only, Laplace's equation in cylindrical coordinates simplifies to

$$\frac{d^2V}{d\phi^2} = 0. \tag{4-42}$$

The general solution of Eq. (4-42) is

$$V = c_1\phi + c_2,$$

which, upon using the boundary conditions in Eqs. (4-41a) and (4-41b), becomes

$$V = \frac{2V_0}{\pi}\phi. \tag{4-43}$$

The current density is

$$\begin{aligned}\mathbf{J} &= \sigma\mathbf{E} = -\sigma\nabla V \\ &= -\mathbf{a}_\phi\sigma\frac{\partial V}{r\partial\phi} = -\mathbf{a}_\phi\frac{2\sigma V_0}{\pi r}.\end{aligned} \tag{4-44}$$

The total current I can be found by integrating $\mathbf{J}$ over the $\phi = \pi/2$ surface at which $d\mathbf{s} = -\mathbf{a}_\phi h\,dr$. We have

$$\begin{aligned}I &= \int_S \mathbf{J}\cdot d\mathbf{s} = \frac{2\sigma V_0}{\pi}h\int_a^b \frac{dr}{r} \\ &= \frac{2\sigma h V_0}{\pi}\ln\frac{b}{a}.\end{aligned} \tag{4-45}$$

Therefore,

$$R = \frac{V_0}{I} = \frac{\pi}{2\sigma h \ln (b/a)}. \tag{4-46}$$

When the given geometry is such that **J** can be determined easily from a total current I, we may start the solution by assuming an I. From I, **J** and $\mathbf{E} = \mathbf{J}/\sigma$ are found. Then the potential difference V_0 is determined from the relation

$$V_0 = -\int \mathbf{E} \cdot d\boldsymbol{\ell},$$

where the integration is from the low-potential terminal to the high-potential terminal. The resistance $R = V_0/I$ is independent of the assumed I, which will be canceled in the process.

EXERCISE 4.5 The radii of the inner and outer conductors of a coaxial cable are a and b respectively, and the medium in-between has a conductivity σ. Find the leakage resistance between the conductors per unit length by first assuming a leakage current I from the inner to the outer conductor, then determining **J**, **E**, V_0, and $R_1 = V_0/I$. Check your result with Eq. (4-39).

REVIEW QUESTIONS

Q.4-12 What is the relation between the conductance and the capacitance formed by two conductors immersed in a lossy dielectric medium that has permittivity ϵ and conductivity σ?

Q.4-13 What is the relation between the resistance and the capacitance formed by two conductors immersed in a lossy dielectric that has dielectric constant ϵ_r and conductivity σ?

REMARKS

The total leakage resistance *between* two parallel conductors of length ℓ is equal to the leakage resistance per unit length *divided by* (not multiplied by) ℓ.

SUMMARY

In this chapter we

- considered two types of steady electric currents: convection currents (not governed by Ohm's law) and conduction currents,
- defined conductivity that led to the point form of Ohm's law,

- introduced the equation of continuity and the concept of relaxation time,
- studied Joule's law and power dissipation,
- obtained the governing equations for steady current density,
- examined the boundary conditions for current density, and
- discussed methods for resistance calculations.

PROBLEMS

P.4-1 A d-c voltage of 6 (V) applied to the ends of 1 (km) of a conducting wire of 0.5 (mm) radius results in a current of 1/6 (A). Find

a) the conductivity of the wire,
b) the electric field intensity in the wire,
c) the power dissipated in the wire,
d) the electron drift velocity, assuming electron mobility in the wire to be 1.4×10^{-3} ($\text{m}^2/\text{V} \cdot \text{s}$).

P.4-2 A long, round wire of radius a and conductivity σ is coated with a material of conductivity 0.1σ.

a) What must be the thickness of the coating so that the resistance per unit length of the uncoated wire is reduced by 50%?
b) Assuming a total current I in the coated wire, find **J** and **E** in both the core and the coating material.

P.4-3 Lightning strikes a lossy dielectric sphere—$\epsilon = 1.2\epsilon_0$, $\sigma = 10$ (S/m)—of radius 0.1 (m) at time $t = 0$, depositing uniformly in the sphere a total charge 1 (mC). Determine, for all t,

a) the electric field intensity both inside and outside the sphere,
b) the current density in the sphere.

P.4-4 Refer to Problem P.4-3.

a) Calculate the time it takes for the charge density in the sphere to diminish to 1% of its initial value.
b) Calculate the change in the electrostatic energy stored in the sphere as the charge density diminishes from the initial value to 1% of its value. What happens to this energy?
c) Determine the electrostatic energy stored in the space outside the sphere. Does this energy change with time?

P.4-5 Find the current and the heat dissipated in each of the five resistors in the network shown in Fig. 4-5 if

$$R_1 = \tfrac{1}{3}(\Omega), \quad R_2 = 20(\Omega), \quad R_3 = 30(\Omega), \quad R_4 = 8(\Omega), \quad R_5 = 10(\Omega),$$

and if the source is an ideal d-c voltage generator of 0.7 (V) with its positive polarity at terminal 1. What is the total resistance seen by the source at terminal pair 1–2?

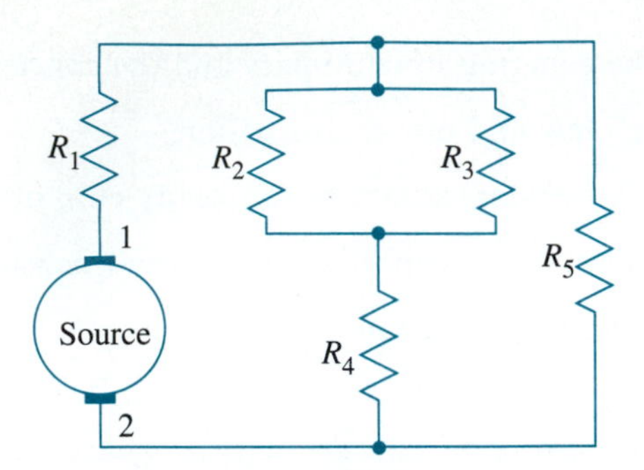

FIGURE 4-5 A network problem (Problem P.4-5).

P.4-6 Two lossy homogeneous dielectric media with dielectric constants $\epsilon_{r1} = 2$, $\epsilon_{r2} = 3$ and conductivities $\sigma_1 = 15\,(\text{mS})$, $\sigma_2 = 10\,(\text{ms})$ are in contact at the $z = 0$ plane. In the $z > 0$ region (medium 1) a uniform electric field $\mathbf{E}_1 = \mathbf{a}_x 20 - \mathbf{a}_z 50$ (V/m) exists. Find (a) $\mathbf{E}_2$ in medium 2, (b) $\mathbf{J}_1$ and $\mathbf{J}_2$, (c) the angles that $\mathbf{J}_1$ and $\mathbf{J}_2$ make with the $z = 0$ plane, and (d) the surface charge density at the interface.

P.4-7 The space between two parallel conducting plates each having an area S is filled with an inhomogeneous ohmic medium whose conductivity varies linearly from σ_1 at one plate ($y = 0$) to σ_2 at the other plate ($y = d$). A d-c voltage V_0 is applied across the plates. Determine

a) the total resistance between the plates, and

b) the surface charge densities on the plates.

P.4-8 A d-c voltage V_0 is applied across a parallel-plate capacitor of area S. The space between the conducting plates is filled with two different lossy dielectrics of thicknesses d_1 and d_2, permittivities ϵ_1 and ϵ_2, and conductivities σ_1 and σ_2, respectively, as shown in Fig. 4-6. Determine

a) the current density between the plates,

b) the electric field intensities in both dielectrics, and

c) the equivalent R-C circuit between terminals a and b.

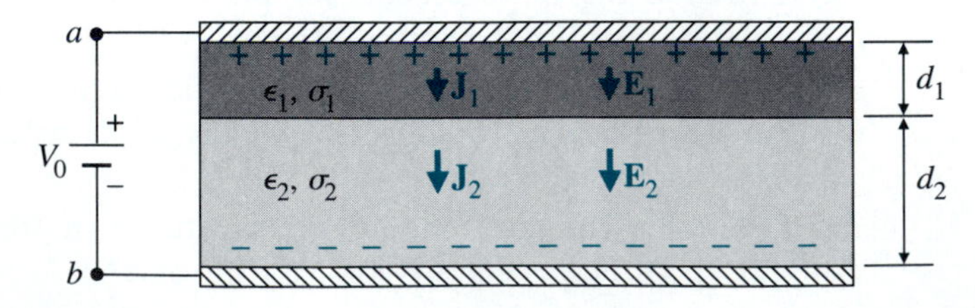

FIGURE 4-6 Parallel-plate capacitor with two lossy dielectrics (Problem P.4-8).

P.4-9 A d-c voltage V_0 is applied across a cylindrical capacitor of length L. The radii of the inner and outer conductors are a and b, respectively. The space between the conductors is filled with two different lossy dielectrics having, respectively, permittivity ϵ_1 and conductivity σ_1 in the region $a < r < c$, and permittivity ϵ_2 and conductivity σ_2 in the region $c < r < b$. Determine

a) the equivalent R-C circuit between the inner and outer conductors, and

b) the current density in each region.

HINT: Use the results in Example 4-3(a).

P.4-10 Refer to the flat conducting quarter-circular washer in Example 4-4 and Fig. 4-4. Find the resistance between the top and bottom flat faces.

P.4-11 Refer to the flat conducting quarter-circular washer in Example 4-4 and Fig. 4-4. Find the resistance between the curved sides.

P.4-12 Find the resistance between two concentric spherical surfaces of radii R_1 and R_2 $(R_1 < R_2)$ if the space between the surfaces is filled with a homogeneous and isotropic material having a conductivity σ.

CHAPTER 5

5-1 OVERVIEW Earlier we discussed the interaction between electric charges at rest by introducing the concept of *electric field*. We saw in Chapter 3 that electric field intensity **E** is the only fundamental vector field quantity required for the study of electrostatics in free space. In a material medium it is convenient to define a second vector field quantity, the electric flux density (or electric displacement) **D**, to account for the effect of polarization. The following two equations form the basis of the electrostatic model:

Two basic equations of the electrostatic model

$$\nabla \cdot \mathbf{D} = \rho_v, \tag{5-1}$$

$$\nabla \times \mathbf{E} = 0. \tag{5-2}$$

The electrical property of the medium determines the relation between **D** and **E**. If the medium is linear and isotropic, we have the simple constitutive relation $\mathbf{D} = \epsilon \mathbf{E}$, where the permittivity ϵ is a scalar.

The phenomenon of magnetism was discovered when pieces of magnetic lodestone were found to exhibit a mysterious attractive power. Since the lodestone pieces were found near the ancient Greek city called *Magnesia*, the derived terms *magnet*, *magnetism*, *magnetization*, and *magnetron* have come into use. We study magnetism by introducing the concept of *magnetic field*. A magnetic field can be caused by a permanent magnet (like the magnetized lodstone), by moving charges, or by a current flow.

Static Magnetic Fields

When a small test charge q is placed in an electric field $\mathbf{E}$, it experiences an ***electric force*** $\mathbf{F}_e$, which is a function of the position of q.

Electric intensity defined in terms of force on a stationary charge

$$\boxed{\mathbf{F}_e = q\mathbf{E} \qquad (\text{N}).} \tag{5-3}$$

When the test charge is in motion in a magnetic field characterized by a ***magnetic flux density*** $\mathbf{B}$,† experiments show that charge q also experiences a ***magnetic force*** $\mathbf{F}_m$ given by

Magnetic flux density defined in terms of force experienced by a moving charge

$$\boxed{\mathbf{F}_m = q\mathbf{u} \times \mathbf{B} \qquad (\text{N}),} \tag{5-4}$$

SI unit for B

where $\mathbf{u}$ (m/s) is the velocity of the moving charge, and $\mathbf{B}$ is measured in webers per square meter (Wb/m^2) or teslas (T).‡ The total ***electromagnetic force*** on a charge q is, then, $\mathbf{F} = \mathbf{F}_e + \mathbf{F}_m$; that is,

Lorentz's force equation

$$\boxed{\mathbf{F} = q(\mathbf{E} + \mathbf{u} \times \mathbf{B}) \qquad (\text{N}),} \tag{5-5}$$

†Magnetic flux density is also called ***magnetic induction***, mainly in physics texts.

‡One weber per square meter or one tesla equals 10^4 gauss in CGS units. The earth's magnetic field is about $\frac{1}{2}$ gauss or 0.5×10^{-4}T. (A weber is the same as a volt-second.)

which is called ***Lorentz's force equation***. Its validity has been unquestionably established by experiments. We may consider $\mathbf{F}_e/q$ for a small q as the definition for electric field intensity $\mathbf{E}$ (as we did in Eq. 3-1), and $\mathbf{F}_m/q = \mathbf{u} \times \mathbf{B}$ as the defining relation for magnetic flux density $\mathbf{B}$. Alternatively, we may consider Lorentz's force equation as a fundamental postulate of our electromagnetic model; it cannot be derived from other postulates.

Charges in motion produce a current that, in turn, creates a magnetic field. Steady currents are accompanied by static magnetic fields, which are the subject of study of this chapter. We begin the study of static magnetic fields in free space by two postulates specifying the divergence and the curl of **B**. From the solenoidal character of **B** a vector magnetic potential is defined, which is shown to obey a vector Poisson's equation. Next we derive the Biot-Savart law, which can be used to determine the magnetic field of a current-carrying circuit. The postulated curl relation leads directly to Ampère's circuital law, which is particularly useful when symmetry exists.

The macroscopic effect of magnetic materials in a magnetic field can be studied by defining a magnetization vector. Here we introduce another vector field quantity, the magnetic field intensity **H**. From the relation between **B** and **H** we define the permeability of the material, and discuss the behavior of magnetic materials. We then examine the boundary conditions of **B** and **H** at the interface of two different magnetic media; define self- and mutual inductances; and discuss magnetic energy, forces, and torques.

5-2 FUNDAMENTAL POSTULATES OF MAGNETOSTATICS IN FREE SPACE

To study magnetostatics (steady magnetic fields) in free space or in nonmagnetic material,† we need only consider the magnetic flux density vector, **B**. *The two fundamental postulates of magnetostatics that specify the divergence and the curl of* **B** *in nonmagnetic media are*

Divergence of B vanishes

$$\nabla \cdot \mathbf{B} = 0, \tag{5-6}$$

and

Curl of static B in nonmagnetic media

$$\nabla \times \mathbf{B} = \mu_0 \mathbf{J}. \quad \text{(in nonmagnetic media)} \tag{5-7}$$

†Except for ferromagnetic materials (nickel, cobalt, iron, or their alloys), the permeability of substances is very close (within 0.01%) to μ_0 for free space. (See table in Appendix B-5.) In this book, for simplicity, when we deal with magnetic fields in nonferromagnetic materials such as air, water, copper, and aluminum, we will consider them to be in free space.

In Eq. (5-7) μ_0 is the permeability of free space:

$$\mu_0 = 4\pi \times 10^{-7} \quad \text{(H/m)}$$

(see Eq. 1-9), and $\mathbf{J}$ is the current density (A/m^2). Since the divergence of the curl of any vector field is zero, we obtain from Eq. (5-7)

$$\nabla \cdot \mathbf{J} = 0, \tag{5-8}$$

which is consistent with Eq. (4-21) for steady currents.

Taking the volume integral of Eq. (5-6) and applying the divergence theorem, we have

There are no magnetic flow sources (no isolated magnetic charges).

$$\boxed{\oint_S \mathbf{B} \cdot d\mathbf{s} = 0,} \tag{5-9}$$

where the surface integral is carried out over the bounding surface of an arbitrary volume. Comparison of Eq. (5-9) with Eq. (3-6) leads us to conclude that there is no magnetic analogue for electric charges. ***There are no magnetic flow sources, and the magnetic flux lines always close upon themselves.*** Equation (5-9) is also referred to as an expression for ***the law of conservation of magnetic flux*** because it states that the total outward magnetic flux through any closed surface is zero.

Net magnetic flux flowing out of any closed surface is zero.

The traditional designation of north and south poles in a permanent bar magnet does not imply that an isolated positive magnetic charge exists at the north pole and a corresponding amount of isolated negative magnetic charge exists at the south pole. Consider the bar magnet with north and south poles in Fig. 5-1(a). If this magnet is cut into two segments, new south and

FIGURE 5-1 Successive division of a bar magnet.

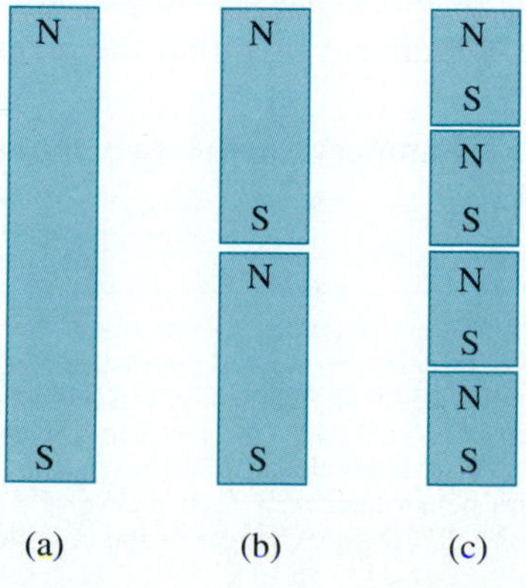

(a)	(b)	(c)
N … S	N … S / N … S	N S / N S / N S / N S

north poles appear, and we have two shorter magnets as in Fig. 5-1(b). If each of the two shorter magnets is cut again into two segments, we have four magnets, each with a north pole and a south pole as in Fig. 5-1(c). This process could be continued until the magnets are of atomic dimensions; but each infinitesimally small magnet would still have a north pole and a south pole. Obviously, then, magnetic poles cannot be isolated. The magnetic flux lines follow closed paths from one end of a magnet to the other end outside the magnet and then continue inside the magnet back to the first end. The designation of north and south poles is in accordance with the fact that the respective ends of a bar magnet freely suspended in the earth's magnetic field will seek the north and south directions.[†]

North and south magnetic poles cannot be isolated.

The integral form of the curl relation in Eq. (5-7) can be obtained by integrating both sides over an open surface and applying Stokes's theorem. We have

$$\int_S (\nabla \times \mathbf{B}) \cdot d\mathbf{s} = \mu_0 \int_S \mathbf{J} \cdot d\mathbf{s},$$

or

$$\boxed{\oint_C \mathbf{B} \cdot d\boldsymbol{\ell} = \mu_0 I,} \qquad \text{(in nonmagnetic media)} \tag{5-10}$$

where the path C for the line integral is the contour bounding the surface S, and I is the total current through S. The sense of tracing C and the direction of current flow follow the right-hand rule. Note that Eq. (5-10) is a derived relation from the curl postulate for $\mathbf{B}$. It is a form of ***Ampère's circuital law***, which states that ***the circulation of the magnetic flux density in a nonmagnetic medium around any closed path is equal to μ_0 times the total current flowing through the surface bounded by the path.*** Ampère's circuital law is very useful in determining the magnetic flux density $\mathbf{B}$ caused by a current I when there is a closed path C around the current such that the magnitude of $\mathbf{B}$ is constant over the path.

Ampère's circuital law in nonmagnetic media

The following is a summary of the two fundamental postulates of magnetostatics in free space:

[†]We note here parenthetically that examination of some prehistoric rock formations has led to the belief that there have been dramatic reversals of the earth's magnetic field every ten million years or so. The earth's magnetic field is thought to be produced by the rolling motions of the molten iron in the earth's outer core, but the exact reasons for the field reversals are still not well understood. The next such reversal is predicted to be only about 2000 years from now. One cannot conjecture all the dire consequences of such a reversal, but among them would be disruptions in global navigation and drastic changes in the migratory patterns of birds.

Two fundamental postulates of magnetostatics in nonmagnetic media

Postulates of Magnetostatics in nonmagnetic media	
Differential Form	Integral Form
$\nabla \cdot \mathbf{B} = 0$	$\oint_S \mathbf{B} \cdot d\mathbf{s} = 0$
$\nabla \times \mathbf{B} = \mu_0 \mathbf{J}$	$\oint_C \mathbf{B} \cdot d\boldsymbol{\ell} = \mu_0 I$

EXAMPLE 5-1

An infinitely long, straight, solid, nonmagnetic conductor with a circular cross section of radius b carries a steady current I. Determine the magnetic flux density both inside and outside the conductor.

SOLUTION

First we note that this is a problem with cylindrical symmetry and that Ampère's circuital law can be used to advantage. If we align the conductor along the z-axis, the magnetic flux density $\mathbf{B}$ will be ϕ-directed and will be constant along any circular path around the z-axis. Figure 5-2(a) shows a cross section of the conductor and the two circular paths of integration, C_1 and C_2, inside and outside, respectively, the current-carrying conductor. Note again that the directions of C_1 and C_2 and the direction of I follow the right-hand rule. (When the fingers of the right hand follow the directions of C_1 and C_2, the thumb of the right hand points to the direction of I.)

a) *Inside the conductor:*

$$\mathbf{B}_1 = \mathbf{a}_\phi B_{\phi 1}, \qquad d\boldsymbol{\ell} = \mathbf{a}_\phi r_1\, d\phi$$

$$\oint_{C_1} \mathbf{B}_1 \cdot d\boldsymbol{\ell} = \int_0^{2\pi} B_{\phi 1} r_1\, d\phi = 2\pi r_1 B_{\phi 1}.$$

The current through the area enclosed by C_1 is

$$I_1 = \frac{\pi r_1^2}{\pi b^2} I = \left(\frac{r_1}{b}\right)^2 I.$$

Therefore, from Ampère's circuital law,

$$\mathbf{B}_1 = \mathbf{a}_\phi B_{\phi 1} = \mathbf{a}_\phi \frac{\mu_0 r_1 I}{2\pi b^2}, \qquad r_1 \le b. \tag{5-11}$$

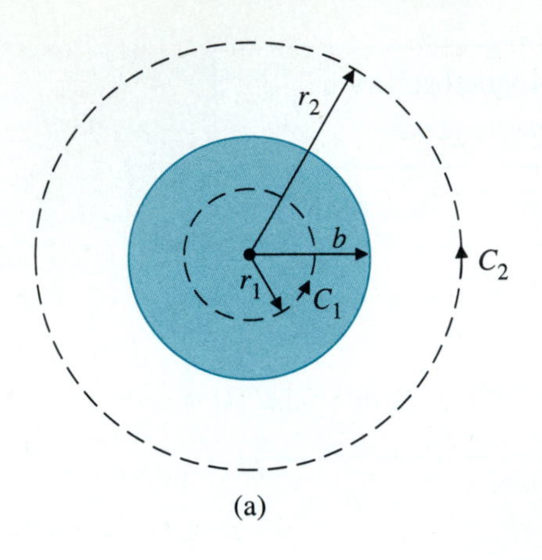

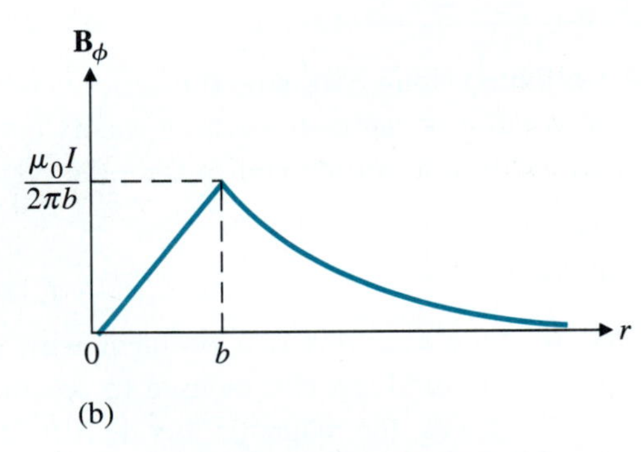

FIGURE 5-2 Magnetic flux density of an infinitely long circular conductor carrying a current I out of paper (Example 5-1).

b) *Outside the conductor:*

$$\mathbf{B}_2 = \mathbf{a}_\phi B_{\phi 2}, \qquad d\boldsymbol{\ell} = \mathbf{a}_\phi r_2 \, d\phi$$

$$\oint_{C_2} \mathbf{B}_2 \cdot d\boldsymbol{\ell} = 2\pi r_2 B_{\phi 2}.$$

Path C_2 outside the conductor encloses the total current I. Hence

$$\mathbf{B}_2 = \mathbf{a}_\phi B_{\phi 2} = \mathbf{a}_\phi \frac{\mu_0 I}{2\pi r_2}, \qquad r_2 \geq b. \tag{5-12}$$

Examination of Eqs. (5-11) and (5-12) reveals that the magnitude of $\mathbf{B}$ increases linearly with r_1 from 0 until $r_1 = b$, after which it decreases inversely with r_2. The variation of B_ϕ versus r is sketched in Fig. 5-2(b).

■ **EXERCISE 5.1** An infinitely long, very thin cylindrical conducting tube of radius b carries a uniform surface current $\mathbf{J}_s = \mathbf{a}_z J_s$ (A/m). Find $\mathbf{B}$ everywhere.

ANS. 0 for $r < b$, $\mathbf{a}_\phi \mu_0 J_s b/r$ for $r > b$.

EXAMPLE 5-2

Determine the magnetic flux density inside a closely wound toroidal coil with an air core having N turns of coil and carrying a current I. The toroid has a mean radius b, and the radius of each turn is a.

SOLUTION

Figure 5-3 depicts the geometry of this problem. Cylindrical symmetry ensures that $\mathbf{B}$ has only a ϕ-component and is constant along any circular path about the axis of the toroid. We construct a circular contour C with radius r as shown. For $(b-a)<r<b+a$, Eq. (5-10) leads directly to

$$\oint \mathbf{B}\cdot d\boldsymbol{\ell} = 2\pi r B_\phi = \mu_0 N I,$$

where we have assumed that the toroid has an air core with permeability μ_0. Therefore,

$$\mathbf{B} = \mathbf{a}_\phi B_\phi = \mathbf{a}_\phi \frac{\mu_0 N I}{2\pi r}, \qquad (b-a) < r < (b+a). \tag{5-13}$$

$\mathbf{B} = 0$ for $r < (b-a)$ and $r > (b+a)$, because the net total current enclosed by a contour constructed in these two regions is zero.

EXERCISE 5.2 Find the magnetic flux density inside a very long cylindrical solenoid with an air core having n turns per meter and carrying a current I.

ANS. $\mu_0 n I$.

FIGURE 5-3 A current-carrying toroidal coil (Example 5-2).

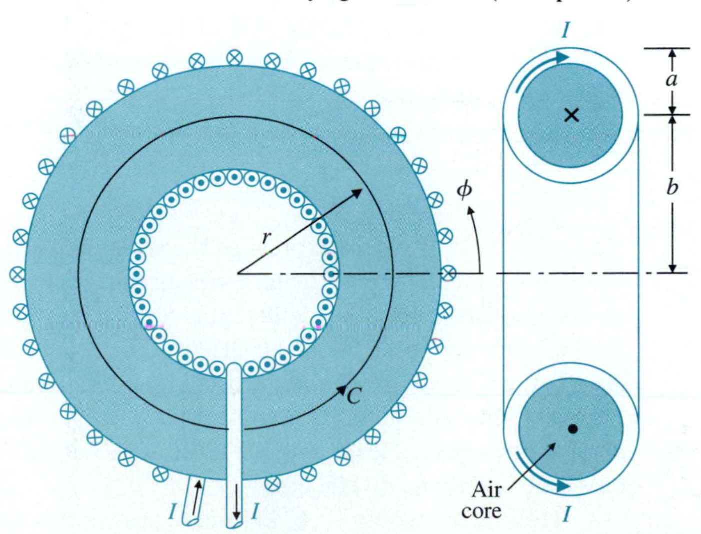

REVIEW QUESTIONS

Q.5-1 What is the expression for the force on a test charge q that moves with velocity $\mathbf{u}$ in a magnetic field of flux density $\mathbf{B}$?

Q.5-2 Verify that tesla (T), the unit for magnetic flux density, is the same as volt-second per square meter ($\text{V} \cdot \text{s/m}^2$).

Q.5-3 Write Lorentz's force equation.

Q.5-4 What are the two fundamental postulates of magnetostatics?

Q.5-5 Which postulate of magnetostatics denies the existence of isolated magnetic charges?

Q.5-6 State the law of conservation of magnetic flux.

Q.5-7 State Ampère's circuital law.

Q.5-8 In what manner does the **B**-field of an infinitely long straight filament carrying a direct current I vary with distance?

REMARKS

1. The magnetic force on charge q moving with a velocity $\mathbf{u}$ in a magnetic field $\mathbf{B}$ is perpendicular to both $\mathbf{u}$ and $\mathbf{B}$; there is no force on q if $\mathbf{u}$ is parallel to $\mathbf{B}$.
2. There are no isolated magnetic charges.
3. Magnetic field is solenoidal, and magnetic flux lines always close upon themselves.

5-3 VECTOR MAGNETIC POTENTIAL

The divergence-free postulate of $\mathbf{B}$ in Eq. (5-6), $\nabla \cdot \mathbf{B} = 0$, assures that $\mathbf{B}$ is solenoidal. As a consequence, $\mathbf{B}$ can be expressed as the curl of another vector field, say $\mathbf{A}$, such that

Partial definition of vector magnetic potential A

$$\boxed{\mathbf{B} = \nabla \times \mathbf{A} \qquad \text{(T).}} \tag{5-14}$$

SI unit for A

The vector field A so defined is called the ***vector magnetic potential.*** Its SI unit is weber per meter (Wb/m). Thus if we can find $\mathbf{A}$ of a current distribution, $\mathbf{B}$ can be obtained from $\mathbf{A}$ by a differential (curl) operation. This is quite similar to the introduction of the scalar electric potential V for the curl-free $\mathbf{E}$ in electrostatics (Section 3-5) and the obtaining of $\mathbf{E}$ from the relation $\mathbf{E} = -\nabla V$. However, the definition of a vector requires the specification of both its curl and its divergence. Hence Eq. (5-14) alone is not sufficient to define $\mathbf{A}$; we must still specify its divergence.

How do we choose $\nabla \cdot \mathbf{A}$? Before we answer this question, let us take the curl of $\mathbf{B}$ in Eq. (5-14) and substitute it in Eq. (5-7). We have

$$\nabla \times \nabla \times \mathbf{A} = \mu_0 \mathbf{J}. \tag{5-15}$$

Here we digress to introduce a formula for the curl curl of a vector.†

$$\nabla \times \nabla \times \mathbf{A} = \nabla(\nabla \cdot \mathbf{A}) - \nabla^2 \mathbf{A}, \tag{5-16a}$$

or

$$\nabla^2 \mathbf{A} = \nabla(\nabla \cdot \mathbf{A}) - \nabla \times \nabla \times \mathbf{A}. \tag{5-16b}$$

Equations (5-16a) or (5-16b) can be regarded as the definition of $\nabla^2 \mathbf{A}$, the Laplacian of $\mathbf{A}$. For *Cartesian coordinates* it can also be verified by direct substitution that

$$\nabla^2 \mathbf{A} = \mathbf{a}_x \nabla^2 A_x + \mathbf{a}_y \nabla^2 A_y + \mathbf{a}_z \nabla^2 A_z. \tag{5-17}$$

Thus, *for Cartesian coordinates* the Laplacian of a vector field $\mathbf{A}$ is another vector field whose components are the Laplacian (the divergence of the gradient) of the corresponding components of $\mathbf{A}$. This, however, is not true for other coordinate systems.

■ **EXERCISE 5.3** Verify Eq. (5-17) in Cartesian coordinates.

We now expand $\nabla \times \nabla \times \mathbf{A}$ in Eq. (5-15) according to Eq. (5-16a) and obtain

$$\nabla(\nabla \cdot \mathbf{A}) - \nabla^2 \mathbf{A} = \mu_0 \mathbf{J}. \tag{5-18}$$

With the purpose of simplifying Eq. (5-18) to the greatest extent possible we choose

Coulomb condition for divergence of A

$$\boxed{\nabla \cdot \mathbf{A} = 0,} \tag{5.19‡}$$

and Eq. (5-18) becomes

Operator form of vector Poisson's equation

$$\boxed{\nabla^2 \mathbf{A} = -\mu_0 \mathbf{J}.} \tag{5-20}$$

This is a ***vector Poisson's equation***. In Cartesian coordinates, Eq. (5-20) is equivalent to three scalar Poisson's equations:

$$\nabla^2 A_x = -\mu_0 J_x, \tag{5-21a}$$

$$\nabla^2 A_y = -\mu_0 J_y, \tag{5-21b}$$

$$\nabla^2 A_z = -\mu_0 J_z. \tag{5-21c}$$

†This formula can be readily verified in Cartesian coordinates by direct substitution.

‡This relation is called ***Coulomb condition*** or ***Coulomb gauge***.

Each of these three equations is mathematically the same as the scalar Poisson's equation, Eq. (3-126) in electrostatics. In free space the equation

$$\nabla^2 V = -\frac{\rho_v}{\epsilon_0}$$

has a particular solution (see Eq. 3-38),

$$V = \frac{1}{4\pi\epsilon_0} \int_{V'} \frac{\rho_v}{R}\, dv'.$$

Hence the solution for Eq. (5-21a) is

$$A_x = \frac{\mu_0}{4\pi} \int_{V'} \frac{J_x}{R}\, dv'.$$

We can write similar solutions for A_y and A_z. Combining the three components, we have the solution for Eq. (5-20):

Finding vector magnetic potential from current density

$$\mathbf{A} = \frac{\mu_0}{4\pi} \int_{V'} \frac{\mathbf{J}}{R}\, dv' \qquad \text{(Wb/m)}. \tag{5-22}$$

Equation (5-22) enables us to find the vector magnetic potential **A** from the volume current density **J**. The magnetic flux density **B** can then be obtained from $\nabla \times \mathbf{A}$ by differentiation.

Vector potential **A** relates to the magnetic flux Φ through a given area S that is bounded by contour C in a simple way:

$$\Phi = \int_S \mathbf{B} \cdot d\mathbf{s}. \tag{5-23}$$

SI unit for magnetic flux

The SI unit for magnetic flux is weber (Wb), which is equivalent to tesla-square meter ($\text{T} \cdot \text{m}^2$). Using Eq. (5-14) and Stokes's theorem, we have

$$\Phi = \int_S (\nabla \times \mathbf{A}) \cdot d\mathbf{s} = \oint_C \mathbf{A} \cdot d\boldsymbol{\ell} \qquad \text{(Wb)}. \tag{5-24}$$

Relation between vector magnetic potential and magnetic flux

Thus, vector magnetic potential **A** does have physical significance in that its line integral around any closed path equals the total magnetic flux passing through the area enclosed by the path.

5-4 THE BIOT-SAVART LAW AND APPLICATIONS

In many applications we are interested in determining the magnetic field due to a current-carrying circuit. For a thin wire with cross-sectional area S, dv' equals $S\, d\ell'$, and the current flow is entirely along the wire. We have

$$\mathbf{J}\, dv' = JS\, d\boldsymbol{\ell}' = I\, d\boldsymbol{\ell}', \tag{5-25}$$

and Eq. (5-22) becomes

Finding vector magnetic potential from current in a closed circuit

$$\mathbf{A} = \frac{\mu_0 I}{4\pi} \oint_{C'} \frac{d\boldsymbol{\ell}'}{R} \quad \text{(Wb/m)}, \tag{5-26}$$

where a circle has been put on the integral sign because the current I must flow in a closed path,† which is designated C'. The magnetic flux density is then

$$\begin{aligned} \mathbf{B} = \nabla \times \mathbf{A} &= \nabla \times \left[\frac{\mu_0 I}{4\pi} \oint_{C'} \frac{d\boldsymbol{\ell}'}{R} \right] \\ &= \frac{\mu_0 I}{4\pi} \oint_{C'} \nabla \times \left(\frac{d\boldsymbol{\ell}'}{R} \right). \end{aligned} \tag{5-27}$$

It is very important to note in Eq. (5-27) that the *unprimed* curl operation implies differentiations with respect to the space coordinates of the *field point*, and that the integral operation is with respect to the *primed source coordinates*. The integrand in Eq. (5-27) can be expanded into two terms by using the following identity (see Eq. 2-115):

$$\nabla \times (f\mathbf{G}) = f\nabla \times \mathbf{G} + (\nabla f) \times \mathbf{G}. \tag{5-28}$$

We have, with $f = 1/R$ and $\mathbf{G} = d\boldsymbol{\ell}'$,

$$\mathbf{B} = \frac{\mu_0 I}{4\pi} \oint_{C'} \left[\frac{1}{R} \nabla \times d\boldsymbol{\ell}' + \left(\nabla \frac{1}{R} \right) \times d\boldsymbol{\ell}' \right]. \tag{5-29}$$

Now, since the unprimed and primed coordinates are independent, $\nabla \times d\boldsymbol{\ell}'$ equals 0, and the first term on the right side of Eq. (5-29) vanishes. The distance R is measured from $d\boldsymbol{\ell}'$ at (x', y', z') to the field point at (x, y, z). Thus we have

$$\frac{1}{R} = [(x-x')^2 + (y-y')^2 + (z-z')^2]^{-1/2};$$

$$\begin{aligned} \nabla\left(\frac{1}{R}\right) &= \mathbf{a}_x \frac{\partial}{\partial x}\left(\frac{1}{R}\right) + \mathbf{a}_y \frac{\partial}{\partial y}\left(\frac{1}{R}\right) + \mathbf{a}_z \frac{\partial}{\partial z}\left(\frac{1}{R}\right) \\ &= -\frac{\mathbf{a}_x(x-x') + \mathbf{a}_y(y-y') + \mathbf{a}_z(z-z')}{[(x-x')^2 + (y-y')^2 + (z-z')^2]^{3/2}} \\ &= -\frac{\mathbf{R}}{R^3} = -\mathbf{a}_R \frac{1}{R^2}, \end{aligned} \tag{5-30}$$

†We are now dealing with direct (non-time-varying) currents that give rise to steady magnetic fields. Circuits containing time-varying sources may send time-varying currents along an open wire and deposit charges at its ends. Antennas are examples.

where $\mathbf{a}_R$ is the unit vector directed *from the source point to the field point*. Substituting Eq. (5-30) in Eq. (5-29), we get

Biot-Savart law for finding magnetic flux density from current in a closed circuit

$$\mathbf{B} = \frac{\mu_0 I}{4\pi} \oint_{C'} \frac{d\boldsymbol{\ell}' \times \mathbf{a}_R}{R^2} \quad \text{(T)}. \tag{5-31}$$

Equation (5-31) is known as ***Biot-Savart law***. It is a formula for determining **B** caused by a current I in a closed path C', and was *derived* from the divergence postulate for **B**. Many books use Biot-Savart law as the starting point for developing magnetostatics, but it is difficult to see the experimental procedure used to establish such a precise and complicated relation as Eq. (5-31). We prefer to derive both Ampère's circuital law and Biot-Savart law from our simple divergence and curl postulates for **B**.

Sometimes it is convenient to write Eq. (5-31) in two steps:

$$\mathbf{B} = \oint_{C'} d\mathbf{B} \quad \text{(T)}, \tag{5-32a}$$

with

$$d\mathbf{B} = \frac{\mu_0 I}{4\pi} \left(\frac{d\boldsymbol{\ell}' \times \mathbf{a}_R}{R^2} \right) \quad \text{(T)}, \tag{5-32b}$$

which is the magnetic flux density due to a current element $I\, d\boldsymbol{\ell}'$. An alternative and sometimes more convenient form for Eq. (5-32b) is

$$d\mathbf{B} = \frac{\mu_0 I}{4\pi} \left(\frac{d\boldsymbol{\ell}' \times \mathbf{R}}{R^3} \right) \quad \text{(T)}. \tag{5-32c}$$

EXAMPLE 5-3

A direct current I flows in a straight wire of length $2L$. Find the magnetic flux density **B** at a point located at a distance r from the wire in the bisecting plane: (a) by determining the vector magnetic potential **A** first, and (b) by applying Biot-Savart law.

SOLUTION

Direct currents exist only in closed circuits. Hence the wire in the present problem must be a part of a current-carrying closed loop. Since we do not

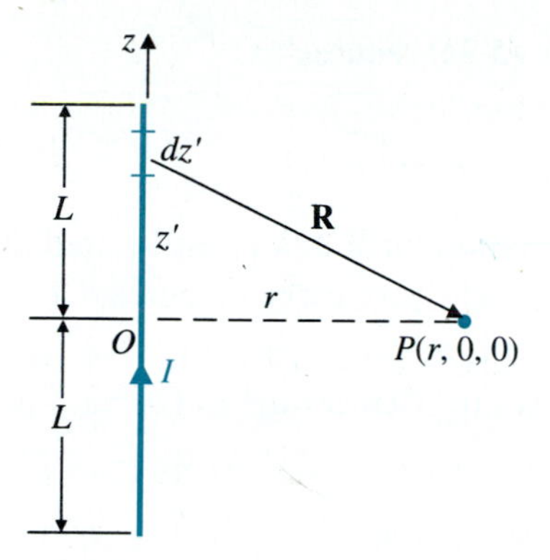

FIGURE 5-4 A current-carrying straight wire (Example 5-3).

know the rest of the circuit, Ampère's circuital law cannot be used to advantage. Refer to Fig. 5-4. The current-carrying line segment is aligned with the z-axis. A typical element on the wire is

$$d\boldsymbol{\ell}' = \mathbf{a}_z\, dz'.$$

The cylindrical coordinates of the field point P are $(r, 0, 0)$.

a) *By finding* **B** *from* $\nabla \times \mathbf{A}$. Substituting $R = \sqrt{z'^2 + r^2}$ into Eq. (5-26) we have

$$\begin{aligned}\mathbf{A} &= \mathbf{a}_z \frac{\mu_0 I}{4\pi} \int_{-L}^{L} \frac{dz'}{\sqrt{z'^2 + r^2}} \\ &= \mathbf{a}_z \frac{\mu_0 I}{4\pi} \left[\ln\left(z' + \sqrt{z'^2 + r^2}\right)\right]\Big|_{-L}^{L} \\ &= \mathbf{a}_z \frac{\mu_0 I}{4\pi} \ln \frac{\sqrt{L^2 + r^2} + L}{\sqrt{L^2 + r^2} - L}.\end{aligned} \tag{5-33}$$

Therefore,

$$\mathbf{B} = \nabla \times \mathbf{A} = \nabla \times (\mathbf{a}_z A_z) = \mathbf{a}_r \frac{1}{r}\frac{\partial A_z}{\partial \phi} - \mathbf{a}_\phi \frac{\partial A_z}{\partial r}.$$

Cylindrical symmetry around the wire assures that $\partial A_z/\partial \phi = 0$. Thus,

$$\begin{aligned}\mathbf{B} &= -\mathbf{a}_\phi \frac{\partial}{\partial r}\left[\frac{\mu_0 I}{4\pi} \ln \frac{\sqrt{L^2 + r^2} + L}{\sqrt{L^2 + r^2} - L}\right] \\ &= \mathbf{a}_\phi \frac{\mu_0 I L}{2\pi r \sqrt{L^2 + r^2}}.\end{aligned} \tag{5-34}$$

When $r \ll L$, Eq. (5-34) reduces to

$$\mathbf{B}_\phi = \mathbf{a}_\phi \frac{\mu_0 I}{2\pi r}, \tag{5-35}$$

which is the expression for $\mathbf{B}$ at a point located at a distance r from an infinitely long, straight wire carrying current I, as given in Eq. (5-12).

b) *By applying Biot-Savart law.* From Fig. 5-4 we see that the distance vector *from* the source element dz' *to* the field point P is

$$\mathbf{R} = \mathbf{a}_r r - \mathbf{a}_z z'$$

$$d\boldsymbol{\ell}' \times \mathbf{R} = \mathbf{a}_z\, dz' \times (\mathbf{a}_r r - \mathbf{a}_z z') = \mathbf{a}_\phi r\, dz'.$$

Substitution in Eq. (5-32c) gives

$$\mathbf{B} = \int d\mathbf{B} = \mathbf{a}_\phi \frac{\mu_0 I}{4\pi} \int_{-L}^{L} \frac{r\, dz'}{(z'^2 + r^2)^{3/2}}$$

$$= \mathbf{a}_\phi \frac{\mu_0 I L}{2\pi r \sqrt{L^2 + r^2}},$$

which is the same as Eq. (5-34).

EXAMPLE 5-4

Find the magnetic flux density at the center of a planar square loop, with side w carrying a direct current I.

SOLUTION

Assume that the loop lies in the xy-plane, as shown in Fig. 5-5. The magnetic flux density at the center of the square loop is equal to four times that caused by a single side of length w. We have, by setting $L = r = w/2$ in Eq. (5-34),

$$\mathbf{B} = \mathbf{a}_z \frac{\mu_0 I}{\sqrt{2}\pi w} \times 4 = \mathbf{a}_z \frac{2\sqrt{2}\mu_0 I}{\pi w}, \tag{5-36}$$

FIGURE 5-5 A square loop carrying current I (Example 5-4).

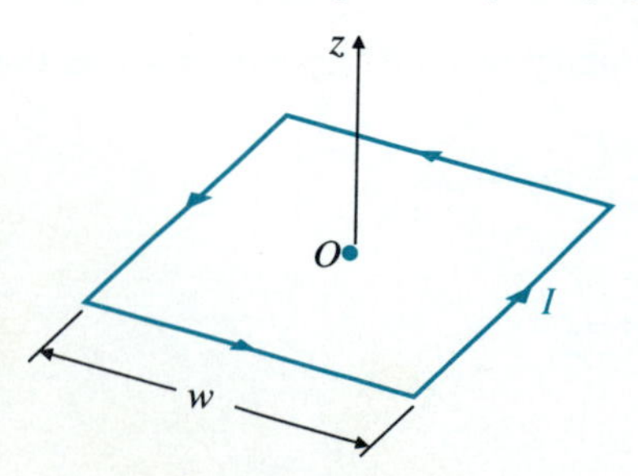

where the direction of **B** and that of the current in the loop follow the right-hand rule.

EXERCISE 5.4 An 8 (cm) × 6 (cm) rectangular conducting loop lies in the xy-plane. A direct current of 5 (A) flows in a clockwise direction viewing from the top. Find **B** at the center of the loop.

ANS. $-\mathbf{a}_z 83.3\,(\mu\text{T})$.

EXAMPLE 5-5

Find the magnetic flux density at a point on the axis of a circular loop of radius b that carries a direct current I.

SOLUTION

We apply Biot-Savart law to the circular loop shown in Fig. 5-6:

$$d\boldsymbol{\ell}' = \mathbf{a}_\phi b\,d\phi',$$
$$\mathbf{R} = \mathbf{a}_z z - \mathbf{a}_r b,$$
$$R = (z^2 + b^2)^{1/2}.$$

Again it is important to remember that **R** is the vector *from* the source element $d\boldsymbol{\ell}'$ *to* the field point P. We have

$$d\boldsymbol{\ell}' \times \mathbf{R} = \mathbf{a}_\phi b\,d\phi' \times (\mathbf{a}_z z - \mathbf{a}_r b)$$
$$= \mathbf{a}_r bz\,d\phi' + \mathbf{a}_z b^2\,d\phi'.$$

Because of cylindrical symmetry, it is easy to see that the $\mathbf{a}_r$-component is canceled by the contribution of the element located diametrically opposite to $d\boldsymbol{\ell}'$, so we need only consider the $\mathbf{a}_z$-component of this cross product.

FIGURE 5-6 A circular loop carrying current I (Example 5-5).

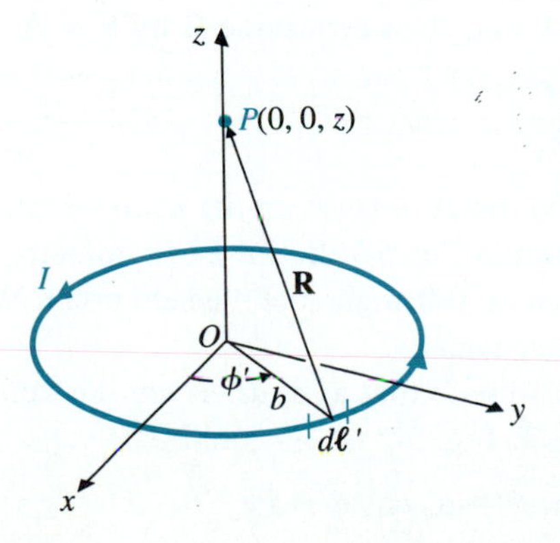

We write, from Eqs. (5-32a) and (5-32c),

$$\mathbf{B} = \frac{\mu_0 I}{4\pi} \int_0^{2\pi} \mathbf{a}_z \frac{b^2\, d\phi'}{(z^2 + b^2)^{3/2}},$$

or

$$\boxed{\mathbf{B} = \mathbf{a}_z \frac{\mu_0 I b^2}{2(z^2 + b^2)^{3/2}} \quad \text{(T)}.} \tag{5-37}$$

EXERCISE 5.5 Refer to Fig. 5-6. Find **B**:

a) at the center of a circular loop of radius 5 (cm) carrying a direct current 2 (A), and

b) at the center of a semi-circular loop of radius 8 (cm) carrying a direct current 4 (A).

ANS. (a) 8π (μT), (b) 5π (μT).

5-5 THE MAGNETIC DIPOLE

We begin this section with an example.

EXAMPLE 5-6

A magnetic dipole

Find the magnetic flux density at a distant point of a small circular loop of radius b that carries a current I (a ***magnetic dipole***).

SOLUTION

We select the center of the loop to be the origin of spherical coordinates, as shown in Fig. 5-7. The source coordinates are primed. We first find the vector magnetic potential **A** and then determine **B** by $\nabla \times \mathbf{A}$:

$$\mathbf{A} = \frac{\mu_0 I}{4\pi} \oint_{C'} \frac{d\boldsymbol{\ell}'}{R_1}, \tag{5-38}$$

where R_1 denotes the distance between the source element $d\boldsymbol{\ell}'$ at P' and the field point P, as shown in Fig. 5-7. Because of symmetry, the magnetic field is obviously independent of the angle ϕ of the field point. We pick $P(R, \theta, \pi/2)$ in the yz-plane for convenience.

It is important to note that $\mathbf{a}_\phi$ at $d\boldsymbol{\ell}'$ is *not* the same as $\mathbf{a}_\phi$ at point P. In fact, $\mathbf{a}_\phi$ at P, shown in Fig. 5-7 is $-\mathbf{a}_x$, and

$$d\boldsymbol{\ell}' = (-\mathbf{a}_x \sin\phi' + \mathbf{a}_y \cos\phi')b\, d\phi'. \tag{5-39}$$

For every $I\,d\boldsymbol{\ell}'$ there is another symmetrically located differential current element on the other side of the y-axis that will contribute an equal amount to $\mathbf{A}$ in the $-\mathbf{a}_x$ direction but will cancel the contribution of $I\,d\boldsymbol{\ell}'$ in the $\mathbf{a}_y$ direction. Equation (5-38) can be written as

$$\mathbf{A} = -\mathbf{a}_x \frac{\mu_0 I}{4\pi} \int_0^{2\pi} \frac{b \sin \phi'}{R_1}\, d\phi',$$

or

$$\mathbf{A} = \mathbf{a}_\phi \frac{\mu_0 I b}{2\pi} \int_{-\pi/2}^{\pi/2} \frac{\sin \phi'}{R_1}\, d\phi'. \tag{5-40}$$

The law of cosines applied to the triangle OPP' gives

$$R_1^2 = R^2 + b^2 - 2bR \cos \psi,$$

where $R \cos \psi$ is the projection of R on the radius OP', which is the same as the projection of OP'' ($OP'' = R \sin \theta$) on OP'. Hence,

$$R_1^2 = R^2 + b^2 - 2bR \sin \theta \sin \phi',$$

and

$$\frac{1}{R_1} = \frac{1}{R}\left(1 + \frac{b^2}{R^2} - \frac{2b}{R} \sin \theta \sin \phi'\right)^{-1/2}.$$

When $R^2 \gg b^2$, b^2/R^2 can be neglected in comparison with 1:

$$\begin{aligned} \frac{1}{R_1} &\cong \frac{1}{R}\left(1 - \frac{2b}{R} \sin \theta \sin \phi'\right)^{-1/2} \\ &\cong \frac{1}{R}\left(1 + \frac{b}{R} \sin \theta \sin \phi'\right). \end{aligned} \tag{5-41}$$

FIGURE 5-7 A small circular loop carrying current I (Example 5-6).

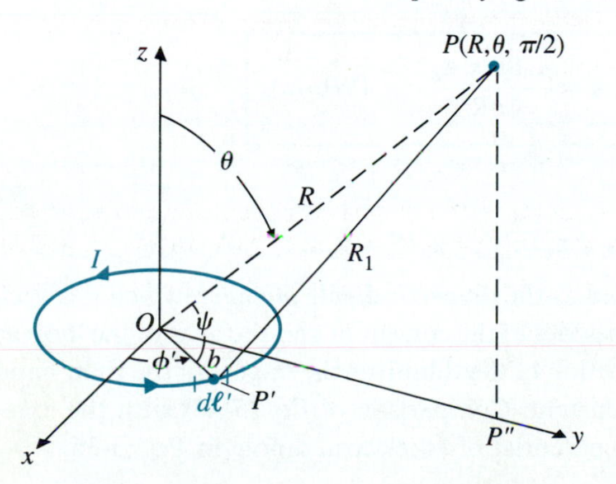

Substitution of Eq. (5-41) in Eq. (5-40) gives

$$\mathbf{A} = \mathbf{a}_\phi \frac{\mu_0 I b}{2\pi R} \int_{-\pi/2}^{\pi/2} \left(1 + \frac{b}{R} \sin\theta \sin\phi'\right) \sin\phi' \, d\phi',$$

which yields

$$\mathbf{A} = \mathbf{a}_\phi \frac{\mu_0 I b^2}{4R^2} \sin\theta. \tag{5-42}$$

The magnetic flux density is $\mathbf{B} = \nabla \times \mathbf{A}$. Equation (2-97) or the formula on the inside of the back cover can be used to find

$$\mathbf{B} = \frac{\mu_0 I b^2}{4R^3} (\mathbf{a}_R \, 2\cos\theta + \mathbf{a}_\theta \sin\theta), \tag{5-43}$$

which is our answer.

At this point we recognize the similarity between Eq. (5-43) and the expression for the electric field intensity *in the far field* of an electrostatic dipole as given in Eq. (3-37). Hence, at distant points the magnetic flux lines of a magnetic dipole (placed in the xy-plane) such as that in Fig. 5-7 will have the same form as the electric field lines of an electric dipole (lying in the z-direction). In the vicinity of the dipoles, however, the flux lines of a magnetic dipole are continuous, whereas the field lines of an electric dipole terminate on the charges, always going from the positive to the negative charge. This is illustrated in Fig. 5-8.

Let us now rearrange the expression of the vector magnetic potential in Eq. (5-42) as

$$\mathbf{A} = \mathbf{a}_\phi \frac{\mu_0 (I\pi b^2)}{4\pi R^2} \sin\theta,$$

or

Finding vector magnetic potential from magnetic dipole moment

$$\boxed{\mathbf{A} = \frac{\mu_0 \mathbf{m} \times \mathbf{a}_R}{4\pi R^2} \qquad \text{(Wb/m)},} \tag{5-44}$$

where

$$\mathbf{m} = \mathbf{a}_z I\pi b^2 = \mathbf{a}_z IS = \mathbf{a}_z m \qquad (\text{A}\cdot\text{m}^2) \tag{5-45}$$

is defined as the ***magnetic dipole moment***, which is a vector whose magnitude is the product of the current in and the area of the loop and whose direction is the direction of the thumb as the fingers of the right hand follow the direction of the current. Comparison of Eq. (5-44) with the expression for the scalar electric potential of an electric dipole in Eq. (3-36),

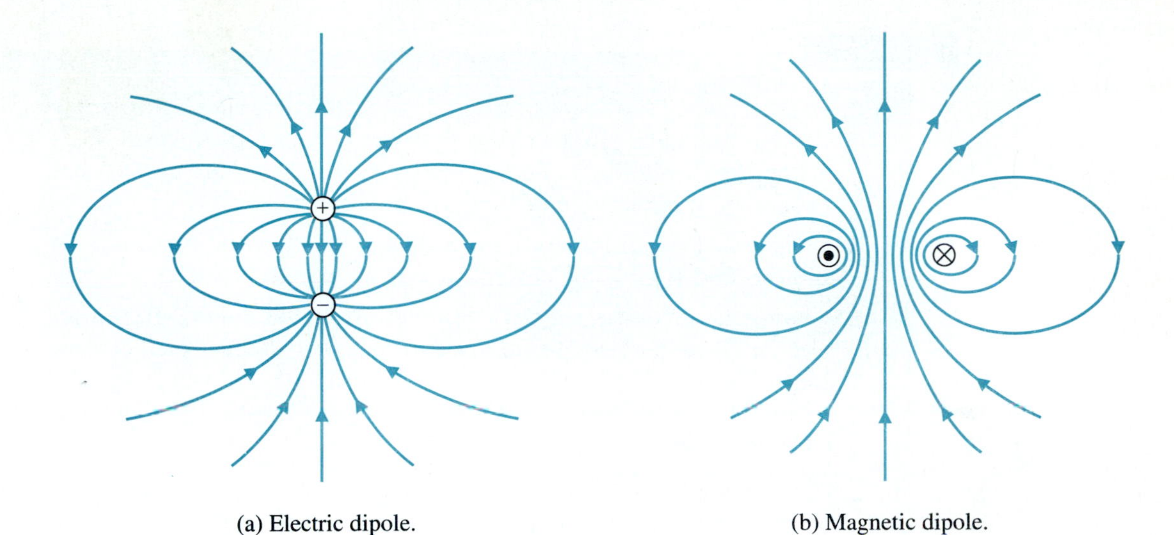

(a) Electric dipole. (b) Magnetic dipole.

FIGURE 5-8 Electric field lines of an electric dipole and magnetic flux lines of a magnetic dipole.

$$V = \frac{\mathbf{p} \cdot \mathbf{a}_R}{4\pi\epsilon_0 R^2} \quad \text{(V)}, \tag{5-46}$$

reveals that, for the two cases, **A** is analogous to V. We call a small current-carrying loop a ***magnetic dipole***.

In a similar manner we can also rewrite Eq. (5-43) as

Magnetic flux density due to a magnetic dipole

$$\mathbf{B} = \frac{\mu_0 m}{4\pi R^3}(\mathbf{a}_R\, 2\cos\theta + \mathbf{a}_\theta \sin\theta) \quad \text{(T)}. \tag{5-47}$$

Except for the change of p to m and ϵ_0 to $1/\mu_0$, Eq. (5-47) has the same form as Eq. (3-37) for the expression for **E** at a distant point of an electric dipole. Although the magnetic dipole in Example 5-6 was taken to be a small circular loop, it can be shown that the same far-zone expressions—Eqs. (5-44) and (5-47)—are obtained when the loop has other shapes, with $m = IS$.

REVIEW QUESTIONS

Q.5-9 Define *vector magnetic potential* **A**. What is its SI unit?

Q.5-10 What is the relation between vector magnetic potential **A** and the magnetic flux through a given area?

Q.5-11 State Biot-Savart law.

Q.5-12 What is a magnetic dipole? Define ***magnetic dipole moment***. What is its SI unit?

REMARKS

1. In determining **B** due to a current distribution, it is simplest to apply Ampère's circuital law if a closed path can be found, over which **B** has a constant magnitude. The geometry of the problem usually has cylindrical symmetry and/or infinite length.
2. If the above condition does not exist, the Biot-Savart law should be used to find **B** from the current in a given circuit.
3. Analogous quantities in the computation of the **E**-field due to an electric dipole and the **B**-field due to a magnetic dipole (small current loop):

Electric dipole	Magnetic dipole
$\mathbf{p} = q\mathbf{d}$	$\mathbf{m} = \mathbf{a}_n IS$
$\mathbf{E}$	$\mathbf{B}$
ϵ_0	$1/\mu_0$

5-6 MAGNETIZATION AND EQUIVALENT CURRENT DENSITIES

According to the elementary atomic model of matter, all materials are composed of atoms, each with a positively charged nucleus and a number of orbiting negatively charged electrons. The orbiting electrons cause circulating currents and form microscopic magnetic dipoles. In addition, both the electrons and the nucleus of an atom rotate (spin) on their own axes with certain magnetic dipole moments. The magnetic dipole moment of a spinning nucleus is usually negligible in comparison to that of an orbiting or spinning electron because of the much larger mass and lower angular velocity of the nucleus.

In the absence of an external magnetic field, the magnetic dipoles of the atoms of most materials (except permanent magnets) have random orientations, resulting in no net magnetic moment. The application of an external magnetic field causes both an alignment of the magnetic moments of the spinning electrons and an induced magnetic moment due to a change in the orbital motion of electrons. To obtain a formula for determining the quantitative change in the magnetic flux density caused by the presence of a magnetic material, we let $\mathbf{m}_k$ be the magnetic dipole moment of an atom. If there are n atoms per unit volume, we define a ***magnetization vector***, $\mathbf{M}$, as

Magnetization vector is the volume density of magnetic dipole moment.

$$\mathbf{M} = \lim_{\Delta v \to 0} \frac{\sum_{k=1}^{n\Delta v} \mathbf{m}_k}{\Delta v} \qquad \text{(A/m)}, \tag{5-48}$$

which is the volume density of magnetic dipole moment. The magnetic dipole moment $d\mathbf{m}$ of an elemental volume dv' is $d\mathbf{m} = \mathbf{M}\,dv'$ that, according to Eq. (5-44), will produce a vector magnetic potential

$$d\mathbf{A} = \frac{\mu_0 \mathbf{M} \times \mathbf{a}_R}{4\pi R^2}\,dv'. \tag{5-49}$$

The total $\mathbf{A}$ is the volume integral of $d\mathbf{A}$ in Eq. (5-49), and the contribution of magnetization to the magnetic flux density $\mathbf{B}$ is $\nabla \times \mathbf{A}$. Equation (5-49) is analogous to the expression of dV in Eq. (3-55), from which we obtained the potential V due to a polarized dielectric medium, and $\mathbf{E}$ from $-\nabla V$.

Similar to the equivalence of $\mathbf{P}$ of induced electric dipoles to a polarization surface charge density $\rho_{ps} = \mathbf{P}\cdot\mathbf{a}_n$ and a polarization volume charge density $\rho_{pv} = -\nabla\cdot\mathbf{P}$ discussed in Subsection 3-6.2, we can prove analytically the equivalence of $\mathbf{M}$ of magnetic dipoles to a ***magnetization surface current density***

Equivalent magnetization surface current density

$$\boxed{\mathbf{J}_{ms} = \mathbf{M} \times \mathbf{a}_n \qquad (\text{A/m}),} \tag{5-50}$$

where $\mathbf{a}_n$ is the unit outward normal to the boundary, and a ***magnetization volume current density***

Equivalent magnetization volume current density

$$\boxed{\mathbf{J}_{mv} = \nabla \times \mathbf{M} \qquad (\text{A/m}^2).} \tag{5-51}$$

A qualitative interpretation can be had by referring to Fig. 5-9, which

FIGURE 5-9 A cross section of a magnetized material.

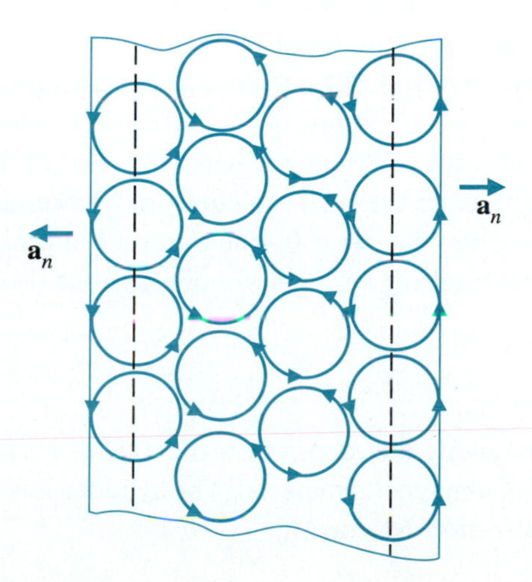

represents a cross section of a magnetized material having a certain thickness. We see that the magnetic dipoles at the surface effectively contribute to a surface current beyond the dashed lines. The magnitude of the surface current is directly proportional to the volume density of the magnetic dipole moment, and the direction of the current on both boundaries is correctly given by $\mathbf{M} \times \mathbf{a}_n$ in the figure, as stipulated in Eq. (5-50).

The equivalent magnetization volume current density $\mathbf{J}_{mv}$, given in Eq. (5-51), is a little difficult to visualize, but we can accept that magnetic-moment density $\mathbf{M}$ produces an internal flux density $\mathbf{B}_i$ which is proportional to $\mathbf{M}$. We may write

$$\mathbf{B}_i = \mu_0 \mathbf{M}, \tag{5-52}$$

or

$$\frac{\mathbf{B}_i}{\mu_0} = \mathbf{M}. \tag{5-53}$$

From Eq. (5-7) we see that

$$\nabla \times \left(\frac{\mathbf{B}_e}{\mu_0}\right) = \mathbf{J}, \tag{5-54}$$

where $\mathbf{B}_e$ denotes the external magnetic flux density due to the free current density $\mathbf{J}$. We write from Eq. (5-53):

$$\nabla \times \frac{\mathbf{B}_i}{\mu_0} = \nabla \times \mathbf{M} = \mathbf{J}_{mv}, \tag{5-55}$$

where $\mathbf{J}_{mv}$ is the equivalent magnetization volume current density. Adding Eqs. (5-54) and (5-55), we have

$$\nabla \times \mathbf{B} = \mu_0(\mathbf{J} + \mathbf{J}_{mv}), \tag{5-56}$$

where $\mathbf{B} = \mathbf{B}_{total} = \mathbf{B}_e + \mathbf{B}_i$. Hence the resultant magnetic flux density in the presence of a magnetized material is changed by an amount $\mathbf{B}_i$. If $\mathbf{M}$ is uniform inside the material, the currents of the neighboring atomic dipoles that flow in opposite directions will cancel everywhere, leaving no net currents in the interior. This is predicted by Eq. (5-51), since the space derivatives (and therefore the curl) of a constant $\mathbf{M}$ vanish. However, if $\mathbf{M}$ has space variations and $\nabla \times \mathbf{M} \neq 0$, the internal atomic currents do not completely cancel, resulting in a net volume current density $\mathbf{J}_{mv}$.

EXAMPLE 5-7

Determine the magnetic flux density on the axis of a uniformly magnetized circular cylinder of a magnetic material. The cylinder has a radius b, length L, and axial magnetization $\mathbf{M} = \mathbf{a}_z M_0$.

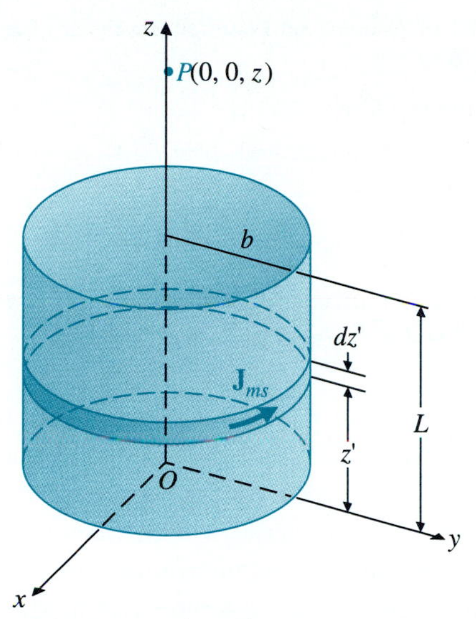

FIGURE 5-10 A uniformly magnetized circular cylinder (Example 5-7).

SOLUTION

In this problem concerning a cylindrical bar magnet, let the axis of the magnetized cylinder coincide with the z-axis of a cylindrical coordinate system, as shown in Fig. 5-10. Since the magnetization $\mathbf{M}$ is a constant within the magnet, $\mathbf{J}_m = \nabla' \times \mathbf{M} = 0$, and there is no equivalent volume current density. The equivalent magnetization surface current density on the side wall is

$$\begin{aligned}\mathbf{J}_{ms} &= \mathbf{M} \times \mathbf{a}'_n = (\mathbf{a}_z M_0) \times \mathbf{a}_r \\ &= \mathbf{a}_\phi M_0.\end{aligned} \tag{5-57}$$

The magnet is then like a cylindrical sheet with a lineal circumferential current density of M_0 (A/m). There is no surface current on the top and bottom faces. To find $\mathbf{B}$ at $P(0, 0, z)$, we consider a differential length dz' with a current $\mathbf{a}_\phi M_0\, dz'$ and use Eq. (5-37) to obtain

$$d\mathbf{B} = \mathbf{a}_z \frac{\mu_0 M_0 b^2\, dz'}{2[(z-z')^2 + b^2]^{3/2}}$$

and

$$\begin{aligned}\mathbf{B} = \int d\mathbf{B} &= \mathbf{a}_z \int_0^L \frac{\mu_0 M_0 b^2\, dz'}{2[(z-z')^2 + b^2]^{3/2}} \\ &= \mathbf{a}_z \frac{\mu_0 M_0}{2}\left[\frac{z}{\sqrt{z^2+b^2}} - \frac{z-L}{\sqrt{(z-L)^2+b^2}}\right].\end{aligned} \tag{5-58}$$

EXERCISE 5.6 A cylindrical magnet of radius 5 (cm) and length 12 (cm) has an axial magnetization $\mathbf{a}_z 130$ (A/cm). Find **B** at

a) the center of the top face,

b) the center of the bottom face, and

c) the center of the magnet.

ANS. (a) and (b) $\mathbf{a}_z 7.54$ (mT), (c) $\mathbf{a}_z 12.55$ (mT).

5-7 MAGNETIC FIELD INTENSITY AND RELATIVE PERMEABILITY

Because the application of an external magnetic field causes both an alignment of the internal dipole moments and an induced magnetic moment in a magnetic material, we expect that the resultant magnetic flux density in the presence of a magnetic material will be different from its value in free space. The macroscopic effect of magnetization can be studied by incorporating the equivalent magnetization volume current density, $\mathbf{J}_{mv}$ in Eq. (5-51), into the basic curl equation, Eq. (5-7). We have

$$\frac{1}{\mu_0}\nabla \times \mathbf{B} = \mathbf{J} + \mathbf{J}_{mv} = \mathbf{J} + \nabla \times \mathbf{M},$$

or

$$\nabla \times \left(\frac{\mathbf{B}}{\mu_0} - \mathbf{M}\right) = \mathbf{J}. \tag{5-59}$$

We now define a new fundamental field quantity, the ***magnetic field intensity*** **H**, such that

Definition of magnetic field intensity H

$$\boxed{\mathbf{H} = \frac{\mathbf{B}}{\mu_0} - \mathbf{M} \qquad \text{(A/m)}.} \tag{5-60}$$

Combining Eqs. (5-59) and (5-60), we obtain the new equation

$$\boxed{\nabla \times \mathbf{H} = \mathbf{J} \qquad (\text{A/m}^2),} \tag{5-61}$$

where $\mathbf{J}\,(\text{A/M}^2)$ is the volume density of *free current*. Equations (5-6) and (5-61) are the two fundamental governing differential equations for magnetostatics. The permeability of the medium does not appear explicitly in these two equations.

The corresponding integral form of Eq. (5-61) is obtained by taking the scalar surface integral of both sides:

$$\int_S (\nabla \times \mathbf{H}) \cdot d\mathbf{s} = \int_S \mathbf{J} \cdot d\mathbf{s}, \tag{5-62}$$

or, according to Stokes's theorem,

Generalized Ampère's circuital law for steady currents–applicable in nonmagnetic as well as magnetic media.

$$\oint_C \mathbf{H} \cdot d\boldsymbol{\ell} = I \qquad \text{(A)}, \tag{5-63}$$

where C is the contour (closed path) bounding the surface S and I is the total free current passing through S. The relative directions of C and current flow I follow the right-hand rule. Equation (5-63) is another form of ***Ampère's circuital law***, which holds in a nonmagnetic as well as a magnetic medium. It states that ***the circulation of the magnetic field intensity around any closed path is equal to the free current flowing through the surface bounded by the path.***

When the magnetic properties of the medium are *linear* and *isotropic*, the magnetization is directly proportional to the magnetic field intensity:

Definition of magnetic susceptibility

$$\mathbf{M} = \chi_m \mathbf{H}, \tag{5-64}$$

where χ_m is a dimensionless quantity called ***magnetic susceptibility***. Substitution of Eq. (5-64) in Eq. (5-60) yields the following constitutive relation:

$$\begin{aligned} \mathbf{B} &= \mu_0(1 + \chi_m)\mathbf{H} \\ &= \mu_0\mu_r\mathbf{H} = \mu\mathbf{H} \qquad (\text{Wb/m}^2), \end{aligned} \tag{5-65}$$

or

$$\mathbf{H} = \frac{1}{\mu}\mathbf{B} \qquad \text{(A/m)}, \tag{5-66}$$

where

$$\mu_r = 1 + \chi_m = \frac{\mu}{\mu_0} \tag{5-67}$$

Relative permeability and absolute permeability of a medium

is another dimensionless quantity known as the ***relative permeability*** of the medium. The parameter $\mu = \mu_0\mu_r$ is the ***absolute permeability*** (or sometimes just ***permeability***) of the medium and is measured in H/m; χ_m, and therefore μ_r, can be a function of space coordinates. For a simple medium—linear, isotropic, and homogeneous—χ_m and μ_r are constants.

Definition of a simple medium

The permeability of most materials is very close to that of free space (μ_0). For ferromagnetic materials such as iron, nickel, and cobalt, μ_r could be very large (50–5000, and up to 10^6 or more for special alloys); the permeability depends not only on the magnitude of H but also on the previous history of the material. Section 5-8 contains some qualitative discussions of the macroscopic behavior of magnetic materials.

5-8 BEHAVIOR OF MAGNETIC MATERIALS

In Eq. (5-64) of the previous section, we described the macroscopic magnetic property of a linear, isotropic medium by defining the magnetic susceptibility χ_m, a dimensionless coefficient of proportionality between magnetization **M** and magnetic field intensity **H**. The relative permeability μ_r is simply $1 + \chi_m$. Magnetic materials can be roughly classified into three main groups in accordance with their μ_r values. A material is said to be

Three types of magnetic material

Diamagnetic, if $\mu_r \lesssim 1$ (χ_m is a very small negative number).

Paramagnetic, if $\mu_r \gtrsim 1$ (χ_m is a very small positive number).

Ferromagnetic, if $\mu_r \gg 1$ (χ_m is a large positive number).

A thorough understanding of microscopic magnetic phenomena requires a knowledge of quantum theory. Here we simply state that diamagnetism arises mainly from the orbital motion of the electrons within an atom, whereas paramagnetism is mainly from the magnetic dipole moments of the spinning electrons. The magnetic susceptibility for most known *diamagnetic* materials (copper, germanium, silver, gold) is of the order of -10^{-5}, and that for *paramagnetic* materials such as aluminum, magnesium, titanium and tungsten is of the order of 10^{-5}.

The magnetization of *ferromagnetic* materials can be many orders of magnitude larger than that of paramagnetic substances. (See Appendix B-5 for typical values of relative permeability.)

Ferromagnetism can be explained in terms of magnetized ***domains***. According to this model, which has been experimentally confirmed, a ferromagnetic material (such as cobalt, nickel, and iron) is composed of many small domains, their linear dimensions ranging from a few microns to about 1 mm. These domains, each containing about 10^{15} or 10^{16} atoms, are fully magnetized in the sense that they contain aligned magnetic dipoles resulting from spinning electrons even in the absence of an applied magnetic field. Quantum theory asserts that strong coupling forces exist between the magnetic dipole moments of the atoms in a domain, holding the dipole moments in parallel. Between adjacent domains there is a transition region about 100 atoms thick called a ***domain wall***. In an unmagnetized state the magnetic moments of the adjacent domains in a ferromagnetic material have different directions, as exemplified in Fig. 5-11 by the polycrystalline specimen shown. Viewed as a whole, the random nature of the orientations in the various domains results in no net magnetization.

When an external magnetic field is applied to a ferromagnetic material, the walls of those domains having magnetic moments aligned with the applied field move in such a way as to make the volumes of those domains grow at the expense of other domains. As a result, magnetic flux density is

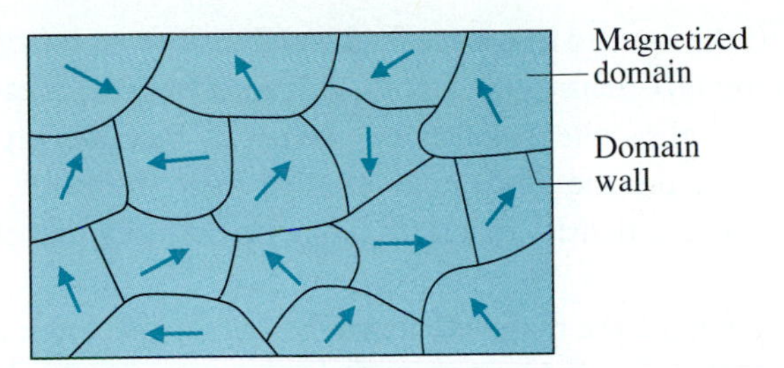

FIGURE 5-11 Domain structure of a polycrystalline ferromagnetic specimen.

increased. For weak applied fields, say up to point P_1 on the B–H magnetization curve in Fig. 5-12 domain-wall movements are reversible. But when an applied field becomes stronger (past P_1), domain-wall movements are no longer reversible, and domain rotation toward the direction of the applied field will also occur. For example, if an applied field is reduced to zero at point P_2, the B–H relationship will not follow the solid curve P_2P_1O, but will go down from P_2 to P_2', along the broken curve in the figure. This phenomenon of magnetization lagging behind the field producing it is called ***hysteresis***, which is derived from a Greek word meaning "to lag." As the applied field becomes even much stronger (past P_2 to P_3), domain-wall motion and domain rotation will cause essentially a total alignment of the microscopic magnetic moments with the applied field, at which point the magnetic material is said to have reached *saturation*. The curve $OP_1P_2P_3$ on the B–H plane is called the ***normal magnetization curve***.

Phenomenon of hysteresis

FIGURE 5-12 Hysteresis loops in the B–H plane for ferromagnetic material.

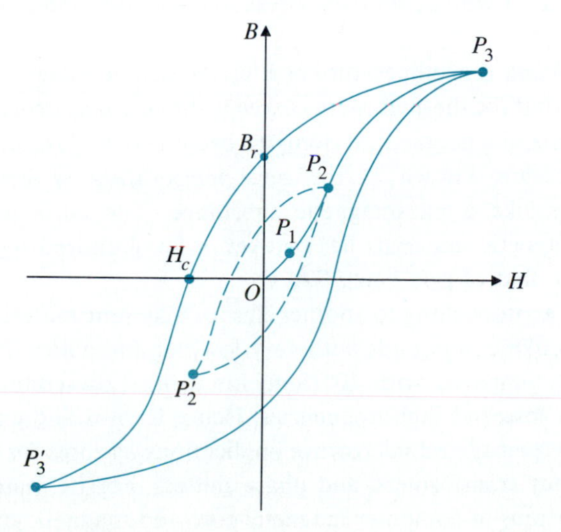

If the applied magnetic field is reduced to zero from the value at P_3, the magnetic flux density does not go to zero but assumes the value at B_r. This value is called the ***residual*** or ***remanent flux density*** (in Wb/m^2) and is dependent on the maximum applied field intensity. The existence of a remanent flux density in a ferromagnetic material makes permanent magnets possible.

Permanent magnets have remanent flux density.

To make the magnetic flux density of a specimen zero, it is necessary to apply a magnetic field intensity H_c in the opposite direction. This required H_c is called *coercive force*, but a more appropriate name is ***coercive field intensity*** (in A/m). Like B_r, H_c also depends on the maximum value of the applied magnetic field intensity.

Ferromagnetic materials for use in electric generators, motors, and transformers should have a large magnetization for a very small applied field; they should have tall, narrow hysteresis loops. As the applied magnetic field intensity varies periodically between $\pm H_{\text{max}}$, the hysteresis loop is traced once per cycle. The area of the hysteresis loop corresponds to energy loss (***hysteresis loss***) per unit volume per cycle. Hysteresis loss is the energy lost in the form of heat in overcoming the friction encountered during domain-wall motion and domain rotation. Ferromagnetic materials, which have tall, narrow hysteresis loops with small loop areas, are referred to as "soft" materials; they are usually well-annealed materials with very few dislocations and impurities so that the domain walls can move easily.

Hysteresis loss

"Soft" and "hard" ferromagnetic materials

Good permanent magnets, on the other hand, should show a high resistance to demagnetization. This requires that they be made with materials that have large coercive field intensities H_c and hence fat hysteresis loops. These materials are referred to as "hard" ferromagnetic materials. The coercive field intensity of hard ferromagnetic materials (such as Alnico alloys) can be 10^5 (A/m) or more, whereas that for soft materials is usually 50 (A/m) or less.

When the temperature of a ferromagnetic material is raised to such an extent that the thermal energy exceeds the coupling energy of magnetic dipole moments, the magnetized domains become disorganized. Above this critical temperature, known as the ***curie temperature***, a ferromagnetic material behaves like a paramagnetic substance. The curie temperature of most ferromagnetic materials lies between a few hundred to a thousand degrees Celcius, that of iron being 770°C.

Definition of curie temperature

Ferrites belong to another class of magnetic materials. Some ferrites are ceramiclike compounds with very low conductivities (for instance, 10^{-4} to 1 (S/m) compared with 10^7 (S/m) for iron). Low conductivity limits eddy-current losses at high frequencies. Hence ferrites find extensive uses in such high-frequency and microwave applications as cores for FM antennas, high-frequency transformers, and phase shifters. Ferrite material also has broad applications in computer magnetic-core and magnetic-disk memory devices.

Characteristics of ferrites

5-9 BOUNDARY CONDITIONS FOR MAGNETOSTATIC FIELDS

In order to solve problems concerning magnetic fields in regions having media with different physical properties, it is necessary to study the conditions (boundary conditions) that **B** and **H** vectors must satisfy at the interfaces of different media. Using techniques similar to those employed in Section 3-8 to obtain the boundary conditions for electrostatic fields, we can derive magnetostatic boundary conditions by applying the two fundamental governing equations, Eqs. (5-6) and (5-61) to a small pillbox and a small closed path, respectively, which include the interface. From the divergenceless nature of the **B** field in Eq. (5-6) we may conclude directly, as in Eq. (4-34), that ***the normal component of* B *is continuous across an interface***; that is,

The normal component of B is continuous across an interface.

$$\boxed{B_{1n} = B_{2n} \qquad \text{(T)}.} \tag{5-68}$$

For linear and isotropic media, $\mathbf{B}_1 = \mu_1 \mathbf{H}_1$ and $\mathbf{B}_2 = \mu_2 \mathbf{H}_2$, Eq. (5-68) becomes

$$\boxed{\mu_1 H_{1n} = \mu_2 H_{2n}.} \tag{5-69}$$

The tangential component of magnetic field is not continuous if there is a surface current along the interface. An expression of the boundary condition for the tangential components of **H** can be derived by applying Eq. (5-63) to a closed path *abcda* about the interface of two media, as shown in Fig. 5-13. We have, in letting the sides $bc = da = \Delta h$ approach zero,

$$\int_{abcda} \mathbf{H}_i \cdot d\boldsymbol{\ell} = \mathbf{H}_1 \cdot \Delta\mathbf{w} + \mathbf{H}_2 \cdot (-\Delta\mathbf{w}) = \mathbf{J}_{sn},$$

FIGURE 5-13 Closed path about the interface of two media for determining the boundary condition of H_t.

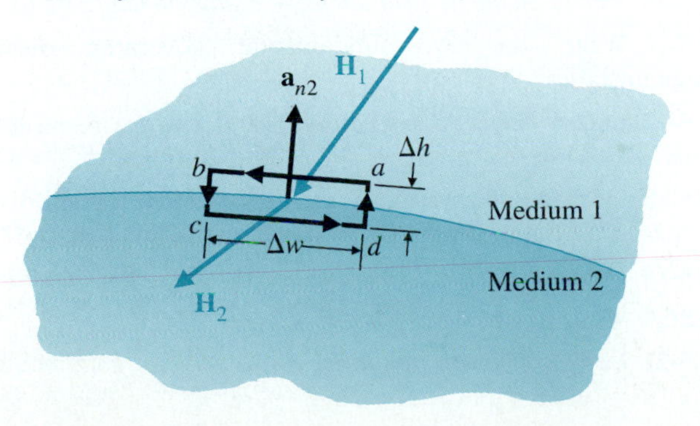

or

$$H_{1t} - H_{2t} = J_{sn} \qquad \text{(A/m)}, \tag{5-70}$$

where J_{sn} is the surface current density on the interface normal to the contour *abcda*. The direction of J_{sn} is that of the thumb when the fingers of the right hand follow the direction of the path. In Fig. 5-13, the positive direction of J_{sn} for the chosen path is out of the paper. The more general form for Eq. (5-70) is

Boundary condition for tangential component of H

$$\mathbf{a}_{n2} \times (\mathbf{H}_1 - \mathbf{H}_2) = \mathbf{J}_s \qquad \text{(A/m)}, \tag{5-71}$$

where $\mathbf{a}_{n2}$ is the *outward unit normal from medium* 2 at the interface.

When the conductivities of both media are finite, currents are specified by volume current densities and free surface currents are not defined on the interface. Hence $\mathbf{J}_s$ equals zero, and ***the tangential component of* H *is continuous across the boundary of almost all physical media; it is discontinuous only when an interface with an ideal perfect conductor or a superconductor is assumed.*** Thus, for magnetostatic fields we normally have

$$\mathbf{H}_{1t} = \mathbf{H}_{2t}. \tag{5-72}$$

EXERCISE 5.7 The magnetic field intensity $\mathbf{H}_1$ in medium 1 having a permeability μ_1 makes an angle α_1 with the normal at an interface with medium 2 having a permeability μ_2. Find the relation between the angle α_2 (that $\mathbf{H}_2$ makes with the normal) and α_1.

ANS. $\tan \alpha_2 / \tan \alpha_1 = \mu_2/\mu_1$.

REVIEW QUESTIONS

Q.5-13 Define *magnetization vector*. What is its SI unit?

Q.5-14 What is meant by "equivalent magnetization current densities"? What are the SI units for $\nabla \times \mathbf{M}$ and $\mathbf{M} \times \mathbf{a}_n$?

Q.5-15 Define *magnetic field intensity vector*. What is its SI unit?

Q.5-16 Write the two fundamental governing differential equations for magnetostatics.

Q.5-17 Define *magnetic susceptibility* and *relative permeability*. What are their SI units?

Q.5-18 Does the magnetic field intensity due to a current distribution depend on the properties of the medium? Does the magnetic flux density?

Q.5-19 Define *diamagnetic*, *paramagnetic*, and *ferromagnetic* materials.

Q.5-20 What is a hysteresis loop?

Q.5-21 Define *remanent flux density* and *coercive field intensity*.

Discuss the difference between soft and hard ferromagnetic materials.

What is *curie temperature*?

What are the boundary conditions for magnetostatic fields at an interface between two different magnetic media?

REMARKS

1. Cylindrical permanent magnets having uniform magnetization are like cylindrical sheets with a constant circumferential surface current.
2. Do not confuse the relative permeability, μ_r, of a medium with its (absolute) permeability, μ. μ_r equals $1 + \chi_m$ and is a dimensionless quantity, while the SI unit for μ is (H/m).
3. μ_r for nonferromagnetic materials can be taken as unity. Ferromagnetic materials (nickel, cobalt, iron, and their alloys) have very large μ_r and are nonlinear. (**B** is not proportional to **H**.)

5-10 INDUCTANCES AND INDUCTORS

Consider two neighboring closed loops, C_1 and C_2 bounding surfaces S_1 and S_2, respectively, as shown in Fig. 5-14. If a current I_1 flows in C_1, a magnetic field $\mathbf{B}_1$ will be created. Some of the magnetic flux due to $\mathbf{B}_1$ will link with C_2—that is, will pass through the surface S_2 bounded by C_2. Let us designate this mutual flux Φ_{12}. We have

$$\Phi_{12} = \int_{S_2} \mathbf{B}_1 \cdot d\mathbf{s}_2 \qquad \text{(Wb)}. \tag{5-73}$$

FIGURE 5-14 Two magnetically coupled loops.

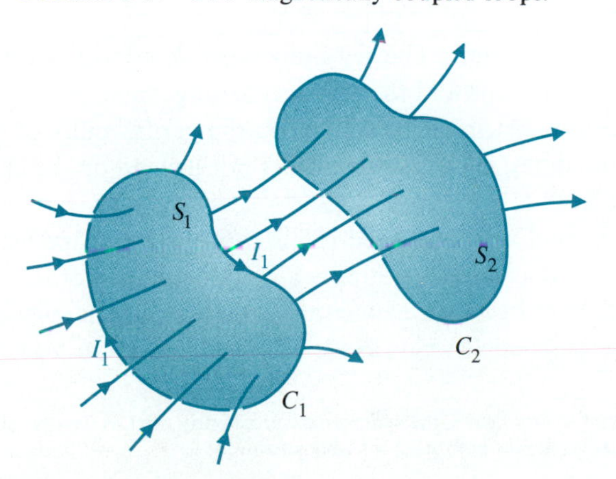

From Biot-Savart law, Eq. (5-31), we see that B_1 is directly proportional to I_1; hence Φ_{12} is also proportional to I_1. We write

$$\Phi_{12} = L_{12}I_1, \tag{5-74}$$

where the proportionality constant L_{12} is called the ***mutual inductance***† between loops C_1 and C_2, with SI unit henry (H). In case C_2 has N_2 turns, the ***flux linkage*** Λ_{12} due to Φ_{12} is

$$\Lambda_{12} = N_2\Phi_{12} \qquad \text{(Wb)}. \tag{5-75}$$

Equation (5-74) then generalizes to

$$\Lambda_{12} = L_{12}I_1 \qquad \text{(Wb)}, \tag{5-76}$$

or

$$\boxed{L_{12} = \frac{\Lambda_{12}}{I_1} = \frac{N_2}{I_1}\int_{S_2} \mathbf{B}_1 \cdot d\mathbf{s}_2 \qquad \text{(H)}.} \tag{5-77}$$

Mutual inductance

The ***mutual inductance between two circuits*** *is then the magnetic flux linkage with one circuit per unit current in the other*. In Eq. (5-77) it is implied that the permeability of the medium does not change with I_1. In other words, Eq. (5-74) and hence Eq. (5-77) apply only to *linear* media.

Some of the magnetic flux produced by I_1 links only with C_1 itself, and not with C_2. The total flux linkage with C_1 caused by I_1 is

$$\Lambda_{11} = N_1\Phi_{11} > N_1\Phi_{12}. \tag{5-78}$$

Self-inductance

*The **self-inductance** of loop C_1 is defined as the magnetic flux linkage per unit current in the loop itself*; that is,

$$\boxed{L_{11} = \frac{\Lambda_{11}}{I_1} = \frac{N_1}{I_1}\int_{S_1} \mathbf{B}_1 \cdot d\mathbf{s}_1 \qquad \text{(H)},} \tag{5-79}$$

for a linear medium. The self-inductance of a loop or circuit depends on the geometrical shape and the physical arrangement of the conductor constituting the loop or circuit, as well as on the permeability of the medium. With a linear medium, self-inductance does not depend on the current in the loop or circuit.

A conductor arranged in an appropriate shape (such as a conducting wire wound as a coil) to supply a certain amount of self-inductance is called an ***inductor***. Just as a capacitor can store electric energy, an inductor can storage magnetic energy, as we shall see in Section 5-11. When we deal with

†In circuit theory books, the symbol M is frequently used to denote mutual inductance. We use L_{12} here, as M has been used for magnetization.

only one loop or coil, there is no need to carry the subscripts in Eq. (5-79), and ***inductance*** without an adjective will be taken to mean self-inductance. The procedure for determining the self-inductance of an inductor is as follows:

Procedure for determining self-inductance

1. Choose an appropriate coordinate system for the given geometry.
2. Assume a current I in the conducting wire.
3. Find $\mathbf{B}$ from I by Ampère's circuital law, Eq. (5-10), if symmetry exists; if not, Biot-Savart law, Eq. (5-31) must be used.
4. Find the flux linking with each turn, Φ, from $\mathbf{B}$ by integration:

$$\Phi = \int_S \mathbf{B} \cdot d\mathbf{s},$$

where S is the area over which $\mathbf{B}$ exists and links with the assumed current.

5. Find the flux linkage Λ by multiplying Φ by the number of turns.
6. Find L by taking the ratio $L = \Lambda/I$.

Only a slight modification of this procedure is needed to determine the mutual inductance L_{12} between two circuits. After choosing an appropriate coordinate system, proceed as follows: Assume $I_1 \rightarrow$ Find $\mathbf{B}_1 \rightarrow$ Find Φ_{12} by integrating $\mathbf{B}_1$ over surface $S_2 \rightarrow$ Find flux linkage $\Lambda_{12} = N_2\Phi_{12} \rightarrow$ Find $L_{12} = \Lambda_{12}/I_1$.

EXAMPLE 5-8

N turns of wire are tightly wound on a toroidal frame of a rectangular cross section with dimensions as shown in Fig. 5-15. Assuming the permeability of the medium to be μ_0, find the self-inductance of the toroidal coil.

SOLUTION

It is clear that the cylindrical coordinate system is appropriate for this problem because the toroid is symmetrical about its axis. Assuming a current I in the conducting wire, we find, by applying Eq. (5-10) to a circular path with radius $r\,(a < r < b)$:

$$\mathbf{B} = \mathbf{a}_\phi B_\phi,$$

$$d\boldsymbol{\ell} = \mathbf{a}_\phi r\, d\phi,$$

$$\oint_C \mathbf{B} \cdot d\boldsymbol{\ell} = \int_0^{2\pi} B_\phi r\, d\phi = 2\pi r B_\phi.$$

This result is obtained because both B_ϕ and r are constant around the circular path C. Since the path encircles a total current NI, we have

$$2\pi r B_\phi = \mu_0 NI$$

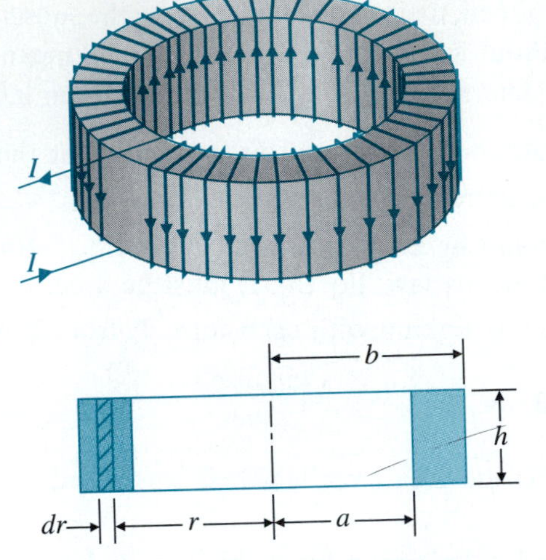

FIGURE 5-15 A closely wound toroidal coil (Example 5-8).

and

$$B_\phi = \frac{\mu_0 NI}{2\pi r}, \tag{5-80}$$

which is the same as Eq. (5-13) in Example 5-2 for a toroidal coil with a circular cross section.

Next we find

$$\Phi = \int_S \mathbf{B} \cdot d\mathbf{s} = \int_S \left(\mathbf{a}_\phi \frac{\mu_0 NI}{2\pi r}\right) \cdot (\mathbf{a}_\phi h\, dr)$$

$$= \frac{\mu_0 NIh}{2\pi} \int_a^b \frac{dr}{r} = \frac{\mu_0 NIh}{2\pi} \ln \frac{b}{a}.$$

The flux linkage Λ is $N\Phi$ or

$$\Lambda = \frac{\mu_0 N^2 Ih}{2\pi} \ln \frac{b}{a}.$$

Finally, we obtain

Inductance of an N-turn closely wound toroidal coil

$$\boxed{L = \frac{\Lambda}{I} = \frac{\mu_0 N^2 h}{2\pi} \ln \frac{b}{a} \quad \text{(H)}.} \tag{5-81}$$

We note that the self-inductance is not a function of I (for a constant medium permeability) and that it is proportional to the square of the number of turns. The qualification that the coil be closely wound on the toroid is to minimize the linkage flux around the individual turns of the wire.

EXAMPLE 5-9

Find the inductance per unit length of a very long solenoid having n turns per unit length. The permeability of the core is μ.

SOLUTION

The magnetic flux density inside a very long solenoid can be obtained from Eq. (5-80) by regarding the solenoid as a toroidal coil with an infinite radius. In such a case the dimensions of the cross section of the core are very small in comparison with the radius, and the magnetic flux density inside the solenoid is approximately constant. We have, from Eq. (5-80),

$$B = \mu\left(\frac{N}{2\pi r}\right)I = \mu n I, \tag{5-82}$$

where n is the number of turns per unit length. Thus,

$$B = \mu n I,$$

which is constant inside the solenoid. Hence,

$$\Phi = BS = \mu n S I, \tag{5-83}$$

where S is the cross-sectional area of the solenoid. The flux linkage per unit length† is

$$\Lambda' = n\Phi = \mu n^2 S I. \tag{5-84}$$

Therefore the inductance per unit length is

Inductance per unit length of a long solenoid

$$\boxed{L' = \mu n^2 S \qquad \text{(H/m)}.} \tag{5-85}$$

Equation (5-85) is an approximate formula, based on the assumption that the length of the solenoid is very much greater than the linear dimensions of its cross section. A more accurate derivation for the magnetic flux density and flux linkage per unit length near the ends of a finite solenoid will show that they are less than the values given, respectively, by Eqs. (5-82) and (5-84). Hence the total inductance of a finite solenoid is somewhat less than the values of L', as given in Eq. (5-85) multiplied by the length.

EXAMPLE 5-10

An air coaxial transmission line has a solid inner conductor of radius a and a very thin outer conductor of inner radius b. Determine the inductance per unit length of the line.

†We use a prime to indicate quantities that are per unit length.

SOLUTION

Refer to Fig. 5-16. Assume that a current I flows in the inner conductor and returns via the outer conductor in the other direction. Because of the cylindrical symmetry, $\mathbf{B}$ has only a ϕ-component. Assume also that the current I is uniformly distributed over the cross section of the inner conductor. We find the values of $\mathbf{B}$ first.

a) *Inside the inner conductor,*

$0 \le r \le a.$

From Eq. (5-11),

$$\mathbf{B}_1 = \mathbf{a}_\phi B_{\phi 1} = \mathbf{a}_\phi \frac{\mu_0 r I}{2\pi a^2}. \tag{5-86}$$

b) *Between the inner and outer conductors,*

$a \le r \le b.$

From Eq. (5-12),

$$\mathbf{B}_2 = \mathbf{a}_\phi B_{\phi 2} = \mathbf{a}_\phi \frac{\mu_0 I}{2\pi r}. \tag{5-87}$$

Now consider an annular ring in the inner conductor between radii r and $r + dr$. The current in a unit length of this annular ring is linked by the flux that can be obtained by integrating Eqs. (5-86) and (5-87). We have

$$\begin{aligned} d\Phi' &= \int_r^a B_{\phi 1}\, dr + \int_a^b B_{\phi 2}\, dr \\ &= \frac{\mu_0 I}{2\pi a^2} \int_r^a r\, dr + \frac{\mu_0 I}{2\pi} \int_a^b \frac{dr}{r} \\ &= \frac{\mu_0 I}{4\pi a^2}(a^2 - r^2) + \frac{\mu_0 I}{2\pi} \ln \frac{b}{a}. \end{aligned} \tag{5-88}$$

But the current in the annular ring is only a fraction $(2\pi r\, dr/\pi a^2 = 2r\, dr/a^2)$

FIGURE 5-16 Two views of a coaxial transmission line (Example 5-10).

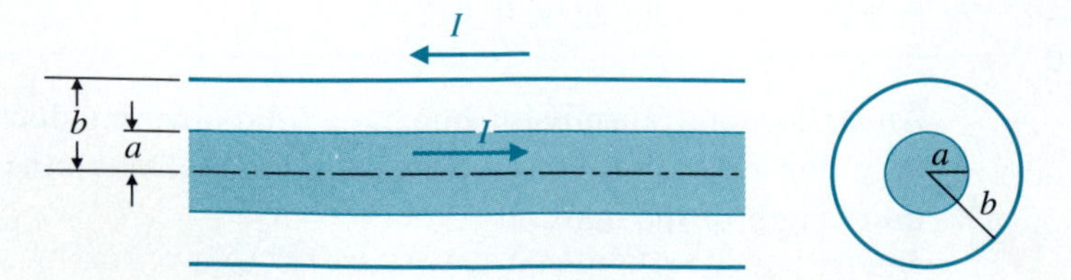

of the total current I. Hence the flux linkage for this annular ring is

$$d\Lambda' = \frac{2r\,dr}{a^2}\,d\Phi'. \tag{5-89}$$

The total flux linkage per unit length is

$$\begin{aligned}\Lambda' &= \int_{r=0}^{r=a} d\Lambda' \\ &= \frac{\mu_0 I}{\pi a^2}\left[\frac{1}{2a^2}\int_0^a (a^2 - r^2)r\,dr + \left(\ln\frac{b}{a}\right)\int_0^a r\,dr\right] \\ &= \frac{\mu_0 I}{2\pi}\left(\frac{1}{4} + \ln\frac{b}{a}\right).\end{aligned}$$

The inductance of a unit length of the coaxial transmission line is therefore

Inductance per unit length of a coaxial transmission line

$$L' = \frac{\Lambda'}{I} = \frac{\mu_0}{8\pi} + \frac{\mu_0}{2\pi}\ln\frac{b}{a} \qquad \text{(H/m)}. \tag{5-90}$$

The first term $\mu_0/8\pi$ arises from the flux linkage internal to the solid inner conductor; it is known as the ***internal inductance*** per unit length of the inner conductor. The second term comes from the linkage of the flux that exists between the inner and the outer conductors; this term is known as the ***external inductance*** per unit length of the coaxial line. If the inner conductor is a thin hollow tube, then the $\mu_0/8\pi$ term does not exist; we would only have external inductance.

EXAMPLE 5-11

Calculate the internal and external inductances per unit length of a transmission line consisting of two long parallel conducting wires of radius a that carry currents in opposite directions. The axes of the wires are separated by a distance d, which is much larger than a.

SOLUTION

The internal self-inductance per unit length of each wire is, from Eq. (5-90) $\mu_0/8\pi$. So for two wires we have

$$L_i' = 2 \times \frac{\mu_0}{8\pi} = \frac{\mu_0}{4\pi} \qquad \text{(H/m)}. \tag{5-91}$$

To find the external self-inductance per unit length, we first calculate the magnetic flux linking with a unit length of the transmission line for an

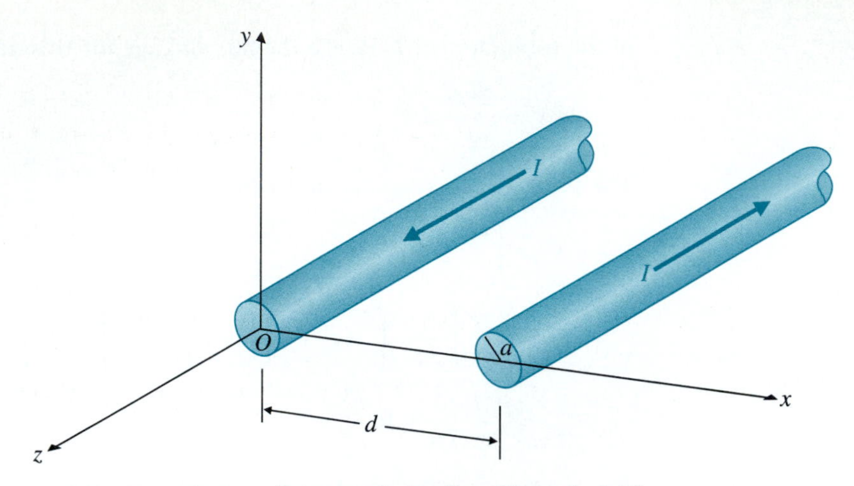

FIGURE 5-17 A two-wire transmission line (Example 5-11).

assumed current I in the wires. In the xz-plane where the two wires lie, as in Fig. 5-17, the contributing **B** vectors due to the equal and opposite currents in the two wires have only a y-component:

$$B_{y1} = \frac{\mu_0 I}{2\pi x}, \tag{5-92}$$

$$B_{y2} = \frac{\mu_0 I}{2\pi(d-x)}. \tag{5-93}$$

The flux linkage per unit length is then

$$\begin{aligned}\Phi' &= \int_a^{d-a} (B_{y1} + B_{y2})\,dx \\ &= \int_a^{d-a} \frac{\mu_0 I}{2\pi}\left[\frac{1}{x} + \frac{1}{d-x}\right]dx \\ &= \frac{\mu_0 I}{\pi} \ln\left(\frac{d-a}{a}\right) \cong \frac{\mu_0 I}{\pi} \ln\frac{d}{a} \qquad \text{(Wb/m)}.\end{aligned}$$

Therefore,

$$L'_e = \frac{\Phi'}{I} = \frac{\mu_0}{\pi} \ln\frac{d}{a} \qquad \text{(H/m)}, \tag{5-94}$$

and the total self-inductance per unit length of the two-wire line is

Inductance per unit length of a parallel-wire transmission line

$$L' = L'_i + L'_e = \frac{\mu_0}{\pi}\left(\frac{1}{4} + \ln\frac{d}{a}\right) \qquad \text{(H/m)}. \tag{5-95}$$

It can be proved formally that the mutual inductance L_{12} between two circuits C_1 and C_2 obtained from the magnetic flux linking C_2 due to a unit current in C_1 is the same as the mutual inductance L_{21} obtained from the magnetic flux linking C_1 due to a unit current in C_2; that is, $L_{12} = L_{21}$. Therefore, as a first step in working a problem of determining mutual inductance, we should examine the given geometry and use the simpler of the two ways.

EXAMPLE 5-12

Determine the mutual inductance between a conducting rectangular loop and a very long straight wire as shown in Fig. 5-18.

SOLUTION

In the present problem we see that it is quite simple to assume a current in the long straight wire, write the magnetic flux density, and find the flux linking with the rectangular loop. However, it would be more involved to find the magnetic flux density and the flux linking with the straight wire due to an assumed current in the rectangular loop.

Let us designate the long straight wire as circuit 1 and the rectangular loop as circuit 2. Magnetic flux density $\mathbf{B}_1$ due to a current I_1 in the wire is, by

FIGURE 5-18 A conducting rectangular loop and a long straight wire (Example 5-12).

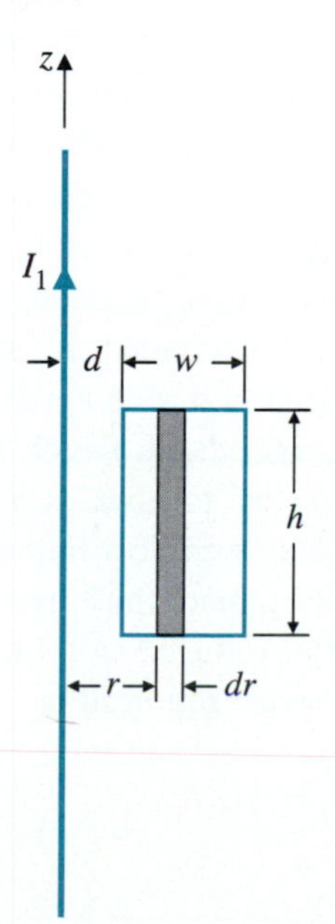

applying Ampère's circuital law,

$$\mathbf{B}_1 = \mathbf{a}_\phi \frac{\mu_0 I_1}{2\pi r}. \tag{5-96}$$

The flux linkage $\Lambda_{12} = \Phi_{12}$ is

$$\Lambda_{12} = \int_{S_2} \mathbf{B}_1 \cdot d\mathbf{s}_2, \tag{5-97}$$

where $d\mathbf{s}_2 = \mathbf{a}_\phi h\,dr$. Combination of Eqs. (5-96) and (5-97) gives

$$\begin{aligned}\Lambda_{12} &= \frac{\mu_0 I_1}{2\pi} h \int_d^{d+w} \frac{dr}{r} \\ &= \frac{\mu_0 h I_1}{2\pi} \ln\left(1 + \frac{w}{d}\right).\end{aligned} \tag{5-98}$$

Hence the mutual inductance is

$$L_{12} = \frac{\Lambda_{12}}{I_1} = \frac{\mu_0 h}{2\pi} \ln\left(1 + \frac{w}{d}\right) \quad \text{(H)}. \tag{5-99}$$

5-11 MAGNETIC ENERGY

So far we have discussed self- and mutual inductances in static terms. However, we know that resistanceless inductors appear as short-circuits to steady (d-c) currents; it is obviously necessary that we consider alternating currents when the effects of inductances on circuits and magnetic fields are of interest. A general consideration of time-varying electromagnetic fields (electrodynamics) will be deferred until the next chapter.

In Section 3-10 we discussed the fact that work is required to assemble a group of charges and that the work is stored as electric energy. We certainly expect that work also needs to be expended in sending currents into conducting loops and that it will be stored as magnetic energy. Consider a single closed loop with a self-inductance L_1 in which the current is initially zero. A current generator is connected to the loop, which increases the current i_1 from zero to I_1. From physics we know that an electromotive force (emf) will be induced in the loop that opposes the current change. An amount of work must be done to overcome this induced emf. Let $v_1 = L_1\, di_1/dt$ be the voltage across the inductance. The work required is

$$W_1 = \int v_1 i_1\, dt = L_1 \int_0^{I_1} i_1\, di_1 = \tfrac{1}{2} L_1 I_1^2. \tag{5-100}$$

which is stored as *magnetic energy*.

Now consider two closed loops C_1 and C_2 carrying currents i_1 and i_2, respectively. The currents are initially zero and are to be increased to I_1 and I_2, respectively. To find the amount of work required, we first keep $i_2 = 0$ and increase i_1 from zero to I_1. This requires a work W_1 in loop C_1, as given in Eq. (5-100); no work is done in loop C_2, since $i_2 = 0$. Next we keep i_1 at I_1 and increase i_2 from zero to I_2. Because of mutual coupling, some of the magnetic flux due to i_2 will link with loop C_1, giving rise to an induced emf that must be overcome by a voltage $v_{21} = \pm L_{21}\, di_2/dt$ in order to keep i_1 constant at its value I_1. The work involved is

$$W_{21} = \int v_{21} I_1\, dt = \pm L_{21} I_1 \int_0^{I_2} di_2 = \pm L_{21} I_1 I_2. \tag{5-101}$$

In Eq. (5-101), the plus sign applies if I_1 and I_2 in C_1 and C_2 are such that their magnetic fields strengthen each other and the minus sign applies if their magnetic fields oppose each other.

At the same time a work W_{22} must be done in loop C_2 in order to counteract the induced emf as i_2 is increased from 0 to I_2.

$$W_{22} = \tfrac{1}{2} L_2 I_2^2. \tag{5-102}$$

The total amount of work done in raising the currents in loops C_1 and C_2 from zero to I_1 and I_2, respectively, is then the sum of W_1, W_{21}, and W_{22}:

Magnetic energy stored in two coupled current-carrying loops

$$\boxed{W_2 = \tfrac{1}{2} L_1 I_1^2 \pm L_{21} I_1 I_2 + \tfrac{1}{2} L_2 I_2^2,} \tag{5-103}$$

which is the energy stored in the magnetic field of the two coupled current-carrying loops.

For a current I flowing in a single inductor with inductance L, the stored magnetic energy is

Magnetic energy stored in an inductance

$$\boxed{W_m = \tfrac{1}{2} L I^2 \qquad \text{(J)}.} \tag{5-104}$$

■ **EXERCISE 5.8** Express the stored magnetic energy in terms of flux linkage Φ and current I in an inductor having an inductance L.

ANS. $\Phi I/2$.

5-11.1 MAGNETIC ENERGY IN TERMS OF FIELD QUANTITIES

When we discussed electrostatic energy in Subsection 3-10.1, we found it convenient to express W_e in terms of field quantities $\mathbf{E}$ and $\mathbf{D}$, as was done in Eqs. (3-105) and (3-106). From our work so far we notice the following analogous relations between the quantities in electrostatics and those in magnetostatics:

Electrostatics	Magnetostatics
E	**B**
D	**H**
ϵ	$1/\mu$

It turns out that we could correctly write magnetic energy W_m in a linear medium in terms of **B** and **H** from Eq. (3-105) using the above analogy. Thus,

Magnetic energy in terms of B and H

$$W_m = \frac{1}{2}\int_{V'} \mathbf{H}\cdot\mathbf{B}\,dv \qquad \text{(J)}. \tag{5-105}$$

On using the constitutive relation $\mathbf{H} = \mathbf{B}/\mu$ for a linear medium, we can write

Magnetic energy in terms of B and μ

$$W_m = \frac{1}{2}\int_{V'} \frac{B^2}{\mu}\,dv \qquad \text{(J)}. \tag{5-106}$$

A separate formal derivation of Eqs. (5-105) and (5-106) will not be included here. We will deal with electric and magnetic energies again in Section 7-5 when we discuss flow of electromagnetic power.

If we define a ***magnetic energy density***, w_m, such that its volume integral equals the total magnetic energy

$$W_m = \int_{V'} w_m\,dv, \tag{5-107}$$

we can write w_m as

$$w_m = \tfrac{1}{2}\mathbf{H}\cdot\mathbf{B} \qquad (\text{J/m}^3), \tag{5-108a}$$

or

$$w_m = \frac{B^2}{2\mu} \qquad (\text{J/m}^3). \tag{5-108b}$$

By using Eq. (5-104) in conjunction with Eq. (5-105) or Eq. (5-106), we can often determine self-inductance more easily from stored magnetic energy calculated in terms of **B** and/or **H**, than from flux linkage. We have

$$L = \frac{2W_m}{I^2} \qquad \text{(H)}. \tag{5-109}$$

EXAMPLE 5-13

By using stored magnetic energy, determine the inductance per unit length of an air coaxial transmission line that has a solid inner conductor of radius a and a very thin outer conductor of inner radius b.

SOLUTION

This is the same problem as that in Example 5-10, in which the self-inductance was determined through a consideration of flux linkages. Refer again to Fig. 5-16. Assume that a uniform current I flows in the inner conductor and returns in the outer conductor. The magnetic energy per unit length stored in the inner conductor is, from Eqs. (5-86) and (5-106),

$$\begin{aligned} W'_{m1} &= \frac{1}{2\mu_0}\int_0^a B_{\phi 1}^2 2\pi r\, dr \\ &= \frac{\mu_0 I^2}{4\pi a^4}\int_0^a r^3\, dr = \frac{\mu_0 I^2}{16\pi} \qquad \text{(J/m)}. \end{aligned} \tag{5-110}$$

The magnetic energy per unit length stored in the region between the inner and outer conductors is, from Eq. (5-87) and (5-106),

$$\begin{aligned} W'_{m2} &= \frac{1}{2\mu_0}\int_a^b B_{\phi 2}^2 2\pi r\, dr \\ &= \frac{\mu_0 I^2}{4\pi}\int_a^b \frac{1}{r}\, dr = \frac{\mu_0 I^2}{4\pi}\ln\frac{b}{a} \qquad \text{(J/m)}. \end{aligned} \tag{5-111}$$

Therefore, from Eq. (5-109) we have

$$\begin{aligned} L' &= \frac{2}{I^2}(W'_{m1} + W'_{m2}) \\ &= \frac{\mu_0}{8\pi} + \frac{\mu_0}{2\pi}\ln\frac{b}{a} \qquad \text{(H/m)}, \end{aligned} \tag{5-112}$$

which is the same as Eq. (5-90). The procedure used in this solution is comparatively simpler than that used in Example 5-10.

■ **EXERCISE 5.9**

A current I flows in the N-turn toroidal coil in Fig. 5-15.

a) Obtain an expression for the stored magnetic energy.

b) Use Eq. (5-109) to determine its self-inductance and check your result with Eq. (5-81).

ANS. (a) $[\mu_0 (NI)^2 h \ln(b/a)]/4\pi$ (J).

REVIEW QUESTIONS

Q.5-25 Define (a) the mutual inductance between two circuits, and (b) the self-inductance of a single coil.

Q.5-26 What is meant by the internal inductance of a conductor?

Q.5-27 Write the expression for the stored magnetic energy of two coupled current-carrying loops.

Q.5-28 Write the expression for stored magnetic energy in terms of field quantities.

REMARKS

1. The self-inductance of wire-wound solenoidal and toroidal coils is proportional to the *square* of the number of turns.
2. The internal inductance of straight, thin, conducting tubes is approximately zero, and that of straight, solid, nonferromagnetic conductors is $\mu_0/8\pi$ (H/m).
3. The mutual inductance between two coupled circuits has the property that $L_{12} = L_{21}$.
4. Consideration of inductances necessarily involves alternating currents (a.c.) because at d.c. resistanceless inductors appear as short circuits.
5. At very high frequencies, current distribution in conductors is not uniform and tends to concentrate on the surface (due to *skin effect*, which will be discussed in Chapter 7). This phenomenon should be taken into consideration in inductance calculations.

5-12 MAGNETIC FORCES AND TORQUES

Earlier we noted that a charge q moving with a velocity $\mathbf{u}$ in a magnetic field with flux density $\mathbf{B}$ experiences a magnetic force $\mathbf{F}_m$ given by Eq. (5-4), which is repeated below:

$$\mathbf{F}_m = q\mathbf{u} \times \mathbf{B} \qquad \text{(N)}. \tag{5-113}$$

In this section we will discuss various aspects of forces and torques on current-carrying circuits in static magnetic fields.

5-12.1 FORCES AND TORQUES ON CURRENT-CARRYING CONDUCTORS

Let us consider an element of conductor $d\boldsymbol{\ell}$ with a cross-sectional area S. If there are N charge carriers (electrons) per unit volume moving with a velocity $\mathbf{u}$ in the direction of $d\boldsymbol{\ell}$, then the magnetic force on the differential element is, according to Eq. (5-113),

$$d\mathbf{F}_m = -NeS|d\boldsymbol{\ell}|\mathbf{u} \times \mathbf{B}$$
$$= -NeS|\mathbf{u}|\, d\boldsymbol{\ell} \times \mathbf{B}, \tag{5-114}$$

where e is the electronic charge. The two expressions in Eq. (5-114) are equivalent, since $\mathbf{u}$ and $d\boldsymbol{\ell}$ have the same direction. Now, since $-NeS|\mathbf{u}|$ equals the current in the conductor, we can write Eq. (5-114) as

$$\boxed{d\mathbf{F}_m = I\, d\boldsymbol{\ell} \times \mathbf{B} \qquad (\text{N}).} \tag{5-115}$$

The magnetic force on a complete (closed) circuit of contour C that carries a current I in a magnetic field $\mathbf{B}$ is then

Magnetic force on a current-carrying circuit in a magnetic field

$$\boxed{\mathbf{F}_m = I \oint_C d\boldsymbol{\ell} \times \mathbf{B} \qquad (\text{N}).} \tag{5-116}$$

When we have two circuits carrying currents I_1 and I_2, respectively, the situation is that of one current-carrying circuit in the magnetic field of the other. In the presence of the magnetic flux $\mathbf{B}_{12}$, which was caused by the current I_1 in C_1, the force $\mathbf{F}_{12}$ on circuit C_2 can be written as

$$\mathbf{F}_{12} = I_2 \oint_{C_2} d\boldsymbol{\ell}_2 \times \mathbf{B}_{12}, \tag{5-117a}$$

where $\mathbf{B}_{12}$ is, from the Biot-Savart law in Eq. (5-31),

$$\mathbf{B}_{12} = \frac{\mu_0 I_1}{4\pi} \oint_{C_1} \frac{d\boldsymbol{\ell}_1 \times \mathbf{a}_{R_{12}}}{R_{12}^2}. \tag{5-117b}$$

Combining Eqs. (5-117a) and (5-117b), we obtain

Ampère's law of force between two current-carrying circuits

$$\boxed{\mathbf{F}_{12} = \frac{\mu_0}{4\pi} I_2 I_1 \oint_{C_2} \oint_{C_1} \frac{d\boldsymbol{\ell}_2 \times (d\boldsymbol{\ell}_1 \times \mathbf{a}_{R_{12}})}{R_{12}^2} \qquad (\text{N}),} \tag{5-118}$$

which is ***Ampère's law of force*** between two current-carrying circuits. It is an inverse-square relationship and should be compared with Coulomb's law of force in Eq. (3-13) between two stationary charges. We see that the force formula for two current-carrying circuits is much more complicated than that for two stationary charges. In actual computation it is convenient to break the formidable-looking Eq. (5-118) into two steps as represented by Eqs. (5-117a) and (5-117b).

The force $\mathbf{F}_{21}$ on circuit C_1 due to magnetic flux caused by current I_2 in C_2 is obtained from Eq. (5-118) by simply interchanging 1 and 2 in the subscripts. Newton's third law governing action and reaction assures that $\mathbf{F}_{21} = -\mathbf{F}_{12}$.

EXAMPLE 5-14

Determine the force per unit length between two infinitely long, thin, parallel conducting wires carrying currents I_1 and I_2 in the same direction. The wires are separated by a distance d.

SOLUTION

Let the wires lie in the yz-plane and the left-hand wire be designated circuit 1, as shown in Fig. 5-19. This problem is a straightforward application of Eq. (5-117a). Using $\mathbf{F}'_{12}$ to denote the force per unit length on wire 2, we have

$$\mathbf{F}'_{12} = I_2(\mathbf{a}_z \times \mathbf{B}_{12}), \tag{5-119}$$

where $\mathbf{B}_{12}$, the magnetic flux density at wire 2, set up by the current I_1 in wire 1, is constant over wire 2. Because the wires are assumed to be infinitely long and cylindrical symmetry exists, it is not necessary to use Eq. (5-117b) for the determination of $\mathbf{B}_{12}$. We apply Ampère's circuital law and write, from Eq. (5-12),

$$\mathbf{B}_{12} = -\mathbf{a}_x \frac{\mu_0 I_1}{2\pi d}. \tag{5-120}$$

Substitution of Eq. (5-120) in Eq. (5-119) yields

$$\mathbf{F}'_{12} = -\mathbf{a}_y \frac{\mu_0 I_1 I_2}{2\pi d} \quad \text{(N/m)}. \tag{5-121}$$

Force of attraction (repulsion) between wires carrying currents in the same direction (in opposite directions)

We see that the force on wire 2 pulls it toward wire 1. Hence the force between two wires carrying *currents in the same direction* is one of *attraction* (unlike the force between two charges of the same polarity, which is one of repulsion).

FIGURE 5-19 Force between two parallel current-carrying wires (Example 5-14).

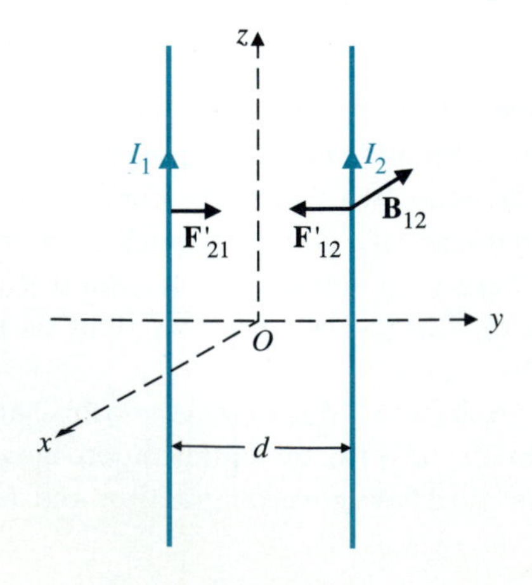

■ **EXERCISE 5.10** Assume that a current I_2 flows in the rectangular loop in Fig. 5-18 in the clockwise direction. Determine the net force on the loop.

ANS. $-\mathbf{a}_r \mu_0 I_1 I_2 hw/2\pi d(d + w)$.

Let us now consider a small circular loop of radius b and carrying a current I in a uniform magnetic field of flux density $\mathbf{B}$. It is convenient to resolve $\mathbf{B}$ into two components, $\mathbf{B} = \mathbf{B}_\perp + \mathbf{B}_\parallel$, where $\mathbf{B}_\perp$ and $\mathbf{B}_\parallel$ are perpendicular and parallel, respectively, to the plane of the loop. As illustrated in Fig. 5-20(a), the perpendicular component $\mathbf{B}_\perp$ tends to expand the loop (or contract it if the direction of I is reversed), but exerts no net force to move the loop. The parallel component $\mathbf{B}_\parallel$ produces an upward force $d\mathbf{F}_1$ (out from the paper) on element $d\boldsymbol{\ell}_1$ and a downward force (into the paper) $d\mathbf{F}_2 = -d\mathbf{F}_1$ on the symmetrically located element $d\boldsymbol{\ell}_2$, as shown in Fig. 5-20(b). Although the net force on the entire loop caused by $\mathbf{B}_\parallel$ is also zero, a torque exists that tends to rotate the loop about the x-axis in such a way as to *align* the magnetic field (due to I) with the external $\mathbf{B}_\parallel$ field. The differential torque produced by $d\mathbf{F}_1$ and $d\mathbf{F}_2$ is

$$\begin{aligned} d\mathbf{T} &= \mathbf{a}_x(dF)2b \sin\phi \\ &= \mathbf{a}_x(I\,d\ell\,B_\parallel \sin\phi)2b \sin\phi \\ &= \mathbf{a}_x 2Ib^2 B_\parallel \sin^2\phi\,d\phi, \end{aligned} \tag{5-122}$$

where $dF = |d\mathbf{F}_1| = |d\mathbf{F}_2|$ and $d\ell = |d\boldsymbol{\ell}_1| = |d\boldsymbol{\ell}_2| = b\,d\phi$. The total torque acting on the loop is then

$$\begin{aligned} \mathbf{T} &= \int d\mathbf{T} = \mathbf{a}_x 2Ib^2 B_\parallel \int_0^\pi \sin^2\phi\,d\phi \\ &= \mathbf{a}_x I(\pi b^2) B_\parallel . \end{aligned} \tag{5-123}$$

FIGURE 5-20 A circular loop in a uniform magnetic field $\mathbf{B} = \mathbf{B}_\perp + \mathbf{B}_\parallel$.

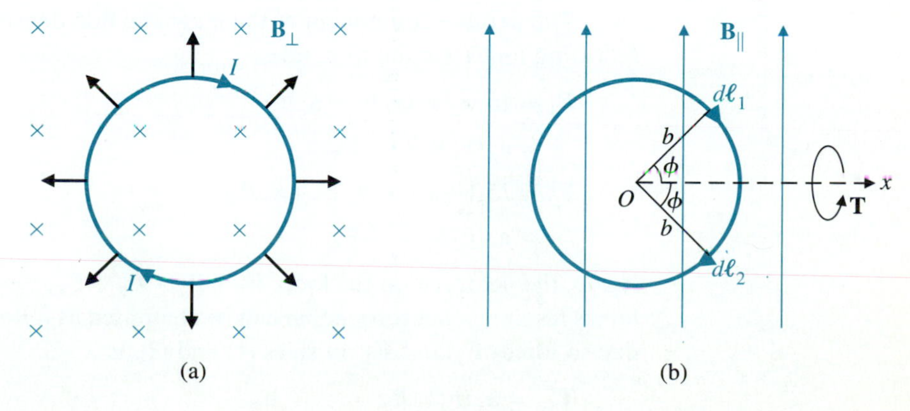

If the definition of the magnetic dipole moment in Eq. (5-45) is used,

$$\mathbf{m} = \mathbf{a}_n I(\pi b^2) = \mathbf{a}_n IS,$$

where $\mathbf{a}_n$ is a unit vector in the direction of the right thumb (normal to the plane of the loop) as the fingers of the right hand follow the direction of the current, we can write Eq. (5-123) as

Torque experienced by a current-carrying circuit in a magnetic field

$$\boxed{\mathbf{T} = \mathbf{m} \times \mathbf{B} \qquad (\text{N}\cdot\text{m}).} \tag{5-124}$$

The vector $\mathbf{B}$ (instead of $\mathbf{B}_{\parallel}$) is used in Eq. (5-124) because $\mathbf{m} \times (\mathbf{B}_{\perp} + \mathbf{B}_{\parallel}) = \mathbf{m} \times \mathbf{B}_{\parallel}$. This is the torque that aligns the microscopic magnetic dipoles in magnetic materials and causes the materials to be magnetized by an applied magnetic field. It should be remembered that Eq. (5-124) does not hold if $\mathbf{B}$ is not uniform over the current-carrying loop.

EXAMPLE 5-15

A rectangular loop in the xy-plane with sides b_1 and b_2 carrying a current I lies in a *uniform* magnetic field $\mathbf{B} = \mathbf{a}_x B_x + \mathbf{a}_y B_y + \mathbf{a}_z B_z$. Determine the force and torque on the loop.

SOLUTION

Resolving $\mathbf{B}$ into perpendicular and parallel components $\mathbf{B}_{\perp}$ and $\mathbf{B}_{\parallel}$, we have

$$\mathbf{B}_{\perp} = \mathbf{a}_z B_z; \tag{5-125a}$$

$$\mathbf{B}_{\parallel} = \mathbf{a}_x B_x + \mathbf{a}_y B_y. \tag{5-125b}$$

Assuming that the current flows in a clockwise direction, as shown in Fig. 5-21, we find that the perpendicular component $\mathbf{a}_z B_z$ results in forces $Ib_1 B_z$ on sides (1) and (3) and forces $Ib_2 B_z$ on sides (2) and (4), all directed toward the center of the loop. The vector sum of these four contracting forces is zero, and no torque is produced.

The parallel component of the magnetic flux density, $\mathbf{B}_{\parallel}$, produces the following forces on the four sides:

$$\begin{aligned} \mathbf{F}_1 &= Ib_1 \mathbf{a}_x \times (\mathbf{a}_x B_x + \mathbf{a}_y B_y) \\ &= \mathbf{a}_z I b_1 B_y = -\mathbf{F}_3; \end{aligned} \tag{5-126a}$$

$$\begin{aligned} \mathbf{F}_2 &= Ib_2(-\mathbf{a}_y) \times (\mathbf{a}_x B_x + \mathbf{a}_y B_y) \\ &= \mathbf{a}_z I b_2 B_x = -\mathbf{F}_4. \end{aligned} \tag{5-126b}$$

Again, the net force on the loop, $\mathbf{F}_1 + \mathbf{F}_2 + \mathbf{F}_3 + \mathbf{F}_4$, is zero. However, these forces result in a net torque that can be computed as follows. The torque $\mathbf{T}_{13}$, due to forces $\mathbf{F}_1$ and $\mathbf{F}_3$ on sides (1) and (3), is

$$\mathbf{T}_{13} = \mathbf{a}_x I b_1 b_2 B_y; \tag{5-127a}$$

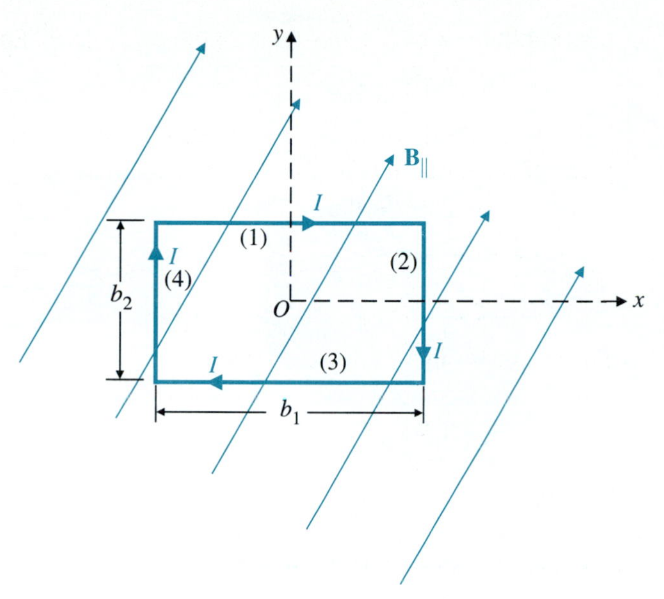

FIGURE 5-21 A rectangular loop in a uniform magnetic field (Example 5-15).

the torque $\mathbf{T}_{24}$, due to forces $\mathbf{F}_2$ and $\mathbf{F}_4$ on sides (2) and (4), is

$$\mathbf{T}_{24} = -\mathbf{a}_y I b_1 b_2 B_x. \tag{5-127b}$$

The total torque on the rectangular loop is then

$$\mathbf{T} = \mathbf{T}_{13} + \mathbf{T}_{24} = I b_1 b_2 (\mathbf{a}_x B_y - \mathbf{a}_y B_x) \qquad (\text{N}\cdot\text{m}). \tag{5-128}$$

Since the magnetic moment of the loop is $\mathbf{m} = -\mathbf{a}_z I b_1 b_2$, the result in Eq. (5-128) is exactly $\mathbf{T} = \mathbf{m} \times (\mathbf{a}_x B_x + \mathbf{a}_y B_y) = \mathbf{m} \times \mathbf{B}$. Hence in spite of the fact that Eq. (5-124) was derived for a circular loop, the torque formula holds also for a rectangular loop. As a matter of fact, it can be proved that Eq. (5-124) holds for a planar loop of any shape as long as it is located in a uniform magnetic field.

5-12.2 DIRECT-CURRENT MOTORS

The principle of operation of direct-current (d-c) motors is based on Eq. (5-124). Figure 5-22(a) shows a schematic diagram of such a motor. The magnetic field $\mathbf{B}$ is produced by a field current I_f in a winding around the pole pieces. When a current I is sent through the rectangular loop, a torque results that makes the loop rotate in a clockwise direction as viewed from the $+x$-direction. This is illustrated in Fig. 5-22(b). A split ring with brushes is necessary so that the currents in the two legs of the coil reverse their directions every half of a turn in order to maintain the torque $\mathbf{T}$ always in the

Principle of operation of d-c motors

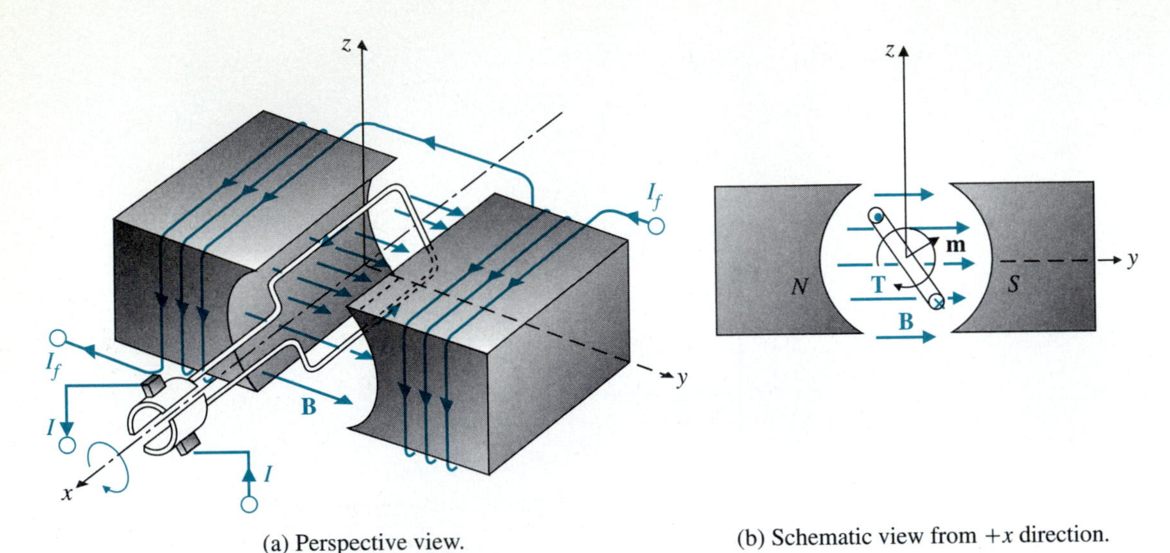

(a) Perspective view. (b) Schematic view from $+x$ direction.

FIGURE 5-22 Illustrating the principle of operation of d-c motor.

same direction; the magnetic moment **m** of the loop must have a positive z-component.

To obtain a smooth and efficient operation, an actual d-c motor has many such rectangular loops wound and distributed around an armature with a ferromagnetic core. The ends of each loop are attached to a pair of conducting bars arranged on a small cylindrical drum called a ***commutator***. The commutator has twice as many parallel conducting bars insulated from one another as there are loops.

5-12.3 FORCES AND TORQUES IN TERMS OF STORED MAGNETIC ENERGY

All current-carrying conductors and circuits experience magnetic forces when situated in a magnetic field. They are held in place only if mechanical forces, equal and opposite to the magnetic forces, exist. Except for special symmetrical cases (such as the case of the two infinitely long, current-carrying, parallel conducting wires in Example 5-14), determining the magnetic forces between current-carrying circuits by Ampère's law of force is often a tedious task. We now examine an alternative method of finding magnetic forces and torques based on the ***principle of virtual displacement***. This principle was used in Subsection 3-10.2 to determine electrostatic forces between charged conductors.

If we assume that no changes in flux linkages result from a virtual differential displacement $d\boldsymbol{\ell}$ of one of the current-carrying circuits, there will be no induced emf's, and the sources will supply no energy to the system. The mechanical work, $\mathbf{F}_\Phi \cdot d\boldsymbol{\ell}$, done *by the system* is at the expense of a decrease in the stored magnetic energy, W_m. Here, $\mathbf{F}_\Phi$ denotes the force under the constant-flux condition. We have

$$\mathbf{F}_\Phi \cdot d\boldsymbol{\ell} = -dW_m = -(\nabla W_m)\cdot d\boldsymbol{\ell}, \tag{5-129}$$

from which it follows that

Determining magnetic force on a current-carrying circuit, by method of virtual displacement

$$\boxed{\mathbf{F}_\Phi = -\nabla W_m \qquad \text{(N).}} \tag{5-130}$$

In three-dimensional space, the vector equation (5-130) is actually three equations. For instance, in Cartesian coordinates the force in the x-direction is

$$(F_\Phi)_x = -\frac{\partial W_m}{\partial x}. \tag{5-131}$$

Similar expressions can be written for the other directions.

If the circuit is constrained to rotate about an axis, say the z-axis, the mechanical work done by the system will be $(T_\Phi)_z\, d\phi$, and

Torque about a given axis on a current-carrying circuit in a magnetic field, by method of virtual displacement

$$\boxed{(T_\Phi)_z = -\frac{\partial W_m}{\partial \phi} \qquad \text{(N}\cdot\text{m),}} \tag{5-132}$$

which is the z-component of the torque acting on the circuit under the condition of constant flux linkages.

EXAMPLE 5-16

Consider the electromagnet in Fig. 5-23 in which a current I in an N-turn coil produces a flux Φ in the magnetic circuit. The cross-sectional area of the core is S. Determine the lifting force on the armature.

FIGURE 5-23 An electromagnet (Example 5-16).

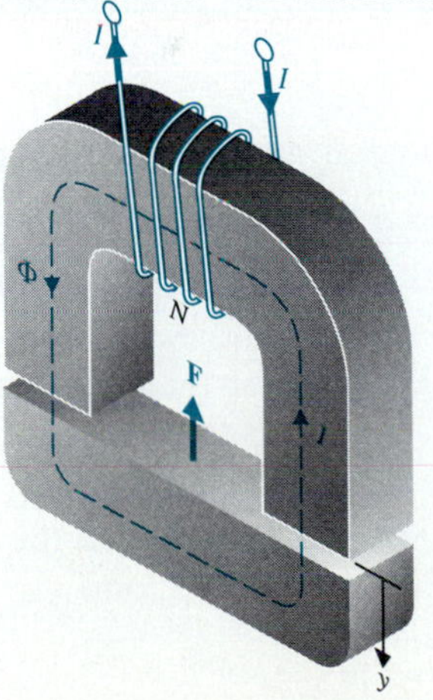

SOLUTION

Let the armature take a virtual displacement dy (a differential increase in y) and the source be adjusted to keep the flux Φ constant. A displacement of the armature changes only the length of the air gaps; consequently, the displacement changes only the magnetic energy stored in the two air gaps. We have, from Eq. (5-106),

$$\begin{aligned} dW_m = d(W_m)_{\substack{\text{air}\\ \text{gap}}} &= 2\left(\frac{B^2}{2\mu_0}\, S\, dy\right) \\ &= \frac{\Phi^2}{\mu_0 S}\, dy. \end{aligned} \tag{5-133}$$

An increase in the air-gap length (a positive dy) increases the stored magnetic energy if Φ is constant. Using Eq. (5-130), we obtain the force in the y-direction:

$$\mathbf{F}_\Phi = \mathbf{a}_y(F_\Phi)_y = \mathbf{a}_y\left(-\frac{\partial W_m}{\partial y}\right) = -\mathbf{a}_y\frac{\Phi^2}{\mu_0 S} \qquad \text{(N)}. \tag{5-134}$$

Here the negative sign indicates that the force tends to reduce the air-gap length; that is, it is a force of attraction.

REVIEW QUESTIONS

Q.5-29 Give the integral expression for the force on a closed circuit that carries a current I in a magnetic field $\mathbf{B}$.

Q.5-30 Write the formula expressing the torque on a current-carrying circuit in a magnetic field.

Q.5-31 Explain the principle of operation of d-c motors.

Q.5-32 What is the relation between the force and the stored magnetic energy in a system of current-carrying circuits under the condition of constant flux linkages?

REMARKS

1. The magnetic force between two current-carrying wires is one of *attraction* if the currents are in the *same* direction, and one of *repulsion* if the currents are in *opposite* directions.
2. The torque on a current-carrying loop (including microscopic magnetic dipoles) is in a direction that tends to *align* the magnetic moment of the loop with the applied magnetic field.
3. The torque formula $\mathbf{T} = \mathbf{m} \times \mathbf{B}$ holds only if the external magnetic field is *uniform* over the current-carrying loop.

SUMMARY

A charge in motion in a region where electric and magnetic fields exist experiences both an electric force and a magnetic force. The total electromagnetic force is given by the Lorentz's force equation. After introducing the formula for the magnetic force on a moving charge in a magnetic field, we

- presented the two fundamental postulates of magnetostatics in free space that specify the divergence and the curl of **B**,
- derived Ampère's circuital law, which enabled us to determine the magnetic flux density due to a current distribution under conditions of symmetry,
- introduced the concept of magnetic vector potential,
- derived Biot-Savart law for determining **B** caused by a current flowing in a closed path,
- discussed the microscopic effect of induced dipole moments by finding the equivalent magnetization current densities,
- defined magnetic field intensity, **H**, and relative permeability,
- compared the behavior of different magnetic materials,
- found the boundary conditions for static magnetic fields,
- defined self-inductance and mutual inductance, and explained the procedure for their determination,
- found the formula for stored magnetic energy,
- discussed the forces and torques on current-carrying circuits in magnetic fields, and
- explained the principle of operation of direct-current motors.

PROBLEMS

P.5-1 A point charge Q with a velocity $\mathbf{u} = \mathbf{a}_x u_0$ enters a region having a uniform magnetic field $\mathbf{B} = \mathbf{a}_x B_x + \mathbf{a}_y B_y + \mathbf{a}_z B_z$. What **E** field should exist in the region so that the charge proceeds without a change of velocity?

P.5-2 Find the total magnetic flux through a circular toroid with a rectangular cross section of height h. The inner and outer radii of the toroid are a and b, respectively. A current I flows in N turns of closely wound wire around the toroid. Determine the percentage of error if the flux is found by multiplying the cross-sectional area by the flux density at the mean radius. What is the error if $b/a = 5$?

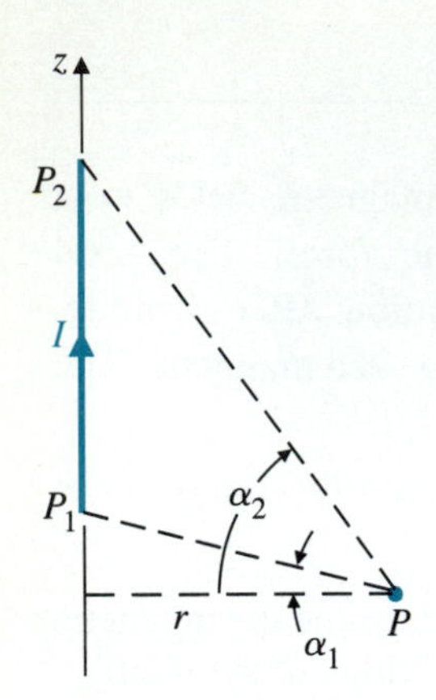

FIGURE 5-24 A finite straight conductor carrying a current I (Problem P.5-3).

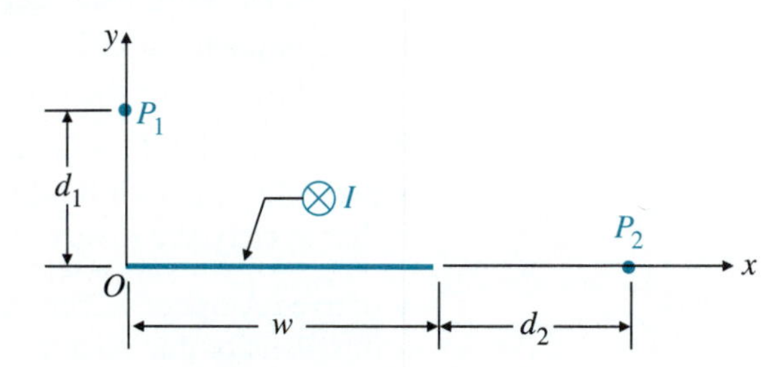

FIGURE 5-25 A thin conducting sheet carrying a current I (for Problems P.5-4 and P.5-5).

P.5-3 A direct current I flows in a straight filamentary conductor P_1P_2.

a) Prove that $\mathbf{B}$ at a point P, whose location is specified by the perpendicular distance r and the two angles α_1 and α_2 shown in Fig. 5-24 is

$$\mathbf{B}_P = \mathbf{a}_\phi \frac{\mu I}{4\pi r}(\sin \alpha_2 - \sin \alpha_1). \tag{5-135}$$

b) Verify that Eq. (5-135) reduces to Eq. (5-35) when the wire is infinitely long.

P.5-4 A current I flows lengthwise in a very long, thin conducting sheet of width w, as shown in Fig. 5-25. Assuming that the current flows into the paper, determine the magnetic flux density $\mathbf{B}_1$ at point $P_1(0, d)$.

P.5-5 Refer to Problem P.5-4 and Fig. 5-25. Find the magnetic flux density $\mathbf{B}_2$ at point $P_2(w + d_2, 0)$.

P.5-6 A current I flows in the inner conductor of an infinitely long coaxial line and returns via the outer conductor. The radius of the inner conductor is a, and the inner and outer radii of the outer conductor are b and c, respectively. Find the magnetic flux density $\mathbf{B}$ for all regions and plot $|\mathbf{B}|$ versus r.

P.5-7 A thin conducting wire of length $3w$ forms a planar equilateral triangle. A direct current I flows in the wire. Find the magnetic flux density at the center of the triangle.

P.5-8 Refer to Fig. 5-26. Determine the magnetic flux density at a point P on the axis of a solenoid with radius b and length L, and with a current I in its N turns of closely wound coil. Show that the result reduces to that given in Eq. (5-82) when L approaches infinity. *Hint:* Use Eq. (5-37).

P.5-9 A direct current I flows in an infinitely long wire of a radius 2(mm) along the z-axis.

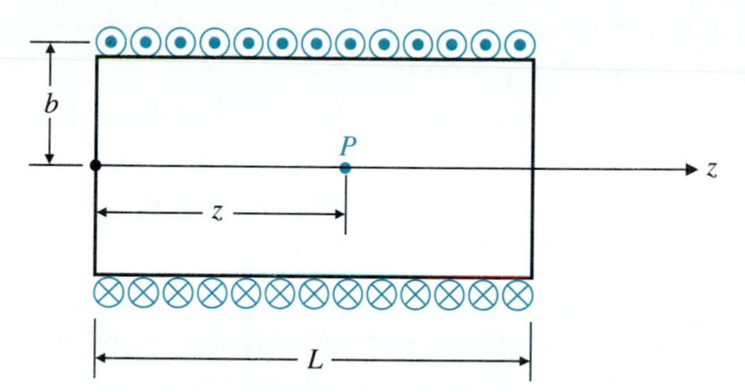

FIGURE 5-26 A solenoid with circular cross section (Problem P.5-8).

a) Obtain the vector magnetic potential **A** at $r > 2$(mm) from the expression of **B** in Eq. (5-12). Choose the reference zero potential at wire surface.

b) If $I = 10$(A), determine from **A** the total amount of magnetic flux passing through a square loop specified by $z = \pm 0.3$(m) and $y = 0.1$(m) and 0.7(m).

P.5-10 A d-c surface current with density $\mathbf{a}_x J_{s0}$ flows in an infinite conducting sheet coinciding with the xy-plane.

a) Determine the magnetic flux density **B** at $(0, 0, z)$ and at $(0, 0, -z)$.

b) Find the vector magnetic potential **A** at $(0, 0, z)$ from **B**. Choose the reference zero potential at an arbitrary point $z = z_0$.

P.5-11 A very large slab of material of thickness d lies perpendicularly to a uniform magnetic field of intensity $\mathbf{H}_0 = \mathbf{a}_z H_0$. Ignoring edge effect, determine the magnetic field intensity in the slab:

a) if the slab material has a permeability μ, and

b) if the slab is a permanent magnet having a magnetization vector $\mathbf{M}_i = \mathbf{a}_z M_i$.

P.5-12 A circular rod of magnetic material with permeability μ is inserted coaxially in a very long air solenoid. The radius of the rod, a, is less than the inner radius, b, of the solenoid. The solenoid's winding has n turns per unit length and carries a current I.

a) Find the values of **B**, **H**, and **M** inside the solenoid for $r < a$ and for $a < r < b$.

b) What are the equivalent magnetization current densities $\mathbf{J}_{mv}$ and $\mathbf{J}_{ms}$ for the magnetized rod?

P.5-13 A ferromagnetic sphere of radius b is magnetized uniformly with a magnetization $\mathbf{M} = \mathbf{a}_z M_0$.

a) Determine the equivalent magnetization current densities $\mathbf{J}_{mv}$ and $\mathbf{J}_{ms}$.

b) Determine the magnetic flux density at the center of the sphere.

P.5-14 Consider a plane boundary ($y = 0$) between air (region 1, $\mu_{r1} = 1$) and iron (region 2, $\mu_{r2} = 5000$).

a) Assuming $\mathbf{B}_1 = \mathbf{a}_x 2 - \mathbf{a}_y 10$ (mT), find $\mathbf{B}_2$ and the angle that $\mathbf{B}_2$ makes with the interface.

b) Assuming $\mathbf{B}_2 = \mathbf{a}_x 10 + \mathbf{a}_y 2$ (mT), find $\mathbf{B}_1$ and the angle that $\mathbf{B}_1$ makes with the normal to the interface.

P.5-15 Determine the self-inductance of a toroidal coil of N turns of wire wound on an air frame with mean radius r_o and a circular cross section of radius b. Obtain an approximate expression assuming $b \ll r_o$.

P.5-16 Determine the mutual inductance between a very long, straight wire and a conducting equilateral triangular loop, as shown in Fig. 5-27.

P.5-17 Find the mutual inductance between two coplanar rectangular loops with parallel sides, as shown in Fig. 5-28. Assume that $h_1 \gg h_2 (h_2 > w_2 > d)$.

P.5-18 Calculate the force per unit length on each of three equidistant, infinitely long, parallel wires 10 (cm) apart, each carrying a current of 25 (A) in the same direction. A cross section of the arrangement is shown in Fig. 5-29. Specify the direction of the force.

P.5-19 The cross section of a long thin metal strip and a parallel wire is shown in Fig. 5-30. Equal and opposite currents I flow in the conductors. Find the force per unit length on the conductors.

P.5-20 The bar AA' in Fig. 5-31 serves as a conducting path (such as the blade

FIGURE 5-27 A long, straight wire and a conducting equilateral triangular loop (Problem P.5-16).

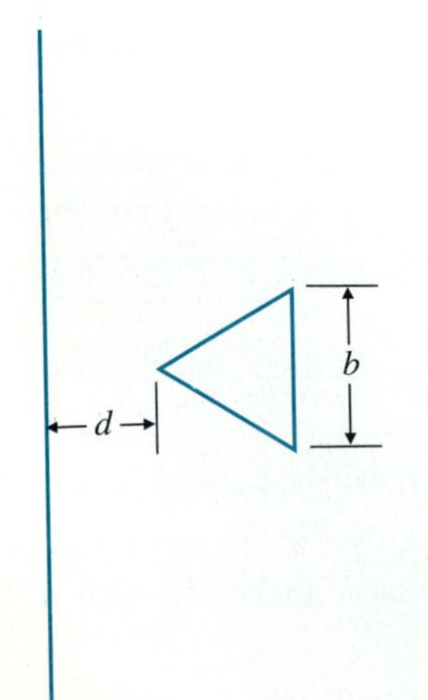

FIGURE 5-28 Two coplanar rectangular loops, $h_1 \gg h_2$ (Problem P.5-17).

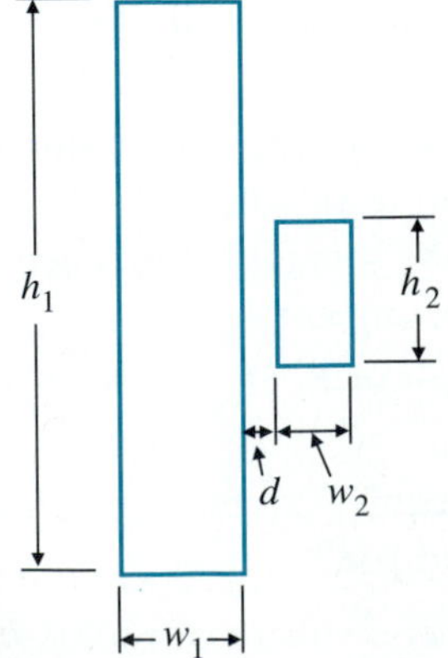

FIGURE 5-29 Three equidistant, infinitely long, current-carrying wires (Problem P.5-18).

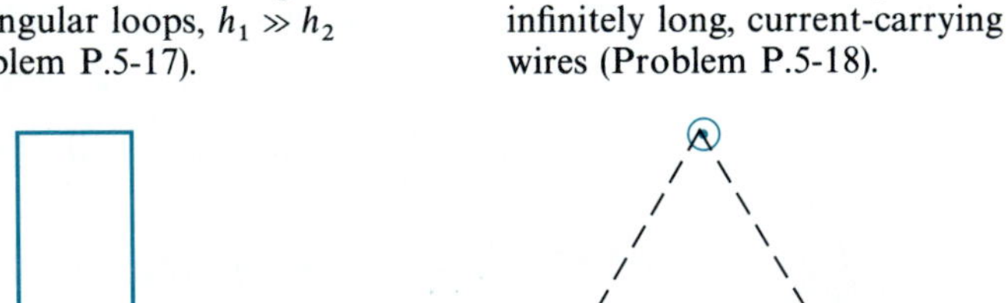

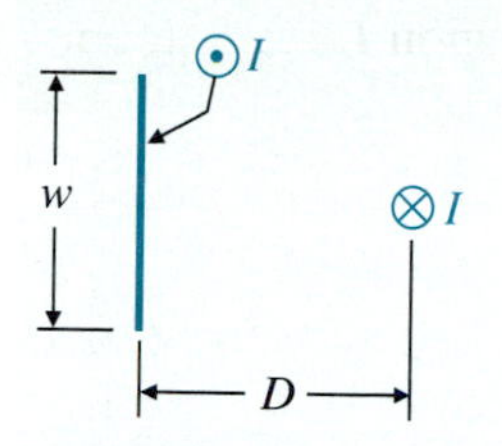

FIGURE 5-30 Cross section of parallel strip and wire conductor (Problem P.5-19).

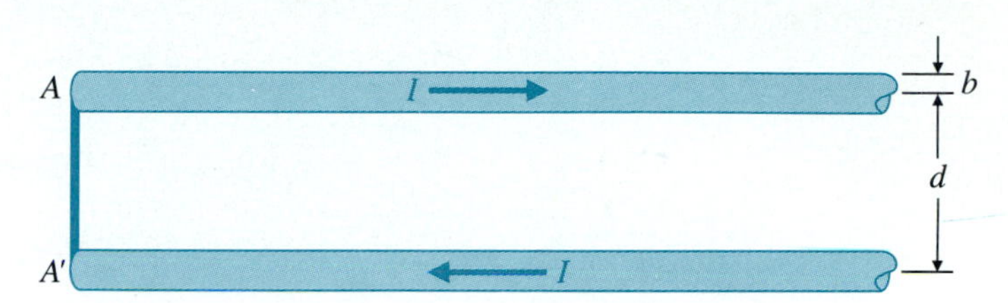

FIGURE 5-31 Force on end conducting bar (Problem P.5-20).

of a circuit breaker) for the current I in two very long parallel lines. The lines have a radius b and are spaced at a distance d apart. Find the direction and the magnitude of the magnetic force on the bar.

P.5-21 A d-c current $I = 10\,(\text{A})$ flows in a triangular loop in the xy-plane as in Fig. 5-32. Assuming a uniform magnetic flux density $\mathbf{B} = \mathbf{a}_y 6\,(\text{mT})$ in the region, find the forces and torque on the loop. The dimensions are in (cm).

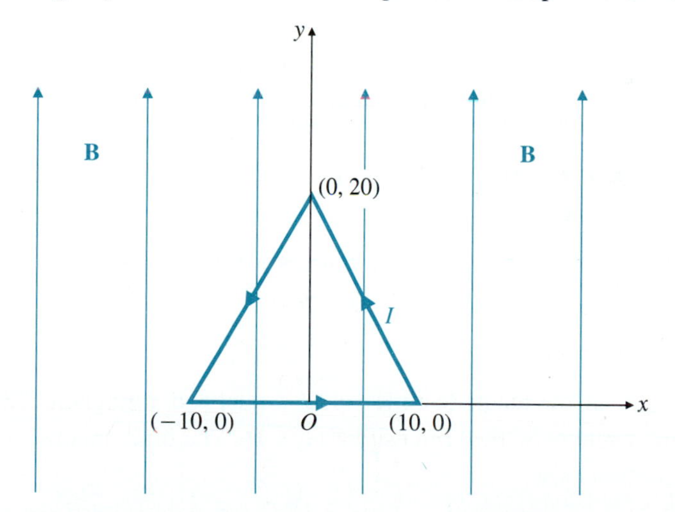

FIGURE 5-32 A triangular loop in a uniform magnetic field (Problem P.5-21).

P.5-22 A small circular turn of wire of radius r_1 that carries a steady current I_1 is placed at the center of a much larger turn of wire of radius r_2 ($r_2 \gg r_1$) that carries a steady current I_2 in the same direction. The angle between the normals of the two circuits is θ and the small circular wire is free to turn about its diameter. Determine the magnitude and the direction of the torque on the small circular wire.

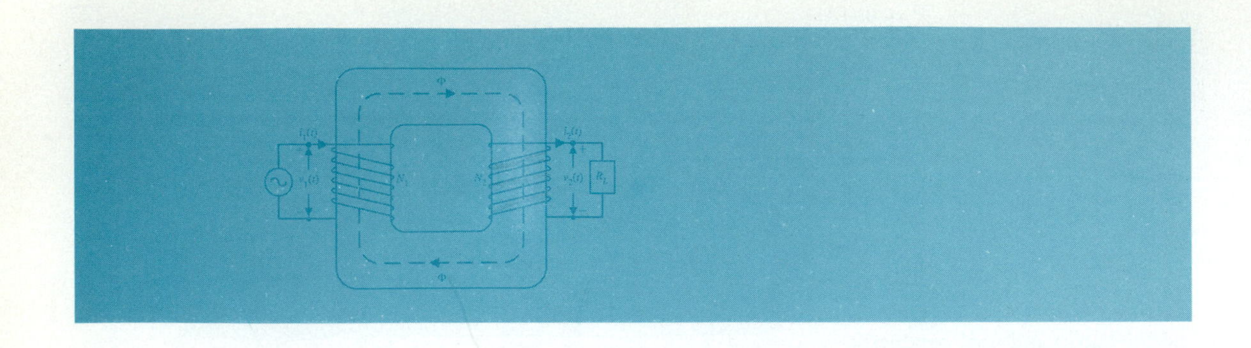

CHAPTER 6

6-1 OVERVIEW

So far we have only dealt with fields that do not change with time. In constructing the electrostatic model we defined an electric field intensity vector, **E**, and an electric flux density (electric displacement) vector, **D**. The fundamental governing differential equations are

Fundamental governing equations for electrostatic model

$$\nabla \times \mathbf{E} = 0, \qquad (3\text{-}4)(6\text{-}1)$$

$$\nabla \cdot \mathbf{D} = \rho_v. \qquad (3\text{-}63)(6\text{-}2)$$

For linear and isotropic (not necessarily homogeneous) media, **E** and **D** are related by the constitutive relation

Constitutive relation for the electric model

$$\mathbf{D} = \epsilon \mathbf{E}. \qquad (3\text{-}67)(6\text{-}3)$$

For the magnetostatic model we defined a magnetic flux density vector, **B**, and a magnetic field intensity vector, **H**. The fundamental governing differential equations are

Fundamental governing equations for magnetostatic model

$$\nabla \cdot \mathbf{B} = 0, \qquad (5\text{-}6)(6\text{-}4)$$

$$\nabla \times \mathbf{H} = \mathbf{J}. \qquad (5\text{-}61)(6\text{-}5)$$

The constitutive relation for **B** and **H** in linear and isotropic media is

Constitutive relation for the magnetic model

$$\mathbf{H} = \frac{1}{\mu} \mathbf{B}. \qquad (5\text{-}66)(6\text{-}6)$$

We observe that **E** and **D** in the electrostatic model are not related to **B** and **H** in the magnetostatic model. In a conducting medium, static electric and

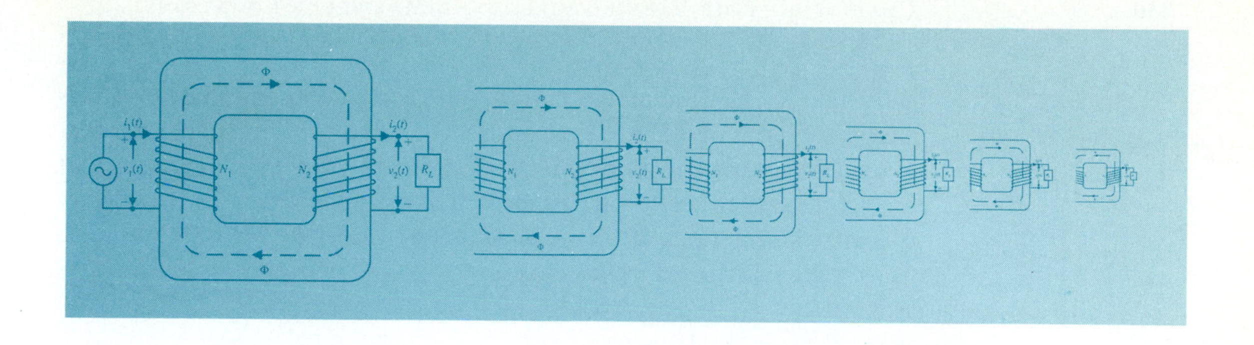

Time-Varying Fields and Maxwell's Equations

Electromagneto-static field

magnetic fields may both exist and form an ***electromagnetostatic field***. A static electric field in a conducting medium causes a steady current to flow that, in turn, gives rise to a static magnetic field. However, the electric field can be completely determined from the static electric charges or potential distributions. The magnetic field is a consequence; it does not enter into the calculation of the electric field.

Static models are simple, but they are inadequate for explaining time-varying electromagnetic phenomena. Static electric and magnetic fields do not give rise to waves that propagate and carry energy and information. Waves are the essence of electromagnetic action at a distance. In this chapter we will see that a changing magnetic field induces an electric field, and vice versa. Under time-varying conditions it is necessary to construct an electromagnetic model in which the electric field vectors **E** and **D** are properly related to the magnetic field vectors **B** and **H**.

We will begin with a fundamental postulate that modifies the $\nabla \times \mathbf{E}$ equation in Eq. (6-1) and leads to Faraday's law of electromagnetic induction. The concepts of transformer emf and motional emf will be discussed. With the new postulate we will also need to modify the $\nabla \times \mathbf{H}$ equation in order to make the governing equations consistent with the law of conservation of charge. The two modified curl equations together with the two divergence equations (6-2) and (6-4), are known as Maxwell's equations and form the foundation of electromagnetic theory. The governing equations for electrostatics and magnetostatics are special forms of Maxwell's equations when all quantities are independent of time. Maxwell's equations can be combined to yield wave equations that predict the existence of electromagnetic waves propagating with the velocity of light. The solutions of the wave equations, especially for time-harmonic fields, will be discussed in this chapter.

6-2 FARADAY'S LAW OF ELECTROMAGNETIC INDUCTION

A major advance in electromagnetic theory was made by Michael Faraday, who, in 1831, discovered experimentally that a current was induced in a conducting loop when the magnetic flux linking the loop changed. The quantitative relationship between the induced emf and the rate of change of flux linkage, based on experimental observation, is known as ***Faraday's law***. It is an experimental law and can be considered as a postulate. However, we do not take the experimental relation concerning a finite loop as the starting point for developing the theory of electromagnetic induction. Instead, we follow our approach in Chapter 3 for electrostatics and in Chapter 5 for magnetostatics by stating a fundamental postulate and developing from it the integral forms of Faraday's law.

The fundamental postulate for electromagnetic induction is

Fundamental postulate for electromagnetic induction

$$\boxed{\nabla \times \mathbf{E} = -\frac{\partial \mathbf{B}}{\partial t}.} \tag{6-7}$$

Equation (6-7) expresses a point-function relationship; that is, it applies to every point in space, whether it be in free space or in a material medium. *The electric field intensity in a region of time-varying magnetic flux density is therefore nonconservative and cannot be expressed as the negative gradient of a scalar potential.*

Electric field intensity in a time-varying magnetic field is not conservative.

Taking the surface integral of both sides of Eq. (6-7) over an open surface and applying Stokes's theorem, we obtain

$$\oint_C \mathbf{E} \cdot d\boldsymbol{\ell} = -\int_S \frac{\partial \mathbf{B}}{\partial t} \cdot d\mathbf{s}. \tag{6-8}$$

Equation (6-8) is valid for any surface S with a bounding contour C, whether or not a physical circuit exists around C. Of course, in a field with no time variation, $\partial \mathbf{B}/\partial t = 0$, Eqs. (6-7) and (6-8) reduce to Eqs. (6-1) and (3-7), respectively, for electrostatics.

In the following subsections we discuss separately the cases of a stationary circuit in a time-varying magnetic field, a moving conductor in a static magnetic field, and a moving circuit in a time-varying magnetic field.

6-2.1 A STATIONARY CIRCUIT IN A TIME-VARYING MAGNETIC FIELD

For a stationary circuit with a contour C and surface S, Eq. (6-8) can be written as

$$\oint_C \mathbf{E} \cdot d\boldsymbol{\ell} = -\frac{d}{dt} \int_S \mathbf{B} \cdot d\mathbf{s}. \tag{6-9}$$

If we define

$$\mathscr{V} = \oint_C \mathbf{E} \cdot d\boldsymbol{\ell} = \text{emf induced in circuit with contour } C \quad \text{(V)} \tag{6-10}$$

and

$$\Phi = \int_S \mathbf{B} \cdot d\mathbf{s} = \text{magnetic flux crossing surface } S \quad \text{(Wb)}, \tag{6-11}$$

then Eq. (6-9) becomes

$$\boxed{\mathscr{V} = -\frac{d\Phi}{dt} \quad \text{(V)}.} \tag{6-12}$$

Faraday's law of electromagnetic induction

Lenz's law of transformer emf

Equation (6-12) states that ***the electromotive force induced in a stationary closed circuit is equal to the negative rate of increase of the magnetic flux linking the circuit***. This is a statement of ***Faraday's law of electromagnetic induction***. The negative sign in Eq. (6-12) is an assertion that the induced emf will cause a current to flow in the closed loop in such a direction as to oppose the change in the linking magnetic flux. This assertion is known as ***Lenz's law***. The emf induced in a stationary loop caused by a time-varying magnetic field is a ***transformer emf***.

EXAMPLE 6-1

A circular loop of N turns of conducting wire lies in the xy-plane with its center at the origin of a magnetic field specified by $\mathbf{B} = \mathbf{a}_z B_0 \cos(\pi r/2b) \sin \omega t$, where b is the radius of the loop and ω is the angular frequency. Find the emf induced in the loop.

SOLUTION

The problem specified a stationary loop in a time-varying magnetic field; hence Eq. (6-12) can be used directly to find the induced emf, $\mathscr{V}$. The magnetic flux linking each turn of the circular loop is

$$\begin{aligned}\Phi &= \int_S \mathbf{B}\cdot d\mathbf{s}\\ &= \int_0^b \left[\mathbf{a}_z B_0 \cos\left(\frac{\pi r}{2b}\right)\sin\omega t\right]\cdot(\mathbf{a}_z 2\pi r\, dr)\\ &= \frac{8b^2}{\pi}\left(\frac{\pi}{2}-1\right)B_0 \sin\omega t.\end{aligned}$$

Since there are N turns, the total flux linkage is $N\Phi$, and we obtain

$$\begin{aligned}\mathscr{V} &= -N\frac{d\Phi}{dt}\\ &= -\frac{8N}{\pi}b^2\left(\frac{\pi}{2}-1\right)B_0\omega\cos\omega t \qquad \text{(V)}.\end{aligned}$$

As the phase of $\cos\omega t$ leads that of $\sin\omega t$ by 90°, we see that the phase of the induced emf lags behind that of the flux linkage by 90°.

EXERCISE 6.1 Find the emf induced in the N-turn circular loop of radius b in Example 6-1 if the magnetic flux density is $\mathbf{B} = \mathbf{a}_z B_0(b-r)\cos\omega t$. What is the phase relationship between the induced emf and the magnetic field?

ANS. $\mathscr{V} = \frac{\pi}{3}\omega b^3 B_0 \sin\omega t$, lags 90° behind **B** in phase.

6-2.2 TRANSFORMERS

Functions of a transformer

A transformer is an alternating-current (a-c) device that transforms voltages, currents, and impedances. It usually consists of two or more coils coupled magnetically through a common ferromagnetic core, such as that sketched in Fig. 6-1. For the closed path in the magnetic circuit traced by magnetic flux Φ, we have

$$N_1 i_1 - N_2 i_2 = \mathscr{R}\Phi, \tag{6-13}$$

where N_1, N_2 and i_1, i_2 are the numbers of turns and the currents in the primary and secondary circuits, respectively. The left side of Eq. (6-13) is the closed line integral $\oint \mathbf{H}\cdot d\boldsymbol{\ell}$ around the transformer core, a consequence of Ampère's circuital law as expressed in Eq. (5-63), and represents the net *magnetomotive force* (mmf—SI unit: ampere-turn). We have noted, in accordance with Lenz's law, that the induced mmf in the secondary circuit, $N_2 i_2$, opposes the flow of magnetic flux created by the mmf in the primary circuit, $N_1 i_1$. On the right side of Eq. (6-13) $\mathscr{R}$ denotes the *reluctance* of the

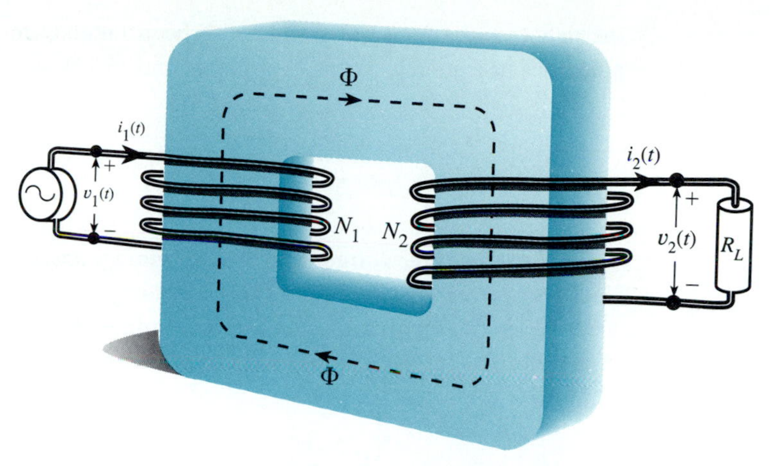

FIGURE 6-1 Schematic diagram of a transformer.

magnetic circuit, which depends on the geometry and is inversely proportional to the permeability of the core material.† Equation (6-13) for a magnetic circuit is analogous to an expression of Kirchhoff's voltage law for a d-c electric circuit, which says that the net emf around a closed circuit is equal to the sum of the resistances times the current. Here $\mathscr{R}$ and Φ are analogous to resistance and current respectively.

Conditions of an ideal transformer

For ideal transformers we assume there is no leakage flux, and that $\mu \rightarrow \infty$, $\mathscr{R} = 0$. Equation (6-13) becomes

$$\boxed{\frac{i_1}{i_2} = \frac{N_2}{N_1}.} \tag{6-14}$$

Current relation for an ideal transformer

Equation (6-14) states that ***the ratio of the currents in the primary and secondary windings of an ideal transformer is equal to the inverse ratio of the numbers of turns***. Faraday's law tells us that

$$v_1 = N_1 \frac{d\Phi}{dt} \tag{6-15}$$

and

$$v_2 = N_2 \frac{d\Phi}{dt}, \tag{6-16}$$

†The calculation of the reluctance of a magnetic circuit is similar to that of a resistance in an electric circuit, but the result can only be approximate because of the existence of leakage flux and the nonuniformity of the cross-sectional area of the transformer core. (Leakage flux is that part of magnetic flux that does not link with the entire magnetic circuit.) We will not elaborate on the determination of $\mathscr{R}$ here. The SI unit for reluctance is reciprocal henry (H^{-1}).

the proper signs for v_1 and v_2 having been taken care of by the designated polarities in Fig. 6-1. From Eqs. (6-15) and (6-16) we have

$$\frac{v_1}{v_2} = \frac{N_1}{N_2}. \tag{6-17}$$

Voltage relation for an ideal transformer

Thus, ***the ratio of the voltages across the primary and secondary windings of an ideal transformer is equal to the turns ratio.***

When the secondary winding is terminated in a load resistance R_L, as shown in Fig. 6-1, the effective load seen by the source connected to primary winding is

$$(R_1)_{\text{eff}} = \frac{v_1}{i_1} = \frac{(N_1/N_2)v_2}{(N_2/N_1)i_2}, \tag{6-18}$$

or

Resistance transformation by an ideal transformer

$$(R_1)_{\text{eff}} = \left(\frac{N_1}{N_2}\right)^2 R_L, \tag{6-19}$$

which is the load resistance multiplied by the square of the turns ratio. For a sinusoidal source $v_1(t)$ and a load impedance Z_L, the effective load seen by the source is $(N_1/N_2)^2 Z_L$, an impedance transformation. We have

Impedance transformation by an ideal transformer

$$(Z_1)_{\text{eff}} = \left(\frac{N_1}{N_2}\right)^2 Z_L. \tag{6-20}$$

We note from Section 5-10 that the inductance of a coil is proportional to the permeability of the medium. Thus, the assumption of an infinite μ for an ideal transformer also implies infinite inductances.

Why an exact analysis of a real transformer is difficult

For real transformers we have the following real-life conditions: the existence of leakage flux, noninfinite inductances, nonzero winding resistances, and the presence of hysteresis and eddy-current losses. The nonlinear nature of the ferromagnetic core (the dependence of permeability on magnetic field intensity) further compounds the difficulty of an exact analysis of real transformers.

Defintion of eddy current

Principle of induction heating

When time-varying magnetic flux flows in the ferromagnetic core, an induced emf will result in accordance with Faraday's law. This induced emf will produce local currents in the conducting core normal to the magnetic flux. These currents are called ***eddy currents***. Eddy currents produce ohmic power loss and cause local heating. As a matter of fact, this is the principle of induction heating. Induction furnaces have been built to produce high

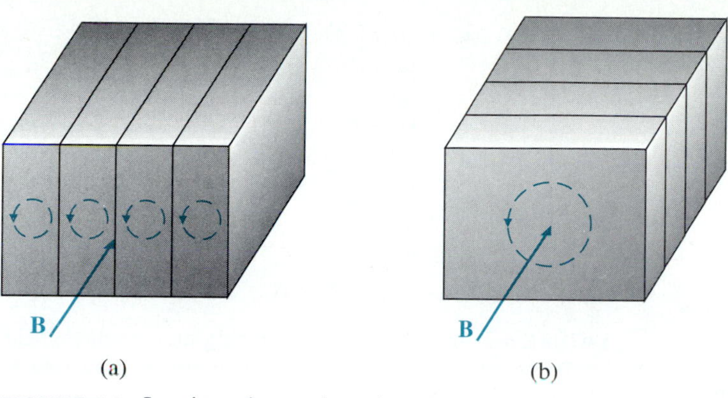

FIGURE 6-2 Laminated cores in a time-varying magnetic field.

enough temperatures to melt metals. In transformers this eddy-current power loss is undesirable and can be reduced by using core materials that have high permeability but low conductivity (high μ and low σ). Ferrites are such materials. For low-frequency, high-power applications, an economical way for reducing eddy-current power loss is to use laminated cores; that is, to make transformer cores out of stacked ferromagnetic (iron) sheets, each electrically insulated from its neighbors by thin varnish or oxide coatings. The insulating coatings should be parallel to the direction of the magnetic flux, as shown in Fig. 6-2(a), so that eddy currents normal to the flux are restricted to the laminated sheets. The arrangement in Fig. 6-2(b) with insulated layers normal to the magnetic flux clearly has little effect in reducing eddy-current power loss. It can be proved that the total eddy-current power loss decreases as the number of laminations increases. The amount of power-loss reduction depends on the shape and size of the cross section as well as on the method of lamination.

Core lamination for reducing eddy-current power loss

EXERCISE 6.2 A resistance of 75 (Ω) is to be converted into 300 (Ω) by means of an ideal transformer. What turns-ratio should the transformer have?

ANS. 2:1.

6-2.3 A MOVING CONDUCTOR IN A STATIC MAGNETIC FIELD

When a conductor moves with a velocity **u** in a static (non-time-varying) magnetic field **B** as shown in Fig. 6-3, a force $\mathbf{F}_m = q\mathbf{u} \times \mathbf{B}$ will cause the freely movable electrons in the conductor to drift toward one end of the conductor and leave the other end positively charged. This separation of the positive and negative charges creates a Coulombian force of attraction. The charge-separation process continues until the electric and magnetic forces balance each other and a state of equilibrium is reached. At equilibrium,

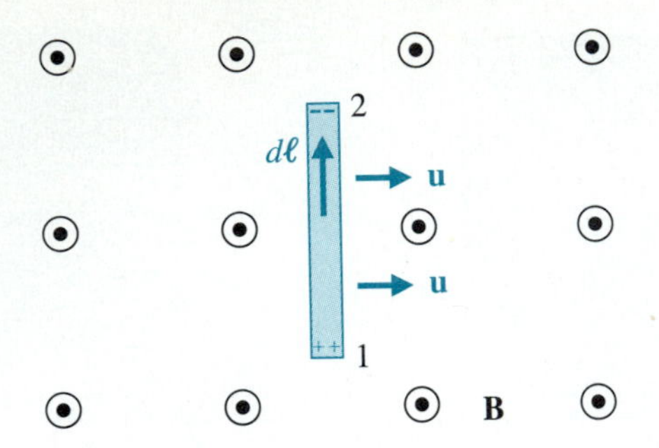

FIGURE 6-3 A conducting bar moving in a magnetic field.

which is reached very rapidly, the net force on the free charges in the moving conductor is zero.

To an observer moving with the conductor there is no apparent motion, and the magnetic force per unit charge $\mathbf{F}_m/q = \mathbf{u} \times \mathbf{B}$ can be interpreted as an induced electric field acting along the conductor and producing a voltage

$$V_{21} = \int_1^2 (\mathbf{u} \times \mathbf{B}) \cdot d\boldsymbol{\ell}. \tag{6-21}$$

If the moving conductor is a part of a closed circuit C, then the emf generated around the circuit is

$$\mathcal{V}' = \oint_C (\mathbf{u} \times \mathbf{B}) \cdot d\boldsymbol{\ell} \qquad \text{(V)}. \tag{6-22}$$

Definition of motional emf

This is referred to as a ***flux-cutting emf*** or a ***motional emf***. Obviously, only the part of the circuit that moves in a direction not parallel to (and hence, figuratively, "cutting") the magnetic flux will contribute to $\mathcal{V}'$ in Eq. (6-22).

EXAMPLE 6-2

A metal bar slides over a pair of conducting rails in a uniform magnetic field $\mathbf{B} = \mathbf{a}_z B_0$ with a constant velocity $\mathbf{u}$, as shown in Fig. 6-4.

a) Determine the open-circuit voltage V_0 that appears across terminals 1 and 2.

b) Assuming that a resistance R is connected between the terminals, find the electric power dissipated in R.

c) Show that this electric power is equal to the mechanical power required to move the sliding bar with a velocity $\mathbf{u}$. Neglect the electric resistance of the metal bar and of the conducting rails. Neglect also the mechanical friction at the contact points.

SOLUTION

a) The moving bar generates a flux-cutting emf. We use Eq. (6-22) to find the open-circuit voltage V_0:

$$V_0 = V_1 - V_2 = \oint_C (\mathbf{u} \times \mathbf{B}) \cdot d\boldsymbol{\ell}$$

$$= \int_{2'}^{1'} (\mathbf{a}_x u \times \mathbf{a}_z B_0) \cdot (\mathbf{a}_y \, d\ell)$$

$$= -uB_0 h \qquad \text{(V)}. \tag{6-23}$$

b) When a resistance R is connected between terminals 1 and 2, a current $I = uB_0 h/R$ will flow from terminal 2 to terminal 1, so the electric power, P_e, dissipated in R is

$$P_e = I^2 R = \frac{(uB_0 h)^2}{R} \qquad \text{(W)}. \tag{6-24}$$

c) The mechanical power, P_m, required to move the sliding bar is

$$P_m = \mathbf{F} \cdot \mathbf{u} \qquad \text{(W)}, \tag{6-25}$$

where $\mathbf{F}$ is the mechanical force required to counteract the magnetic force, $\mathbf{F}_m$, which the magnetic field exerts on the current-carrying metal bar. From Eq. (5-116) we have

$$\mathbf{F}_m = I \int_{2'}^{1'} d\boldsymbol{\ell} \times \mathbf{B} = -\mathbf{a}_x I B_0 h \qquad \text{(N)}. \tag{6-26}$$

The negative sign in Eq. (6-26) arises because current I flows in a direction opposite to that of $d\boldsymbol{\ell}$. Hence,

$$\mathbf{F} = -\mathbf{F}_m = \mathbf{a}_x I B_0 h = \mathbf{a}_x u B_0^2 h^2 / R \qquad \text{(N)}. \tag{6-27}$$

FIGURE 6-4 A metal bar sliding over conducting rails (Example 6-2).

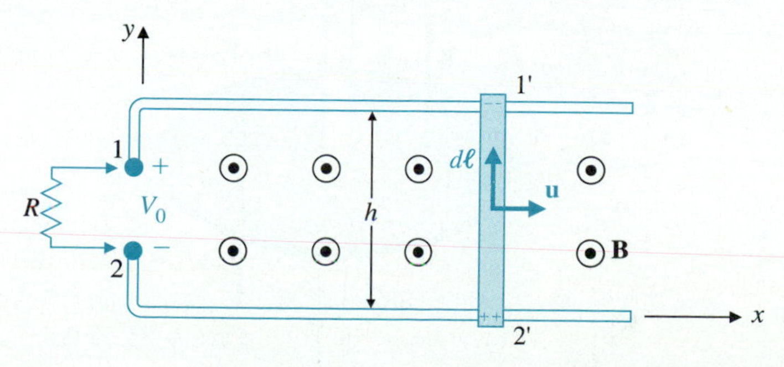

Substitution of Eq. (6-27) in Eq. (6-25) proves that $P_m = P_e$, which upholds the principle of conservation of energy.

EXAMPLE 6-3

Faraday disk generator

The ***Faraday disk generator*** consists of a circular metal disk rotating with a constant angular velocity ω in a uniform and constant magnetic field of flux density $\mathbf{B} = \mathbf{a}_z B_0$ that is parallel to the axis of rotation. Brush contacts are provided at the axis and on the rim of the disk, as depicted in Fig. 6-5. Determine the open-circuit voltage of the generator if the radius of the disk is b.

SOLUTION

Let us consider the circuit 122′341′1. Of the part 2′34 that moves with the disk, only the straight portion 34 "cuts" the magnetic flux. We have, from Eq. (6-22),

$$\begin{aligned} V_0 &= \oint (\mathbf{u} \times \mathbf{B}) \cdot d\boldsymbol{\ell} \\ &= \int_3^4 [(\mathbf{a}_\phi r\omega) \times \mathbf{a}_z B_0] \cdot (\mathbf{a}_r dr) \\ &= \omega B_0 \int_b^0 r\, dr = -\frac{\omega B_0 b^2}{2} \qquad \text{(V)}, \end{aligned} \tag{6-28}$$

which is the emf of the Faraday disk generator. To measure V_0, we must use a voltmeter of a very high resistance so that no appreciable current flows in the circuit to modify the externally applied magnetic field.

FIGURE 6-5 Faraday disk generator (Example 6-3).

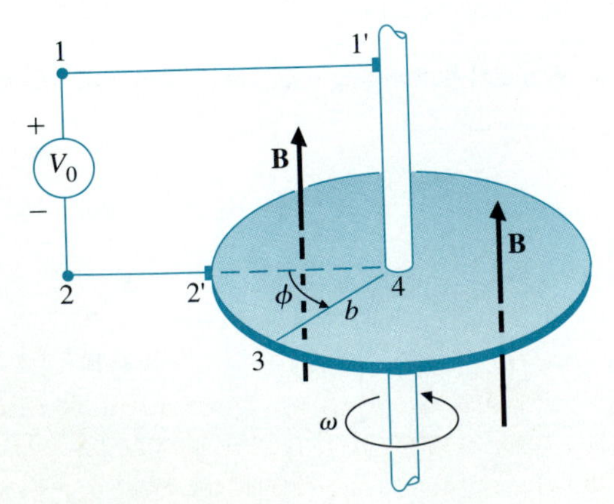

6-2.4 A MOVING CIRCUIT IN A TIME-VARYING MAGNETIC FIELD

When a charge q moves with a velocity $\mathbf{u}$ in a region where both an electric field $\mathbf{E}$ and a magnetic field $\mathbf{B}$ exist, the electromagnetic force $\mathbf{F}$ on q, as measured by a laboratory observer, is given by Lorentz's force equation:

Lorentz's force equation

$$\mathbf{F} = q(\mathbf{E} + \mathbf{u} \times \mathbf{B}). \tag{5-5)(6-29}$$

To an observer moving with q, there is no apparent motion, and the force on q can be interpreted as caused by an electric field $\mathbf{E}'$, where

$$\mathbf{E}' = \mathbf{E} + \mathbf{u} \times \mathbf{B}, \tag{6-30}$$

or

$$\mathbf{E} = \mathbf{E}' - \mathbf{u} \times \mathbf{B}. \tag{6-31}$$

Hence, when a conducting circuit with contour C and surface S moves with a velocity $\mathbf{u}$ in a field $(\mathbf{E}, \mathbf{B})$, we use Eq. (6-31) in Eq. (6-8) to obtain

General form of Faraday's law

$$\oint_C \mathbf{E}' \cdot d\boldsymbol{\ell} = -\int_S \frac{\partial \mathbf{B}}{\partial t} \cdot d\mathbf{s} + \oint_C (\mathbf{u} \times \mathbf{B}) \cdot d\boldsymbol{\ell} \qquad \text{(V)}. \tag{6-32}$$

Equation (6-32) is the general form of ***Faraday's law*** for a moving circuit in a time-varying magnetic field. The line integral on the left side is the emf induced in the moving frame of reference. The first term on the right side represents the transformer emf due to the time variation of $\mathbf{B}$, and the second term represents the motional emf due to the motion of the circuit in $\mathbf{B}$.

If we designate the left side of Eq. (6-32) by

$$\begin{aligned}\mathscr{V}' &= \oint_C \mathbf{E}' \cdot d\boldsymbol{\ell} \\ &= \text{emf induced in circuit } C \text{ measured in the moving frame,}\end{aligned} \tag{6-33}$$

it can be proved in general that Eq. (6-32) is equivalent to

Another form of Faraday's law

$$\begin{aligned}\mathscr{V}' &= -\frac{d}{dt}\int_S \mathbf{B} \cdot d\mathbf{s} \\ &= -\frac{d\Phi}{dt} \qquad \text{(V)},\end{aligned} \tag{6-34}$$

which is of the same form as Eq. (6-12). Of course, if a circuit is not in motion, $\mathscr{V}'$ reduces to $\mathscr{V}$. Hence, Faraday's law that the emf induced in a closed circuit equals the negative time-rate of increase of the magnetic flux linking a circuit applies to a stationary circuit as well as a moving one. Either Eq. (6-32) or Eq. (6-34) can be used to evaluate the induced emf in the general case.

EXAMPLE 6-4

Determine the open-circuit voltage of the Faraday disk generator in Example 6-3 by using Eq. (6-34).

SOLUTION

We solved the problem in Example 6-3 by using Eq. (6-22), which is Eq. (6-32) with $\partial \mathbf{B}/\partial t = 0$. In order to use Eq. (6-34), we first find the magnetic flux linking the circuit 122′341′1 in Fig. 6-5, which is the flux passing through the wedge-shaped area 2′342′.

$$\begin{aligned}\Phi &= \int_S \mathbf{B}\cdot d\mathbf{s} = B_0 \int_0^b \int_0^{\omega t} r\, d\phi\, dr \\ &= B_0(\omega t)\frac{b^2}{2}\end{aligned} \tag{6-35}$$

and

$$V_0 = \mathscr{V}' = -\frac{d\Phi}{dt} = -\frac{\omega B_0 b^2}{2}, \tag{6-36}$$

which is the same as Eq. (6-28).

■ **EXERCISE 6.3** Determine the open-circuit voltage that appears across terminals 1 and 2 in Example 6-2 by using Eq. (6-34).

ANS. $-uB_0h$.

EXAMPLE 6-5

An h by w rectangular conducting loop is situated in a changing magnetic field $\mathbf{B} = \mathbf{a}_y B_0 \sin \omega t$. The normal of the loop initially makes an angle α with $\mathbf{a}_y$, as shown in Fig. 6-6. Find the induced emf in the loop: (a) when the loop is at rest, and (b) when the loop rotates with an angular velocity ω about the x-axis.

SOLUTION

a) When the loop is at rest, we use Eq. (6-12):

$$\begin{aligned}\Phi &= \int \mathbf{B}\cdot d\mathbf{s} \\ &= (\mathbf{a}_y B_0 \sin \omega t)\cdot(\mathbf{a}_n hw) \\ &= B_0 hw \sin \omega t \cos \alpha.\end{aligned}$$

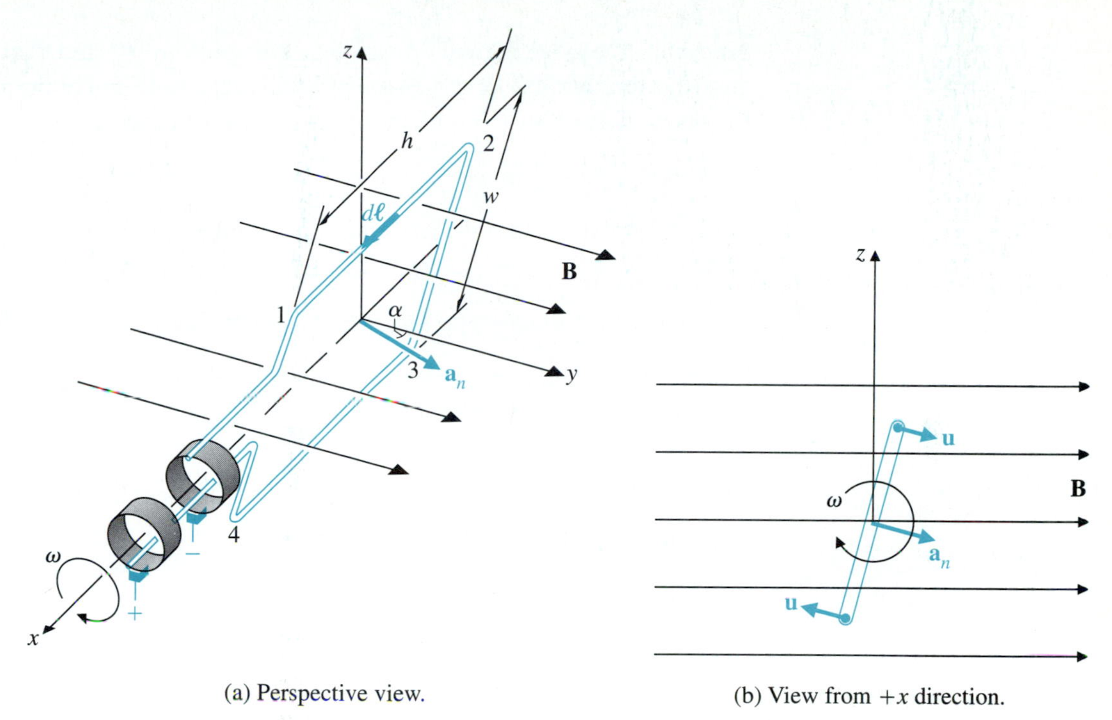

(a) Perspective view.

(b) View from $+x$ direction.

FIGURE 6-6 A rectangular conducting loop rotating in a changing magnetic field (Example 6-5).

Therefore,

$$\mathscr{V}_a = -\frac{d\Phi}{dt} = -B_0 S\omega \cos \omega t \cos \alpha, \tag{6-37}$$

where $S = hw$ is the area of the loop. The relative polarities of the terminals are as indicated. If the circuit is completed through an external load, $\mathscr{V}_a$ will produce a current that will oppose the change in Φ.

b) When the loop rotates about the x-axis, both terms in Eq. (6-32) contribute: the first term contributes the transformer emf $\mathscr{V}_a$ in Eq. (6-37), and the second term contributes a motional emf $\mathscr{V}_a'$, where

$$\begin{aligned}
\mathscr{V}_a' &= \oint_C (\mathbf{u} \times \mathbf{B}) \cdot d\boldsymbol{\ell} \\
&= \int_2^1 \left[\left(\mathbf{a}_n \frac{w}{2} \omega\right) \times (\mathbf{a}_y B_0 \sin \omega t)\right] \cdot (\mathbf{a}_x \, dx) \\
&\quad + \int_4^3 \left[\left(-\mathbf{a}_n \frac{w}{2} \omega\right) \times (\mathbf{a}_y B_0 \sin \omega t)\right] \cdot (\mathbf{a}_x \, dx) \\
&= 2\left(\frac{w}{2} \omega B_0 \sin \omega t \sin \alpha\right) h.
\end{aligned}$$

Note that the sides 23 and 41 do not contribute to $\mathscr{V}_a'$ and that the contributions of sides 12 and 34 are of equal magnitude and in the same direction. If $\alpha = 0$ at $t = 0$, then $\alpha = \omega t$, and we can write

$$\mathscr{V}_a' = B_0 S\omega \sin \omega t \sin \omega t. \tag{6-38}$$

The total emf induced or generated in the rotating loop is the sum of $\mathscr{V}_a$ in Eq. (6-37) and $\mathscr{V}_a'$ in Eq. (6-38):

$$\mathscr{V}_t' = -B_0 S\omega(\cos^2 \omega t - \sin^2 \omega t) = -B_0 S\omega \cos 2\omega t, \tag{6-39}$$

which has an angular frequency 2ω. Thus the arrangement in Fig. 6-6 is a generator of second harmonics.

We can determine the total induced emf $\mathscr{V}_t'$ by applying Eq. (6-34) directly. At any time t, the magnetic flux linking the loop is

$$\begin{aligned}\Phi(t) &= \mathbf{B}(t) \cdot [\mathbf{a}_n(t)S] = B_0 S \sin \omega t \cos \alpha \\ &= B_0 S \sin \omega t \cos \omega t = \tfrac{1}{2} B_0 S \sin 2\omega t.\end{aligned}$$

Hence,

$$\begin{aligned}\mathscr{V}_t' &= -\frac{d\Phi}{dt} = -\frac{d}{dt}\left(\frac{1}{2} B_0 S \sin 2\omega t\right), \\ &= -B_0 S\omega \cos 2\omega t\end{aligned}$$

as before.

REVIEW QUESTIONS

Q.6-1 What constitutes an *electromagnetostatic field*? In what ways are **E** and **B** related in a conducting medium under static conditions?

Q.6-2 Write the fundamental postulate for electromagnetic induction.

Q.6-3 State Lenz's law.

Q.6-4 Write the expression for transformer emf.

Q.6-5 How do the primary-to-secondary voltage and current ratios in an ideal transformer depend on the turns ratio of its windings?

Q.6-6 What are *eddy currents*?

Q.6-7 What is the principle of inductiong heating?

Q.6-8 Why are materials having high permeability and low conductivity preferred as transformer cores?

Q.6-9 Why are the cores of power transformers laminated?

Q.6-10 Write the general form of Faraday's law.

Q.6-11 What is a Faraday disk generator?

REMARKS

1. **E** in a region of time-varying magnetic field is *not* conservative, and cannot be expressed as the gradient of a scalar potential alone.
2. A time-varying magnetic field linking with a circuit induces an emf in the circuit.
3. Transformers are inherently a-c devices.
4. An ideal transformer assumes infinite permeability for its core and infinite inductances for its windings.
5. The insulated laminations of a transformer core to reduce eddy-current power loss should be parallel to the direction of the magnetic flux.

6-3 MAXWELL'S EQUATIONS

The fundamental postulate for electromagnetic induction assures us that a time-varying magnetic field gives rise to an electric field. This assurance has been amply verified by numerous experiments. The $\nabla \times \mathbf{E} = 0$ equation (6-1) must therefore be replaced by Eq. (6-7) in the time-varying case. We now have the following collection of two curl equations, (6-7) and (6-5), and two divergence equations, (6-2) and (6-4):

$$\nabla \times \mathbf{E} = -\frac{\partial \mathbf{B}}{\partial t}, \tag{6-7)(6-40a}$$

$$\nabla \times \mathbf{H} = \mathbf{J}, \tag{6-5)(6-40b}$$

$$\nabla \cdot \mathbf{D} = \rho_v, \tag{6-2)(6-40c}$$

$$\nabla \cdot \mathbf{B} = 0. \tag{6-4)(6-40d}$$

In addition, we know that the principle of conservation of charge must be satisfied at all times. The mathematical expression of charge conservation is the equation of continuity:

$$\nabla \cdot \mathbf{J} = -\frac{\partial \rho_v}{\partial t}. \tag{4-20)(6-41}$$

The crucial question here is whether the set of four equations in (6-40a, b, c, and d) is now consistent with the requirement specified by Eq. (6-41) in a time-varying situation. That the answer is in the negative is immediately obvious by simply taking the divergence of Eq. (6-40b),

$$\nabla \cdot (\nabla \times \mathbf{H}) = 0 = \nabla \cdot \mathbf{J}. \tag{6-42}$$

Equation (6-42) follows from the null identity, Eq. (2-109), which asserts that the divergence of the curl of any well-behaved vector field is zero. Since Eq. (6-41) indicates that $\nabla \cdot \mathbf{J}$ does not vanish in a time-varying situation, Eq. (6-40b) is, in general, not true.

How should Eqs. (6-40a, b, c, and d) be modified so that they are consistent with Eq. (6-41)? First of all, a term $\partial\rho_v/\partial t$ must be added to the right side of Eq. (6-42):

$$\nabla \cdot (\nabla \times \mathbf{H}) = 0 = \nabla \cdot \mathbf{J} + \frac{\partial \rho_v}{\partial t}. \tag{6-43}$$

Using Eq. (6-40c) in Eq. (6-43), we have

$$\nabla \cdot (\nabla \times \mathbf{H}) = \nabla \cdot \left(\mathbf{J} + \frac{\partial \mathbf{D}}{\partial t}\right),$$

which implies that

$$\boxed{\nabla \times \mathbf{H} = \mathbf{J} + \frac{\partial \mathbf{D}}{\partial t}.} \tag{6-44}$$

Equation (6-44) indicates that a time-varying electric field will give rise to a magnetic field, even in the absence of a free current flow (i.e., even when $\mathbf{J} = 0$). The additional term $\partial\mathbf{D}/\partial t$ is necessary to make Eq. (6-44) consistent with the principle of conservation of charge.

Time rate of change of D gives rise to a displacement current density.

It is easy to verify that $\partial\mathbf{D}/\partial t$ has the dimension of a current density (SI unit: A/m^2). The term $\partial\mathbf{D}/\partial t$ is called ***displacement current density***, and its introduction in the $\nabla \times \mathbf{H}$ equation was one of the major contributions of James Clerk Maxwell (1831–1879). In order to be consistent with the equation of continuity in a time-varying situation, both of the curl equations in (6-1) and (6-5) must be generalized. The set of four consistent equations to replace the inconsistent equations in Eqs. (6-40a, b, c, and d) is

Differential (operator) form of Maxwell's equations

$$\nabla \times \mathbf{E} = -\frac{\partial \mathbf{B}}{\partial t}, \tag{6-45a}$$

$$\nabla \times \mathbf{H} = \mathbf{J} + \frac{\partial \mathbf{D}}{\partial t}, \tag{6-45b}$$

$$\nabla \cdot \mathbf{D} = \rho_v, \tag{6-45c}$$

$$\nabla \cdot \mathbf{B} = 0. \tag{6-45d}$$

They are known as ***Maxwell's equations***. Note that ρ_v in Eq. (6-45c) is the volume density of *free charges*, and $\mathbf{J}$ in Eq. (6-45b) is the density of *free currents*, which may comprise both convection current ($\rho_v\mathbf{u}$) and conduction current ($\sigma\mathbf{E}$). These four equations, together with the equation of continuity in Eq. (6-41) and Lorentz's force equation in Eq. (5-5), form the foundation of electromagnetic theory. These equations can be used to explain and predict *all* macroscopic electromagnetic phenomena.

The four Maxwell's equations are not all independent.

Although the four Maxwell's equations in Eqs. (6-45a, b, c, and d) are consistent, they are not all independent. As a matter of fact, the two divergence equations, Eqs. (6-45c and d), can be derived from the two curl equations, Eqs. (6-45a and b), by making use of the equation of continuity, Eq. (6-41). See Exercise 6.4 below.

EXERCISE 6.4 Take the divergence of the two curl equations (6-45a) and (6-45b) and derive the two divergence equations (6-45c) and (6-45d) with the aid of the equation of continuity (6-41).

6-3.1 INTEGRAL FORM OF MAXWELL'S EQUATIONS

The four Maxwell's equations in (6-45a, b, c, and d) are differential equations that are valid at every point in space. In explaining electromagnetic phenomena in a physical environment we must deal with finite objects of specified shapes and boundaries. It is convenient to convert the differential forms into their integral-form equivalents. We take the surface integral of both sides of the curl equations in Eqs. (6-45a) and (6-45b) over an open surface S with a contour C and apply Stokes's theorem to obtain

Integral form of Maxwell's equations

$$\oint_C \mathbf{E} \cdot d\boldsymbol{\ell} = -\int_S \frac{\partial \mathbf{B}}{\partial t} \cdot d\mathbf{s} \tag{6-46a}$$

and

$$\oint_C \mathbf{H} \cdot d\boldsymbol{\ell} = \int_S \left(\mathbf{J} + \frac{\partial \mathbf{D}}{\partial t} \right) \cdot d\mathbf{s}. \tag{6-46b}$$

Taking the volume integral of both sides of the divergence equations in Eqs. (6-45c) and (6-45d) over a volume V with a *closed* surface S and using divergence theorem, we have

$$\oint_S \mathbf{D} \cdot d\mathbf{s} = \int_V \rho_v \, dv \tag{6-46c}$$

and

$$\oint_S \mathbf{B} \cdot d\mathbf{s} = 0. \tag{6-46d}$$

TABLE **6-1** MAXWELL'S EQUATIONS AND THEIR SIGNIFICANCE

Differential Form	Integral Form	Significance
$\nabla \times \mathbf{E} = -\frac{\partial \mathbf{B}}{\partial t}$	$\oint_C \mathbf{E} \cdot d\boldsymbol{\ell} = -\frac{d\Phi}{dt}$	Faraday's law
$\nabla \times \mathbf{H} = \mathbf{J} + \frac{\partial \mathbf{D}}{\partial t}$	$\oint_C \mathbf{H} \cdot d\boldsymbol{\ell} = I + \int_S \frac{\partial \mathbf{D}}{\partial t} \cdot d\mathbf{s}$	Ampère's circuital law
$\nabla \cdot \mathbf{D} = \rho_v$	$\oint_S \mathbf{D} \cdot d\mathbf{s} = Q$	Gauss's law
$\nabla \cdot \mathbf{B} = 0$	$\oint_S \mathbf{B} \cdot d\mathbf{s} = 0$	No isolated magnetic charge

The set of four equations in (6-46a, b, c, and d) is the integral form of Maxwell's equations. Both the differential and integral forms of Maxwell's equations as well as their significance are listed in Table 6-1 for easy reference.

EXAMPLE 6-6

An a-c source of amplitude V_0 and angular frequency ω, $v_c = V_0 \sin \omega t$, is connected across a parallel-plate capacitor C_1, as shown in Fig. 6-7. (a) Verify that the displacement current in the capacitor is the same as the conduction current in the wires. (b) Determine the magnetic field intensity at a distance r from the wire.

SOLUTION

a) The conduction current in the connecting wire is

$$i_C = C_1 \frac{dv_C}{dt} = C_1 V_0 \omega \cos \omega t.$$

For a parallel-plate capacitor with an area A, plate separation d, and a dielectric medium of permittivity ϵ, the capacitance is

$$C_1 = \epsilon \frac{A}{d}.$$

With a voltage v_C appearing between the plates, the uniform electric field intensity E in the dielectric is equal to (neglecting fringing effects) $E = v_C/d$, whence

$$D = \epsilon E = \epsilon \frac{V_0}{d} \sin \omega t.$$

The displacement current is then

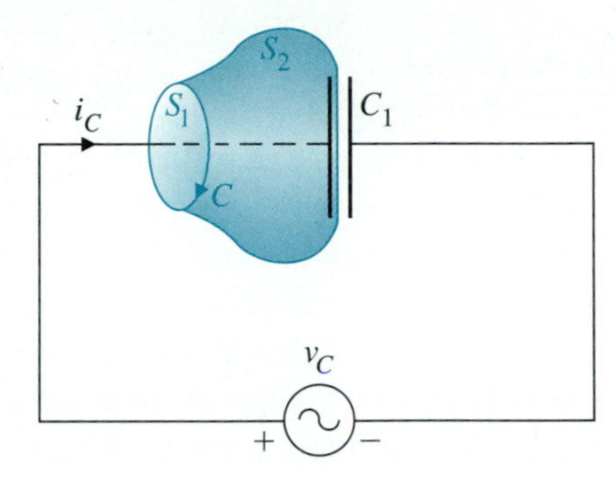

FIGURE 6-7 A parallel-plate capacitor connected to an a-c voltage source (Example 6-6).

$$i_D = \int_A \frac{\partial \mathbf{D}}{\partial t} \cdot d\mathbf{s} = \left(\epsilon \frac{A}{d}\right) V_0 \omega \cos \omega t$$

$$= C_1 V_0 \omega \cos \omega t = i_C,$$

which is what we wanted to verify.

b) The magnetic field intensity at a distance r from the conducting wire can be found by applying the generalized Ampère's circuital law, Eq. (6-46b), to contour C in Fig. 6-7. Two typical open surfaces with rim C may be chosen: (1) a planar disk surface S_1, or (2) a curved surface S_2 passing through the dielectric medium. Symmetry around the wire ensures a constant H_ϕ along the contour C. The line integral on the left side of Eq. (6-46b) is

$$\oint_C \mathbf{H} \cdot d\boldsymbol{\ell} = 2\pi r H_\phi.$$

For the surface S_1, only the first term on the right side of Eq. (6-46b) is nonzero because no charges are deposited along the wire and, consequently, $\mathbf{D} = 0$.

$$\int_{S_1} \mathbf{J} \cdot d\mathbf{s} = i_C = C_1 V_0 \omega \cos \omega t.$$

Since the surface S_2 passes through the dielectric medium, no conduction current flows through S_2. If the second surface integral were not there, the right side of Eq. (6-46b) would be zero. This would result in a contradiction. The inclusion of the displacement-current term by Maxwell eliminates this contradiction. As we have shown in part (a), $i_D = i_C$. Hence we obtain the same result whether surface S_1 or surface S_2 is chosen. Equating the two previous integrals, we find that

$$H_\phi = \frac{C_1 V_0}{2\pi r} \omega \cos \omega t.$$

6-3.2 ELECTROMAGNETIC BOUNDARY CONDITIONS

In order to solve electromagnetic problems involving contiguous regions of different constitutive parameters, it is necessary to know the boundary conditions that the field vectors **E**, **D**, **B**, and **H** must satisfy at the interfaces. Boundary conditions are derived by applying the integral form of Maxwell's equations (6-46a, b, c, and d) to a small region at an interface of two media in manners similar to those used in obtaining the boundary conditions for static electric and magnetic fields. The integral equations are assumed to hold for regions containing discontinuous media. (You should review the procedures followed in Sections 3-8 and 5-9.) In general, the application of the integral form of a curl equation to a flat closed path at a boundary with top and bottom sides in the two touching media yields the boundary condition for the tangential components. On the other hand, the application of the integral form of a divergence equation to a shallow pillbox at an interface with top and bottom faces in the two contiguous media gives the boundary condition for the normal components.

The boundary conditions for the tangential components of **E** and **H** are obtained from Eqs. (6-46a) and (6-46b), respectively:

Boundary condition for tangential component of E

$$E_{1t} = E_{2t} \qquad \text{(V/m)}; \tag{6-47a}$$

Boundary condition for tangential component of H

$$\mathbf{a}_{n2} \times (\mathbf{H}_1 - \mathbf{H}_2) = \mathbf{J}_s \qquad \text{(A/m)}. \tag{6-47b}$$

We note that Eqs. (6-47a) and (6-47b) for the time-varying case are exactly the same as Eq. (3-72) for static electric fields and Eq. (5-71) for static magnetic fields, respectively, in spite of the existence of the time-varying terms in Eqs. (6-46a) and (6-46b).

Similarly, the boundary conditions for the normal components of **D** and **B** are obtained from Eqs. (6-46c) and (6-46d):

Boundary condition for normal component of D

$$D_{1n} - D_{2n} = \rho_s \qquad \text{(C/m}^2\text{)}, \tag{6-47c}$$

where the reference normal direction is outward from medium 2; and

Boundary condition for normal component of B

$$B_{1n} = B_{2n} \qquad \text{(T)}. \tag{6-47d}$$

These are the same as, respectively, Eq. (3-75b) for static electric fields and Eq.

(5-68) for static magnetic fields because we start from the same divergence equations.

In the following we summarize the boundary conditions of two important special cases.

A. *Interface between two lossless media*
A lossless linear medium can be specified by a permittivity ϵ and a permeability μ, with $\sigma = 0$. There are usually no free charges and no surface currents at the interface between two lossless media. We set $\rho_s = 0$ and $\mathbf{J}_s = 0$ in Eqs. (6-47c) and (6-47b), and obtain the boundary conditions listed in Table 6-2. We see that in this case E_t, H_t, D_n and B_n are all continuous at the interface.

B. *Interface between a dielectric and a perfect conductor*
A perfect conductor is one with an infinite conductivity. In the physical world we have an abundance of "good conductors" such as silver, copper, gold, and aluminium, whose conductivities are of the order of 10^7 (S/m). (See the table in Appendix B-4.) In order to simplify the analytical solution of field problems, good conductors are often considered perfect conductors in regard to boundary conditions. In the *interior* of a perfect conductor the electric field is zero (otherwise, it would produce an infinite current density), and any charges the conductor will have will reside on the surface only. The interrelationship between (**E**, **D**) and (**B**, **H**) through Maxwell's equations ensures that **B** and **H** are also zero in the *interior* of a conductor in a *time-varying situation.* Consider an interface between a lossless dielectric (medium 1) and a perfect conductor (medium 2). In medium 2, $\mathbf{E}_2 = 0$, $\mathbf{H}_2 = 0$, $\mathbf{D}_2 = 0$, and $\mathbf{B}_2 = 0$. The general boundary conditions in Eqs. (6-47a, b, c, and d) reduce to those listed in Table 6-3. In this case, E_t and B_n are continuous, but H_t and D_n are discontinuous by an amount equal to the surface current density J_s and surface charge density ρ_s, respectively. It is important to note that, when we apply Eqs. (6-49b) and (6-49c), the reference normal is outward from the surface of the conductor (medium 2).

TABLE **6-2** BOUNDARY CONDITIONS BETWEEN TWO LOSSLESS MEDIA

$$E_{1t} = E_{2t} \longrightarrow \frac{D_{1t}}{D_{2t}} = \frac{\epsilon_1}{\epsilon_2} \tag{6-48a}$$

$$H_{1t} = H_{2t} \longrightarrow \frac{B_{1t}}{B_{2t}} = \frac{\mu_1}{\mu_2} \tag{6-48b}$$

$$D_{1n} = D_{2n} \longrightarrow \epsilon_1 E_{1n} = \epsilon_2 E_{2n} \tag{6-48c}$$

$$B_{1n} = B_{2n} \longrightarrow \mu_1 H_{1n} = \mu_2 H_{2n} \tag{6-48d}$$

TABLE **6-3** BOUNDARY CONDITIONS BETWEEN A DIELECTRIC (MEDIUM 1) AND A PERFECT CONDUCTOR (MEDIUM 2) (TIME-VARYING CASE)

On the Side of Medium 1	On the Side of Medium 2	
$E_{1t} = 0$	$E_{2t} = 0$	(6-49a)
$\mathbf{a}_{n2} \times \mathbf{H}_1 = \mathbf{J}_s$	$H_{2t} = 0$	(6-49b)
$D_{1n} = \rho_s$	$D_{2n} = 0$	(6-49c)
$B_{1n} = 0$	$B_{2n} = 0$	(6-49d)

■ **EXERCISE 6.5** Assume that the $y = 0$ plane separates air in the upper half-space ($y > 0$) from a good conductor and that a surface charge density $\rho_s = C_1 \sin \beta x$ and a surface current density $\mathbf{J}_s = \mathbf{a}_x C_2 \cos \beta x$ exist on the interface. (C_1, C_2, and β are constants.) Determine **E** and **H** in the air at the interface.

ANS. $\mathbf{E} = \mathbf{a}_y (C_1/\epsilon_0) \sin \beta x$, $\mathbf{H} = \mathbf{a}_z C_2 \cos \beta x$.

REVIEW QUESTIONS

Q.6-12 Write the differential form of Maxwell's equations.

Q.6-13 Explain the significance of *displacement current*.

Q.6-14 Write the integral form of Maxwell's equations, and identify each equation with the proper experimental law.

Q.6-15 State the boundary conditions for E_t and for B_n.

Q.6-16 State the boundary conditions for H_t and for D_n.

Q.6-17 Why is the **E** field immediately outside of a perfect conductor perpendicular to the conductor surface?

Q.6-18 Why is the **H** field immediately outside of a perfect conductor tangential to the conductor surface?

Q.6-19 Can a static magnetic field exist in the interior of a perfect conductor? Explain. Can a time-varying magnetic field? Explain.

REMARKS

1. A changing magnetic field produces an electric field, and a changing electric field produces a displacement current that contributes to the magnetic field. In time-varying situations the electric and magnetic fields are coupled through Maxwell's curl equations.
2. Not all four Maxwell's equations are independent.
3. The tangential component of **E** and the normal component of **B** are continuous at the interface of any two media.

6-4 POTENTIAL FUNCTIONS

In Section 5-3 the concept of the vector magnetic potential $\mathbf{A}$ was introduced because of the solenoidal nature of $\mathbf{B}(\nabla \cdot \mathbf{B} = 0)$:

$$\boxed{\mathbf{B} = \nabla \times \mathbf{A} \qquad \text{(T).}} \tag{6-50}$$

If Eq. (6-50) is substituted in the differential form of Faraday's law, Eq. (6-7), we get

$$\nabla \times \mathbf{E} = -\frac{\partial}{\partial t}(\nabla \times \mathbf{A}), \tag{6-51}$$

or

$$\nabla \times \left(\mathbf{E} + \frac{\partial \mathbf{A}}{\partial t}\right) = 0. \tag{6-52}$$

Since the sum of the two vector quantities in the parentheses of Eq. (6-52) is curl-free, it can be expressed as the gradient of a scalar. To be consistent with the definition of the scalar electric potential V in Eq. (3-26) for electrostatics, we write

$$\mathbf{E} + \frac{\partial \mathbf{A}}{\partial t} = -\nabla V,$$

from which we obtain

In time-varying case, E is a function of both scalar electric potential V and vector magnetic potential A.

$$\boxed{\mathbf{E} = -\nabla V - \frac{\partial \mathbf{A}}{\partial t} \qquad \text{(V/m).}} \tag{6-53}$$

In the static case, $\partial \mathbf{A}/\partial t = 0$, and Eq. (6-53) reduces to $\mathbf{E} = -\nabla V$. Hence $\mathbf{E}$ can be determined from V alone, and $\mathbf{B}$ from $\mathbf{A}$ by Eq. (6-50). For time-varying fields, $\mathbf{E}$ depends on both V and $\mathbf{A}$; that is, an electric field intensity can result both from accumulations of charge through the $-\nabla V$ term and from time-varying magnetic fields through the $-\partial \mathbf{A}/\partial t$ term. Inasmuch as $\mathbf{B}$ also depends on $\mathbf{A}$, $\mathbf{E}$ and $\mathbf{B}$ are coupled.

Let us substitute Eqs. (6-50) and (6-53) into Eq. (6-45b) and make use of the constitutive relations $\mathbf{H} = \mathbf{B}/\mu$ and $\mathbf{D} = \epsilon \mathbf{E}$. We have

$$\nabla \times \nabla \times \mathbf{A} = \mu \mathbf{J} + \mu\epsilon \frac{\partial}{\partial t}\left(-\nabla V - \frac{\partial \mathbf{A}}{\partial t}\right), \tag{6-54}$$

where a homogeneous medium has been assumed. Recalling the vector identity for $\nabla \times \nabla \times \mathbf{A}$ in Eq. (5-16a), we can write Eq. (6-54) as

$$\nabla(\nabla \cdot \mathbf{A}) - \nabla^2 \mathbf{A} = \mu \mathbf{J} - \nabla\left(\mu\epsilon \frac{\partial V}{\partial t}\right) - \mu\epsilon \frac{\partial^2 \mathbf{A}}{\partial t^2},$$

or

$$\nabla^2 \mathbf{A} - \mu\epsilon \frac{\partial^2 \mathbf{A}}{\partial t^2} = -\mu \mathbf{J} + \nabla\left(\nabla \cdot \mathbf{A} + \mu\epsilon \frac{\partial V}{\partial t}\right). \tag{6-55}$$

Now, the definition of a vector requires the specification of both its curl and its divergence. Although the curl of **A** is designated **B** in Eq. (6-50), we are still at liberty to choose the divergence of **A**. We let

Lorentz condition for potentials

$$\boxed{\nabla \cdot \mathbf{A} + \mu\epsilon \frac{\partial V}{\partial t} = 0,} \tag{6-56}$$

which makes the second term on the right side of Eq. (6-55) vanish and reduces Eq. (6-55) to a simplest possible form. We then have

Wave equation for vector potential A

$$\boxed{\nabla^2 \mathbf{A} - \mu\epsilon \frac{\partial^2 \mathbf{A}}{\partial t^2} = -\mu \mathbf{J}.} \tag{6-57}$$

Equation (6-57) is the ***nonhomogeneous wave equation for vector potential* A**. It is called a wave equation because its solutions represent waves traveling with a velocity equal to $1/\sqrt{\mu\epsilon}$. This will become clear in Subsection 6-4.1 when the solution of wave equations is discussed. The relation between **A** and V in Eq. (6-56) is called the ***Lorentz condition (or Lorentz gauge) for potentials***. It reduces to the condition $\nabla \cdot \mathbf{A} = 0$ in Eq. (5-19) for static fields.

A corresponding wave equation for the scalar potential V is

Wave potential for scalar potential *V*

$$\boxed{\nabla^2 V - \mu\epsilon \frac{\partial^2 V}{\partial t^2} = -\frac{\rho_v}{\epsilon},} \tag{6-58}$$

which is the ***nonhomogeneous wave equation for scalar potential V***. Hence the Lorentz condition in Eq. (6-56) uncouples the wave equations for **A** and for V. Note the similarities of Eqs. (6-57) and (6-58), and the analogy among the quantities: $\mathbf{A} \sim V$, $\mathbf{J} \sim \rho_v$, and $\mu \sim 1/\epsilon$.

■ **EXERCISE 6.6** Derive the wave equation (6-58) for V by using Eq. (6-53) in Eq. (6-40c).

REVIEW QUESTIONS

Q.6-20 How are the scalar electric potential V and the vector magnetic potential $\mathbf{A}$ defined?

Q.6-21 What is a wave equation?

REMARKS

Time-varying potential functions V and $\mathbf{A}$ satisfy the wave equation.

6-4.1 SOLUTION OF WAVE EQUATIONS

We now consider the solution of the nonhomogeneous wave equation (6-58) for scalar potential V due to a charge distribution ρ_v in a restricted region. Let an elemental point charge $\rho_v\, dv'$ be situated at the origin at time t. At a distance R far away from the origin we can assume spherical symmetry (that is, V depends only on R and t, not on θ or ϕ). Using Eq. (3-129), we can write Eq. (6-58) as

$$\frac{1}{R^2}\frac{\partial}{\partial R}\left(R^2\frac{\partial V}{\partial R}\right) - \mu\epsilon\frac{\partial^2 V}{\partial t^2} = 0. \tag{6-59}$$

We now introduce a new variable

$$V(R,t) = \frac{1}{R}U(R,t), \tag{6-60}$$

which simplifies Eq. (6-59) to

$$\frac{\partial^2 U}{\partial R^2} - \mu\epsilon\frac{\partial^2 U}{\partial t^2} = 0. \tag{6-61}$$

Equation (6-61) is a one-dimensional homogeneous wave equation. It can be verified by direct substitution (see Problem P.6-11) that *any* twice-differentiable function of $(t - R\sqrt{\mu\epsilon})$ is a solution of Eq. (6-61). We write

$$U(R, t) = f(t - R\sqrt{\mu\epsilon}). \tag{6-62}$$

The function at new distance $R + \Delta R$ at a later time $t + \Delta t$ is

$$U(R + \Delta R,\, t + \Delta t) = f[t + \Delta t - (R + \Delta R)\sqrt{\mu\epsilon}]$$

which equals $f(t - R\sqrt{\mu\epsilon})$ and retains its form if $\Delta t = \Delta R\sqrt{\mu\epsilon} = \Delta R/u_p$. The

Velocity of wave propagation in a medium with constitutive parameters ϵ and μ is $1/\sqrt{\mu\epsilon}$.

quantity

$$u_p = \frac{1}{\sqrt{\mu\epsilon}} \tag{6-63}$$

is the ***velocity of wave propagation***, a characteristic of the medium. In air, it is equal to $c = 1/\sqrt{\mu_0\epsilon_0}$. From Eqs. (6-60) and (6-62) we have

$$V(R, t) = \frac{1}{R} f(t - R/u_p). \tag{6-64}$$

To determine what the specific function $f(t - R/u_p)$ must be, we note from Eq. (3-29) that for a static point charge $\rho(t)\,\Delta v'$ at the origin,

$$\Delta V(R) = \frac{\rho_v(t)\,\Delta v'}{4\pi\epsilon R}. \tag{6-65}$$

Comparison of Eqs. (6-64) and (6-65) enables us to identify

$$\Delta f(t - R/u_p) = \frac{\rho_v(t - R/u_p)\,\Delta v'}{4\pi\epsilon}. \tag{6-66}$$

The potential due to a charge distribution over a volume V' is then

Finding retarded scalar potential V from charge distribution

$$\boxed{V(R, t) = \frac{1}{4\pi\epsilon} \int_{V'} \frac{\rho_v(t - R/u_p)}{R}\, dv' \qquad \text{(V)}.} \tag{6-67}$$

Equation (6-67) indicates that the scalar potential at a distance R from the source at time t depends on the value of the charge density at an *earlier* time $(t - R/u_p)$. For this reason, $V(R, t)$ in Eq. (6-67) is called the ***retarded scalar potential***.

The solution of the nonhomogeneous wave equation, Eq. (6-57), for vector magnetic potential **A** can proceed in exactly the same way as that for V. The vector equation, Eq. (6-57), for **A** can be decomposed into three scalar equations, each similar to Eq. (6-58) for V. The ***retarded vector potential*** is thus given by

Finding retarded vector potential A from current distribution

$$\boxed{\mathbf{A}(R, t) = \frac{\mu}{4\pi} \int_{V'} \frac{\mathbf{J}(t - R/u_p)}{R}\, dv' \qquad \text{(Wb/m)}.} \tag{6-68}$$

Circuit theory ignores time-retardation effect

The electric and magnetic fields derived from **A** and V by differentiation will obviously also be functions of $(t - R/u_p)$ and therefore retarded in time. It takes time for electromagnetic waves to travel and for the effects of time-varying charges and currents to be felt at distant points. In circuit theory this time-retardation effect is ignored and instant response is assumed.

■ **EXERCISE 6.7** A radar signal sent from earth to the moon is received back on the earth after a delay of 2.562 (s). Determine the distance between the surfaces of earth and the moon at that moment in kilometers and in miles.

ANS. 3.843×10^5 (km), or 238,844 miles.

■ **EXERCISE 6.8** What is the distance equivalent to one light year?

ANS. 9.46×10^{12} (km), or 5.88 trillion miles.

REVIEW QUESTIONS

Q.6-22 What does *retarded potential* mean in electromagnetics?

Q.6-23 In what ways do the retardation time and the velocity of wave propagation depend on the constitutive parameters of the medium?

REMARKS

1. The response at a distance of changing charge and current distributions is *not instantaneous*, but is *delayed* in time.
2. The velocity of wave propagation is a characteristic of the medium and is *independent of frequency*.

6-5 TIME-HARMONIC FIELDS

Maxwell's equations and all the equations derived from them so far in this chapter hold for electromagnetic quantities with an arbitrary time-dependence. The actual type of time functions that the field quantities assume depends on the source functions ρ_v and **J**. In engineering, sinusoidal time functions occupy a unique position. They are easy to generate; arbitrary periodic time functions can be expanded into Fourier series of harmonic sinusoidal components; and transient nonperiodic functions can be expressed as Fourier integrals. Hence for source functions with an arbitrary time dependence, electrodynamic fields can be determined in terms of those caused by the various frequency components of the source functions. The application of the principle of superposition (adding the results due to the various frequencies) will give us the total fields. In this section we examine *time-harmonic* (steady-state sinusoidal) field relationships.

Time-harmonic fields are sinusoidal fields.

6-5.1 THE USE OF PHASORS—A REVIEW

For time-harmonic fields it is convenient to use a phasor notation. At this time we digress briefly to review the use of phasors. Conceptually, it is simpler to discuss a scalar phasor. The instantaneous (time-dependent) expression of a sinusoidal scalar quantity, such as a current i, can be written as either a cosine or a sine function. *If* we choose a cosine function as the *reference* (which is usually dictated by the functional form of the excitation), then all derived results will refer to the cosine function. The specification of a sinusoidal quantity requires the knowledge of three parameters: amplitude, frequency, and phase. For example,

$$i(t) = I_0 \cos(\omega t + \phi), \tag{6-69}$$

where I_0 is the amplitude; ω is the angular frequency (rad/s)—ω is always equal to $2\pi f$, f being the frequency in hertz; and ϕ is the phase angle referred to the cosine function. We can also write $i(t)$ in Eq. (6-69) as a sine function if we wish: $i(t) = I_0 \sin(\omega t + \phi')$, with $\phi' = \phi + \pi/2$. Thus it is important to decide at the outset whether our reference is a cosine or a sine function, then to stick to that decision throughout a problem.

To work directly with an instantaneous expression such as the cosine function is inconvenient when differentiations or integrations of $i(t)$ are involved because they lead to both sine (first-order differentiation or integration) and cosine (second-order differentiation or integration) functions and because it is tedious to combine sine and cosine functions. For instance, the loop equation for a series RLC circuit with an applied voltage $v(t) = V_0 \cos \omega t$ is

$$L\frac{di}{dt} + Ri + \frac{1}{C}\int i\,dt = v(t). \tag{6-70}$$

If we write the resulting current $i(t)$ in the form of Eq. (6-69), Eq. (6-70) yields

$$\begin{aligned} I_0\left[-\omega L \sin(\omega t + \phi) + R\cos(\omega t + \phi) + \frac{1}{\omega C}\sin(\omega t + \phi)\right] \\ = V_0 \cos \omega t. \end{aligned} \tag{6-71}$$

Obviously, complicated mathematical manipulations are required in order to determine the unknown I_0 and ϕ from Eq. (6-71).

It is much simpler to use exponential functions† by writing the applied voltage as

$$\begin{aligned} v(t) &= V_0 \cos \omega t = \mathscr{R}e[(V_0 e^{j0})e^{j\omega t}] \\ &= \mathscr{R}e(V_s e^{j\omega t}) \end{aligned} \tag{6-72}$$

and $i(t)$ in Eq. (6-69) as

$$\begin{aligned} i(t) &= \mathscr{R}e[(Ie^{j\phi})e^{j\omega t}] \\ &= \mathscr{R}e(I_s e^{j\omega t}), \end{aligned} \tag{6-73}$$

† $e^{j\omega t} = \cos \omega t + j \sin \omega t$, $\cos \omega t = \mathscr{R}e(e^{j\omega t})$, $\sin \omega t = \mathscr{I}m(e^{j\omega t})$.

where $\mathcal{R}e$ means "the real part of." In Eqs. (6-72) and (6-73),

$$V_s = V_0 e^{j0} = V_0 \tag{6-74}$$

$$I_s = I_0 e^{j\phi} \tag{6-75}$$

Phasors: polar forms of complex quantities containing amplitude and phase information

are (scalar) ***phasors*** that contain amplitude and phase information but are *independent of t*. The phasor V_s in Eq. (6-74) with zero phase angle is the reference phasor. From Eq. (6-73) we have

$$\frac{di}{dt} = \mathcal{R}e(j\omega I_s e^{j\omega t}), \qquad \text{and} \tag{6-76}$$

$$\int i\,dt = \mathcal{R}e\left(\frac{I_s}{j\omega} e^{j\omega t}\right). \tag{6-77}$$

Substitution of Eqs. (6-72) through (6-77) in Eq. (6-70) yields

$$\left[R + j\left(\omega L - \frac{1}{\omega C}\right)\right] I_s = V_s, \tag{6-78}$$

from which the current phasor I_s can be solved easily. Note that the time-dependent factor $e^{j\omega t}$ disappears from Eq. (6-78) because it is present in every term after the substitution in Eq. (6-70) and is therefore canceled. After I_s has been determined, the instantaneous current response $i(t)$ can be found from Eq. (6-73) by multiplying I_s by $e^{j\omega t}$ and taking the real part of the product.

Converting phasors to instantaneous sinusoidal time functions

If the applied voltage had been given as a *sine function* such as $v(t) = V_0 \sin \omega t$, the series RLC-circuit problem would be solved in terms of phasors in exactly the same way; only the instantaneous expressions would be obtained by taking the *imaginary part* of the product of the phasors with $e^{j\omega t}$. The complex phasors represent the magnitudes and the phase shifts of the quantities in the solution of time-harmonic problems.

EXAMPLE 6-7

Write the phasor expression I_s for the following current functions using a cosine reference.

a) $i(t) = -I_0 \cos(\omega t - 30°)$, and

b) $i(t) = I_0 \sin(\omega t + 0.2\pi)$.

SOLUTION

For cosine reference we write

$$i(t) = \mathcal{R}e(I_s e^{j\omega t}).$$

a) $$i(t) = -I_0 \cos(\omega t - 30°)$$
$$= \mathcal{R}e[(-I_0 e^{-j30°})e^{j\omega t}].$$

Thus, $I_s = -I_0 e^{-j30°} = -I_0 e^{-j\pi/6} = I_0 e^{j5\pi/6}$.

b) $i(t) = I_0 \sin(\omega t + 0.2\pi)$
$= \mathcal{R}e[(I_0 e^{j0.2\pi}) e^{-j\pi/2} \cdot e^{j\omega t}],$

where the $e^{-j\pi/2}$ factor is needed because the phase of $\sin \omega t$ lags behind that of $\cos \omega t$ by 90° or $\pi/2$ (rad). We have

$$I_s = (I_0 e^{j0.2\pi}) e^{-j\pi/2} = I_0 e^{-j0.3\pi}.$$

■ **EXERCISE 6.9** Find the phasor expressions I'_s for the current functions in Example 6-7 using a sine reference, $i(t) = \mathcal{I}m(I'_s e^{j\omega t})$.

ANS. (a) $-I_0 e^{j\pi/3}$, or $I_0 e^{-j2\pi/3}$. (b) $I_0 e^{j0.2\pi}$.

EXAMPLE 6-8

Write the instantaneous expressions $v(t)$ for the following phasors using a cosine reference:

a) $V_s = V_0 e^{j\pi/4}$, and
b) $V_s = 3 - j4$.

SOLUTION

a) $v(t) = \mathcal{R}e[V_s e^{j\omega t}]$
$= \mathcal{R}e[(V_0 e^{j\pi/4}) e^{j\omega t}]$
$= V_0 \cos(\omega t + \pi/4).$

b) $V_s = 3 - j4 = \sqrt{3^2 + 4^2}\, e^{j\tan^{-1}(-4/3)}$
$= 5e^{-j53.1^\circ}.$

Thus,

$$v(t) = \mathcal{R}e[(5e^{-j53.1^\circ}) e^{j\omega t}]$$
$$= 5\cos(\omega t - 53.1^\circ).$$

■ **EXERCISE 6.10** Write the phasor expression V_s for the voltage $v(t) = 10\cos(\omega t - 45^\circ)$.

ANS. $10e^{-j\pi/4}$.

■ **EXERCISE 6.11** Write the instantaneous expression $v(t)$ for the phasor $V_s = 4 + j3$ using a cosine reference.

ANS. $5\cos(\omega t + 36.9^\circ)$.

■ **EXERCISE 6.12** Write the instantaneous expression $v(t)$ for the phasor $V_s = 4 + j3$ using a sine reference.

ANS. $5\sin(\omega t + 126.9^\circ)$.

REVIEW QUESTIONS

Q.6-24 What is a *phasor*? Are phasors functions of t? Functions of ω?

Q.6-25 What is the difference between a phasor and a vector?

REMARKS

1. Phasors are complex quantities (expressed in polar form) representing the magnitude and the phase of sinusoidal functions.
2. Phase angles can be expressed either in radians or in degrees. *Do not forget* the ° sign when expressed in degrees.
3. *Never* mix factors containing j with instantaneous time functions. Expressions such as $j\cos\omega t$, $e^{j\phi}\sin\omega t$, and $(1-j)i(t)$ are incorrect.

6-5.2 TIME-HARMONIC ELECTROMAGNETICS

Field vectors that vary with space coordinates and are sinusoidal functions of time can similarly be represented by vector phasors that depend on space coordinates but not on time. As an example, we can write a time-harmonic **E** field *referring to* $\cos\omega t$[†] *as*

$$\mathbf{E}(x, y, z; t) = \mathcal{R}e[\mathbf{E}(x, y, z)e^{j\omega t}], \tag{6-79}$$

where $\mathbf{E}(x, y, z)$ is a ***vector phasor*** that contains information on direction, magnitude, and phase. From Eqs. (6-76) and (6-77) we see that, if $\mathbf{E}(x, y, z; t)$ is to be represented by the vector phasor $\mathbf{E}(x, y, z)$, then $\partial\mathbf{E}(x, y, z; t)/\partial t$ and $\int\mathbf{E}(x, y, z; t)\,dt$ would be represented by vector phasors $j\omega\mathbf{E}(x, y, z)$ and $\mathbf{E}(x, y, z)/j\omega$, respectively. Higher-order differentiations and integrations with respect to t would be represented by multiplications and divisions, respectively, of the phasor $\mathbf{E}(x, y, z)$ by higher powers of $j\omega$.

We now write time-harmonic Maxwell's equations (6-45a, b, c, and d) in terms of vector field phasors ($\mathbf{E}, \mathbf{H}$) and source phasors ($\rho_v, \mathbf{J}$) in a simple (linear, isotropic, and homogeneous) medium as follows.

Time-harmonic Maxwell's equations in terms of phasors

$$\nabla\times\mathbf{E} = -j\omega\mu\mathbf{H}, \tag{6-80a}$$

$$\nabla\times\mathbf{H} = \mathbf{J} + j\omega\epsilon\mathbf{E}, \tag{6-80b}$$

$$\nabla\cdot\mathbf{E} = \rho_v/\epsilon, \tag{6-80c}$$

$$\nabla\cdot\mathbf{H} = 0. \tag{6-80d}$$

The space-coordinate arguments and the subscript s indicating a phasor quantity have been omitted for simplicity. The fact that the same notations

[†] If the time reference is not explicitly specified, it is customarily taken as $\cos\omega t$.

are used for the phasors as are used for their corresponding time-dependent quantities should create little confusion because we will deal almost exclusively with time-harmonic fields (and therefore with phasors) in the rest of this book. When there is a need to distinguish an instantaneous quantity from a phasor, the time dependence of the instantaneous quantity will be indicated explicitly by the inclusion of a t in its argument.

REVIEW QUESTION

Q.6-26 Discuss the advantages of using phasors in electromagnetics.

REMARKS

1. Phasor quantities are not functions of t.
2. Instantaneous time functions cannot contain complex numbers.
3. Any electromagnetic expression containing j must necessarily be a relation of phasors.

In terms of phasors we can write time-harmonic wave equation (6-58) for scalar potential V as

$$\nabla^2 V - \mu\epsilon(j\omega)^2 V = -\frac{\rho_v}{\epsilon},$$

or

Helmholtz's equation for scalar potential V

$$\boxed{\nabla^2 V + k^2 V = -\frac{\rho_v}{\epsilon},} \tag{6-81}$$

where

$$k = \omega\sqrt{\mu\epsilon} = \frac{\omega}{u_p}, \tag{6-82a}$$

or

Definition of wave number

$$k = \frac{2\pi f}{u_p} = \frac{2\pi}{\lambda} \tag{6-82b}$$

is called the ***wavenumber***. It is a measure of the number of wavelengths in a cycle. Similarly, the phasor form of time-harmonic wave equation (6-57) for vector potential $\mathbf{A}$ is

Helmholtz's equation for vector potential A

$$\boxed{\nabla^2 \mathbf{A} + k^2 \mathbf{A} = -\mu\mathbf{J}.} \tag{6-83}$$

Equations (6-81) and (6-83) are referred to as ***nonhomogeneous Helmholtz's equations.***

■ **EXERCISE 6.13** Write the phasor form for the Lorentz condition for potentials, Eq. (6-56).

REVIEW QUESTION

Q.6-27 Define *wavenumber*. What is its SI unit?

REMARKS

1. Helmholtz's equations are time-harmonic wave equations in terms of phasors.
2. Wavenumber depends on medium characteristics as well as on the frequency of the wave, but is always equal to 2π divided by the wavelength.

The solution of the nonhomogeneous Helmholtz's equation (6-81) for V can be obtained from Eq. (6-67). The potential $V(R, t)$ involves a time advance R/u_p for ρ_v, which is equivalent to a phase lead of $\omega(R/u_p)$ or kR. In phasor notation this requires a multiplying factor e^{-jkR}. Hence the *phasor form* for Eq. (6-67) is

Phasor form for retarded scalar potential

$$\boxed{V(R) = \frac{1}{4\pi\epsilon}\int_{V'} \frac{\rho_v e^{-jkR}}{R}\, dv' \qquad \text{(V)}.} \tag{6-84}$$

Similarly, the phasor solution of Eq. (6-83) for $\mathbf{A}$ is

Phasor form for retarded vector potential

$$\boxed{\mathbf{A}(R) = \frac{\mu}{4\pi}\int_{V'} \frac{\mathbf{J}e^{-jkR}}{R}\, dv' \qquad \text{(Wb/m)}.} \tag{6-85}$$

These are the expressions for the retarded scalar and vector potentials due to time-harmonic sources. Now the Taylor-series expansion for the exponential factor e^{-jkR} is

$$e^{-jkR} = 1 - jkR + \frac{k^2R^2}{2} + \cdots. \tag{6-86}$$

Thus, if

$$kR = 2\pi\frac{R}{\lambda} \ll 1, \tag{6-87}$$

or if the distance R is very small in comparison to the wavelength λ, e^{-jkR} can be approximated by 1. Equations (6-84) and (6-85) then simplify to the static expressions in Eqs. (3-39) and (5-22), respectively.

Procedure for determining instantaneous electric and magnetic fields

The formal procedure for determining the electric and magnetic fields due to time-harmonic charge and current distributions is as follows:

1. Find phasors $V(R)$ and $\mathbf{A}(R)$ from Eqs. (6-84) and (6-85).
2. Find phasors $\mathbf{E}(R) = -\nabla V - j\omega\mathbf{A}$ and $\mathbf{B}(R) = \nabla \times \mathbf{A}$.
3. Find instantaneous $\mathbf{E}(R, t) = \mathcal{R}e[\mathbf{E}(R)e^{j\omega t}]$ and $\mathbf{B}(R, t) = \mathcal{R}e[\mathbf{B}(R)e^{j\omega t}]$ for a cosine reference.

The degree of difficulty of a problem depends on how difficult it is to perform the integrations in Step 1. We will use this procedure to determine the radiation properties of antennas in Chapter 10.

EXAMPLE 6-9

Given that the electric field intensity of an electromagnetic wave in a nonconducting dielectric medium with permittivity $\epsilon = 9\epsilon_0$ and permeability μ_0 is

$$\mathbf{E}(z, t) = \mathbf{a}_y 5 \cos(10^9 t - \beta z) \qquad \text{(V/m)}, \tag{6-88}$$

find the magnetic field intensity $\mathbf{H}$ and the value of β.

SOLUTION

The given $\mathbf{E}(z, t)$ in Eq. (6-88) is a harmonic time function with angular frequency $\omega = 10^9$ (rad/s). Using phasors with a cosine reference, we have

$$\mathbf{E}(z) = \mathbf{a}_y 5e^{-j\beta z}. \tag{6-89}$$

The magnetic field intensity can be calculated from Eq. (6-80a).

$$\begin{aligned}
\mathbf{H}(z) &= -\frac{1}{j\omega\mu_0}\nabla \times \mathbf{E} \\
&= -\frac{1}{j\omega\mu_0}\begin{vmatrix} \mathbf{a}_x & \mathbf{a}_y & \mathbf{a}_z \\ \dfrac{\partial}{\partial x} & \dfrac{\partial}{\partial y} & \dfrac{\partial}{\partial z} \\ 0 & 5e^{-j\beta z} & 0 \end{vmatrix} = -\frac{1}{j\omega\mu_0}\left(-\mathbf{a}_x\frac{\partial}{\partial z}E_y\right) \\
&= -\frac{1}{j\omega\mu_0}(\mathbf{a}_x j\beta 5e^{-j\beta z}) = -\mathbf{a}_x\frac{\beta}{\omega\mu_0}5e^{-j\beta z}.
\end{aligned} \tag{6-90}$$

In order to determine β, we use Eq. (6-80b). For a nonconducting medium, $\sigma = 0$, $\mathbf{J} = 0$. Thus,

$$\begin{aligned}
\mathbf{E}(z) &= \frac{1}{j\omega\epsilon}\nabla \times \mathbf{H} = \frac{1}{j\omega\epsilon}\left(\mathbf{a}_y\frac{\partial}{\partial z}H_x\right) \\
&= \mathbf{a}_y\frac{\beta^2}{\omega^2\mu_0\epsilon}5e^{-j\beta z}.
\end{aligned} \tag{6-91}$$

Equating Eqs. (6-89) and (6-91), we require

$$\beta = \omega\sqrt{\mu_0\epsilon} = 3\omega\sqrt{\mu_0\epsilon_0} = \frac{3\omega}{c}$$

$$= \frac{3 \times 10^9}{3 \times 10^8} = 10 \qquad \text{(rad/m)}.$$

From Eq. (6-90) we obtain

$$\mathbf{H}(z) = -\mathbf{a}_x \frac{5(10)}{(10^9)(4\pi 10^{-7})} e^{-j10z}$$

$$= -\mathbf{a}_x 0.0398 e^{-j10z}. \tag{6-92}$$

The phasor $\mathbf{H}(z)$ in Eq. (6-92) corresponds to the following instantaneous time function:

$$\mathbf{H}(z, t) = -\mathbf{a}_x 0.0398 \cos(10^9 t - 10z) \qquad \text{(A/m)}. \tag{6-93}$$

REVIEW QUESTIONS

Q.6-28 Write the phasor expression of scalar electric potential $V(R)$ in terms of charge distribution ρ_v.

Q.6-29 Write the phasor expression of vector magnetic potential $\mathbf{A}(R)$ in terms of current distribution $\mathbf{J}$.

REMARKS

The electric and magnetic field intensities of an electromagnetic wave in a given medium are definitely related. They must satisfy Maxwell's equations, and their amplitudes and phases cannot be specified independently.

6-5.3 THE ELECTROMAGNETIC SPECTRUM

In *source-free nonconducting media* characterized by ϵ and $\mu(\sigma = 0)$ Maxwell's equations (6-45a, b, c, and d) reduce to

Maxwell's equations in source-free nonconducting media

$$\nabla \times \mathbf{E} = -\mu \frac{\partial \mathbf{H}}{\partial t}, \tag{6-94a}$$

$$\nabla \times \mathbf{H} = \epsilon \frac{\partial \mathbf{E}}{\partial t}, \tag{6-94b}$$

$$\nabla \cdot \mathbf{E} = 0, \tag{6-94c}$$

$$\nabla \cdot \mathbf{H} = 0. \tag{6-94d}$$

Equations (6-94a, b, c, and d) are first-order differential equations in the two variables **E** and **H**. They can be combined to give a second-order equation in **E** or **H** alone. To do this, we take the curl of Eq. (6-94a) and use Eq. (6-94b):

$$\nabla \times \nabla \times \mathbf{E} = -\mu \frac{\partial}{\partial t}(\nabla \times \mathbf{H}) = -\mu\epsilon \frac{\partial^2 \mathbf{E}}{\partial t^2}.$$

Now $\nabla \times \nabla \times \mathbf{E} = \nabla(\nabla \cdot \mathbf{E}) - \nabla^2 \mathbf{E} = -\nabla^2 \mathbf{E}$ because of Eq. (6-94c). Hence we have

$$\nabla^2 \mathbf{E} - \mu\epsilon \frac{\partial^2 \mathbf{E}}{\partial t^2} = 0; \tag{6-95}$$

or, since $u_p = 1/\sqrt{\mu\epsilon}$,

Homogeneous wave equation for E

$$\boxed{\nabla^2 \mathbf{E} - \frac{1}{u_p^2} \frac{\partial^2 \mathbf{E}}{\partial t^2} = 0.} \tag{6-96}$$

In an entirely similar way we can also obtain an equation in **H**:

Homogeneous wave equation for H

$$\boxed{\nabla^2 \mathbf{H} - \frac{1}{u_p^2} \frac{\partial^2 \mathbf{H}}{\partial t^2} = 0.} \tag{6-97}$$

Equations (6-96) and (6-97) are ***homogeneous vector wave equations***.

We can see that in Cartesian coordinates Eqs. (6-96) and (6-97) can each be decomposed into three one-dimensional, homogeneous, scalar wave equations. Each component of **E** and of **H** will satisfy an equation exactly like Eq. (6-61), whose solutions represent waves. We will extensively discuss wave behavior in various environments in the next chapter.

For time-harmonic fields it is convenient to use phasors. Thus Eq. (6-96) becomes

$$\nabla^2 \mathbf{E}_s + \frac{\omega^2}{u_p^2} \mathbf{E}_s = 0,$$

or

Homogeneous Helmholtz's equation for phasor $\mathbf{E}_s$

$$\boxed{\nabla^2 \mathbf{E}_s + k^2 \mathbf{E}_s = 0,} \tag{6-98}^\dagger$$

in view of Eq. (6-82a). Similarly, Eq. (6-97) leads to

Homogeneous Helmholtz's equation for phasor $\mathbf{H}_s$

$$\boxed{\nabla^2 \mathbf{H}_s + k^2 \mathbf{H}_s = 0.} \tag{6-99}^\dagger$$

†Subscript s has been added to emphasize that $\mathbf{E}_s$ and $\mathbf{H}_s$ are phasors and are not the same as the time-dependent **E** and **H** in Eqs. (6-96) and (6-97).

The solutions of Eqs. (6-98) and (6-99) represent propagating waves, which will be the subject of study of the next chapter.

REVIEW QUESTION

Q.6-30 Explain why there may be nonvanishing solutions for electric and magnetic fields in source-free regions.

REMARKS

1. In source-free nonconducting media **E** and **H** satisfy the same homogeneous wave equation, in Eqs. (6-96) and (6-97).
2. For time-harmonic waves in source-free media phasors $\mathbf{E}_s$ and $\mathbf{H}_s$ satisfy the same homogeneous Helmholtz's equation, in Eqs. (6-98) and (6-99).
3. Maxwell's equations, and therefore the wave and Helmholtz's equations, impose no limit on the frequency of the waves.

The electromagnetic spectrum and applications

The electromagnetic spectrum that has been investigated experimentally extends from very low power frequencies through radio, television, microwave, infrared, visible light, ultraviolet, X-ray, and gamma (γ)-ray frequencies exceeding 10^{24} (Hz). Figure 6-8 shows the electromagnetic spectrum divided into frequency and wavelength ranges on logarithmic scales according to application and natural occurrence.

The visible-light spectrum

Wireless transmission

Microwave radar bands

The term "microwave" is somewhat nebulous and imprecise; it could mean electromagnetic waves above a frequency of 1 (GHz) and all the way up to the lower limit of the infrared band, encompassing UHF, SHF, EHF, and mm-wave regions. The wavelength range of visible light is from deep red at 720 (nm) to violet at 380 (nm), or from 0.72 (μm) to 0.38 (μm), corresponding to a frequency range of from 4.2×10^{14} (Hz) to 7.9×10^{14} (Hz). The bands used for radar, satellite communication, navigation aids, television (TV), FM and AM radio, citizen's band radio (CB), sonar, and others are also noted. Frequencies below the VLF range are seldom used for wireless transmission because huge antennas would be needed for efficient radiation of electromagnetic waves and because of the very low data rate at these low frequencies. There have been proposals to use these frequencies for strategic global communication with submarines submerged in conducting seawater. In radar work it has been found convenient to assign alphabet names to the different microwave frequency bands. They are listed in Table 6-4.

In the next chapter we shall discuss the characteristics of plane electromagnetic waves and examine their behavior as they propagate across discontinuous boundaries.

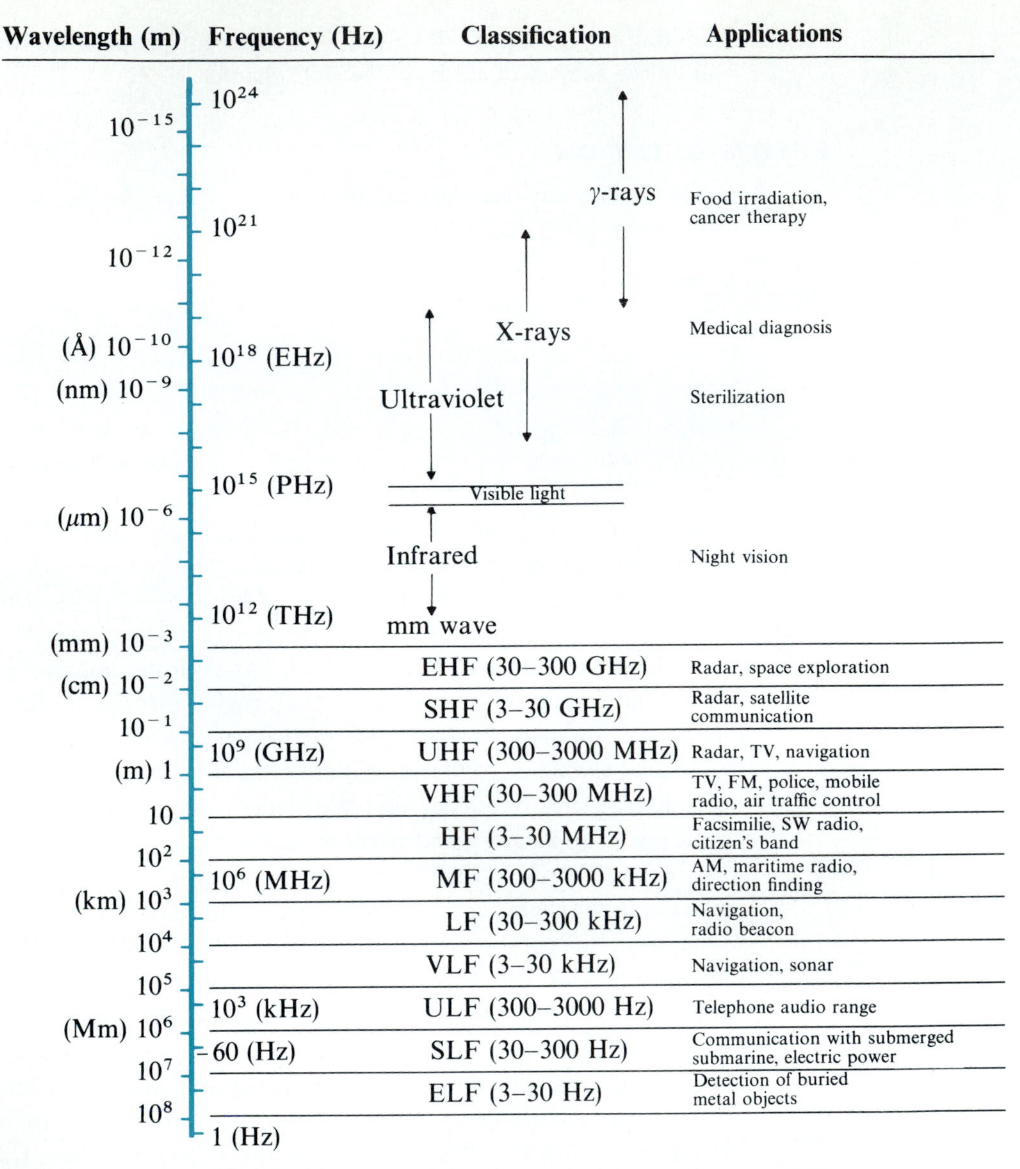

FIGURE 6-8 Spectrum of electromagnetic waves.

REVIEW QUESTIONS

Q.6-31 What is the wavelength range of visible light?

Q.6-32 Why are frequencies below the VLF range rarely used for wireless transmission?

TABLE 6-4 RADAR BAND DESIGNATIONS FOR MICROWAVE FREQUENCIES

Band	Frequency Range (GHz)	Wavelength Range (cm)
U	40–60	0.75–0.50
Ka	26.5–40	1.13–0.75
K	18–26.5	1.67–1.13
Ku	12.4–18	2.42–1.67
X	8–12.4	3.75–2.42
C	4–8	7.5–3.75
S	2–4	15–7.5
L	1–2	30–15

REMARKS

1. Electromagnetic waves of *all frequencies* propagate in a given medium with the same velocity, $u_p = 1/\sqrt{\mu\epsilon}$.
2. The operating frequency of microwave ovens is around 2.45 (GHz).

SUMMARY

In time-varying situations electric and magnetic fields are coupled and the postulates that we introduced in previous chapters for static fields no longer suffice. In this chapter we

- added a fundamental postulate for electromagnetic induction,
- introduced Faraday's law that relates quantitatively the emf induced in a circuit to the time rate of change of flux linkage,
- explained that the induced emf may be decomposed into two parts: a transformer emf and a motional (flux-cutting) emf,
- discussed the characteristics of ideal transformers,
- obtained a set of four Maxwell's equations (two divergence and two curl equations) that are consistent with the equation of continuity,
- considered the general boundary conditions for field vectors at the interface of contiguous regions of different constitutive parameters,
- expressed electric and magnetic field intensities in terms of a scalar electric potential function V and a vector magnetic potential function $\mathbf{A}$,

- derived nonhomogeneous wave equations for V and $\mathbf{A}$,
- introduced the concept of retarded potentials,
- converted wave equations into Helmholtz's equations for time-harmonic fields, and
- discussed the electromagnetic spectrum in source-free space.

PROBLEMS

P.6-1 Express the transformer emf induced in a stationary loop in terms of time-varying vector potential $\mathbf{A}$.

P.6-2 The circuit in Fig. 6-9 is situated in a magnetic field

$$\mathbf{B} = \mathbf{a}_x 3 \cos (5\pi 10^7 t - \tfrac{1}{3}\pi y) \qquad (\mu\text{T}).$$

Assuming $R = 15\,(\Omega)$, find the current i.

P.6-3 A stationary rectangular conducting loop of width w and height h is situated near a very long wire carrying a current i_1 as in Fig. 6-10. Assume $i_1 = I_1 \sin \omega t$ and the self-inductance of the rectangular loop to be L. Find the induced current i_2 in the loop.

HINT: Use phasors.

P.6-4 In Fig. 6-10 assume a constant current $i_1 = I_0$, but that the rectangular loop moves away with a constant velocity $\mathbf{u} = \mathbf{a}_y u_0$. Determine i_2 when the loop is at a position as shown.

P.6-5 A 10(cm) by 10(cm) square conducting loop having a resistance $R = 0.5\,(\Omega)$ rotates in a constant magnetic field $\mathbf{B} = \mathbf{a}_y 0.04\,(\text{T})$ with an

FIGURE 6-9 A circuit in a time-varying magnetic field (Problem P.6-2).

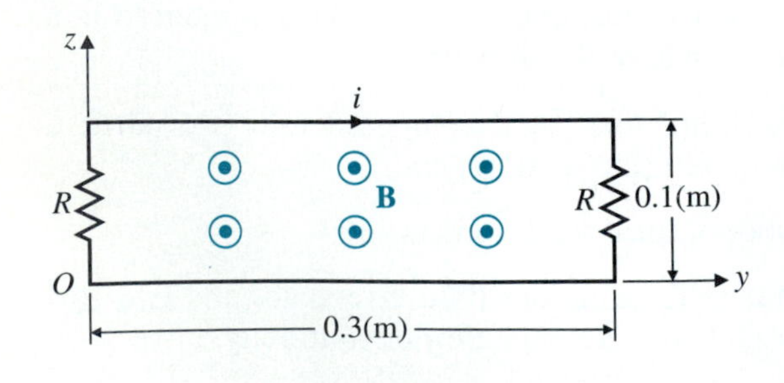

FIGURE 6-10 A rectangular loop near a long current-carrying wire (for Problems P.6-3 and P.6-4).

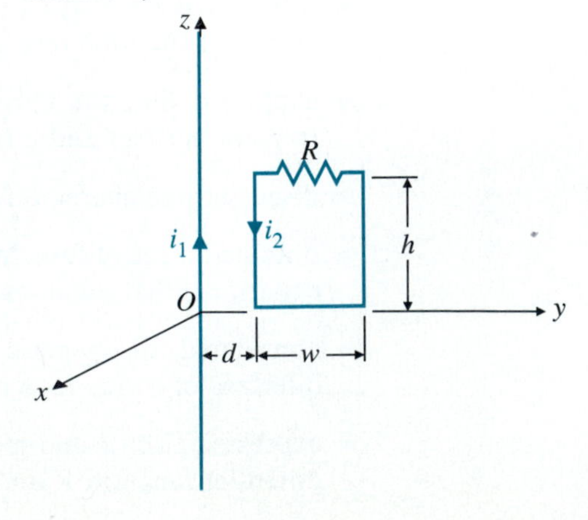

angular frequency $\omega = 100\pi$ (rad/s) about one of its legs, as shown in Fig. 6-11. Assuming that the loop initially lies in the xz-plane, find the induced current i

a) if the self-inductance of the loop is neglected, and

b) if the self-inductance of the loop is 3.5 (mH).

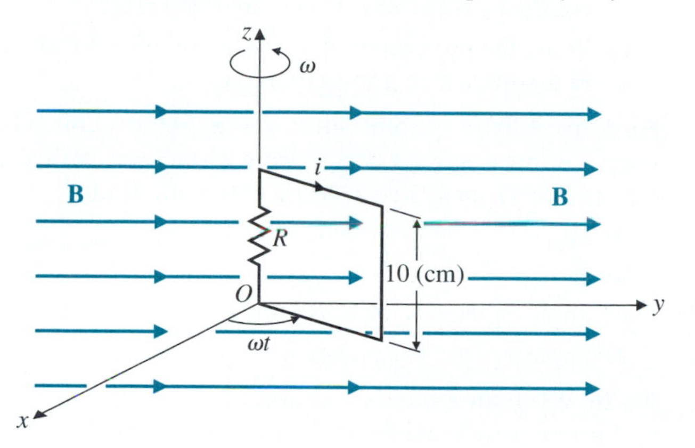

FIGURE 6-11 A rotating square loop in a constant magnetic field (Problem P.6-5).

P.6-6 A conducting sliding bar oscillates over two parallel conducting rails in a sinusoidally varying magnetic field

$$\mathbf{B} = \mathbf{a}_z 5 \cos \omega t \qquad \text{(mT)},$$

as shown in Fig. 6-12. The position of the sliding bar is given by $x = 0.35(1 - \cos \omega t)$ (m), and the rails are terminated in a resistance $R = 0.2\,(\Omega)$. Find i.

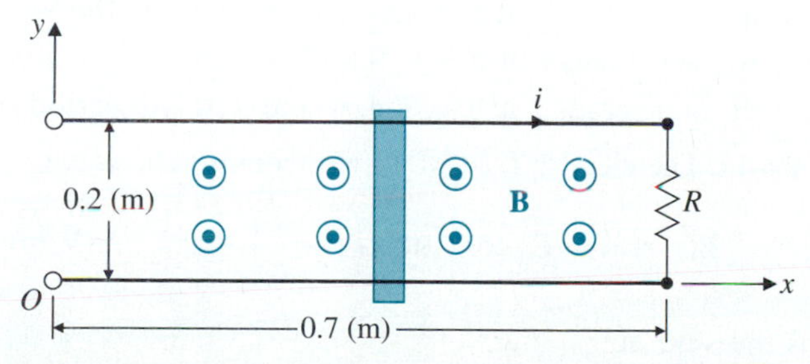

FIGURE 6-12 A conducting bar sliding over parallel rails in a time-varying magnetic field (Problem P.6-6).

P.6-7 Determine the frequency at which a time-harmonic electric field intensity causes a conduction current density and a displacement current density of equal magnitude in

a) seawater with $\epsilon_r = 72$ and $\sigma = 4$ (S/m), and

b) moist soil with $\epsilon_r = 2.5$ and $\sigma = 10^{-3}$ (S/m).

P.6-8 Calculations concerning the electromagnetic effect of currents in a good conductor usually neglect the displacement current even at microwave frequencies.

a) Assuming $\epsilon_r = 1$ and $\sigma = 5.70 \times 10^7$ (S/m) for copper, compare the magnitude of the displacement current density with that of the conduction current density at 100 (GHz).

b) Write the governing differential equation for magnetic field intensity **H** in a source-free good conductor.

P.6-9 An infinite current sheet $\mathbf{J} = \mathbf{a}_x 5$ (A/m) coinciding with the xy-plane separates air (region 1, $z > 0$) from a medium with $\mu_{r2} = 2$ (region 2, $z < 0$). Given that $\mathbf{H}_1 = \mathbf{a}_x 30 + \mathbf{a}_y 40 + \mathbf{a}_z 20$ (A/m), find

a) $\mathbf{H}_2$,

b) $\mathbf{B}_2$,

c) angle α_1 that $\mathbf{B}_1$ makes with the z-axis, and

d) angle α_2 that $\mathbf{B}_2$ makes with the z-axis.

P.6-10 Write the boundary conditions that exist at the interface of free space and a magnetic material of infinite (an approximation) permeability.

P.6-11 Prove by direct substitution that any twice differentiable function of $(t - R\sqrt{\mu\epsilon})$ is a solution of the homogeneous wave equation, Eq. (6-61).

P.6-12 Write the component scalar equations for the set of four Maxwell's equations (6-80a, b, c, and d):

a) in Cartesian coordinates if the field phasors are functions of z only, and

b) in spherical coordinates if the field phasors are functions of R only.

P.6-13 Assuming $E(z, t) = 50 \cos(2\pi 10^9 t - kz)$ (V/m) in air, sketch the following waveforms and label the absissas:

a) $E(t)$ versus t at $z = 100.125\lambda$, where λ is the wavelength,

b) $E(t)$ versus t at $z = -100.125\lambda$, and

c) $E(z)$ versus z at $t = T/4$, where T is the period of the wave.

P.6-14 The electric field of an electromagnetic wave

$$\mathbf{E}(z, t) = \mathbf{a}_x E_0 \cos\left[10^8\pi\left(t - \frac{z}{c}\right) + \psi\right] \qquad \text{(V/m)}$$

is the sum of

$$\mathbf{E}_1(z, t) = \mathbf{a}_x 0.03 \sin 10^8\pi\left(t - \frac{z}{c}\right) \qquad \text{(V/m)}$$

and

$$\mathbf{E}_2(z, t) = \mathbf{a}_x 0.04 \cos\left[10^8\pi\left(t - \frac{z}{c}\right) - \frac{\pi}{3}\right]. \qquad \text{(V/m).}$$

Find E_0 and ψ.

P.6-15 The magnetic field intensity of an electromagnetic wave

$$\mathbf{H}(R, t) = \mathbf{a}_\phi H_0 \cos(\omega t - kR) \qquad \text{(A/m)}$$

is the sum of

$$\mathbf{H}_1(R, t) = \mathbf{a}_\phi 10^{-4} \sin(\omega t - kR) \qquad \text{(A/m)}$$

and

$$\mathbf{H}_2(z, t) = \mathbf{a}_\phi 2 \times 10^{-4} \cos(\omega t - kR + \alpha) \qquad \text{(A/m)}.$$

Find H_0 and α.

P.6-16 Starting from the phasor Maxwell's equations (6-80a, b, c, and d) in simple media with time-harmonic charge and current distributions obtain

a) the nonhomogeneous Helmholtz's equation for **E**, and

b) the nonhomogeneous Helmholtz's equation for **H**.

P.6-17 A short conducting wire carrying a time-harmonic current is a source of electromagnetic waves. Assuming that a uniform current $i(t) = I_0 \cos \omega t$ flows in a very short wire $d\ell$ placed along the z-axis,

a) determine the phasor retarded vector potential **A** at a distance R in spherical coordinates, and

b) find the magnetic field intensity **H** from **A**.

P.6-18 A 60-(MHz) electromagnetic wave exists in an air-dielectric coaxial cable having an inner conductor with radius a and an outer conductor with inner radius b. Assuming perfect conductors, and the phasor form of the electric field intensity to be

$$\mathbf{E} = \mathbf{a}_r \frac{E_0}{r} e^{-jkz} \text{(V/m)}, \qquad a < r < b,$$

a) find k,

b) find **H** from the $\nabla \times \mathbf{E}$ equation, and

c) find the surface current densities on the inner and outer conductors.

P.6-19 It is known that the electric field intensity of a spherical wave in free space is

$$\mathbf{E}(R, \theta; t) = \mathbf{a}_\theta \frac{10^{-3}}{R} \sin\theta \cos(2\pi 10^9 t - kR) \qquad \text{(V/m)}.$$

Determine the magnetic field intensity $\mathbf{H}(R, \theta; t)$ and the value of k.

P.6-20 Given that

$$\mathbf{E}(x, z; t) = \mathbf{a}_y 0.1 \sin(10\pi x) \cos(6\pi 10^9 t - \beta z) \qquad \text{(V/m)}$$

in air, find $\mathbf{H}(x, z; t)$ and β.

P.6-21 Given that

$$\mathbf{H}(x, z; t) = \mathbf{a}_y 2 \cos(15\pi x) \sin(6\pi 10^9 t - \beta z) \qquad \text{(A/m)}$$

in air, find $\mathbf{E}(x, z; t)$ and β.

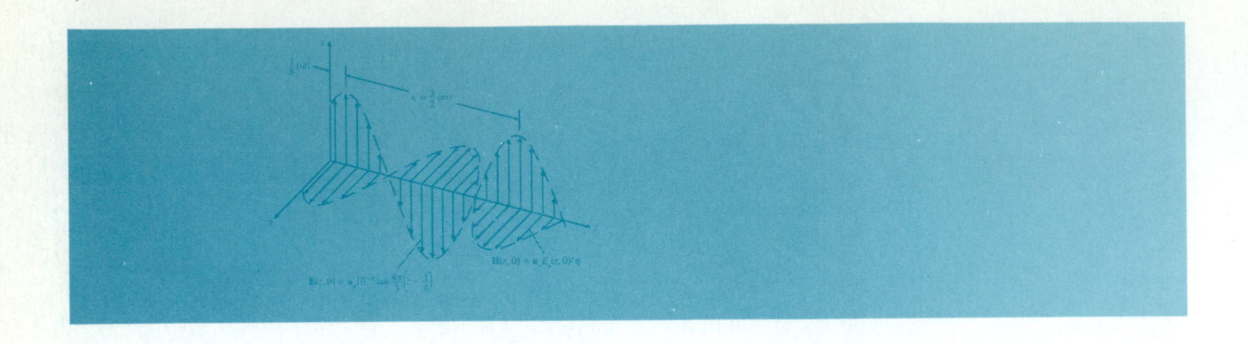

CHAPTER 7

7-1 OVERVIEW In Section 6-5 we showed that in a source-free nonconducting simple medium, Maxwell's equations can be combined to yield homogeneous vector wave equations in **E** and in **H**. These two equations, Eqs. (6-96) and (6-97), have exactly the same form. For example, the source-free wave equation for **E** is

$$\nabla^2\mathbf{E} - \frac{1}{u_p^2}\frac{\partial^2\mathbf{E}}{\partial t^2} = 0. \tag{6-96)(7-1}$$

In Cartesian coordinates, Eq. (7-1) can be decomposed into three one-dimensional, homogeneous, scalar wave equations. Each component equation will be of the form of Eq. (6-61), whose solution represents a wave propagating in the medium with a velocity

$$u_p = \frac{1}{\sqrt{\mu\epsilon}}. \tag{6-63)(7-2}$$

Since time-varying **E** and **H** are coupled through Maxwell's curl equations (6-94a) and (6-94b), the consequence is an electromagnetic wave, which we use to explain electromagnetic action at a distance. The study of the behavior of waves that have a one-dimensional spatial dependence is the main concern of this chapter.

We begin the chapter with a study of the propagation of time-harmonic plane-wave fields in an unbounded homogeneous medium. Medium parameters such as intrinsic impedance, attenuation constant, and phase constant will be introduced. The meaning of ***skin depth***, the depth of wave penetration into a

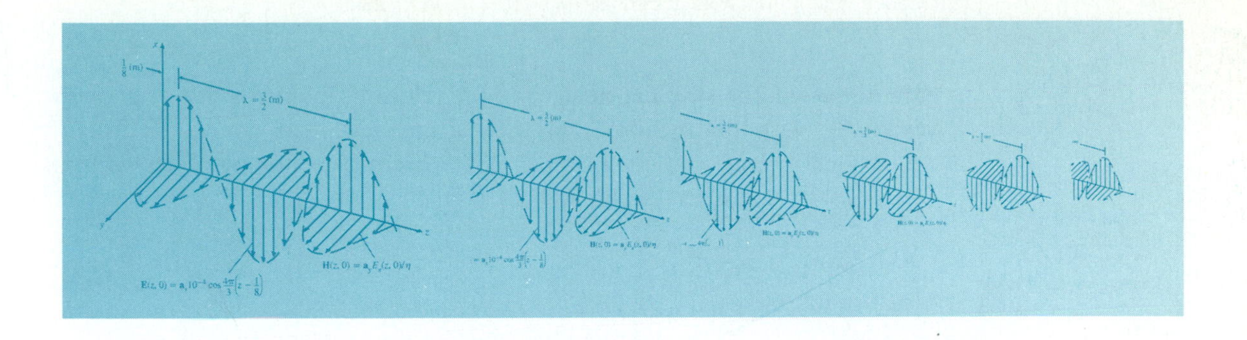

Plane Electromagnetic Waves

good conductor, will be explained. Electromagnetic waves carry with them electromagnetic power. The concept of ***Poynting vector***, a power flux density, will be discussed.

We will study the behavior of a plane wave incident obliquely on a plane boundary. The laws governing the reflection and refraction of plane waves as well as the conditions for no reflection and for total reflection will be examined.

Definition of uniform plane wave

A ***uniform plane wave*** is a particular solution of Maxwell's equations with **E** assuming the same direction, same magnitude, and same phase in infinite planes perpendicular to the direction of propagation (similarly for **H**). Strictly speaking, a uniform plane wave does not exist in practice because a source infinite in extent would be required to create it, and practical wave sources are always finite in extent. But, if we are far enough away from a source, the ***wavefront*** (surface of constant phase) becomes almost spherical; and a very small portion of the surface of a giant sphere is very nearly a plane. The characteristics of uniform plane waves are particularly simple, and their study is of fundamental theoretical, as well as practical, importance.

Definition of Wavefront

7-2 Plane Waves in Lossless Media

In this and future chapters we focus our attention on wave behavior in the sinusoidal steady state. Sinusoidal quantities will be represented by phasors without a subscript s, for simplicity. In situations where instantaneous time functions are discussed, the time dependence of the relevant quantities will be shown explicitly with the symbol t in their arguments. The source-free wave equations in nonconducting simple media becomes a homogeneous vector Helmholtz's equation (see Eq. 6-98):

Homogeneous vector Helmholtz's equation in E in nonconducting simple media

$$\boxed{\nabla^2\mathbf{E} + k^2\mathbf{E} = 0,} \tag{7-3}$$

where k is the wavenumber. In a medium characterized by ϵ and μ, we have, from Eq. (6-82a),

$$\boxed{k = \omega\sqrt{\mu\epsilon} = \frac{\omega}{u_p} \qquad \text{(rad/m).}} \tag{7-4}$$

In Cartesian coordinates, Eq. (7-3) is equivalent to three scalar Helmholtz's equations, one each in the components E_x, E_y, and E_z. Writing it for the component E_x, we have

$$\left(\frac{\partial^2}{\partial x^2} + \frac{\partial^2}{\partial y^2} + \frac{\partial^2}{\partial z^2} + k^2\right)E_x = 0. \tag{7-5}$$

Consider a uniform plane wave characterized by a uniform E_x (uniform magnitude and constant phase) over plane surfaces perpendicular to z; that is,

$$\partial^2 E_x/\partial x^2 = 0 \qquad \text{and} \qquad \partial^2 E_x/\partial y^2 = 0.$$

Equation (7-5) simplifies to

$$\frac{d^2E_x}{dz^2} + k^2E_x = 0, \tag{7-6}$$

which is an ordinary differential equation because E_x, a phasor, depends only on z.

The solution of Eq. (7-6) is readily seen to be

$$\begin{aligned} E_x(z) &= E_x^+(z) + E_x^-(z) \\ &= E_0^+ e^{-jkz} + E_0^- e^{jkz}, \end{aligned} \tag{7-7}$$

where E_0^+ and E_0^- are arbitrary constants that must be determined by boundary conditions.

Let us now examine the first phasor term on the right side of Eq. (7-7) and write

$$\mathbf{E}(z) = \mathbf{a}_x E_x^+(z) = \mathbf{a}_x E_0^+ e^{-jkz}. \tag{7-8}$$

For a cosine reference, the instantaneous expression for $\mathbf{E}$ in Eq. (7-8) is

$$\begin{aligned} \mathbf{E}(z, t) &= \mathbf{a}_x E_x^+(z, t) = \mathbf{a}_x \mathcal{R}e[E_x^+(z)e^{j\omega t}] \\ &= \mathbf{a}_x \mathcal{R}e[E_0^+ e^{j(\omega t - kz)}] = \mathbf{a}_x E_0^+ \cos(\omega t - kz). \end{aligned} \tag{7-9}$$

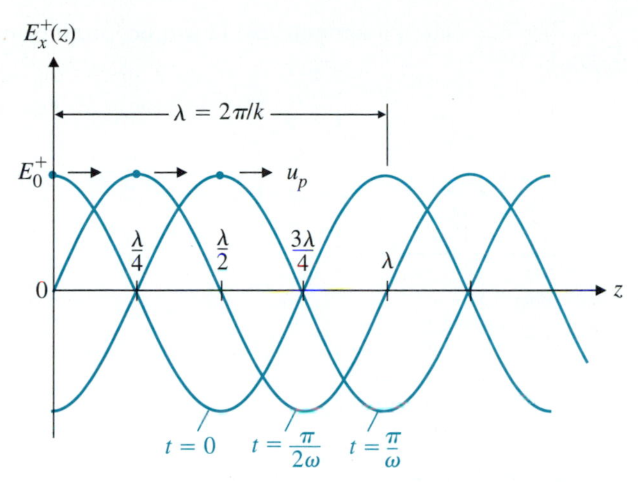

FIGURE 7-1 Wave traveling in positive z direction $E_x^+(z, t) = E_0^+ \cos(\omega t - kz)$, for several values of t.

Characteristics of a traveling wave

Equation (7-9) has been plotted in Fig. 7-1 for several values of t. At $t = 0$, $E_x^+(z, 0) = E_0^+ \cos kz$ is a cosine curve with an amplitude E_0^+. At successive times the curve effectively *travels in the positive z direction.* We have, then, a ***traveling wave***. If we fix our attention on a particular point (a point of a particular phase) on the wave, we set $\cos(\omega t - kz)$ = a constant, or

$$\omega t - kz = \text{A constant phase},$$

from which we obtain

$$u_p = \frac{dz}{dt} = \frac{\omega}{k} = \frac{1}{\sqrt{\mu\epsilon}}. \tag{7-10}$$

Definition of phase velocity

Equation (7-10) assures us that the velocity of propagation of an equiphase front (the ***phase velocity***) is equal to the velocity of light. Wavenumber k bears a definite relation to the wavelength.

$$\boxed{k = \omega\sqrt{\mu\epsilon} = \frac{2\pi}{\lambda},} \tag{7-11}$$

as has been noted in Eqs. (6-82a and b).

We can see, without replotting, that the second phasor term on the right side of Eq. (7-7), $E_0^- e^{jkz}$, represents a cosinusoidal wave traveling in the $-z$ direction with the same velocity u_p. If we are concerned only with the wave traveling in the $+z$ direction, we set $E_0^- = 0$. However, if there are discontinuities in the medium, reflected waves traveling in the opposite direction must also be considered, as we will see later in this chapter.

The associated magnetic field **H** can be found from the $\nabla \times \mathbf{E}$ equation (6-80a):

$$\nabla \times \mathbf{E} = \begin{vmatrix} \mathbf{a}_x & \mathbf{a}_y & \mathbf{a}_z \\ 0 & 0 & \dfrac{\partial}{\partial z} \\ E_x^+(z) & 0 & 0 \end{vmatrix} = -j\omega\mu(\mathbf{a}_x H_x^+ + \mathbf{a}_y H_y^+ + \mathbf{a}_z H_z^+),$$

which leads to

$$H_x^+ = 0, \tag{7-12a}$$

$$H_y^+ = \frac{1}{-j\omega\mu}\frac{\partial E_x^+(z)}{\partial z}, \tag{7-12b}$$

$$H_z^+ = 0. \tag{7-12c}$$

Thus H_y^+ is the only nonzero component of **H** corresponding to the **E** in Eq. (7-8), and since

$$\frac{\partial E_x^+(z)}{\partial z} = \frac{\partial}{\partial z}(E_0^+ e^{-jkz}) = -jkE_x^+(z),$$

Eq. (7-12b) yields

$$\begin{aligned} \mathbf{H} &= \mathbf{a}_y H_y^+(z) = \mathbf{a}_y \frac{k}{\omega\mu} E_x^+(z) \\ &= \mathbf{a}_y \frac{1}{\eta} E_x^+(z). \end{aligned} \tag{7-13}$$

We have introduced a new quantity, η, in Eq. (7-13): $\eta = \omega\mu/k$, or

Intrinsic impedance of medium

$$\boxed{\eta = \sqrt{\frac{\mu}{\epsilon}} \quad (\Omega),} \tag{7-14}$$

which is called the ***intrinsic impedance*** of the medium. In air $\eta_0 = \sqrt{\mu_0/\epsilon_0} = 120\pi = 377\,(\Omega)$. $H_y^+(z)$ is in phase with $E_x^+(z)$, and we can write the instantaneous expression for **H** as

$$\begin{aligned} \mathbf{H}(z, t) &= \mathbf{a}_y H_y^+(z, t) = \mathbf{a}_y \mathscr{R}e[H_y^+(z)e^{j\omega t}] \\ &= \mathbf{a}_y \frac{E_0^+}{\eta} \cos(\omega t - kz). \end{aligned} \tag{7-15}$$

EXERCISE 7.1 Starting from the phasor expression $\mathbf{E} = \mathbf{a}_x E_x^-(z) = \mathbf{a}_x E_0^- e^{jkz}$ for a uniform plane wave traveling in the $-z$ direction, determine the associated $\mathbf{H} = \mathbf{a}_y H_y^-(z)$. Evaluate the ratio $E_x^-(z)/H_y^-(z)$.

ANS. $E_x/H_y^- = -\eta$.

REVIEW QUESTIONS

Q.7-1 Define *uniform plane wave.*

Q.7-2 What is a *wavefront*?

Q.7-3 What is a traveling wave?

Q.7-4 Define *phase velocity.*

Q.7-5 Define *intrinsic impedance* of a medium. What is the value of the intrinsic impedance of free space?

REMARKS

1. The phase velocity of an electromagnetic wave in a lossless medium is independent of its frequency or its direction of travel.
2. The ratio of the magnitudes of **E** and **H** for a uniform plane wave is the intrinsic impedance of the medium.
3. The direction of the **E**-field crosses into the direction of the **H**-field gives the direction of wave propagation; the three directions are mutually perpendicular.

EXAMPLE 7-1

A uniform plane wave with $\mathbf{E} = \mathbf{a}_x E_x$ propagates in a lossless simple medium ($\epsilon_r = 4$, $\mu_r = 1$, $\sigma = 0$) in the $+z$-direction. Assume that E_x is sinusoidal with a frequency 100 (MHz) and has a maximum value of $+10^{-4}$ (V/m) at $t = 0$ and $z = \frac{1}{8}$ (m).

a) Write the instantaneous expression for **E** for any t and z.

b) Write the instantaneous expression for **H**.

c) Determine the locations where E_x is a positive maximum when $t = 10^{-8}$ (s).

SOLUTION

First we find k:

$$k = \omega\sqrt{\mu\epsilon} = \frac{\omega}{c}\sqrt{\mu_r\epsilon_r}$$

$$= \frac{2\pi 10^8}{3 \times 10^8}\sqrt{4} = \frac{4\pi}{3} \qquad \text{(rad/m)}.$$

a) Using $\cos \omega t$ as the reference, we find the instantaneous expression for **E** to be

$$\mathbf{E}(z, t) = \mathbf{a}_x E_x = \mathbf{a}_x 10^{-4} \cos(2\pi 10^8 t - kz + \psi).$$

Since E_x equals $+10^{-4}$ when the argument of the cosine function equals zero—that is, when

$$2\pi 10^8 t - kz + \psi = 0,$$

we have, at $t = 0$ and $z = \frac{1}{8}$,

$$\psi = kz = \left(\frac{4\pi}{3}\right)\left(\frac{1}{8}\right) = \frac{\pi}{6} \qquad \text{(rad)}.$$

Thus,

$$\begin{aligned}\mathbf{E}(z, t) &= \mathbf{a}_x 10^{-4} \cos\left(2\pi 10^8 t - \frac{4\pi}{3} z + \frac{\pi}{6}\right) \\ &= \mathbf{a}_x 10^{-4} \cos\left[2\pi 10^8 t - \frac{4\pi}{3}\left(z - \frac{1}{8}\right)\right] \qquad \text{(V/m)}.\end{aligned}$$

This expression shows a shift of $\frac{1}{8}$(m) in the $+z$-direction and could have been written down directly from the statement of the problem.

b) The instantaneous expression for **H** is

$$\mathbf{H} = \mathbf{a}_y H_y = \mathbf{a}_y \frac{E_x}{\eta},$$

where

$$\eta = \sqrt{\frac{\mu}{\epsilon}} = \frac{\eta_0}{\sqrt{\epsilon_r}} = 60\pi \qquad (\Omega).$$

Hence,

$$\mathbf{H}(z, t) = \mathbf{a}_y \frac{10^{-4}}{60\pi} \cos\left[2\pi 10^8 t - \frac{4\pi}{3}\left(z - \frac{1}{8}\right)\right] \qquad \text{(A/m)}.$$

c) At $t = 10^{-8}$, we equate the argument of the cosine function to $\pm 2n\pi$ in order to make E_x a positive maximum:

$$2\pi 10^8 (10^{-8}) - \frac{4\pi}{3}\left(z_m - \frac{1}{8}\right) = \pm 2n\pi,$$

from which we get

$$z_m = \frac{13}{8} \pm \frac{3}{2} n \quad \text{(m)}, \qquad n = 0, 1, 2, \ldots; \qquad z_m > 0.$$

Examining this result more closely, we note that the wavelength in the given medium is

$$\lambda = \frac{2\pi}{k} = \frac{3}{2} \qquad \text{(m)}.$$

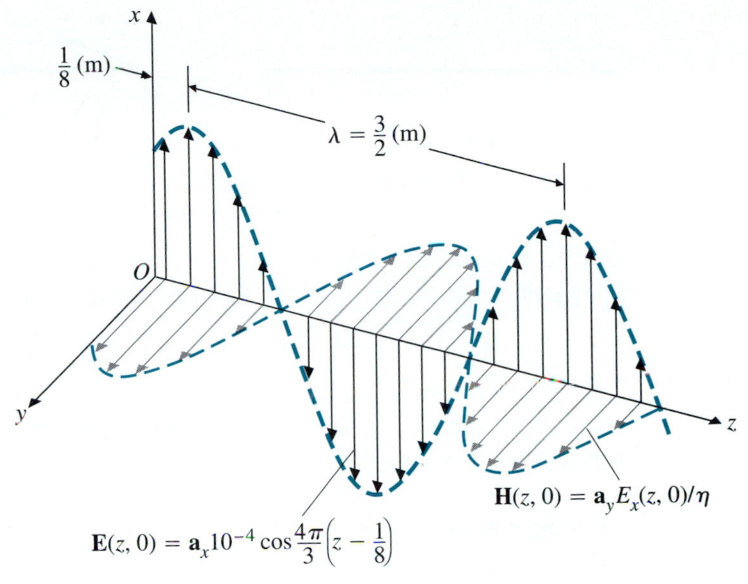

FIGURE 7-2 **E** and **H** fields of a uniform plane wave at $t = 0$ (Example 7-1).

Hence the positive maximum value of E_x occurs at

$$z_m = \frac{13}{8} \pm n\lambda \qquad \text{(m)}.$$

The **E** and **H** fields are shown in Fig. 7-2 as functions of z for the reference time $t = 0$.

7-2.1 DOPPLER EFFECT

Meaning of Doppler effect

When there is relative motion between a time-harmonic source and a receiver, the frequency of the wave detected by the receiver tends to be different from that emitted by the source. This phenomenon is known as ***Doppler effect***.† The Doppler effect manifests itself in acoustics as well as in electromagnetics. Perhaps you have experienced the changes in the pitch of a fast-moving locomotive whistle. In the following we give an explanation of the Doppler effect.

Let us assume that the source (transmitter) T of a time-harmonic wave of a frequency f moves with a velocity **u** at an angle θ relative to the direct

†C. Doppler (1803–1853).

FIGURE 7-3 Illustrating the Doppler effect.

line to a stationary receiver R, as illustrated in Fig. 7-3(a). The electromagnetic wave emitted by T in air at a reference time $t = 0$ will reach R at

$$t_1 = \frac{r_0}{c}. \tag{7-16}$$

At a later time $t = \Delta t$, T has moved to the new position T', and the wave emitted by T' at that time will reach R at

$$\begin{aligned} t_2 &= \Delta t + \frac{r'}{c} \\ &= \Delta t + \frac{1}{c}[r_0^2 - 2r_0(u\,\Delta t)\cos\theta + (u\,\Delta t)^2]^{1/2}. \end{aligned} \tag{7-17}$$

If $(u\,\Delta t)^2 \ll r_0^2$, Eq. (7-17) becomes

$$t_2 \cong \Delta t + \frac{r_0}{c}\left(1 - \frac{u\,\Delta t}{r_0}\cos\theta\right).$$

Hence the time elapsed at R, $\Delta t'$, corresponding to Δt at T is

$$\begin{aligned} \Delta t' &= t_2 - t_1 \\ &= \Delta t\left(1 - \frac{u}{c}\cos\theta\right), \end{aligned} \tag{7-18}$$

which is not equal to Δt.

If Δt represents a period of the time-harmonic source—that is, if $\Delta t = 1/f$—then the frequency of the received wave at R is

$$\begin{aligned} f' &= \frac{1}{\Delta t'} = \frac{f}{\left(1 - \frac{u}{c}\cos\theta\right)} \\ &\cong f\left(1 + \frac{u}{c}\cos\theta\right) \end{aligned} \tag{7-19}$$

for the usual case of $(u/c)^2 \ll 1$. Equation (7-19) clearly indicates that the frequency perceived at R is higher than the transmitted frequency when T moves toward R. Conversely, the perceived frequency is lower than the transmitted frequency when T moves away from R. Similar results are obtained if R moves and T is stationary.

Principle of operation of radar speed monitor

The Doppler effect is the basis of operation of the (Doppler) radar used by police to check the speed of a moving vehicle. The frequency shift of the received wave reflected by a moving vehicle is proportional to the speed of the vehicle and can be detected and displayed on a hand-held unit. The Doppler effect is also the cause of the so-called ***red shift*** of the light spectrum emitted by a receding distant star in astronomy. As the star *moves away* at a high speed from an observer on earth, the received frequency shifts toward the *lower frequency* (*red*) end of the spectrum.

Red shift phenomenon in astronomy

■ **EXERCISE 7.2**

A train traveling at 130 (km/hr) toward an observer at an angle 20° from the line of sight. It gives out a 800 (Hz) whistle. What is the frequency perceived by the observer? (The velocity of sound is approximately 340 m/s.)

ANS. 880 (Hz).

REVIEW QUESTION

Q.7-6 What is Doppler effect?

REMARKS

The frequency of a wave detected by a receiver is higher (lower) than that emitted by a transmitter if the transmitter moves toward (away from) the receiver.

7-2.2 TRANSVERSE ELECTROMAGNETIC WAVES

We have seen that a uniform plane wave characterized by $\mathbf{E} = \mathbf{a}_x E_x$ propagating in the $+z$-direction has associated with it a magnetic field $\mathbf{H} = \mathbf{a}_y H_y$. Thus $\mathbf{E}$ and $\mathbf{H}$ are perpendicular to each other, and both are *transverse* to the direction of propagation. It is a particular case of a ***transverse electromagnetic*** **(TEM)** ***wave***. We now examine the propagation of a uniform plane wave along an arbitrary direction that does not necessarily coincide with a coordinate axis.

TEM wave

Instead of the $\mathbf{E}(z)$ in Eq. (7-8), let us consider

$$\mathbf{E}(x, z) = \mathbf{a}_y E_0 e^{-jk_x x - jk_z z}, \tag{7-20}$$

which represents the y-directed electric intensity of a uniform plane wave propagating in both $+x$ and $+z$ directions. If we define a ***wavenumber vector***, $\mathbf{k}$, as

Expression of wavenumber vector in Cartesian coordinates

$$\mathbf{k} = \mathbf{a}_x k_x + \mathbf{a}_z k_z = \mathbf{a}_k k, \tag{7-21}$$

and a radius vector $\mathbf{R}$ from the origin to an arbitrary point

$$\mathbf{R} = \mathbf{a}_x x + \mathbf{a}_y y + \mathbf{a}_z z, \tag{7-22}$$

Eq. (7-20) can be written succinctly as

$$\mathbf{E} = \mathbf{a}_y E_0 e^{-j\mathbf{k}\cdot\mathbf{R}} = \mathbf{a}_y E_0 e^{-jk\mathbf{a}_k\cdot\mathbf{R}}. \tag{7-23}$$

The situation is illustrated in Fig. 7-4. The relation

$$\mathbf{a}_k \cdot \mathbf{R} = \text{Length } \overline{OP} \qquad \text{(a constant)}$$

is the equation of a plane (locus of the endpoints of radius vector **R**) normal to $\mathbf{a}_k$, the direction of propagation, and it is a plane of constant phase and uniform amplitude.

The magnetic field **H** associated with the electric field in Eq. (7-23) is, from Eq. (6-80a),

$$\mathbf{H} = -\frac{1}{j\omega\mu}\nabla \times \mathbf{E} = \frac{E_0}{\omega\mu}(-\mathbf{a}_x k_z + \mathbf{a}_z k_x)e^{-jk_x x - jk_z z}. \tag{7-24}$$

Equation (7-24) can be put in a more general form:

Finding H from E of a uniform plane wave

$$\boxed{\mathbf{H} = \frac{k}{\omega\mu}\mathbf{a}_k \times \mathbf{E} = \frac{1}{\eta}\mathbf{a}_k \times \mathbf{E}.} \tag{7-25}$$

Thus, if **E** of a uniform plane wave propagating in a given direction is known, the associated **H** can be easily found from Eq. (7-25).

EXERCISE 7.3

a) Write the phasor expression of the z-directed electric field of a uniform plane wave in air having an amplitude E_0, frequency f, and traveling in the $-y$ direction.

b) Write the expression of the associated magnetic field.

ANS. (a) $\mathbf{E} = \mathbf{a}_z E_0 e^{j2\pi f y/c}$.

FIGURE 7-4 Radius vector and wave normal to a phase front of a uniform plane wave.

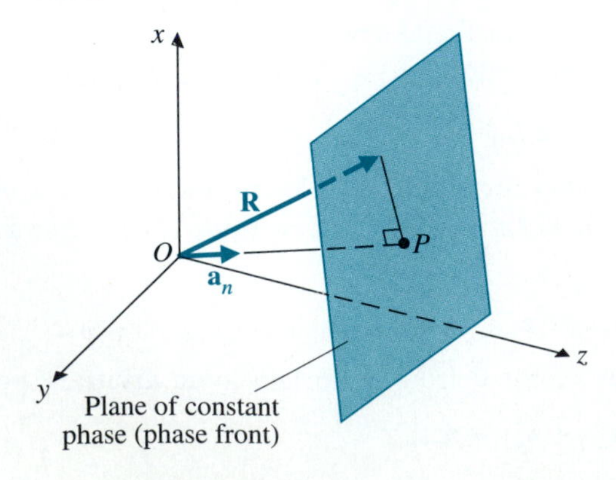

7-2.3 POLARIZATION OF PLANE WAVES

Meaning of polarization of a uniform wave

The ***polarization*** of a uniform plane wave describes the time-varying behavior of the electric field intensity vector at a given point in space. For example, if the **E** vector of a plane wave is fixed in the x direction ($\mathbf{E} = \mathbf{a}_x E_x$, where E_x may be positive or negative), the wave is said to be ***linearly polarized*** in the x-direction. A separate description of magnetic-field behavior is not necessary, inasmuch as the direction of **H** is definitely related to that of **E**.

In some cases the direction of **E** of a plane wave at a given point may change with time. Consider the superposition of two linearly polarized waves: one polarized in the x-direction, and the other polarized in the y-direction and lagging 90° (or $\pi/2$ rad) in time phase. In phasor notation we have

$$\begin{aligned}\mathbf{E}(z) &= \mathbf{a}_x E_1(z) + \mathbf{a}_y E_2(z)\\ &= \mathbf{a}_x E_{10} e^{-jkz} - \mathbf{a}_y j E_{20} e^{-jkz},\end{aligned} \tag{7-26}$$

where E_{10} and E_{20} are real numbers denoting the amplitudes of the two linearly polarized waves. The instantaneous expression for **E** is

$$\begin{aligned}\mathbf{E}(z, t) &= \mathcal{R}e\{[\mathbf{a}_x E_1(z) + \mathbf{a}_y E_2(z)]e^{j\omega t}\}\\ &= \mathbf{a}_x E_{10} \cos(\omega t - kz) + \mathbf{a}_y E_{20} \cos\left(\omega t - kz - \frac{\pi}{2}\right).\end{aligned}$$

In examining the direction change of **E** at a given point as t changes, it is convenient to set $z = 0$. We have

$$\begin{aligned}\mathbf{E}(0, t) &= \mathbf{a}_x E_1(0, t) + \mathbf{a}_y E_2(0, t)\\ &= \mathbf{a}_x E_{10} \cos \omega t + \mathbf{a}_y E_{20} \sin \omega t.\end{aligned} \tag{7-27}$$

As ωt increases from 0 through $\pi/2$, π, and $3\pi/2$—completing the cycle at 2π—the tip of the vector $\mathbf{E}(0, t)$ will traverse an elliptical locus in the counterclockwise direction. Analytically, we have

$$\cos \omega t = \frac{E_1(0, t)}{E_{10}}$$

and

$$\begin{aligned}\sin \omega t &= \frac{E_2(0, t)}{E_{20}}\\ &= \sqrt{1 - \cos^2 \omega t} = \sqrt{1 - \left[\frac{E_1(0, t)}{E_{10}}\right]^2},\end{aligned}$$

which leads to the following equation for an ellipse:

$$\left[\frac{E_2(0, t)}{E_{20}}\right]^2 + \left[\frac{E_1(0, t)}{E_{10}}\right]^2 = 1. \tag{7-28}$$

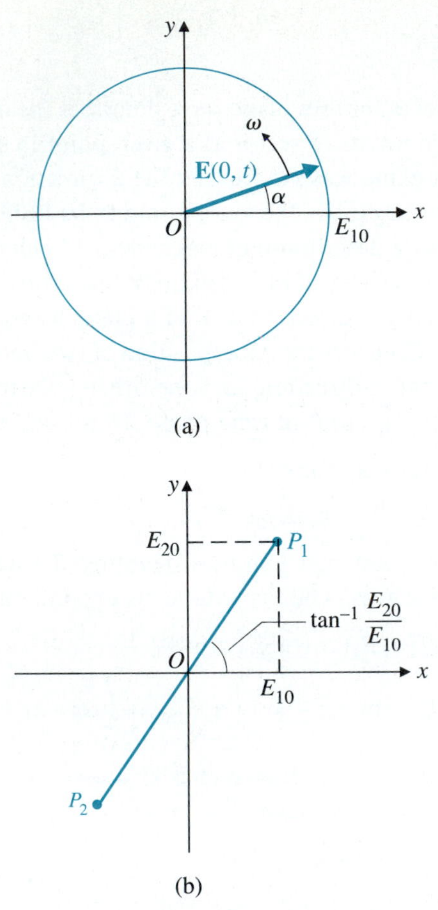

FIGURE 7-5 Polarization diagrams for sum of two linearly polarized waves in space quadrature at $z=0$: (a) circular polarization, $E(0,t)=E_{10}(\mathbf{a}_x \cos \omega t + \mathbf{a}_y \sin \omega t)$; (b) linear polarization, $\mathbf{E}(0,t)=(\mathbf{a}_x E_{10}+\mathbf{a}_y E_{20}) \cos \omega t$.

Meaning of elliptically and circularly polarized waves

Hence **E**, which is the sum of two linearly polarized waves in both space and time quadrature, is ***elliptically polarized*** if $E_{20} \neq E_{10}$, and is ***circularly polarized*** if $E_{20}=E_{10}$. A typical polarization circle is shown in Fig. 7-5(a).

When $E_{20}=E_{10}$, the instantaneous angle α that **E** makes with the x-axis at $z=0$ is

$$\alpha = \tan^{-1}\frac{E_2(0,t)}{E_1(0,t)} = \omega t, \tag{7-29}$$

which indicates that **E** rotates at a uniform rate with an angular velocity ω in a *counterclockwise* direction. When the fingers of the right hand follow the direction of the rotation of **E**, the thumb points to the direction of

Right-hand (or positive) circularly polarized wave

propagation of the wave. This is a ***right-hand*** or ***positive circularly polarized wave***.

If we start with an $E_2(z)$, which *leads* $E_1(z)$ by 90° ($\pi/2$ rad) in time phase, Eqs. (7-26) and (7-27) will be, respectively,

$$\mathbf{E}(z) = \mathbf{a}_x E_{10} e^{-jkz} + \mathbf{a}_y j E_{20} e^{-jkz} \tag{7-30}$$

and

$$\mathbf{E}(0, t) = \mathbf{a}_x E_{10} \cos \omega t - \mathbf{a}_y E_{20} \sin \omega t. \tag{7-31}$$

Left-hand (or negative) circularly polarized wave

Comparing Eq. (7-31) with Eq. (7-27), we see that **E** will still be elliptically polarized. If $E_{20} = E_{10}$, **E** will be circularly polarized, and its angle measured from the x-axis at $z = 0$ will now be $-\omega t$, indicating that **E** will rotate with an angular velocity ω in a *clockwise* direction; this is a ***left-hand*** or ***negative circularly polarized wave***.

Linearly polarized wave

If $E_2(z)$ and $E_1(z)$ are in space quadrature but in time phase, the instantaneous expression for **E** at $z = 0$ is

$$\mathbf{E}(0, t) = (\mathbf{a}_x E_{10} + \mathbf{a}_y E_{20}) \cos \omega t. \tag{7-32}$$

The tip of the $\mathbf{E}(0, t)$ will be at the point P_1 in Fig. 7-5(b) when $\omega t = 0$. Its magnitude will decrease toward zero as ωt increases toward $\pi/2$. After that, $\mathbf{E}(0, t)$ starts to increase again, in the opposite direction, toward the point P_2 where $\omega t = \pi$. We say that the sum **E** is linearly polarized along a line that makes an angle $\tan^{-1}(E_{20}/E_{10})$ with the x-axis.

In the general case, $E_2(z)$ and $E_1(z)$, which are in space quadrature, can have unequal amplitudes ($E_{20} \neq E_{10}$) and can differ in phase by an arbitrary amount (not zero or an integral multiple of $\pi/2$). Their sum **E** will be elliptically polarized.

EXAMPLE 7-2

Prove that a linearly polarized plane wave can be resolved into a right-hand circularly polarized wave and a left-hand circularly polarized wave of equal amplitude.

SOLUTION

Consider a linearly polarized plane wave propagating in the $+z$-direction. We can assume, with no loss of generality, that **E** is polarized in the x-direction. In phasor notation we have

$$\mathbf{E}(z) = \mathbf{a}_x E_0 e^{-jkz}. \tag{7-33}$$

But this can be written as

$$\mathbf{E}(z) = \mathbf{E}_{rc}(z) + \mathbf{E}_{lc}(z), \tag{7-34}$$

where

$$\mathbf{E}_{rc}(z) = \frac{E_0}{2}(\mathbf{a}_x - j\mathbf{a}_y)e^{-jkz} \tag{7-34a}$$

and

$$\mathbf{E}_{lc}(z) = \frac{E_0}{2}(\mathbf{a}_x + j\mathbf{a}_y)e^{-jkz}. \tag{7-34b}$$

A linearly polarized plane wave can be resolved into two circularly polarized waves of equal magnitude.

From previous discussions we recognize that $\mathbf{E}_{rc}(z)$ in Eq. (7-34a) and $\mathbf{E}_{lc}(z)$ in Eq. (7-34b) represent right-hand and left-hand circularly polarized waves, respectively, each having an amplitude $E_0/2$. The statement of this problem is therefore proved. The converse statement that the sum of two oppositely rotating circularly polarized waves of equal amplitude is a linearly polarized wave is, of course, also true.

EXERCISE 7.4 Describe the polarization of a wave whose electric intensity is represented by $\mathbf{E}(x, t) = (\mathbf{a}_y E_{10} - \mathbf{a}_z E_{20}) \sin(\omega t - kx)$.

REVIEW QUESTIONS

Q.7-7 What is a TEM wave?

Q.7-8 Is the intrinsic impedance of a lossless medium a function of frequency? Explain.

Q.7-9 What is meant by the *polarization* of a wave? When is a wave linearly polarized? Circularly polarized?

Q.7-10 Two orthogonal linearly polarized waves of the same frequency are combined. State the conditions under which the resultant will be (a) another linearly polarized wave, (b) a circularly polarized wave, and (c) an elliptically polarized wave.

REMARKS

1. For a TEM wave in an unbounded lossless medium: (a) **E** and **H** are in phase, and (b) $|\mathbf{E}| = \eta|\mathbf{H}|$.
2. The **E**-field from AM broadcast stations is linearly polarized with the **E**-field perpendicular to the ground. For maximum reception the receiving antenna should be in a vertical position.
3. The **E**-field of television signals is linearly polarized in the horizontal direction. Note the horizontal position of TV receiving antennas on roof tops.
4. The wave radiated by FM broadcast stations are generally circularly polarized.

7-3 PLANE WAVES IN LOSSY MEDIA

So far we have considered wave propagation in source-free ($\rho_v = 0, \mathbf{J} = 0$) lossless simple media. If a medium is conducting ($\sigma \neq 0$), a current $\mathbf{J} = \sigma\mathbf{E}$ will flow because of the existence of $\mathbf{E}$. The time-harmonic $\nabla \times \mathbf{H}$ equation (6-80b) should then be changed to

$$\nabla \times \mathbf{H} = (\sigma + j\omega\epsilon)\mathbf{E} = j\omega\left(\epsilon + \frac{\sigma}{j\omega}\right)\mathbf{E}$$

$$= j\omega\epsilon_c\mathbf{E} \tag{7-35}$$

with

Complex permittivity of lossy media

$$\epsilon_c = \epsilon - j\frac{\sigma}{\omega} \qquad \text{(F/m)}. \tag{7-36}$$

The other three equations, Eqs. (6-80a, c, and d), are unchanged. Hence, all the previous equations for nonconducting media will apply to conducting media if ϵ is replaced by the ***complex permittivity*** ϵ_c in Eq. (7-36).

As we discussed in Subsection 3-6.2, when an external time-varying electric field is applied to material bodies, small displacements of bound charges result, giving rise to a volume density of polarization. This polarization vector will vary with the same frequency as that of the applied field. As the frequency increases, the inertia of the charged particles tends to prevent the particle displacements from keeping in phase with the field changes, leading to a frictional damping mechanism that causes power loss because work must be done to overcome the damping forces. This phenomenon of out-of-phase polarization can be characterized by a complex electric susceptibility and hence a complex permittivity. If, in addition, the material body or medium has an appreciable amount of free charge carriers such as the electrons in a conductor, the electrons and holes in a semiconductor, or the ions in an electrolyte, there will also be ohmic losses. In treating such media it is customary to include the effects of both the damping and the ohmic losses in the imaginary part of a complex permittivity ϵ_c:

$$\epsilon_c = \epsilon' - j\epsilon'' \qquad \text{(F/m)}, \tag{7-37}$$

where both ϵ' and ϵ'' may be functions of frequency. Alternatively, we may define an equivalent conductivity representing all losses and write

$$\sigma = \omega\epsilon'' \qquad \text{(S/m)}. \tag{7-38}$$

Combination of Eqs. (7-37) and (7-38) gives Eq. (7-36).

Loss tangent of a lossy medium

The ratio ϵ''/ϵ' is called a ***loss tangent*** because it is a measure of the power loss in the medium:

$$\tan\delta_c = \frac{\epsilon''}{\epsilon'} \cong \frac{\sigma}{\omega\epsilon}. \tag{7-39}$$

The quantity δ_c in Eq. (7-39) may be called the ***loss angle***.

Difference between a good conductor and a good insulator

A medium is said to be a ***good conductor*** if $\sigma \gg \omega\epsilon$, and a ***good insulator*** if $\omega\epsilon \gg \sigma$. Thus, a material may be a good conductor at low frequencies but may have the properties of a lossy dielectric at very high frequencies. For example, a moist ground has a dielectric constant ϵ_r and a conductivity σ that are in the neighborhood of 10 and 10^{-2} (S/m), respectively. The loss tangent $\sigma/\omega\epsilon$ of the moist ground then equals 1.8×10^4 at 1 (kHz), making it a relatively good conductor. At 10 (GHz), $\sigma/\omega\epsilon$ becomes 1.8×10^{-3}, and the moist ground behaves more like an insulator.

EXAMPLE 7-3

A sinusoidal electric intensity of amplitude 250 (V/m) and frequency 1 (GHz) exists in a lossy dielectric medium that has a relative permittivity of 2.5 and a loss tangent of 0.001. Find the average power dissipated in the medium per cubic meter.

SOLUTION

First we must find the effective conductivity of the lossy medium:

$$\tan \delta_c = 0.001 = \frac{\sigma}{\omega\epsilon_0\epsilon_r},$$

$$\sigma = 0.001(2\pi 10^9)\left(\frac{10^{-9}}{36\pi}\right)(2.5)$$

$$= 1.39 \times 10^{-4} \text{ (S/m)}.$$

The average power dissipated per unit volume is

$$p = \tfrac{1}{2}JE = \tfrac{1}{2}\sigma E^2$$

$$= \tfrac{1}{2} \times (1.39 \times 10^{-4}) \times 250^2 = 4.34 \qquad (\text{W/m}^3).$$

EXERCISE 7.5 A microwave oven cooks food by irradiating the food with microwave power generated by a magnetron. Assuming the dielectric constant of a beef steak to be 40 with a loss tangent of 0.35 at an operating frequency 2.45 (GHz), calculate the average power per cubic meter. (Neglect ***skin effect***, which will be discussed in Subsection 7-3.2.)

ANS. 59.6 (kW/m^3).

In view of the above discussion, the study of the time-harmonic behavior in lossy media can proceed from Eq. (7-3) by simply replacing the real k by a complex wavenumber k_c:

$$k_c = \omega\sqrt{\mu\epsilon_c}. \tag{7-40}$$

We need to examine the solution of the following homogeneous Helmholtz's equation:

$$\nabla^2\mathbf{E} + k_c^2\mathbf{E} = 0. \tag{7-41}$$

In an effort to conform with the conventional notation used in transmission-line theory, it is customary to define a ***propagation constant***, γ, such that

Relation between propagation constant and wavenumber

$$\boxed{\gamma = jk_c = j\omega\sqrt{\mu\epsilon_c} \qquad (\text{m}^{-1}).} \tag{7-42}$$

Since γ is complex, we write, with the help of Eq. (7-36),

$$\gamma = \alpha + j\beta = j\omega\sqrt{\mu\epsilon}\left(1 + \frac{\sigma}{j\omega\epsilon}\right)^{1/2}, \tag{7-43}$$

or, from Eq. (7-37),

$$\gamma = \alpha + j\beta = j\omega\sqrt{\mu\epsilon'}\left(1 - j\frac{\epsilon''}{\epsilon'}\right)^{1/2}, \tag{7-44}$$

where α and β are the real and imaginary parts of γ, respectively. Their physical significance will be explained presently. For a lossless medium, $\sigma = 0\,(\epsilon'' = 0,\ \epsilon = \epsilon')$, $\alpha = 0$, and $\beta = k = \omega\sqrt{\mu\epsilon}$.

With Eq. (7-42), Eq. (7-41) becomes

$$\nabla^2\mathbf{E} - \gamma^2\mathbf{E} = 0. \tag{7-45a}$$

For a uniform plane wave propagating in the $+z$ direction and characterized by $\mathbf{E} = \mathbf{a}_x E_x$ and $\mathbf{H} = \mathbf{a}_y H_y$, Eq. (7-45a) reduces to

$$\frac{d^2E_x}{dz^2} = \gamma^2 E_x. \tag{7-45b}$$

The solution of Eq. (7-45b) is

$$E_x = E_0 e^{-\gamma z} = E_0 e^{-\alpha z} e^{-j\beta z}, \tag{7-46}$$

Attenuation constant and its SI unit

Phase constant and its SI unit

where both α and β are positive quantities. The first factor, $e^{-\alpha z}$, decreases as z increases and thus is an attenuation factor, and α is called an ***attenuation constant***. The SI unit of the attenuation constant is neper per meter (Np/m).† The second factor, $e^{-j\beta z}$, is a phase factor; β is called a ***phase constant*** and is expressed in radians per meter (rad/m). The phase constant expresses the amount of phase shift that occurs as the wave travels one meter.

■ **EXERCISE 7.6** Assuming that the amplitude of the electric intensity of a plane wave propagating in a certain lossy medium is 1 (mV/m) at P_1 and 0.8 (mV/m) at P_2 50 (m) away, find

†Neper is a dimensionless quantity. If $\alpha = 1$ (Np/m), then a unit wave amplitude decreases to a magnitude e^{-1} (=0.368) as it travels a distance of 1 (m). An attenuation of 1 (Np/m) equals $20\log_{10}e = 8.69$ (dB/m).

a) the total attenuation between points P_1 and P_2 both in nepers and in decibels,
b) α in (Np/m) and in (dB/m).

ANS. (a) 0.223 (Np), 1.94 (dB), (b) 0.00446 (Np/m), 0.0388 (dB/m).

7-3.1 LOW-LOSS DIELECTRICS

A low-loss dielectric is a good but imperfect insulator with a nonzero equivalent conductivity, such that $\epsilon'' \ll \epsilon'$ or $\sigma/\omega\epsilon \ll 1$. Under this condition, γ in Eq. (7-44) can be approximated by using the binomial expansion:

Propagation constant of a lossy dielectric

$$\gamma = \alpha + j\beta \cong j\omega\sqrt{\mu\epsilon'}\left[1 - j\frac{\epsilon''}{2\epsilon'} + \frac{1}{8}\left(\frac{\epsilon''}{\epsilon'}\right)^2\right],$$

from which we obtain the attenuation constant

$$\alpha = \mathscr{R}e(\gamma) \cong \frac{\omega\epsilon''}{2}\sqrt{\frac{\mu}{\epsilon'}} \quad \text{(Np/m)} \tag{7-47}$$

and the phase constant

$$\beta = \mathscr{I}m(\gamma) \cong \omega\sqrt{\mu\epsilon'}\left[1 + \frac{1}{8}\left(\frac{\epsilon''}{\epsilon'}\right)^2\right] \quad \text{(rad/m)}. \tag{7-48}$$

It is seen from Eq. (7-47) that the attenuation constant of a low-loss dielectric is a positive quantity and is approximately directly proportional to the frequency. The phase constant in Eq. (7-48) deviates only very slightly from the value $\omega\sqrt{\mu\epsilon}$ for a perfect (lossless) dielectric.

The intrinsic impedance of a low-loss dielectric is a complex quantity.

Intrinsic impedance of a lossy dielectric

$$\begin{aligned}\eta_c &= \sqrt{\frac{\mu}{\epsilon'}}\left(1 - j\frac{\epsilon''}{\epsilon'}\right)^{-1/2} \\ &\cong \sqrt{\frac{\mu}{\epsilon'}}\left(1 + j\frac{\epsilon''}{2\epsilon'}\right) \quad (\Omega).\end{aligned} \tag{7-49}$$

Since the intrinsic impedance is the ratio of E_x and H_y for a uniform plane wave, the electric and magnetic field intensities in a lossy dielectric are thus not in time phase, as they are in a lossless medium.

The phase velocity u_p is obtained from the ratio ω/β. Using Eq. (7-48), we have

$$u_p = \frac{\omega}{\beta} \cong \frac{1}{\sqrt{\mu\epsilon'}}\left[1 - \frac{1}{8}\left(\frac{\epsilon''}{\epsilon'}\right)^2\right] \quad \text{(m/s)}, \tag{7-50}$$

which is slightly lower than its value when the medium is lossless.

REVIEW QUESTIONS

Q.7-11 What makes the permittivity of a dielectric medium a complex quantity?

Q.7-12 Define *loss tangent* of a medium.

Q.7-13 What is the relation between *propagation constant* and wavenumber?

Q.7-14 Define *attenuation constant* and *phase constant* of a wave propagating in a medium. What are their SI units?

REMARKS

1. The electric and magnetic fields for uniform plane waves in a *lossy* medium are in space quadrature and have different time phases.
2. Both α and β are real quantities and both are, in general, functions of frequency.
3. Attenuation of wave amplitude in nepers is the natural logarithm of the ratio of the amplitude at the beginning point to that at the end point.
4. 1 (Np) = 8.69 (dB).

7-3.2 GOOD CONDUCTORS

A good conductor is a medium for which $\sigma/\omega\epsilon \gg 1$. Under this condition it is convenient to use Eq. (7-43) and neglect 1 in comparison with $\sigma/\omega\epsilon$. We write

$$\gamma \cong j\omega\sqrt{\mu\epsilon}\sqrt{\frac{\sigma}{j\omega\epsilon}} = \sqrt{j}\sqrt{\omega\mu\sigma} = \frac{1+j}{\sqrt{2}}\sqrt{\omega\mu\sigma},$$

or

Propagation constant of a good conductor

$$\gamma = \alpha + j\beta \cong (1+j)\sqrt{\pi f\mu\sigma}, \tag{7-51}$$

where we have used the relations

$$\sqrt{j} = (e^{j\pi/2})^{1/2} = e^{j\pi/4} = (1+j)/\sqrt{2}$$

and $\omega = 2\pi f$. Equation (7-51) indicates that α and β for a good conductor are approximately equal and both increase as $\sqrt{f}$ and $\sqrt{\sigma}$. For a good conductor,

Attenuation constant and phase constant of a good conductor are equal.

$$\boxed{\alpha = \beta = \sqrt{\pi f\mu\sigma}.} \tag{7-52}$$

Relation between intrinsic impedance and attenuation constant of a good conductor

The intrinsic impedance of a good conductor is

$$\eta_c = \sqrt{\frac{\mu}{\epsilon_c}} \cong \sqrt{\frac{j\omega\mu}{\sigma}} = (1+j)\sqrt{\frac{\pi f \mu}{\sigma}} = (1+j)\frac{\alpha}{\sigma} \qquad (\Omega), \tag{7-53}$$

which has a phase angle of 45°. Hence the magnetic field intensity lags behind the electric field intensity by 45°.

The phase velocity in a good conductor is

Phase velocity in a good conductor

$$u_p = \frac{\omega}{\beta} \cong \sqrt{\frac{2\omega}{\mu\sigma}} \qquad \text{(m/s)}, \tag{7-54}$$

which is proportional to $\sqrt{f}$ and $1/\sqrt{\sigma}$. Consider copper as an example:

$$\sigma = 5.80 \times 10^7 \qquad \text{(S/m)},$$

$$\mu = 4\pi \times 10^{-7} \qquad \text{(H/m)},$$

$$u_p = 720\,\text{(m/s)} \qquad \text{at} \qquad 3\,\text{(MHz)},$$

which is many orders of magnitude slower than the velocity of light in air. The wavelength of a plane wave in a good conductor is

Wavelength in a good conductor

$$\lambda = \frac{2\pi}{\beta} = \frac{u_p}{f} = 2\sqrt{\frac{\pi}{f\mu\sigma}} \qquad \text{(m)}. \tag{7-55}$$

For copper at 3 (MHz), $\lambda = 0.24$ (mm). As a comparison, a 3-(MHz) electromagnetic wave in air has a wavelength of 100 (m).

At very high frequencies the attenuation constant α for a good conductor, as given by Eq. (7-52), tends to be very large. For copper at 3 (MHz),

$$\alpha = \sqrt{\pi(3 \times 10^6)(4\pi \times 10^{-7})(5.80 \times 10^7)} = 2.62 \times 10^4 \qquad \text{(Np/m)}.$$

Since the attenuation factor is $e^{-\alpha z}$, the amplitude of a wave will be attenuated by a factor of $e^{-1} = 0.368$ when it travels a distance $\delta = 1/\alpha$. For copper at 3 (MHz) this distance is $(1/2.62) \times 10^{-4}$ (m), or 0.038 (mm). At 10 (GHz) it is only 0.66 (μm)—a very small distance indeed. Thus a high-frequency electromagnetic wave is attenuated very rapidly as it propagates in a good conductor. The distance δ through which the amplitude of a traveling plane wave decreases by a factor of e^{-1} or 0.368 is called the ***skin depth*** or the ***depth of penetration*** of a conductor:

Skin depth

Finding skin depth from conductivity and permeability of conductor and frequency

$$\boxed{\delta = \frac{1}{\alpha} = \frac{1}{\sqrt{\pi f \mu \sigma}} \qquad \text{(m)}.} \tag{7-56}$$

Since $\alpha = \beta$ for a good conductor, δ can also be written as

$$\boxed{\delta = \frac{1}{\beta} = \frac{\lambda}{2\pi} \qquad \text{(m)}.} \tag{7-57}$$

At microwave frequencies the skin depth or depth of penetration of a good conductor is so small that fields and currents can be considered as, for all practical purposes, confined in a very thin layer (that is, in the *skin*) of the conductor surface.

EXAMPLE 7-4

The electric field intensity of a linearly polarized uniform plane wave propagating in the $+z$-direction in seawater is $\mathbf{E} = \mathbf{a}_x 100 \cos(10^7 \pi t)$ (V/m) at $z = 0$. The constitutive parameters of seawater are $\epsilon_r = 72$, $\mu_r = 1$, and $\sigma = 4$ (S/m).

a) Determine the attenuation constant, phase constant, intrinsic impedance, phase velocity, wavelength, and skin depth.

b) Find the distance at which the amplitude of $\mathbf{E}$ is 1% of its value at $z = 0$.

c) Write the expressions for $\mathbf{E}(z, t)$ and $\mathbf{H}(z, t)$ at $z = 0.8$ (m) as functions of t.

SOLUTION

$$\omega = 10^7 \pi \qquad \text{(rad/s)},$$

$$f = \frac{\omega}{2\pi} = 5 \times 10^6 \text{(Hz)} = 5 \text{(MHz)},$$

$$\frac{\sigma}{\omega\epsilon} = \frac{\sigma}{\omega\epsilon_0\epsilon_r} = \frac{4}{10^7\pi\left(\dfrac{1}{36\pi} \times 10^{-9}\right)72} = 200 \gg 1.$$

Hence we can use the formulas for good conductors.

a) Attenuation constant:

$$\alpha = \sqrt{\pi f \mu \sigma} = \sqrt{5\pi 10^6 (4\pi 10^{-7})4} = 8.89 \qquad \text{(Np/m)}.$$

Phase constant:

$$\beta = \sqrt{\pi f \mu \sigma} = 8.89 \qquad \text{(rad/m)}.$$

Intrinsic impedance:

$$\eta_c = (1 + j)\sqrt{\frac{\pi f \mu}{\sigma}}$$

$$= (1 + j)\sqrt{\frac{\pi(5 \times 10^6)(4\pi \times 10^{-7})}{4}} = \pi e^{j\pi/4} \qquad (\Omega).$$

Phase velocity:

$$u_p = \frac{\omega}{\beta} = \frac{10^7\pi}{8.89} = 3.53 \times 10^6 \qquad \text{(m/s)}.$$

Wavelength:

$$\lambda = \frac{2\pi}{\beta} = \frac{2\pi}{8.89} = 0.707 \qquad \text{(m)}.$$

Skin depth:

$$\delta = \frac{1}{\alpha} = \frac{1}{8.89} = 0.112 \qquad \text{(m)}.$$

b) Distance z_1 at which the amplitude of wave decreases to 1% of its value at $z = 0$:

$$e^{-\alpha z_1} = 0.01 \qquad \text{or} \qquad e^{\alpha z_1} = \frac{1}{0.01} = 100,$$

$$z_1 = \frac{1}{\alpha} \ln 100 = \frac{4.605}{8.89} = 0.518 \qquad \text{(m)}.$$

c) In phasor notation,

$$\mathbf{E}(z) = \mathbf{a}_x 100 e^{-\alpha z} e^{-j\beta z}.$$

The instantaneous expression for **E** is

$$\begin{aligned}\mathbf{E}(z, t) &= \mathcal{R}e[\mathbf{E}(z)e^{j\omega t}] \\ &= \mathcal{R}e[\mathbf{a}_x 100 e^{-\alpha z} e^{j(\omega t - \beta z)}] = \mathbf{a}_x 100 e^{-\alpha z} \cos(\omega t - \beta z).\end{aligned}$$

At $z = 0.8$ (m) we have

$$\begin{aligned}\mathbf{E}(0.8, t) &= \mathbf{a}_x 100 e^{-0.8\alpha} \cos(10^7\pi t - 0.8\beta) \\ &= \mathbf{a}_x 0.082 \cos(10^7\pi t - 7.11) \\ &= \mathbf{a}_x 0.082 \cos(10^7\pi t - 47.5^\circ)^\dagger \qquad \text{(V/m)}.\end{aligned}$$

We know that a uniform plane wave is a TEM wave with $\mathbf{E} \perp \mathbf{H}$ and that both are normal to the direction of wave propagation $\mathbf{a}_z$. Thus $\mathbf{H} = \mathbf{a}_y H_y$. To find $\mathbf{H}(z, t)$, the instantaneous expression of **H** as a function of t, *we must not make the mistake of writing* $H_y(z, t) = E_x(z, t)/\eta_c$ because this would be mixing real time functions $E_x(z, t)$ and $H_z(z, t)$ with a complex quantity η_c. Phasor quantities $E_x(z)$ and $H_y(z)$ must be used. That is,

$$H_y(z) = \frac{E_x(z)}{\eta_c},$$

† 7.11 (rad) $= 7.11 \times (180/\pi) = 407.4^\circ$, which is equivalent to $407.4^\circ - 360^\circ = 47.5^\circ$ in phase relations.

from which we obtain the relation between instantaneous quantities

$$H_y(z,t) = \mathcal{R}e\left[\frac{E_x(z)}{\eta_c}e^{j\omega t}\right].$$

For the present problem we have, in phasors,

$$H_y(0.8) = \frac{100e^{-0.8\alpha}e^{-j0.8\beta}}{\pi e^{j\pi/4}} = \frac{0.082e^{-j7.11}}{\pi e^{j\pi/4}} = 0.026e^{-j1.61}.$$

Note that *both* angles must be in radians before combining. The instantaneous expression for **H** at $z = 0.8$ (m) is then

$$\begin{aligned}\mathbf{H}(0.8,t) &= \mathbf{a}_y 0.026 \cos(10^7\pi t - 1.61)\\ &= \mathbf{a}_y 0.026 \cos(10^7\pi t - 92.3^\circ) \qquad \text{(A/m)}.\end{aligned}$$

We can see that a 5-(MHz) plane wave attenuates very rapidly in seawater and becomes negligibly weak a very short distance from the source. (Field amplitudes at a depth of 0.8 (m) are reduced to $0.082/100 = 0.00082$ times their value at the surface.) Even at very low frequencies, long-distance communication with a submerged submarine is very difficult.

■ **EXERCISE 7.7** Determine the frequency at which the skin depth in seawater is ten meters. Calculate the corresponding wavelength in seawater and compare it to that in air.

ANS. 633 (Hz), 62.8 (m), 474 (km).

REVIEW QUESTIONS

Q.7-15 What distinguishes a good conductor from a good insulator at a given frequency?

Q.7-16 What is meant by the *skin depth* of a conductor?

REMARKS

1. The attenuation constant and the phase constant of a good conductor are numerically equal.
2. The intrinsic impedance of a good conductor has a phase angle of 45°.
3. The skin depth of a good conductor is numerically equal to the reciprocal of its attenuation constant, and is inversely proportional to $\sqrt{f}$ and $\sqrt{\sigma}$.
4. The skin depth of good conductors is less than 1 (μm) at 10 (GHz).

7-4 GROUP VELOCITY

Definition of phase velocity

In Eq. (7-10) we defined the phase velocity, u_p, of a single-frequency plane wave as the velocity of propagation of an equiphase wavefront. The relation between u_p and the phase constant, β, is

$$u_p = \frac{\omega}{\beta} \qquad \text{(m/s)}. \tag{7-58}$$

For plane waves in a lossless medium, $\beta = k = \omega\sqrt{\mu\epsilon}$ is a linear function of ω. As a consequence, the phase velocity $u_p = 1/\sqrt{\mu\epsilon}$ is a constant that is independent of frequency. However, in some cases (such as wave propagation in a lossy dielectric, as discussed previously, or along a transmission line, or in a waveguide) the phase constant is not a linear function of ω; waves of different frequencies will propagate with different phase velocities. Because all information-bearing signals consist of a band of frequencies, waves of the component frequencies will travel with different phase velocities, causing a distortion in the signal wave shape. The signal "disperses." The phenomenon of signal distortion caused by a dependence of the phase velocity on frequency is called ***dispersion***. In view of Eqs. (7-50) and (7-39), we conclude that a lossy dielectric is a ***dispersive medium***.

Dispersion

An information-bearing signal normally has a small spread of frequencies (sidebands) around a high carrier frequency. Such a signal comprises a "group" of frequencies and forms a wave packet. A ***group velocity*** is the velocity of propagation of the wave-packet envelope (of a group of frequencies).

Definition of Group velocity

Consider the simplest case of a wave packet that consists of two traveling waves having equal amplitude and slightly different angular frequencies $\omega_0 + \Delta\omega$ and $\omega_0 - \Delta\omega\,(\Delta\omega \ll \omega_0)$. The phase constants, being functions of frequency, will also be slightly different. Let the phase constants corresponding to the two frequencies be $\beta_0 + \Delta\beta$ and $\beta_0 - \Delta\beta$. We have

$$\begin{aligned} E(z, t) &= E_0 \cos\,[(\omega_0 + \Delta\omega)t - (\beta_0 + \Delta\beta)z] \\ &\quad + E_0 \cos\,[(\omega_0 - \Delta\omega)t - (\beta_0 - \Delta\beta)z] \\ &= 2E_0 \cos\,(t\,\Delta\omega - z\,\Delta\beta)\cos\,(\omega_0 t - \beta_0 z). \end{aligned} \tag{7-59}$$

Since $\Delta\omega \ll \omega_0$, the expression in Eq. (7-59) represents a rapidly oscillating wave having an angular frequency ω_0 and an amplitude that varies slowly with an angular frequency $\Delta\omega$. This wave pattern is depicted in Fig. 7-6.

The wave inside the envelope propagates with a phase velocity found by setting $\omega_0 t - \beta_0 z = \text{Constant}$:

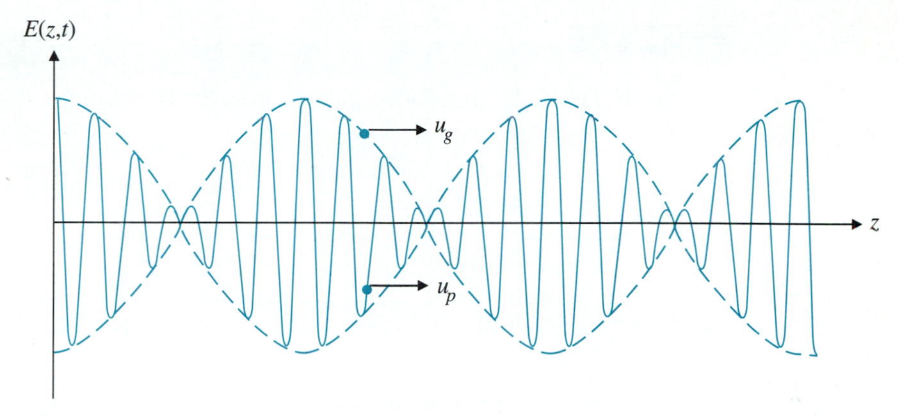

FIGURE 7-6 Sum of two time-harmonic traveling waves of equal amplitude and slightly different frequencies at a given t.

$$u_p = \frac{dz}{dt} = \frac{\omega_0}{\beta_0}.$$

The velocity of the envelope (the ***group velocity*** u_g) can be determined by setting the argument of the first cosine factor in Eq. (7-58) equal to a constant:

$$t\,\Delta\omega - z\,\Delta\beta = \text{Constant},$$

from which we obtain

$$u_g = \frac{dz}{dt} = \frac{\Delta\omega}{\Delta\beta} = \frac{1}{\Delta\beta/\Delta\omega}.$$

In the limit that $\Delta\omega \to 0$, we have the formula for computing the group velocity in a dispersive medium:

Formula for group velocity in dispersive media

$$\boxed{u_g = \frac{1}{d\beta/d\omega} \quad \text{(m/s)}.} \tag{7-60}$$

This is the velocity of a point on the envelope of the wave packet, as shown in Fig. 7-6, and is identified as the velocity of the narrow-band signal. Group velocity in a dispersive medium may be higher or lower than phase velocity. A medium is said to exhibit normal dispersion if $u_g < u_p$, and to exhibit anomalous dispersion if $u_g > u_p$. When $u_g = u_p$, there is no dispersion.

REVIEW QUESTIONS

Q.7-17 What is meant by the *dispersion* of a signal? Give an example of a dispersive medium.

Q.7-18 Define *group velocity*. In what ways is group velocity different from phase velocity?

REMARKS

1. Information-bearing signals propagate with no distortion only in nondispersive media.
2. A medium is nondispersive if β is a linear function of (directly proportional to) ω.

7-5 FLOW OF ELECTROMAGNETIC POWER AND THE POYNTING VECTOR

Electromagnetic waves carry with them electromagnetic power. Energy is transported through space to distant receiving points by electromagnetic waves. We will now derive a relation between the rate of such energy transfer and the electric and magnetic field intensities associated with a traveling electromagnetic wave.

We begin with the curl equations:

$$\nabla \times \mathbf{E} = -\frac{\partial \mathbf{B}}{\partial t}, \tag{6-45a)(7-61}$$

$$\nabla \times \mathbf{H} = \mathbf{J} + \frac{\partial \mathbf{D}}{\partial t}. \tag{6-45b)(7-62}$$

The following identity of vector operations can be verified in a straightforward manner by using Cartesian coordinates:

$$\nabla \cdot (\mathbf{E} \times \mathbf{H}) = \mathbf{H} \cdot (\nabla \times \mathbf{E}) - \mathbf{E} \cdot (\nabla \times \mathbf{H}). \tag{7-63}$$

Substitution of Eqs. (7-61) and (7-62) in Eq. (7-63) yields

$$\nabla \cdot (\mathbf{E} \times \mathbf{H}) = -\mathbf{H} \cdot \frac{\partial \mathbf{B}}{\partial t} - \mathbf{E} \cdot \frac{\partial \mathbf{D}}{\partial t} - \mathbf{E} \cdot \mathbf{J}. \tag{7-64}$$

In a simple medium, whose constitutive parameters ϵ, μ, and σ do not change with time, we have

$$\mathbf{H} \cdot \frac{\partial \mathbf{B}}{\partial t} = \mathbf{H} \cdot \frac{\partial(\mu \mathbf{H})}{\partial t} = \frac{1}{2} \frac{\partial(\mu \mathbf{H} \cdot \mathbf{H})}{\partial t} = \frac{\partial}{\partial t}\left(\frac{1}{2} \mu H^2\right),$$

$$\mathbf{E} \cdot \frac{\partial \mathbf{D}}{\partial t} = \mathbf{E} \cdot \frac{\partial(\epsilon \mathbf{E})}{\partial t} = \frac{1}{2} \frac{\partial(\epsilon \mathbf{E} \cdot \mathbf{E})}{\partial t} = \frac{\partial}{\partial t}\left(\frac{1}{2} \epsilon E^2\right),$$

$$\mathbf{E} \cdot \mathbf{J} = \mathbf{E} \cdot (\sigma \mathbf{E}) = \sigma E^2.$$

Equation (7-64) can then be written as

$$\nabla \cdot (\mathbf{E} \times \mathbf{H}) = -\frac{\partial}{\partial t}\left(\frac{1}{2} \epsilon E^2 + \frac{1}{2} \mu H^2\right) - \sigma E^2, \tag{7-65}$$

which is a point-function relationship. An integral form of Eq. (7-65) is obtained by integrating both sides over the volume of concern:

$$\oint_S (\mathbf{E} \times \mathbf{H})\cdot d\mathbf{s} = -\frac{\partial}{\partial t}\int_V \left(\frac{1}{2}\epsilon E^2 + \frac{1}{2}\mu H^2\right) dv - \int_V \sigma E^2\, dv, \tag{7-66}$$

where the divergence theorem has been applied to convert the volume integral of $\nabla\cdot(\mathbf{E} \times \mathbf{H})$ to the closed surface integral of $(\mathbf{E} \times \mathbf{H})$.

We recognize that the first and second terms on the right side of Eq. (7-66) represent the time-rate of change of the energy stored in the electric and magnetic fields, respectively. [Compare with Eqs. (3-106) and (5-106).] The last term is the ohmic power dissipated in the volume as a result of the flow of conduction current density $\sigma\mathbf{E}$ in the presence of the electric field $\mathbf{E}$. Hence we may interpret the right side of Eq. (7-66) as the *rate of decrease* of the electric and magnetic energies stored, subtracted by the ohmic power dissipated as heat in the volume V. To be consistent with the law of conservation of energy, this must equal the power (rate of energy) *leaving* the volume through its surface. Thus the quantity $(\mathbf{E} \times \mathbf{H})$ is a vector representing the power flow per unit area. Define

$$\boxed{\mathscr{P} = \mathbf{E} \times \mathbf{H} \qquad (\text{W/m}^2).} \tag{7-67}$$

Definition of Poynting vector

Poynting's theorem

Quantity $\mathscr{P}$ is known as the ***Poynting vector***, which is a power density vector associated with an electromagnetic field. The assertion that the surface integral of $\mathscr{P}$ over a closed surface, as given by the left side of Eq. (7-66), equals the power leaving the enclosed volume is referred to as ***Poynting's theorem***. This assertion is not limited to plane waves.

Equation (7-66) may be written in another form:

$$-\oint_S \mathscr{P}\cdot d\mathbf{s} = \frac{\partial}{\partial t}\int_V (w_e + w_m)\, dv + \int_V p_\sigma\, dv, \tag{7-68}$$

where

$$w_e = \tfrac{1}{2}\epsilon E^2 = \tfrac{1}{2}\epsilon \mathbf{E}\cdot\mathbf{E}^* = \text{Electric energy density}, \tag{7-69}$$

$$w_m = \tfrac{1}{2}\mu H^2 = \tfrac{1}{2}\mu \mathbf{H}\cdot\mathbf{H}^* = \text{Magnetic energy density}, \tag{7-70}$$

$$p_\sigma = \sigma E^2 = J^2/\sigma = \sigma\mathbf{E}\cdot\mathbf{E}^* = \mathbf{J}\cdot\mathbf{J}^*/\sigma = \text{Ohmic power density}. \tag{7-71}$$

In words, Eq. (7-68) states that the total power flowing *into* a closed surface at any instant equals the sum of the rates of increase of the stored electric and magnetic energies and the ohmic power dissipated within the enclosed volume. An asterisk on a quantity denotes the complex conjugate of that quantity.

EXAMPLE 7-5

Find the Poynting vector on the surface of a long, straight conducting wire (of radius b and conductivity σ) that carries a direct current I. Verify Poynting's theorem.

SOLUTION

Since we have a d-c situation, the current in the wire is uniformly distributed over its cross-sectional area. Let us assume that the axis of the wire coincides with the z-axis. Figure 7-7 shows a segment of length ℓ of the long wire. We have

$$\mathbf{J} = \mathbf{a}_z \frac{I}{\pi b^2}$$

and

$$\mathbf{E} = \frac{\mathbf{J}}{\sigma} = \mathbf{a}_z \frac{I}{\sigma \pi b^2}.$$

FIGURE 7-7 Illustrating Poynting's theorem (Example 7-5).

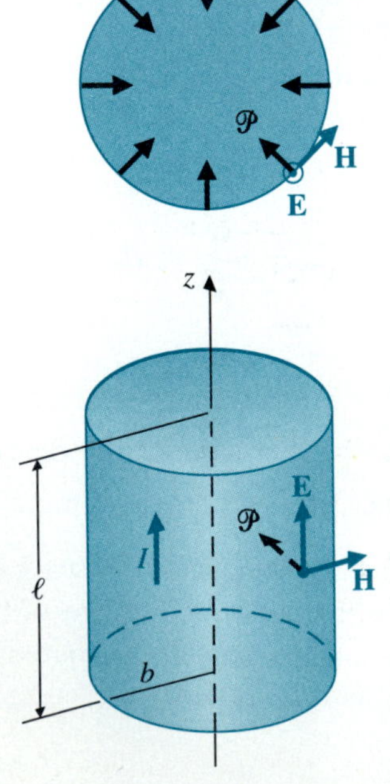

On the surface of the wire,

$$\mathbf{H} = \mathbf{a}_\phi \frac{I}{2\pi b}.$$

Thus the Poynting vector at the surface of the wire is

$$\begin{aligned}\mathscr{P} = \mathbf{E} \times \mathbf{H} &= (\mathbf{a}_z \times \mathbf{a}_\phi)\frac{I^2}{2\sigma\pi^2 b^3} \\ &= -\mathbf{a}_r \frac{I^2}{2\sigma\pi^2 b^3},\end{aligned}$$

which is directed everywhere into the wire surface.

To verify Poynting's theorem, we integrate $\mathscr{P}$ over the wall of the wire segment in Fig. 7-7:

$$\begin{aligned}-\oint_S \mathscr{P}\cdot d\mathbf{s} = -\oint_S \mathscr{P}\cdot \mathbf{a}_r\, ds &= \left(\frac{I^2}{2\sigma\pi^2 b^3}\right) 2\pi b\ell \\ &= I^2\left(\frac{\ell}{\sigma\pi b^2}\right) = I^2 R,\end{aligned}$$

where the formula for the resistance of a straight wire in Eq. (4-16), $R = \ell/\sigma S$, has been used. The above result affirms that the negative surface integral of the Poynting vector is exactly equal to the I^2R ohmic power loss in the conducting wire. Hence Poynting's theorem is verified.

7-5.1 INSTANTANEOUS AND AVERAGE POWER DENSITIES

In dealing with time-harmonic electromagnetic waves we have found it convenient to use phasor notation. The instantaneous value of a quantity is then the real part of the product of the phasor quantity and $e^{j\omega t}$ when $\cos \omega t$ is used as the reference. For example, for the phasor

$$\mathbf{E}(z) = \mathbf{a}_x E_x(z) = \mathbf{a}_x E_0 e^{-(\alpha + j\beta)z}, \tag{7-72}$$

the instantaneous expression is

Writing the instantaneous expression from a phasor

$$\begin{aligned}\mathbf{E}(z, t) = \mathscr{R}e[\mathbf{E}(z)e^{j\omega t}] &= \mathbf{a}_x E_0 e^{-\alpha z} \mathscr{R}e[e^{j(\omega t - \beta z)}] \\ &= \mathbf{a}_x E_0 e^{-\alpha z} \cos(\omega t - \beta z).\end{aligned} \tag{7-73}$$

For a uniform plane wave propagating in a lossy medium in the $+z$-direction, the associated magnetic field intensity phasor is

$$\mathbf{H}(z) = \mathbf{a}_y H_y(z) = \mathbf{a}_y \frac{E_0}{|\eta_c|} e^{-\alpha z} e^{-j(\beta z + \theta_\eta)}, \tag{7-74}$$

where θ_η is the phase angle of the intrinsic impedance $\eta_c = |\eta_c| e^{j\theta_\eta}$ of the medium. The corresponding instantaneous expression for $\mathbf{H}(z)$ is

$$\mathbf{H}(z, t) = \mathscr{R}e[\mathbf{H}(z)e^{j\omega t}] = \mathbf{a}_y \frac{E_0}{|\eta_c|} e^{-\alpha z} \cos(\omega t - \beta z - \theta_\eta). \tag{7-75}$$

The instantaneous expression for the Poynting vector or power density vector, from Eqs. (7-72) and (7-74), is

$$\begin{aligned}\mathscr{P}(z, t) &= \mathbf{E}(z, t) \times \mathbf{H}(z, t) = \mathscr{R}e[\mathbf{E}(z)e^{j\omega t}] \times \mathscr{R}e[\mathbf{H}(z)e^{j\omega t}] \\ &= \mathbf{a}_z \frac{E_0^2}{|\eta_c|} e^{-2\alpha z} \cos(\omega t - \beta z) \cos(\omega t - \beta z - \theta_\eta) \\ &= \mathbf{a}_z \frac{E_0^2}{2|\eta_c|} e^{-2\alpha z} [\cos \theta_\eta + \cos(2\omega t - 2\beta z - \theta_\eta)]. \end{aligned} \tag{7-76}^\dagger$$

As far as the power transmitted by an electromagnetic wave is concerned, its average value is a more significant quantity than its instantaneous value. From Eq. (7-76), we obtain the time-average Poynting vector, $\mathscr{P}_{av}(z)$:

Average power density transmitted by a uniform plane wave in z-direction

$$\mathscr{P}_{av}(z) = \frac{1}{T} \int_0^T \mathscr{P}(z, t)\, dt = \mathbf{a}_z \frac{E_0^2}{2|\eta_c|} e^{-2\alpha z} \cos \theta_\eta \qquad (\text{W/m}^2), \tag{7-77}$$

where $T = 2\pi/\omega$ is the time period of the wave. The second term on the right side of Eq. (7-76) is a cosine function of a double frequency whose average is zero over a fundamental period. For wave propagation in lossless media, $\eta_c \to \eta$ is real, $\sigma = 0$, and $\theta_\eta = 0$, Eq. (7-77) reduces to

$$\mathscr{P}_{av}(z) = \mathbf{a}_z \frac{E_0^2}{2\eta} \qquad (\text{W/m}^2). \tag{7-78}$$

In the general case we may not be dealing with a wave propagating in the z-direction. We write

General formula for average power density in a propagating wave

$$\mathscr{P}_{av} = \tfrac{1}{2}\mathscr{R}e(\mathbf{E} \times \mathbf{H}^*) \qquad (\text{W/m}^2), \tag{7-79}$$

which is a general formula for computing the average power density in a propagating wave.

■ **EXERCISE 7.8** Substitute the phasor expressions of $\mathbf{E}(z)$ and $\mathbf{H}(z)$ given by Eqs. (7-72) and (7-74) in Eq. (7-79) to verify $\mathscr{P}_{av}$ obtained in Eq. (7-77).

† Here we have used the trigonometric identity $\cos \theta_1 \cos \theta_2 = \frac{1}{2}[\cos(\theta_1 - \theta_2) + \cos(\theta_1 + \theta_2)]$.

EXAMPLE 7-6

The phasor expressions of the far field at a distance R from a short vertical current element $I\,d\ell$ located in free space at the origin of a spherical coordinate system are

$$\mathbf{E}(R, \theta) = \mathbf{a}_\theta E_\theta(R, \theta) = \mathbf{a}_\theta \left(j \frac{60\pi I\,d\ell}{\lambda R} \sin\theta \right) e^{-j\beta R} \qquad \text{(V/m)} \tag{7-80}$$

and

$$\mathbf{H}(R, \theta) = \mathbf{a}_\phi \frac{E_\theta(R, \theta)}{\eta_0} = \mathbf{a}_\phi \left(j \frac{I\,d\ell}{2\lambda R} \sin\theta \right) e^{-j\beta R} \qquad \text{(A/m)}, \tag{7-81}$$

where $\lambda = 2\pi/\beta$ is the wavelength.

a) Write the expression for instantaneous Poynting vector.

b) Find the total average power radiated by the current element.

SOLUTION

a) We note that $E_\theta/H_\phi = \eta_0 = 120\pi\,(\Omega)$. The instantaneous Poynting vector is

$$\begin{aligned} \mathscr{P}(R, \theta; t) &= \mathscr{R}e[\mathbf{E}(R, \theta)e^{j\omega t}] \times \mathscr{R}e[\mathbf{H}(R, \theta)e^{j\omega t}] \\ &= (\mathbf{a}_\theta \times \mathbf{a}_\phi) 30\pi \left(\frac{I\,d\ell}{\lambda R}\right)^2 \sin^2\theta \sin^2(\omega t - \beta R) \\ &= \mathbf{a}_R 15\pi \left(\frac{I\,d\ell}{\lambda R}\right)^2 \sin^2\theta [1 - \cos 2(\omega t - \beta R)] \qquad \text{(W/m}^2\text{)}. \end{aligned}$$

b) The average power density vector is, from Eq. (7-79),

$$\mathscr{P}_{av}(R, \theta) = \mathbf{a}_R 15\pi \left(\frac{I\,d\ell}{\lambda R}\right)^2 \sin^2\theta, \tag{7-82}$$

which is seen to equal the time-average value of $\mathscr{P}(R, \theta; t)$ given in part (a) of this solution. The total average power radiated is obtained by integrating $\mathscr{P}_{av}(R, \theta)$ over the surface of the sphere of radius R:

$$\begin{aligned} \text{Total } P_{av} &= \oint_S \mathscr{P}_{av}(R, \theta) \cdot d\mathbf{s} \\ &= \int_0^{2\pi} \int_0^{\pi} \left[15\pi \left(\frac{I\,d\ell}{\lambda R}\right)^2 \sin^2\theta \right] R^2 \sin\theta \, d\theta \, d\phi \\ &= 40\pi^2 \left(\frac{d\ell}{\lambda}\right)^2 I^2 \qquad \text{(W)}, \end{aligned} \tag{7-83}$$

where I is the amplitude ($\sqrt{2}$ times the effective value) of the sinusoidal current in $d\ell$.

EXERCISE 7.9 Refer to Example 7-6. Assuming $I = 5$ (A) and $d\ell = \lambda/20$, determine the power intercepted in the far field at a distance of 9 (m) by a spherical surface facing the current element and defined by $80° \leqslant \theta \leqslant 100°$ and $0° \leqslant \phi \leqslant 20°$.

ANS. 0.354 (W).

REVIEW QUESTIONS

Q.7-19 Define *Poynting vector*. What is the SI unit for this vector?

Q.7-20 State Poynting's theorem.

Q.7-21 For a time-harmonic electromagnetic field, write the expressions in terms of electric and magnetic field intensity vectors for (a) instantaneous Poynting vector, and (b) time-average Poynting vector.

REMARKS

1. The Poynting vector $\mathscr{P}$ is in a direction normal to both **E** and **H**.
2. Poynting's theorem is a manifestation of the principle of conservation of energy.
3. Note that

$$\mathscr{P}(z, t) = \mathscr{R}e[\mathbf{E}(z)e^{j\omega t}] \times \mathscr{R}e[\mathbf{H}(z)e^{j\omega t}] \neq \mathscr{R}e[\mathbf{E}(z) \times \mathbf{H}(z)]e^{j\omega t};$$

that is, it is not correct to cross-multiply **E** and **H** first and then take the real part of the product.

7-6 NORMAL INCIDENCE OF PLANE WAVES AT PLANE BOUNDARIES

Up to this point we have dealt with the propagation of uniform plane waves in an unbounded homogeneous medium. In practice, waves often propagate in bounded regions where several media with different constitutive parameters are present. When an electromagnetic wave traveling in one medium impinges on another medium with a different intrinsic impedance, it experiences a reflection. Unless the second medium is a perfect conductor, a part of the incident power is transmitted into the second medium. In this section we consider the simpler case of the normal incidence of uniform plane waves at a plane boundary. The more general case of oblique incidence will be studied in the next section.

Consider the situation in Fig. 7-8, where the incident wave $(\mathbf{E}_i, \mathbf{H}_i)$ in medium 1 (ϵ_1, μ_1) travels in the $+z$-direction toward medium 2 (ϵ_2, μ_2). The boundary surface is the $z = 0$ plane. Both media are assumed to be lossless. The incident electric and magnetic field intensity phasors are $(\mathbf{a}_{ki} = \mathbf{a}_z)$:

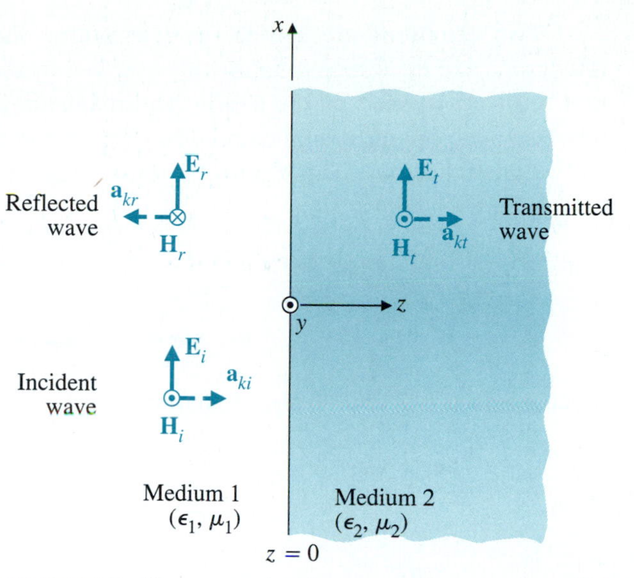

FIGURE 7-8 Plane wave incident normally on a plane dielectric boundary.

$$\mathbf{E}_i(z) = \mathbf{a}_x E_{i0} e^{-j\beta_1 z}, \tag{7-84}$$

$$\mathbf{H}_i(z) = \mathbf{a}_y \frac{E_{i0}}{\eta_1} e^{-j\beta_1 z}. \tag{7-85}$$

Because of the medium discontinuity at $z = 0$, the incident wave is partly reflected back into medium 1 and partly transmitted into medium 2. We have

a) *For the reflected wave* ($\mathbf{E}_r$, $\mathbf{H}_r$): $\mathbf{a}_{kr} = -\mathbf{a}_z$.

$$\mathbf{E}_r(z) = \mathbf{a}_x E_{r0} e^{j\beta_1 z}, \tag{7-86}$$

$$\mathbf{H}_r(z) = (-\mathbf{a}_z) \times \frac{1}{\eta_1} \mathbf{E}_r(z) = -\mathbf{a}_y \frac{E_{r0}}{\eta_1} e^{j\beta_1 z}. \tag{7-87}$$

b) *For the transmitted wave* ($\mathbf{E}_t$, $\mathbf{H}_t$): $\mathbf{a}_{kt} = \mathbf{a}_z$.

$$\mathbf{E}_t(z) = \mathbf{a}_x E_{t0} e^{-j\beta_2 z}, \tag{7-88}$$

$$\mathbf{H}_t(z) = \mathbf{a}_z \times \frac{1}{\eta_2} \mathbf{E}_t(z) = \mathbf{a}_y \frac{E_{t0}}{\eta_2} e^{-j\beta_2 z}, \tag{7-89}$$

where E_{t0} is the magnitude of $\mathbf{E}_t$ at $z = 0$, and β_2 and η_2 are the phase constant and the intrinsic impedance, respectively, of medium 2. Note that the directions of the arrows for $\mathbf{E}_r$ and $\mathbf{E}_t$ in Fig. 7-8 are arbitrarily drawn because E_{r0} and E_{t0} may themselves be positive or negative, depending on the relative magnitudes of the constitutive parameters of the two media.

Two equations are needed for determining the two unknown magnitudes E_{r0} and E_{t0}. These equations are supplied by the boundary conditions that must be satisfied by the electric and magnetic fields. At the dielectric interface $z = 0$ the tangential components (the x-components) of the electric and magnetic field intensities must be continuous. We have

$$\mathbf{E}_i(0) + \mathbf{E}_r(0) = \mathbf{E}_t(0) \qquad \text{or} \qquad E_{i0} + E_{r0} = E_{t0} \tag{7-90}$$

and

$$\mathbf{H}_i(0) + \mathbf{H}_r(0) = \mathbf{H}_t(0) \qquad \text{or} \qquad \frac{1}{\eta_1}(E_{i0} - E_{r0}) = \frac{E_{t0}}{\eta_2}. \tag{7-91}$$

Solving Eqs. (7-90) and (7-91), we obtain

$$E_{r0} = \frac{\eta_2 - \eta_1}{\eta_2 + \eta_1} E_{i0}, \tag{7-92}$$

$$E_{t0} = \frac{2\eta_2}{\eta_2 + \eta_1} E_{i0}. \tag{7-93}$$

Reflection and transmission coefficients

The ratios E_{r0}/E_{i0} and E_{t0}/E_{i0} are called ***reflection coefficient*** and ***transmission coefficient***, respectively. In terms of the intrinsic impedances they are

$$\Gamma = \frac{E_{r0}}{E_{i0}} = \frac{\eta_2 - \eta_1}{\eta_2 + \eta_1} \qquad \text{(Normal incidence)} \tag{7-94}$$

and

$$\tau = \frac{E_{t0}}{E_{i0}} = \frac{2\eta_2}{\eta_2 + \eta_1} \qquad \text{(Normal incidence).} \tag{7-95}$$

The definitions for Γ and τ in Eqs. (7-94) and (7-95) apply even when the media are dissipative—that is, even when η_1 and/or η_2 are complex. Thus, Γ and τ may themselves be complex in the general case. Reflection and transmission coefficients are related by the following equation:

Relation between reflection and transmission coefficients for normal incidence

$$1 + \Gamma = \tau \qquad \text{(Normal incidence).} \tag{7-96}$$

Incident wave plus reflected wave→standing wave

The total field in medium 1 ($\mathbf{E}_1, \mathbf{H}_1$) is the sum of the incident and reflected fields. From Eqs. (7-84) and (7-86), we have

$$\mathbf{E}_1(z) = \mathbf{a}_x E_{i0} e^{-j\beta_1 z}(1 + \Gamma e^{j2\beta_1 z}), \tag{7-97}$$

which is a function of z.$|\mathbf{E}_1(z)|$ will have maximum and minimum values at locations where the factor $(1 + \Gamma e^{j2\beta_1 z})$ is maximum and minimum, respectively. (The magnitude of $e^{-2\beta_1 z}$ is unity.) We have, in fact, a standing wave in medium 1.

The ratio of the maximum value to the minimum value of the electric field intensity of a standing wave is called the ***standing-wave ratio (SWR)***, *S*.

Standing wave ratio

$$S = \frac{|E|_{\max}}{|E|_{\min}} = \frac{1+|\Gamma|}{1-|\Gamma|} \quad \text{(Dimensionless)}. \tag{7-98}$$

An inverse relation of Eq. (7-98) is

Finding the magnitude of reflection coefficient from the standing-wave ratio

$$|\Gamma| = \frac{S-1}{S+1} \quad \text{(Dimensionless)}. \tag{7-99}$$

Range of $|\Gamma|$: 0 to +1

Range of S: 1 to ∞

While the value of Γ ranges from -1 to $+1$, the value of S ranges from 1 to ∞. It is customary to express S on a logarithmic scale. The standing-wave ratio in decibels is $20\log_{10} S$. Thus $S=2$ corresponds to a standing-wave ratio of $20\log_{10}2 = 6.02\,\text{(dB)}$ and $|\Gamma| = (2-1)/(2+1) = \frac{1}{3}$. A standing-wave ratio of $2\,\text{(dB)}$ is equivalent to $S = 1.26$ and $|\Gamma| = 0.115$.

■ **EXERCISE 7.10**

a) Convert $\Gamma = 0.20$ into S in (dB).

b) Convert $S = 3\,\text{(dB)}$ into reflection coefficient $|\Gamma|$.

ANS. (a) 3.52 (dB), (b) 0.17.

The magnetic field intensity in medium 1 is obtained by combining $\mathbf{H}_i(z)$ and $\mathbf{H}_r(z)$ in Eqs. (7-85) and (7-87), respectively:

$$\mathbf{H}_1(z) = \mathbf{a}_y \frac{E_{i0}}{\eta_1}\left(e^{-j\beta_1 z} - \Gamma e^{j\beta_1 z}\right)$$

$$= \mathbf{a}_y \frac{E_{i0}}{\eta_1} e^{-j\beta_1 z}\left(1 - \Gamma e^{j2\beta_1 z}\right). \tag{7-100}$$

This should be compared with $\mathbf{E}_1(z)$ in Eq. (7-97). In a dissipationless medium, Γ is real; and $|\mathbf{H}_1(z)|$ will be a minimum at locations where $|\mathbf{E}_1(z)|$ is a maximum, and vice versa.

In medium 2, $(\mathbf{E}_t, \mathbf{H}_t)$ constitute the transmitted wave propagating in $+z$-direction. From Eqs. (7-88) and (7-95) we have

$$\mathbf{E}_t(z) = \mathbf{a}_x \tau E_{i0} e^{-j\beta_2 z}. \tag{7-101}$$

And from Eq. (7-89) we obtain

$$\mathbf{H}_t(z) = \mathbf{a}_y \frac{\tau}{\eta_2} E_{i0} e^{-j\beta_2 z}. \tag{7-102}$$

EXAMPLE 7-7

A uniform plane wave in a lossless medium with intrinsic impedance η_1 is incident normally onto another lossless medium with intrinsic impedance η_2 through a plane boundary.

a) Obtain the expressions for the time-average power densities in both media.

b) Find the standing-wave ratio in medium 1 if $\eta_2 = 2\eta_1$.

SOLUTION

a) Equation (7-79) provides the formula for computing the time-average power density, or time-average Poynting vector:

$$\mathscr{P}_{av} = \frac{1}{2}\mathscr{R}e(\mathbf{E} \times \mathbf{H}^*). \tag{7-103}$$

In medium 1 we use Eqs. (7-97) and (7-100):

$$\begin{aligned}(\mathscr{P}_{av})_1 &= \mathbf{a}_z \frac{E_{i0}^2}{2\eta_1} \mathscr{R}e[(1+\Gamma e^{j2\beta_1 z})(1-\Gamma e^{-j2\beta_1 z})] \\ &= \mathbf{a}_z \frac{E_{i0}^2}{2\eta_1} \mathscr{R}e[(1-\Gamma^2)+\Gamma(e^{j2\beta_1 z}-e^{-j2\beta_1 z})] \\ &= \mathbf{a}_z \frac{E_{i0}^2}{2\eta_1} \mathscr{R}e[(1-\Gamma^2)+j2\Gamma \sin 2\beta_1 z] \\ &= \mathbf{a}_z \frac{E_{i0}^2}{2\eta_1} (1-\Gamma^2) \qquad (\text{W/m}^2),\end{aligned} \tag{7-104}$$

where Γ is a real number because both media are lossless.

In medium 2 we use Eqs. (7-101) and (7-102) to obtain

$$(\mathscr{P}_{av})_2 = \mathbf{a}_z \frac{E_{i0}^2}{2\eta_2} \tau^2 \qquad (\text{W/m}^2). \tag{7-105}$$

Since we are dealing with lossless media, the power flow in medium 1 must equal that in medium 2; that is,

$$(\mathscr{P}_{av})_1 = (\mathscr{P}_{av})_2, \tag{7-106}$$

or

$$\boxed{1-\Gamma^2 = \frac{\eta_1}{\eta_2}\tau^2.} \tag{7-107}$$

That Eq. (7-107) is true can be readily verified by using Eqs. (7-94) and (7-95).

b) If $\eta_2 = 2\eta_1$,

$$\Gamma = \frac{\eta_2 - \eta_1}{\eta_2 + \eta_1} = \frac{1}{3}.$$

Thus, from Eq. (7-98),

$$S = \frac{1+1/3}{1-1/3} = 2.$$

Expressed in decibels, $S = 20\log_{10}2 = 6.02\,(\text{dB})$.

■ **EXERCISE 7.11** Given $\mathbf{E}_i(z, t) = \mathbf{a}_y 24\cos(10^8 t - \beta z)$ (V/m) in air that impinges normally on a lossless medium with $\epsilon_{r2} = 2.25$, $\mu_{r2} = 1$ in the $z \geqslant 0$ region, find (a) β, Γ, S, τ; (b) $\mathbf{E}_r(z, t)$; (c) $\mathbf{E}_2(z, t)$; (d) $\mathbf{H}_2(z, t)$; and (e) $(\mathscr{P}_{av})_z$.

ANS. (a) 1/3 (rad/m), -0.2, 1.5, 0.8; (b) $-\mathbf{a}_y 4.8\cos(10^8 t + z/3)$ (V/m); (c) $\mathbf{a}_y 19.2\cos(10^8 t - z/3)$ (V/m); (d) $-\mathbf{a}_x 0.0764\cos(10^8 t - z/3)$ (A/m); (e) 0.733 (W/m^2).

7-6.1 NORMAL INCIDENCE ON A GOOD CONDUCTOR

Our discussions of normal incidence of plane waves on a plane boundary so far have assumed lossless media. In practice we often encounter situations where one medium is a good conductor, $\sigma/\omega\epsilon \gg 1$. Examples are metallic reflectors and waveguides. Under such conditions we are generally allowed to use the perfect-conductor approximation ($\sigma \to \infty$) and obtain good results. This approximation simplifies all our formulas.

Consider the incident field vector phasor given in Eqs. (7-84) and (7-85):

$$\mathbf{E}_i(z) = \mathbf{a}_x E_{i0} e^{-j\beta_1 z}, \tag{7-84)(7-108}$$

$$\mathbf{H}_i(z) = \mathbf{a}_y \frac{E_{i0}}{\eta_1} e^{-j\beta_1 z}. \tag{7-85)(7-109}$$

For normal incidence on a plane conducting boundary: $\Gamma = -1$, $\tau = 0$

This wave impinges on a perfectly conducting plane boundary at $z = 0$. Substituting ∞ for σ in Eq. (7-53), we find $\eta_2 = 0$. This is as expected, and the conducting boundary acts as a short circuit. From Eqs. (7-94) and (7-95) we see that $\Gamma = -1$ and $\tau = 0$. Consequently, $\mathbf{E}_{r0} = \Gamma E_{i0} = -E_{i0}$, and $E_{t0} = \tau E_{i0} = 0$. The incident wave is totally reflected with a phase reversal, and no power is transmitted across a perfectly conducting boundary. We have

$$\mathbf{E}_r(z) = -\mathbf{a}_x E_{i0} e^{j\beta_1 z}, \tag{7-110}$$

$$\mathbf{H}_r(z) = -\mathbf{a}_y \times \frac{\mathbf{E}_r(z)}{\eta_1} = \mathbf{a}_y \frac{E_{i0}}{\eta_1} e^{j\beta_1 z}, \tag{7-111}$$

and

$$\begin{aligned}\mathbf{E}_1(z) &= \mathbf{E}_i(z) + \mathbf{E}_r(z) = \mathbf{a}_x E_{i0}(e^{-j\beta_1 z} - e^{j\beta_1 z}) \\ &= -\mathbf{a}_x j2E_{i0} \sin \beta_1 z,\end{aligned} \tag{7-112}$$

$$\mathbf{H}_1(z) = \mathbf{H}_i(z) + \mathbf{H}_r(z) = \mathbf{a}_y \frac{E_{i0}}{\eta_1}(e^{-j\beta_1 z} + e^{j\beta_1 z})$$

$$= \mathbf{a}_y 2\frac{E_{i0}}{\eta_1}\cos\beta_1 z. \tag{7-113}$$

Equations (7-112) and (7-113) show that $\mathbf{E}_1(z)$ and $\mathbf{H}_1(z)$ are in time quadrature ($\mathbf{E}_1$ lags behind $\mathbf{H}_1$ by 90° because of the $-j$ factor). Both represent standing waves, and from Eq. (7-79) we conclude that no average power is associated with the total electromagnetic wave in medium 1.

In order to examine the space-time behavior of the total field in medium 1, we first write the instantaneous expressions corresponding to the electric and magnetic field intensity phasors obtained in Eqs. (7-112) and (7-113):

Total fields $\mathbf{E}_1$ and $\mathbf{H}_1$ exhibit standing waves for normal incidence on a plane conducting boundary.

$$\mathbf{E}_1(z, t) = \mathscr{R}e[\mathbf{E}_1(z)e^{j\omega t}] = \mathbf{a}_x 2E_{i0}\sin\beta_1 z \sin\omega t, \tag{7-114}$$

$$\mathbf{H}_1(z, t) = \mathscr{R}e[\mathbf{H}_1(z)e^{j\omega t}] = \mathbf{a}_y 2\frac{E_{i0}}{\eta_1}\cos\beta_1 z \cos\omega t. \tag{7-115}$$

Both $\mathbf{E}_1(z, t)$ and $\mathbf{H}_1(z, t)$ possess zeros and maxima at fixed distances from the conducting boundary for all t. For a given t, both $\mathbf{E}_1$ and $\mathbf{H}_1$ vary sinusoidally with the distance measured from the boundary plane. (z is negative in medium 1.) The standing waves of $\mathbf{E}_1 = \mathbf{a}_x E_1$ and $\mathbf{H}_1 = \mathbf{a}_y H_1$ are shown in Fig. 7-9 for several values of ωt. We see that $\mathbf{E}_1$ vanishes at the infinitely conducting boundary; it also is zero at points that are multiples of $\lambda_1/2$ from the boundary. The standing wave of $\mathbf{H}_1$ is shifted from that of $\mathbf{E}_1$ by a quarter wavelength ($\lambda_1/4$).

EXAMPLE 7-8

A y-polarized uniform plane wave ($\mathbf{E}_i, \mathbf{H}_i$) with a frequency 100 (MHz) propagates in air in the $+x$ direction and impinges normally on a perfectly conducting plane at $x = 0$. Assuming the amplitude of $\mathbf{E}_i$ to be 6 (mV/m), write the phasor and instantaneous expressions for

a) $\mathbf{E}_i$ and $\mathbf{H}_i$ of the incident wave;

b) $\mathbf{E}_r$ and $\mathbf{H}_r$ of the reflected wave; and

c) $\mathbf{E}_1$ and $\mathbf{H}_1$ of the total wave in air.

SOLUTION

At the given frequency 100 (MHz),

$$\omega = 2\pi f = 2\pi \times 10^8 \qquad \text{(rad/s)},$$

$$\beta_1 = k_0 = \frac{\omega}{c} = \frac{2\pi \times 10^8}{3 \times 10^8} = \frac{2\pi}{3} \qquad \text{(rad/m)},$$

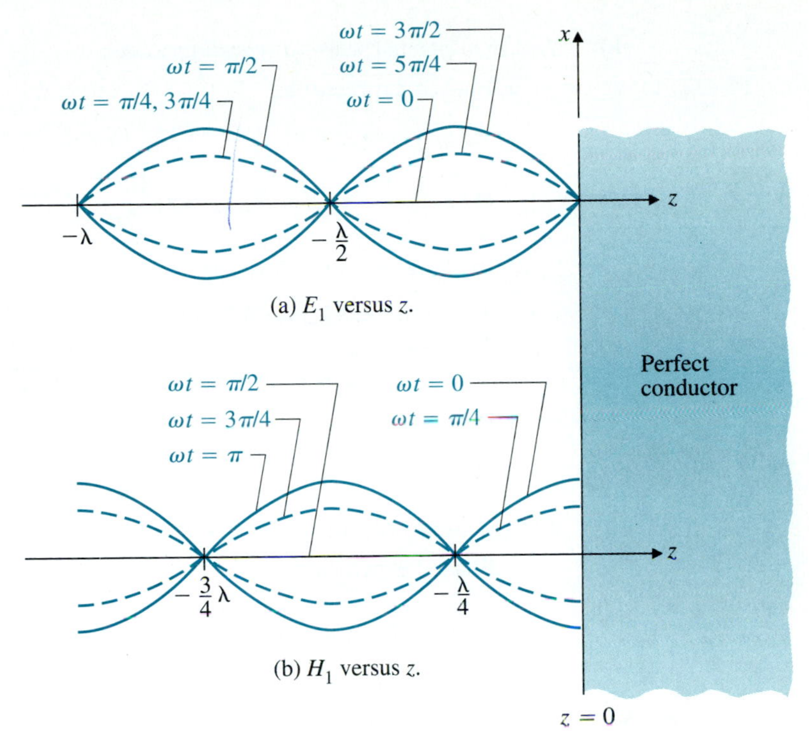

FIGURE 7-9 Standing waves of $\mathbf{E}_1 = \mathbf{a}_x E_1$ and $\mathbf{H}_1 = \mathbf{a}_y H_1$ for several values of ωt.

$$\eta_1 = \eta_0 = \sqrt{\frac{\mu_0}{\epsilon_0}} = 120\pi \qquad (\Omega).$$

a) *For the incident wave (a traveling wave):*

i) Phasor expressions:

$$\mathbf{E}_i(x) = \mathbf{a}_y 6 \times 10^{-3} e^{-j2\pi x/3} \qquad \text{(V/m)},$$

$$\mathbf{H}_i(x) = \frac{1}{\eta_1} \mathbf{a}_x \times \mathbf{E}_i(x) = \mathbf{a}_z \frac{10^{-4}}{2\pi} e^{-j2\pi x/3} \qquad \text{(A/m)}.$$

ii) Instantaneous expressions:

$$\mathbf{E}_i(x, t) = \mathscr{R}e[\mathbf{E}_i(x)e^{j\omega t}]$$
$$= \mathbf{a}_y 6 \times 10^{-3} \cos\left(2\pi \times 10^8 t - \frac{2\pi}{3} x\right) \qquad \text{(V/m)},$$

$$\mathbf{H}_i(x, t) = \mathbf{a}_z \frac{10^{-4}}{2\pi} \cos\left(2\pi \times 10^8 t - \frac{2\pi}{3} x\right) \qquad \text{(A/m)}.$$

b) *For the reflected wave (a traveling wave)*:

i) Phasor expressions:

$$\mathbf{E}_r(x) = -\mathbf{a}_y 6 \times 10^{-3} e^{j2\pi x/3} \qquad \text{(V/m)},$$

$$\mathbf{H}_r(x) = \frac{1}{\eta_1}(-\mathbf{a}_x) \times \mathbf{E}_r(x) = \mathbf{a}_z \frac{10^{-4}}{2\pi} e^{j2\pi x/3} \qquad \text{(A/m)}.$$

ii) Instantaneous expressions:

$$\mathbf{E}_r(x, t) = \mathscr{Re}[\mathbf{E}_r(x)e^{j\omega t}]$$

$$= -\mathbf{a}_y 6 \times 10^{-3} \cos\left(2\pi \times 10^8 t + \frac{2\pi}{3}x\right) \qquad \text{(V/m)},$$

$$\mathbf{H}_r(x, t) = \mathbf{a}_z \frac{10^{-4}}{2\pi} \cos\left(2\pi \times 10^8 t + \frac{2\pi}{3}x\right) \qquad \text{(A/m)}.$$

c) *For the total wave (a standing wave)*:

i) Phasor expressions:

$$\mathbf{E}_1(x) = \mathbf{E}_i(x) + \mathbf{E}_r(x) = -\mathbf{a}_y j12 \times 10^{-3} \sin\left(\frac{2\pi}{3}x\right) \qquad \text{(V/m)},$$

$$\mathbf{H}_1(x) = \mathbf{H}_i(x) + \mathbf{H}_r(x) = \mathbf{a}_z \frac{10^{-4}}{\pi} \cos\left(\frac{2\pi}{3}x\right) \qquad \text{(A/m)}.$$

ii) Instantaneous expressions:

$$\mathbf{E}_1(x, t) = \mathscr{Re}[\mathbf{E}_1(x)e^{j\omega t}]$$

$$= \mathbf{a}_y 12 \times 10^{-3} \sin\left(\frac{2\pi}{3}x\right) \sin(2\pi \times 10^8 t) \qquad \text{(V/m)},$$

$$\mathbf{H}_1(x, t) = \mathbf{a}_z \frac{10^{-4}}{\pi} \cos\left(\frac{2\pi}{3}x\right) \cos(2\pi \times 10^8 t) \qquad \text{(A/m)}.$$

EXERCISE 7.12 For the problem in Example 7-8, find the locations for $|E_1|_{\max}$ and $|H_1|_{\max}$.

ANS. $|E_1|_{\max}$ at $x = -(2n + 1)3/4$ (m), $|H_1|_{\max}$ at $x = -3n/2$ (m), $n = 0, 1, 2, \ldots,$

REVIEW QUESTIONS

Q.7-22 Define *reflection coefficient* and *transmission coefficient*. What is the relationship between them for normal incidence?

Q.7-23 What are the values of the reflection and transmission coefficients at an interface with a perfectly conducting boundary?

Q.7-24 What is a *standing wave*?

Q.7-25 Define *standing-wave ratio*. What is its relationship with reflection coefficient?

Q.7-26 What is the standing-wave ratio of the combined incident and reflected waves from a perfectly conducting boundary at normal incidence?

REMARKS

1. For lossless media, both Γ and τ are real. Γ can be either positive or negative, but τ cannot be negative.
2. For lossy media, both Γ and τ are complex, implying that a phase shift is introduced at the interface upon reflection and transmission.
3. A standing wave is the result of the superposition of an incident wave and a reflected wave.
4. Both Γ and S are dimensionless: $0 \le |\Gamma| \le 1$, and $1 \le S \le \infty$.
5. When $\eta_2 < \eta_1$ $(\Gamma < 0)$, a minimum $|\mathbf{E}_1|$ exists at the interface; when $\eta_2 > \eta_1$ $(\Gamma > 0)$, $|\mathbf{E}_1|$ is a maximum at the interface.

7-7 OBLIQUE INCIDENCE OF PLANE WAVES AT PLANE BOUNDARIES

We now consider the more general case of a uniform plane wave that impinges on a plane boundary obliquely. Refer to Fig. 7-10, where the $z = 0$ plane† is the interface between medium 1 (ϵ_1, μ_1) and medium 2 (ϵ_2, μ_2). The plane containing the normal to the boundary surface and the wavenumber vector $\mathbf{a}_k$ is called the ***plane of incidence***. Three angles are in evidence: the ***angle of incidence*** θ_i, the ***angle of reflection*** θ_r, and the ***angle of refraction*** (or ***angle of transmission***) θ_t representing, respectively, the angles that the incident, reflected, and transmitted waves make with the normal to the boundary. Lines AO, $O'A'$, and $O'B$ are the intersections of the wavefronts (surfaces of constant phase) of the incident, reflected, and transmitted waves respectively, with the plane of incidence. Since both the incident and the reflected waves propagate in medium 1 with the same phase velocity u_{p1}, the distances $\overline{OA'}$ and $\overline{AO'}$ must be equal. Thus,

Plane of incidence

All angles are measured from the normal to the boundary.

$$\overline{OO'} \sin\theta_r = \overline{OO'} \sin\theta_i,$$

or

$$\boxed{\theta_r = \theta_i.} \qquad (7\text{-}116)$$

Equation (7-116) assures us that ***the angle of reflection is equal to the angle of incidence***, which is ***Snell's law of reflection***.

Snell's law of reflection

†There is no loss of generality here, because we can always assign our coordinate system such that the z-axis is perpendicular to the boundary plane.

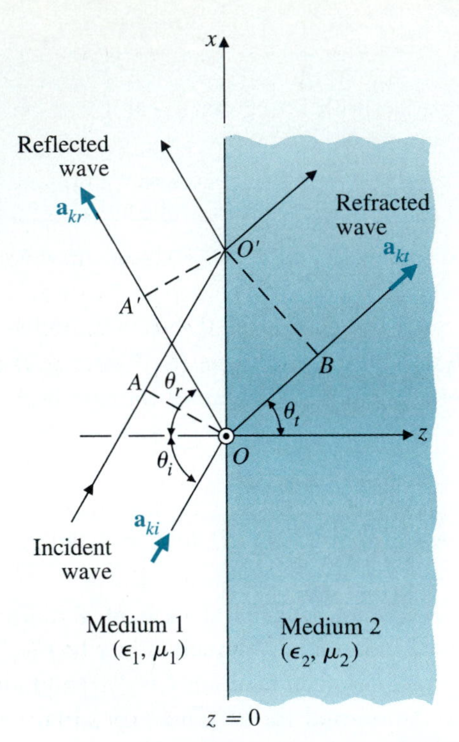

FIGURE 7-10 Uniform plane wave incident obliquely on a plane dielectric boundary.

In medium 2 the time it takes for the transmitted wave to travel from O to B equals the time for the incident wave to travel from A to O'. We have

$$\frac{\overline{OB}}{u_{p2}} = \frac{\overline{AO'}}{u_{p1}},$$

$$\frac{\overline{OB}}{\overline{AO'}} = \frac{\overline{OO'} \sin \theta_t}{\overline{OO'} \sin \theta_i} = \frac{u_{p2}}{u_{p1}},$$

from which we obtain

Snell's law of refraction

$$\boxed{\frac{\sin \theta_t}{\sin \theta_i} = \frac{u_{p2}}{u_{p1}} = \frac{\beta_1}{\beta_2} = \frac{n_1}{n_2},} \tag{7-117}$$

Index of refraction of a medium

where n_1 and n_2 are the indices of refraction for media 1 and 2, respectively. The ***index of refraction*** of a medium is the ratio of the speed of light (electromagnetic wave) in free space to that in the medium; that is, $n = c/u_p$. The relation in Eq. (7-117) is known as ***Snell's law of refraction***.

For media with equal permeability, $\mu_1 = \mu_2$, Eq. (7-117) becomes

Snell's law of refraction for $\mu_1 = \mu_2$

$$\frac{\sin\theta_t}{\sin\theta_i} = \frac{n_1}{n_2} = \frac{\eta_2}{\eta_1} = \sqrt{\frac{\epsilon_1}{\epsilon_2}} = \sqrt{\frac{\epsilon_{r1}}{\epsilon_{r2}}} \qquad (\mu_1 = \mu_2), \tag{7-118}$$

where η_1 and η_2 are the intrinsic impedances of the media.

Note that we have derived here Snell's law of reflection and Snell's law of refraction from a consideration of the ray paths of the incident, reflected, and refracted waves. No mention has been made of the polarization of the waves. Thus Snell's laws are independent of wave polarization.

7-7.1 TOTAL REFLECTION

Let us now examine Snell's law in Eq. (7-118) for $\epsilon_1 > \epsilon_2$—that is, when the wave in medium 1 is incident on a less dense medium 2. In that case, $\theta_t > \theta_i$. Since θ_t increases with θ_i, an interesting situation arises when $\theta_t = \pi/2$, at which angle the refracted wave will glaze along the interface; a further increase in θ_i would result in no refracted wave, and the incident wave is then said to be totally reflected. The angle of incidence θ_c (which corresponds to the threshold of ***total reflection*** $\theta_t = \pi/2$) is called the ***critical angle***. We have, by setting $\theta_t = \pi/2$ in Eq. (7-118),

Phenomenon of total reflection when $\varepsilon_2 < \varepsilon_1$

Critical angle

$$\sin\theta_c = \sqrt{\frac{\epsilon_2}{\epsilon_1}}, \tag{7-119}$$

or

Formula for critical angle

$$\theta_c = \sin^{-1}\sqrt{\frac{\epsilon_2}{\epsilon_1}} = \sin^{-1}\left(\frac{n_2}{n_1}\right) \qquad (\mu_1 = \mu_2). \tag{7-120}$$

This situation is illustrated in Fig. 7-11, where $\mathbf{a}_{ki}$, $\mathbf{a}_{kr}$, and $\mathbf{a}_{kt}$ are unit vectors denoting the directions of propagation of the incident, reflected, and transmitted waves, respectively.

What happens mathematically if θ_i is larger than the critical angle $\theta_c(\sin\theta_i > \sin\theta_c = \sqrt{\epsilon_2/\epsilon_1})$? From Eq. (7-118) we have

$$\sin\theta_t = \sqrt{\frac{\epsilon_1}{\epsilon_2}}\sin\theta_i > 1, \tag{7-121}$$

which does not yield a real solution for θ_t. Although $\sin\theta_t$ in Eq. (7-121) is still real, $\cos\theta_t$ becomes imaginary when $\sin\theta_t > 1$:

$$\cos\theta_t = \sqrt{1-\sin^2\theta_t} = \pm j\sqrt{\frac{\epsilon_1}{\epsilon_2}\sin^2\theta_i - 1}. \tag{7-122}$$

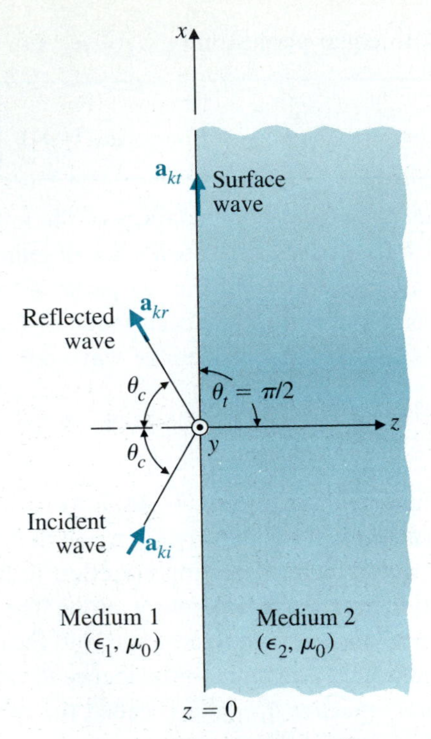

FIGURE 7-11 Plane wave incident at critical angle, $\epsilon_1 > \epsilon_2$.

In medium 2 the unit vector $\mathbf{a}_{kt}$ in the direction of propagation of a typical transmitted (refracted) wave, as shown in Fig. 7-10, is

$$\mathbf{a}_{kt} = \mathbf{a}_x \sin \theta_t + \mathbf{a}_z \cos \theta_t. \tag{7-123}$$

Both $\mathbf{E}_t$ and $\mathbf{H}_t$ vary spatially in accordance with the following factor:

$$e^{-j\beta_2 \mathbf{a}_{kt} \cdot \mathbf{R}} = c^{-j\beta_2(x \sin \theta_t + z \cos \theta_t)}, \tag{7-124}$$

(where $\mathbf{R}$ is a radius vector as in Eq. 7-22.) When Eqs. (7-118) and (7-119) for $\theta_i > \theta_c$ are used, the expression in Eq. (7-124) becomes

$$e^{-\alpha_2 z} e^{-j\beta_{2x} x}, \tag{7-125}$$

where

$$\alpha_2 = \beta_2 \sqrt{(\epsilon_1/\epsilon_2)\sin^2 \theta_i - 1} \tag{7-125a}$$

and

$$\beta_{2x} = \beta_2 \sqrt{\epsilon_1/\epsilon_2} \sin \theta_i. \tag{7-125b}$$

The upper sign in Eq. (7-122) has been abandoned because it would lead to

Evanescent and surface waves

the impossible result of an increasing field as z increases. We can conclude from (7-125) that for $\theta_i > \theta_c$ an ***evanescent wave*** exists *along* the interface (in the x-direction), which is attenuated exponentially (rapidly) in medium 2 in the normal direction (z-direction). This wave is tightly bound to the interface and is called a ***surface wave***. It is illustrated in Fig. 7-11. It is a nonuniform plane wave; no power is transmitted into medium 2 under these conditions. (See Problem P.7-27.)

EXAMPLE 7-9

The permittivity of water at optical frequencies is $1.75\epsilon_0$. It is found that an isotropic light source at a distance d under water yields an illuminated circular area of a radius 5 (m). Determine d.

SOLUTION

The index of refraction of water is $n_w = \sqrt{1.75} = 1.32$. Refer to Fig. 7-12. The radius of illuminated area, $O'P = 5$ (m), corresponds to the critical angle

$$\theta_c = \sin^{-1}\left(\frac{1}{n_w}\right) = \sin^{-1}\left(\frac{1}{1.32}\right) = 49.2^\circ.$$

Hence,

$$d = \frac{\overline{OP}}{\tan\theta_c} = \frac{5}{\tan 49.2^\circ} = 4.32 \qquad \text{(m)}.$$

As illustrated in Fig. 7-12, an incident ray with $\theta_i = \theta_c$ at P results in a reflected ray and a tangential refracted ray. Incident waves for $\theta_i < \theta_c$ are partially reflected back into the water and partially refracted into the air above, and those for $\theta_i > \theta_c$ are totally reflected (the evanescent surface waves are not shown).

FIGURE 7-12 An underwater light source (Example 7-9).

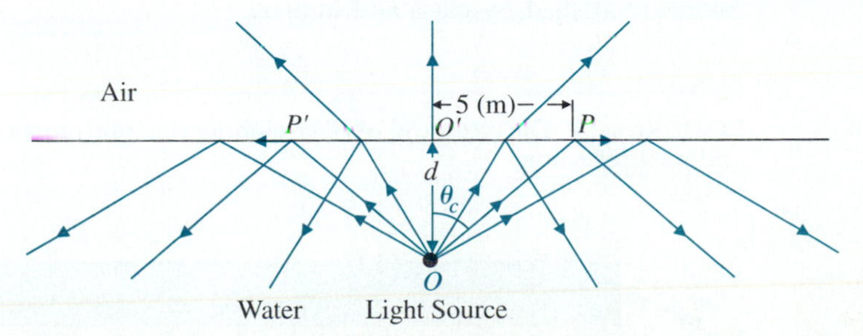

EXAMPLE 7-10

A dielectric rod or fiber of a transparent material can be used to guide light or an electromagnetic wave under the conditions of total internal reflection. Determine the minimum dielectric constant of the guiding medium so that a wave incident on one end at any angle will be confined within the rod until it emerges from the other end.

SOLUTION

Refer to Fig. 7-13. For total internal reflection, θ_1 must be greater than or equal to θ_c for the guiding dielectric medium; that is,

$$\sin \theta_1 \geq \sin \theta_c. \tag{7-126}$$

Since $\theta_1 = \pi/2 - \theta_t$, Eq. (7-126) becomes

$$\cos \theta_t \geq \sin \theta_c. \tag{7-127}$$

From Snell's law of refraction, Eq. (7-118), we have

$$\sin \theta_t = \frac{1}{\sqrt{\epsilon_{r1}}} \sin \theta_i. \tag{7-128}$$

(Note that the roles of ϵ_1 and ϵ_2 in Fig. 7-13 have been interchanged from those in Fig. 7-10.) Substituting Eq. (7-128) in Eq. (7-127) and using Eq. (7-119), we obtain

$$\sqrt{1 - \frac{1}{\epsilon_{r1}} \sin^2 \theta_i} \geq \sin \theta_c = \sqrt{\frac{\epsilon_0}{\epsilon_1}} = \frac{1}{\sqrt{\epsilon_{r1}}},$$

which requires

$$\epsilon_{r1} \geq 1 + \sin^2 \theta_i. \tag{7-129}$$

Since the largest value of the right side of (7-129) is reached when $\theta_i = \pi/2$, we require the dielectric constant of the guiding medium to be at least 2, which corresponds to an index of refraction $n_1 = \sqrt{2}$. This requirement is satisfied by glass and quartz.

Minimum dielectric constant for optical fiber

FIGURE 7-13 Dielectric rod or fiber guiding electromagnetic wave by total internal reflection.

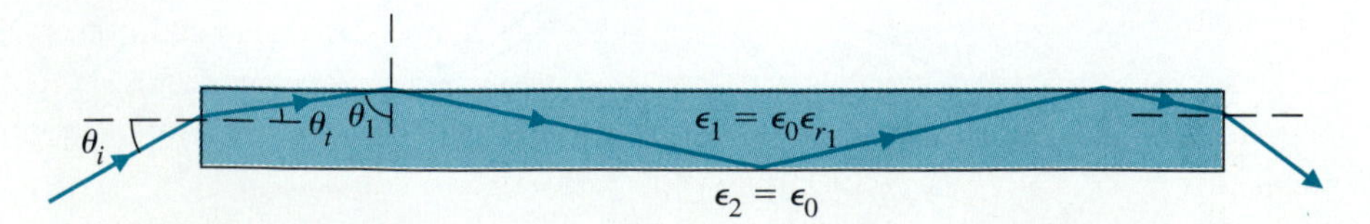

EXERCISE 7.13 A 30-(MHz) uniform plane wave emerges from a lossless dielectric medium ($\epsilon = 2.25\epsilon_0$, $\mu = \mu_0$) into air through a plane boundary at $z = 0$. The angle of incidence is 30°. Find the angle of refraction, and the phase constants both in the dielectric medium and in air.

ANS. 48.6°, 0.94 (rad/m), 0.63 (rad/m).

EXERCISE 7.14 Find the critical angle in Exercise 7.13. Determine the attenuation and phase constants in air if the incident angle is 60°.

ANS. 41.8°, 0.52 (Np/m), 0.82 (rad/m).

REVIEW QUESTIONS

Q.7-27 Define *plane of incidence*.

Q.7-28 State *Snell's law of reflection* in words.

Q.7-29 State *Snell's law of refraction* in terms of the indices of refraction of the media, and in terms of the intrinsic impedances of two contiguous nonmagnetic media.

Q.7-30 Define *critical angle*. What is meant by *total reflection*?

Q.7-31 Define *surface wave*.

REMARKS

1. Snell's laws are independent of wave polarization.
2. Snell's laws are independent of wave frequency if the constitutive parameters of the media are frequency-independent.
3. All angles in Snell's laws are measured from the normal to the interface.
4. Total reflection is possible only when $\epsilon_2 < \epsilon_1$.
5. No power is transmitted across the interface when $\theta_i > \theta_c$.

7-7.2 THE IONOSPHERE

Composition of ionosphere

In the earth's upper atmosphere, roughly from 50 to 500 (km) in altitude, there exist layers of ionized gases called the ***ionosphere***. The ionosphere consists of free electrons and positive ions that are produced when ultraviolet radiation from the sun is absorbed by the atoms and molecules in the upper atmosphere. The charged particles tend to be trapped by the earth's magnetic field. The altitude and character of the ionized layers depend both on the nature of the solar radiation and on the composition of the atmosphere. They change with the sunspot cycle, the season, and the hour of the day in a very complicated way. The electron and ion densities in the individual ionized layers are essentially equal. Ionized gases with equal electron and ion densities are called ***plasmas***.

Plasmas

The ionosphere plays an important role in the propagation of electromagnetic waves and affects telecommunication. Because the electrons are much lighter than the positive ions, they are accelerated more by the electric fields of electromagnetic waves passing through the ionosphere. Analysis has shown that the effect of the ionosphere or plasma on wave propagation can be studied on the basis of an effective permittivity ϵ_p:

Effective permittivity of plasma

$$\epsilon_p = \epsilon_0\left(1 - \frac{\omega_p^2}{\omega^2}\right) = \epsilon_0\left(1 - \frac{f_p^2}{f^2}\right) \quad \text{(F/m)}, \tag{7-130}$$

where ω_p is called the ***plasma angular frequency*** and

$$\omega_p = 2\pi f_p = \sqrt{\frac{Ne^2}{m\epsilon_0}}. \tag{7-131}$$

In Eq. (7-131) N is the number of electrons per unit volume, and e and m are, respectively, the electronic charge and mass.

From Eqs. (7-42) and (7-130) we obtain the propagation constant as

$$\gamma = \alpha + j\beta = j\omega\sqrt{\mu\epsilon_0}\sqrt{1 - \left(\frac{f_p}{f}\right)^2}. \tag{7-132}$$

When $f < f_p$, γ becomes real, indicating an attenuation without propagation. On the other hand, if $f > f_p$, γ is purely imaginary, and electromagnetic waves will propagate unattenuated in the ionosphere (assuming negligible collision losses).

If the value of e, m, and ϵ_0 are substituted into Eq. (7-131), we find a very simple formula for the plasma (cutoff) frequency:

Formula for plasma frequency

$$f_p \cong 9\sqrt{N} \quad \text{(Hz)}. \tag{7-133}$$

As we have mentioned before, N at a given altitude is not a constant; it varies with the time of the day, the season, and other factors. The electron density of the ionosphere ranges from about $10^{10}/\text{m}^3$ in the lowest layer to $10^{12}/\text{m}^3$ in the highest layer. Using these values for N in Eq. (7-133), we find f_p to vary from 0.9 to 9 (MHz). Hence, for communication with a satellite or a space station beyond the ionosphere we must use frequencies much higher than 9 (MHz) to ensure wave penetration through the layer with the largest N at any angle of incidence. Signals with frequencies lower than 0.9 (MHz) cannot penetrate into even the lowest layer of the ionosphere but may propagate very far around the earth by way of multiple reflections at the ionosphere's boundary and the earth's surface. Signals having frequencies between 0.9 and 9 (MHz) will penetrate partially into the lower ionospheric layers but will eventually be turned back where N is large.

EXAMPLE 7-11

When a spacecraft reenters the earth's atmosphere, its speed and temperature ionize the surrounding atoms and molecules and create a plasma. It has been estimated that the electron density is in the neighborhood of 2×10^8 per (cm^3). Discuss the plasma's effect on frequency usage in radio communication between the spacecraft and the mission controllers on earth.

SOLUTION

For

$$N = 2 \times 10^8 \text{ per } (\text{cm}^3)$$
$$= 2 \times 10^{14} \text{ per } (\text{m}^3),$$

Eq. (7-133) gives $f_p = 9 \times \sqrt{2 \times 10^{14}} = 12.7 \times 10^7$ (Hz), or 127 (MHz). Thus, radio communication cannot be established for frequencies below 127 (MHz).

REVIEW QUESTIONS

Q.7-32 What is the composition of the *ionosphere*?

Q.7-33 What is the significance of plasma frequency?

We have remarked previously that Snell's laws and consequently the critical angle for total reflection are independent of the polarization of the incident electric field. However, the formulas for the reflection and transmission coefficients are polarization-dependent. In the following two subsections we discuss the behaviors of perpendicular polarization and parallel polarization separately.

7-7.3 PERPENDICULAR POLARIZATION

Meaning of perpendicular polarization

For oblique incidence with perpendicular polarization, $\mathbf{E}_i$ is perpendicular to the plane of incidence, as illustrated in Fig. 7-14. Noting that

$$\mathbf{a}_{ki} = \mathbf{a}_x \sin\theta_i + \mathbf{a}_z \cos\theta_i, \tag{7-134}$$

we have, from Eqs. (7-23) and (7-25),

$$\mathbf{E}_i(x, z) = \mathbf{a}_y E_{i0} e^{-j\beta_1(x\sin\theta_i + z\cos\theta_i)} \tag{7-135}$$

$$\mathbf{H}_i(x, z) = \frac{E_{i0}}{\eta_1}(-\mathbf{a}_x \cos\theta_i + \mathbf{a}_z \sin\theta_i) e^{-j\beta_1(x\sin\theta_i + z\cos\theta_i)}, \tag{7-136}$$

where β_1 has been used in place of k_1 in a lossless medium.

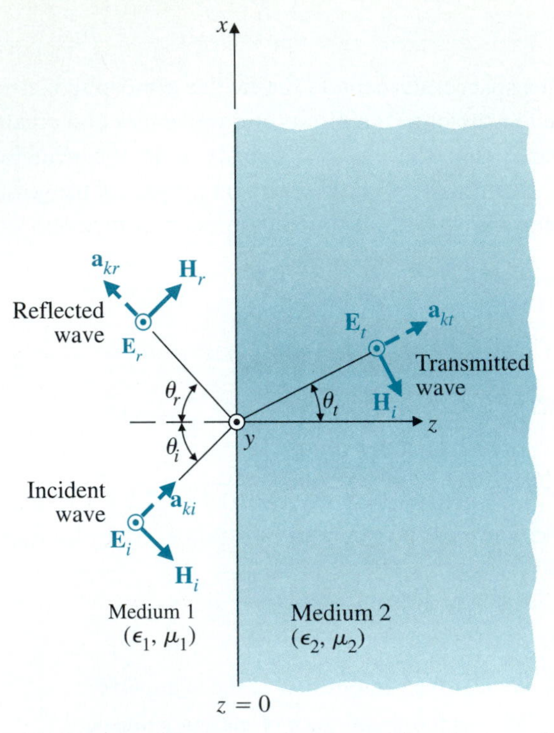

FIGURE 7-14 Plane wave incident obliquely on a plane dielectric boundary (perpendicular polarization).

For the reflected wave,

$$\mathbf{a}_{kr} = \mathbf{a}_x \sin\theta_r - \mathbf{a}_z \cos\theta_r. \tag{7-137}$$

The reflected electric and magnetic fields are

$$\mathbf{E}_r(x, z) = \mathbf{a}_y E_{r0} e^{-j\beta_1(x\sin\theta_r - z\cos\theta_r)} \tag{7-138}$$

$$\mathbf{H}_r(x, z) = \frac{E_{r0}}{\eta_1}(\mathbf{a}_x \cos\theta_r + \mathbf{a}_z \sin\theta_r) e^{-j\beta_1(x\sin\theta_r - z\cos\theta_r)}. \tag{7-139}$$

For the transmitted wave,

$$\mathbf{a}_{kt} = \mathbf{a}_x \sin\theta_t + \mathbf{a}_z \cos\theta_t, \tag{7-140}$$

we have

$$\mathbf{E}_t(x, z) = \mathbf{a}_y E_{t0} e^{-j\beta_2(x\sin\theta_t + z\cos\theta_t)} \tag{7-141}$$

$$\mathbf{H}_t(x, z) = \frac{E_{t0}}{\eta_2}(-\mathbf{a}_x \cos\theta_t + \mathbf{a}_z \sin\theta_t) e^{-j\beta_2(x\sin\theta_t + z\cos\theta_t)}. \tag{7-142}$$

There are four unknown quantities in the above equations, namely, E_{r0}, E_{t0}, θ_r, and θ_t. Their determination follows from the requirements that the tangential components of **E** and **H** be continuous at the boundary $z = 0$.

From $E_{iy}(x,0)+E_{ry}(x,0)=E_{ty}(x,0)$ we have

$$E_{i0}e^{-j\beta_1 x\sin\theta_i}+E_{r0}e^{-j\beta_1 x\sin\theta_r}=E_{t0}e^{-j\beta_2 x\sin\theta_t}. \tag{7-143}$$

Similarly, from $H_{ix}(x,0)+H_{rx}(x,0)=H_{tx}(x,0)$ we require

$$\frac{1}{\eta_1}(-E_{i0}\cos\theta_i e^{-j\beta_1 x\sin\theta_i}+E_{r0}\cos\theta_r e^{-j\beta_1 x\sin\theta_r}) = -\frac{E_{t0}}{\eta_2}\cos\theta_t e^{-j\beta_2 x\sin\theta_t}. \tag{7-144}$$

Because Eqs. (7-143) and (7-144) are to be satisfied *for all* x, all three exponential factors that are functions of x must be equal ("phase-matching"). Thus,

$$\beta_1 x\sin\theta_i=\beta_1 x\sin\theta_r=\beta_2 x\sin\theta_t,$$

which leads to Snell's law of reflection ($\theta_r=\theta_i$) and Snell's law of refraction ($\sin\theta_t/\sin\theta_i=\beta_1/\beta_2=n_1/n_2$). Equations (7-143) and (7-144) can now be written simply as

$$E_{i0}+E_{r0}=E_{t0} \tag{7-145}$$

and

$$\frac{1}{\eta_1}(E_{i0}-E_{r0})\cos\theta_i=\frac{E_{t0}}{\eta_2}\cos\theta_t, \tag{7-146}$$

from which E_{r0} and E_{t0} can be found in terms of E_{i0}. The reflection and transmission coefficients are

Reflection coefficient for perpendicular polarization

$$\boxed{\Gamma_\perp=\frac{E_{r0}}{E_{i0}}=\frac{\eta_2\cos\theta_i-\eta_1\cos\theta_t}{\eta_2\cos\theta_i+\eta_1\cos\theta_t}} \tag{7-147}$$ [†]

and

Transmission coefficient for perpendicular polarization

$$\boxed{\tau_\perp=\frac{E_{t0}}{E_{i0}}=\frac{2\eta_2\cos\theta_i}{\eta_2\cos\theta_i+\eta_1\cos\theta_t}}. \tag{7-148}$$ [†]

When $\theta_i=0$, making $\theta_r=\theta_t=0$, these expressions reduce to those for normal incidence—Eqs. (7-94) and (7-95)—as they should. Furthermore, $\Gamma_\perp$ and $\tau_\perp$ are related in the following way:

Relation between reflection and transmission coefficients for perpendicular polarization

$$\boxed{1+\Gamma_\perp=\tau_\perp,} \tag{7-149}$$

which is similar to Eq. (7-96) for normal incidence.

[†]These are sometimes referred to as ***Fresnel's equations***.

If medium 2 is a perfect conductor, $\eta_2 = 0$. We have $\Gamma_\perp = -1\,(E_{r0} = -E_{i0})$ and $\tau_\perp = 0\,(E_{t0} = 0)$. The tangential **E** field on the surface of the conductor vanishes, and no energy is transmitted across a perfectly conducting boundary.

EXAMPLE 7-12

The instantaneous expression for the electric field of a uniform plane wave in air is

$$\mathbf{E}_i(x, z; t) = \mathbf{a}_y 10 \cos(\omega t + 3x - 4z) \quad \text{(V/m)}.$$

The wave is incident on a perfectly conducting plane boundary at $z = 0$.

a) Find phase constant β_1, angular frequency ω, and angle of incidence, θ_i.

b) Find $\mathbf{E}_r(x, z)$.

c) Discuss the behavior of $\mathbf{E}_1(x, z; t)$.

SOLUTION

a) The phasor expression for $\mathbf{E}_i$ is

$$\mathbf{E}_i(x, z) = \mathbf{a}_y 10 e^{j3x - j4z},$$

which represents a perpendicularly polarized wave propagating in the $-x$- and $+z$-directions. In view of Eqs. (7-20) and (7-21) we have

$$\mathbf{k}_i = \mathbf{a}_{ki} k_i = -\mathbf{a}_x(\beta_1 \sin\theta_i) + \mathbf{a}_z(\beta_1 \cos\theta_i),$$

where we have noted that in the lossless medium 1, $k_i = \beta_1$. Thus,

$$\beta_1 \sin\theta_i = 3, \text{ and}$$

$$\beta_1 \cos\theta_i = 4,$$

from which we obtain

$$\beta_1 = \sqrt{3^2 + 4^2} = 5 \qquad \text{(rad/m), and}$$

$$\theta_i = \tan^{-1}\frac{3}{4} = 36.9^\circ.$$

Also, $\beta_1 = \omega/c$, and $\omega = \beta_1 c = 5 \times (3 \times 10^8) = 1.5 \times 10^9$ (rad/s).

b) For a perfectly conducting interface, $\Gamma_\perp = -1$, $E_{r0} = -E_{i0} = -10$, and

$$\mathbf{E}_r(x, z) = -\mathbf{a}_y 10 e^{j3x + j4z},$$

which propagates in the $-x$- and $-z$-directions.

c)
$$\begin{aligned} \mathbf{E}_1(x, z) &= \mathbf{E}_i(x, z) + \mathbf{E}_r(x, z) \\ &= \mathbf{a}_y 10(e^{-j4z} - e^{j4z})e^{j3x} \\ &= -\mathbf{a}_y 20j(\sin 4z)e^{j3x}, \end{aligned}$$

which corresponds to an instantaneous expression

$$\mathbf{E}_1(x, z; t) = \mathbf{a}_y 20(\sin 4z) \cos(1.5 \times 10^9 t + 3x - \pi/2).$$

Thus, $\mathbf{E}_1(x, z; t)$ is composed of a standing wave in the $-z$-direction and a traveling wave in the $-x$-direction. The standing wave has a value zero at $4z = n\pi$, or $z = n\pi/4$ $(n = 0, 1, 2, \ldots)$. The traveling wave is a nonuniform plane wave since its amplitude is not constant in the z-direction.

7-7.4 PARALLEL POLARIZATION

Meaning of parallel polarization

When a uniform plane wave with parallel polarization is incident obliquely on a plane boundary, $\mathbf{E}_i$ lies in the plane of incidence ($\mathbf{H}_i$ is perpendicular to the plane of incidence), as illustrated in Fig. 7-15. The incident and reflected electric and magnetic field intensity phasors in medium 1 are:

$$\mathbf{E}_i(x, z) = E_{i0}(\mathbf{a}_x \cos\theta_i - \mathbf{a}_z \sin\theta_i)e^{-j\beta_1(x \sin\theta_i + z \cos\theta_i)}, \tag{7-150}$$

FIGURE 7-15 Plane wave incident obliquely on a plane dielectric boundary (parallel polarization).

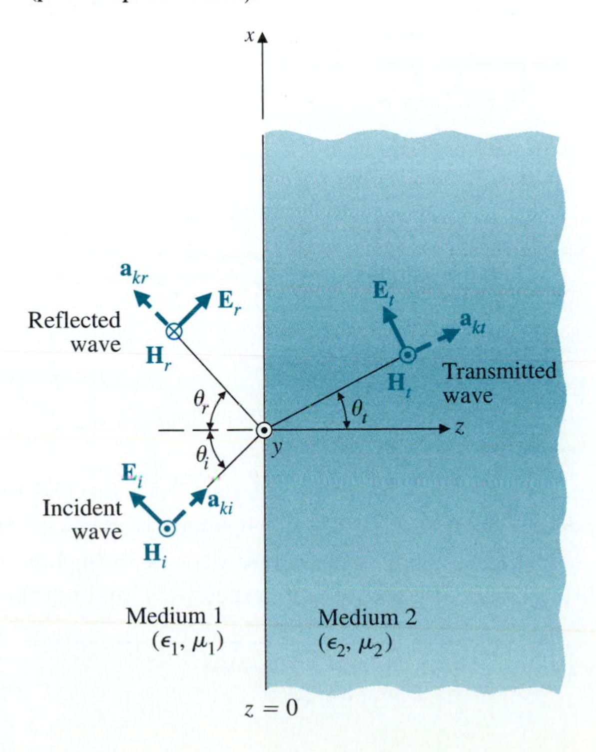

$$\mathbf{H}_i(x, z) = \mathbf{a}_y \frac{E_{i0}}{\eta_1} e^{-j\beta_1(x \sin \theta_i + z \cos \theta_i)}, \tag{7-151}$$

$$\mathbf{E}_r(x, z) = E_{r0}(\mathbf{a}_x \cos \theta_r + \mathbf{a}_z \sin \theta_r) e^{-j\beta_1(x \sin \theta_r - z \cos \theta_r)}, \tag{7-152}$$

$$\mathbf{H}_r(x, z) = -\mathbf{a}_y \frac{E_{r0}}{\eta_1} e^{-j\beta_1(x \sin \theta_r - z \cos \theta_r)}. \tag{7-153}$$

The transmitted electric and magnetic field intensity phasors in medium 2 are

$$\mathbf{E}_t(x, z) = E_{t0}(\mathbf{a}_x \cos \theta_t - \mathbf{a}_z \sin \theta_t) e^{-j\beta_2(x \sin \theta_t + z \cos \theta_t)}, \tag{7-154}$$

$$\mathbf{H}_t(x, z) = \mathbf{a}_y \frac{E_{t0}}{\eta_2} e^{-j\beta_2(x \sin \theta_t + z \cos \theta_t)}. \tag{7-155}$$

Continuity requirements for the tangential components of **E** and **H** at $z = 0$ lead again to Snell's laws of reflection and refraction, as well as to the following two equations:

$$(E_{i0} + E_{r0}) \cos \theta_i = E_{t0} \cos \theta_t, \tag{7-156}$$

$$\frac{1}{\eta_1}(E_{i0} - E_{r0}) = \frac{1}{\eta_2} E_{t0}. \tag{7-157}$$

Solving for E_{r0} and E_{t0} in terms of E_{i0}, we obtain

Reflection coefficient for parallel polarization

$$\Gamma_{\parallel} = \frac{E_{r0}}{E_{i0}} = \frac{\eta_2 \cos \theta_t - \eta_1 \cos \theta_i}{\eta_2 \cos \theta_t + \eta_1 \cos \theta_i} \tag{7-158}^{\dagger}$$

and

Transmission coefficient for parallel polarization

$$\tau_{\parallel} = \frac{E_{t0}}{E_{i0}} = \frac{2\eta_2 \cos \theta_i}{\eta_2 \cos \theta_t + \eta_1 \cos \theta_i}. \tag{7-159}^{\dagger}$$

It is easy to verify that

Reflection and transmission relation between coefficients for parallel polarization

$$1 + \Gamma_{\parallel} = \tau_{\parallel} \left(\frac{\cos \theta_t}{\cos \theta_i} \right). \tag{7-160}$$

Equation (7-160) is seen to be different from Eq. (7-149) for perpendicular polarization except when $\theta_i = \theta_t = 0$, which is the case for normal incidence.

If medium 2 is a perfect conductor ($\eta_2 = 0$), Eqs. (7-158) and (7-159) simplify to $\Gamma_{\parallel} = -1$ and $\tau_{\parallel} = 0$, respectively, making the tangential compo-

† These are also referred to as ***Fresnel's equations***.

nent of the total **E** field on the surface of the conductor vanish, as expected. We note here that the choices of the reference directions of $\mathbf{E}_r$ and $\mathbf{H}_r$ in Figs. 7-14 and 7-15 are all arbitrary. The actual directions of $\mathbf{E}_r$ and $\mathbf{H}_r$ may or may not be the same as those shown, depending on whether $\Gamma_\perp$ in Eq. (7-147) and $\Gamma_\parallel$ in Eq. (7-158) is positive or negative, respectively.

If we plot $|\Gamma_\perp|^2$ and $|\Gamma_\parallel|^2$ versus θ_i, we will find the former always greater than the latter except for $\theta_i = 0$, where they are equal. This means that when an unpolarized wave strikes a plane dielectric interface, the reflected wave will contain more power in the component with perpendicular polarization than that with parallel polarization. A popular application of this fact is the design of Polaroid sunglasses to reduce sun glare. Much of the sunlight received by the eye has been reflected from horizontal surfaces on earth. Because $|\Gamma_\perp|^2 > |\Gamma_\parallel|^2$, the light reaching the eye is predominantly perpendicular to the plane of reflection (same as the plane of incidence), and hence the electric field is parallel to the earth's surface. Polaroid sunglasses are designed to filter out this component.

Principle of Polaroid sunglasses for reducing sun glare

7-7.5 BREWSTER ANGLE OF NO REFLECTION

We note from the expression for reflection coefficient in Eq. (7-158) that the numerator is the difference of two terms. This leads us to inquire whether there is a combination of η_1, η_2, and θ_i that makes $\Gamma_\parallel = 0$ for no reflection. Denoting this particular θ_i by $\theta_{B\parallel}$, we require

$$\eta_2 \cos \theta_t = \eta_1 \cos \theta_{B\parallel}. \tag{7-161}$$

Squaring both sides of Eq. (7-161) and using Eq. (7-117), we obtain

$$\sin^2 \theta_{B\parallel} = \frac{1 - (\eta_2/\eta_1)^2}{1 - (\eta_2\beta_1/\eta_1\beta_2)^2},$$

or

$$\boxed{\sin^2 \theta_{B\parallel} = \frac{1 - \mu_2\epsilon_1/\mu_1\epsilon_2}{1 - (\epsilon_1/\epsilon_2)^2}.} \tag{7-162}$$

The angle $\theta_{B\parallel}$ is known as the ***Brewster angle*** of no reflection for the case of parallel polarization.

Brewster angle of no reflection

$$\boxed{\sin \theta_{B\parallel} = \frac{1}{\sqrt{1 + (\epsilon_1/\epsilon_2)}}, \qquad (\mu_1 = \mu_2).} \tag{7-163}$$

Formula for Brewster angle

An alternative form for Eq. (7-163) is

Alternative formula for Brewster angle

$$\theta_{B\parallel} = \tan^{-1}\sqrt{\frac{\epsilon_2}{\epsilon_1}} = \tan^{-1}\left(\frac{n_2}{n_1}\right), \qquad (\mu_1 = \mu_2). \tag{7-164}$$

EXERCISE 7.15 Verify that Eqs. (7-163) and (7-164) are equivalent.

At this point the reader may wonder why we did not examine the Brewster angle for no reflection for perpendicular polarization. Mathematically we could find a formula for $\theta_{B\perp}$, the angle of incidence θ_i that would make $\Gamma_\perp$ vanish. Setting the numerator of Eq. (7-147) to zero, the condition

$$\eta_2 \cos\theta_{B\perp} = \eta_1 \cos\theta_t, \tag{7-165}$$

in conjunction with Snell's law of reflection, Eq. (7-117), would yield

$$\sin^2\theta_{B\perp} = \frac{1-\mu_1\epsilon_2/\mu_2\epsilon_1}{1-(\mu_1/\mu_2)^2}. \tag{7-166}$$

It is now clear that $\theta_{B\perp}$ does not exist if $\mu_1 = \mu_2$, as usually is the case for wave media.

Because of the difference in the formulas for Brewster angles for perpendicular and parallel polarizations, it is possible to separate these two types of polarization in an unpolarized wave. When an unpolarized wave such as random light is incident upon a boundary at the Brewster angle $\theta_{B\parallel}$ given by Eq. (7-164), only the component with perpendicular polarization will be reflected. Thus a Brewster angle is also referred to as a ***polarizing angle***. Based on this principle, quartz windows set at the Brewster angle at the ends of a laser tube are used to control the polarization of an emitted light beam.

EXAMPLE 7-13

An electromagnetic wave impinges from air on the surface of water, which has a dielectric constant of 80.

a) Determine the Brewster angle for parallel polarization, $\theta_{B\parallel}$, and the corresponding angle of transmission.

b) If the wave has perpendicular polarization and is incident from air on the water surface at $\theta_i = \theta_{B\parallel}$, find the reflection and transmission coefficients.

SOLUTION

a) The Brewster angle of no reflection for parallel polarization can be obtained directly from Eq. (7-163):

$$\theta_{B\parallel} = \sin^{-1}\frac{1}{\sqrt{1+(1/\epsilon_{r2})}}$$

$$= \sin^{-1}\frac{1}{\sqrt{1+(1/80)}} = 81.0^\circ.$$

The corresponding angle of transmission is, from Eq. (7-118),

$$\theta_t = \sin^{-1}\left(\frac{\sin\theta_{B\parallel}}{\sqrt{\epsilon_{r2}}}\right) = \sin^{-1}\left(\frac{1}{\sqrt{\epsilon_{r2}+1}}\right)$$

$$= \sin^{-1}\left(\frac{1}{\sqrt{81}}\right) = 6.38^\circ.$$

b) For an incident wave with perpendicular polarization, we use Eqs. (7-147) and (7-148) to find $\Gamma_\perp$ and $\tau_\perp$ at $\theta_i = 81.0^\circ$ and $\theta_t = 6.38^\circ$ ($\eta_1 = 377\,\Omega$, $\eta_2 = 377/\sqrt{\epsilon_{r2}} = 40.1\,\Omega$):

$$\Gamma_1 = \frac{40.1\cos 81^\circ - 377\cos 6.38^\circ}{40.1\cos 81^\circ + 377\cos 6.38^\circ} = -0.967,$$

$$\tau_\perp = \frac{2 \times 377 \times \cos 81^\circ}{40.1\cos 81^\circ + 377\cos 6.38^\circ} = 0.033.$$

We note that the relation between $\Gamma_\perp$ and $\tau_\perp$ given in Eq. (7-149) is satisfied.

REVIEW QUESTIONS

Q.7-34 Define perpendicular polarization and parallel polarization for oblique incidence of plane waves at a plane boundary.

Q.7-35 Under what conditions will the reflection and transmission coefficients for perpendicular polarization be the same as those for parallel polarization?

Q.7-36 Define *Brewster angle*. When does it exist at an interface of two nonmagnetic media?

Q.7-37 Why is a Brewster angle also called a *polarizing angle*?

REMARKS

1. The reflection and transmission coefficients at an interface depend on the polarization of the incident wave as well as on the constitutive parameters of the media and the angle of incidence.
2. Unlike the critical angle of total reflection, which exists only when $\epsilon_2 < \epsilon_1$, a Brewster angle of no reflection for parallel polarization always exists for two nonmagnetic media whether $\epsilon_2 < \epsilon_1$, or $\epsilon_2 < \epsilon_1$.

SUMMARY

At very large distances from a finite source radiating electromagnetic waves, a small portion of the wavefront is very nearly a plane. Thus the study of uniform plane waves is of particular importance. In this chapter we

- examined the behavior of uniform plane waves in both lossless and lossy media,
- explained Doppler effect when there is relative motion between a time-harmonic source and a receiver,
- discussed the polarization of plane waves and showed the relation between linearly polarized and circularly polarized waves,
- explained the significant of a complex wavenumber and a complex propagation constant in a lossy medium,
- studied the skin effect in conductors and obtained the formula for skin depth,
- introduced the concept of signal dispersion and explained the difference between phase and group velocities,
- discussed the flow of electromagnetic power and Poynting's theorem,
- studied the reflection and refraction of electromagnetic waves at plane boundaries for both normal incidence and oblique incidence,
- derived Snell's laws of reflection and refraction,
- explained the effect of ionosphere on wave propagation, and
- examined the conditions for total reflection and for no reflection.

PROBLEMS

P.7-1 (a) Obtain the wave equations governing the **E** and **H** fields in a source-free conducting medium with constitutive parameters ϵ, μ, and σ. (b) Obtain the corresponding Helmholtz's equations for time-harmonic fields.

P.7-2 A 1-(GHz) Doppler radar on the ground is used to determine the location and speed of an approaching airplane. Assuming that the signal reflected from the airplane at an elevation angle of 15.5° showed a time delay of 0.3 (ms) and a frequency shift of 2.64 (kHz), find the distance, height, and speed of the airplane.

P.7-3 Obtain a general formula that expresses the phasor $\mathbf{E}(\mathbf{R})$ in terms of the phasor $\mathbf{H}(\mathbf{R})$ of a TEM wave and the intrinsic impedance of the medium, where $\mathbf{R}$ is the radius vector.

P.7-4 The instantaneous expression for the magnetic field intensity of a uniform plane wave propagating in the $+y$ direction in air is given by

$$\mathbf{H} = \mathbf{a}_z 4 \times 10^{-6} \cos\left(10^7\pi t - k_0 y + \frac{\pi}{4}\right) \qquad \text{(A/m)}.$$

a) Determine k_0 and the location where H_z vanishes at $t = 3$ (ms).

b) Write the instantaneous expression for $\mathbf{E}$.

P.7-5 The **E**-field of a uniform plane wave propagating in a dielectric medium is given by

$$E(t, z) = \mathbf{a}_x 2 \cos(10^8 t - z/\sqrt{3}) - \mathbf{a}_y \sin(10^8 t - z/\sqrt{3}) \qquad \text{(V/m)}.$$

a) Determine the frequency and wavelength of the wave.

b) What is the dielectric constant of the medium?

c) Describe the polarization of the wave.

d) Find the corresponding **H**-field.

P.7-6 Show that a plane wave with an instantaneous expression for the electric field

$$\mathbf{E}(z, t) = \mathbf{a}_x E_{10} \sin(\omega t - kz) + \mathbf{a}_y E_{20} \sin(\omega t - kz + \psi)$$

is elliptically polarized.

P.7-7 A 3-(GHz), y-polarized uniform plane wave propagates in the $+x$-direction in a nonmagnetic medium having a dielectric constant 2.5 and a loss tangent 0.05.

a) Determine the distance over which the amplitude of the propagating wave will be cut in half.

b) Determine the intrinsic impedance, the wavelength, the phase velocity, and the group velocity of the wave in the medium.

c) Assuming $\mathbf{E} = \mathbf{a}_y 50 \sin(6\pi 10^9 t + \pi/3)$(V/m) at $x = 0$, write the instantaneous expression for **H** for all t and x.

P.7-8 Determine and compare the intrinsic impedance, attenuation constant (in both Np/m and dB/m), and skin depth of copper $[\sigma_{cu} = 5.80 \times 10^7 \text{ (S/m)}]$, and brass $[\sigma_{br} = 1.59 \times 10^7 \text{ (S/m)}]$ at the following frequencies: (a) 1 (MHz), and (b) 1 (GHz).

P.7-9 Given that the skin depth for graphite at 100 (MHz) is 0.16 (mm), determine (a) the conductivity of graphite, and (b) the distance that a 1-(GHz) wave travels in graphite such that its field intensity is reduced by 30 (dB).

P.7-10 There is a continuing discussion on radiation hazards to human health. The following calculations provide a rough comparison.

a) The U.S. standard for personal safety in a microwave environment is that the power density be less than 10 (mW/cm^2). Calculate the corresponding standard in terms of electric field intensity. In terms of magnetic field intensity.

b) It is estimated that the earth receives radiant energy from the sun at a rate of about 1.3 (kW/m^2) on a sunny day. Assuming a monochromatic plane wave (which it is not), calculate the equivalent amplitudes of the electric and magnetic field intensity vectors.

P.7-11 Show that the instantaneous Poynting vector of a circularly polarized plane wave propagating in a lossless medium is a constant that is independent of time and distance.

P.7-12 Assuming that the radiation electric field intensity of an antenna system is

$$\mathbf{E} = \mathbf{a}_\theta E_\theta + \mathbf{a}_\phi E_\phi,$$

find the expression for the average outward power flow per unit area.

P.7-13 From the point of view of electromagnetics, the power transmitted by a lossless coaxial cable can be considered in terms of the Poynting vector inside the dielectric medium between the inner conductor and the outer sheath. Assuming that a d-c voltage V_0 applied between the inner conductor (of radius a) and the outer sheath (of inner radius b) causes a current I to flow to a load resistance, verify that the integration of the Poynting vector over the cross-sectional area of the dielectric medium equals the power $V_0 I$ that is transmitted to the load.

P.7-14 A uniform plane wave in air with $\mathbf{E}_i(x, t) = \mathbf{a}_y 50 \sin(10^8 t - \beta x)$ (V/m) is incident normally on a lossless medium ($\epsilon_r = 2$, $\mu_r = 8$, $\sigma = 0$) in the region $x \geqslant 0$. Find

a) $\mathbf{E}_r$ and $\mathbf{H}_r$,

b) Γ, τ, and S, and

c) $\mathbf{E}_t$ and $\mathbf{H}_t$.

P.7-15 A uniform plane wave propagates in the $+z$-(downward) direction into the ocean ($\epsilon_r = 72$, $\mu_r = 1$, $\sigma = 4$ S/m). The magnetic field at the ocean surface ($z = 0$) is $\mathbf{H}(0, t) = \mathbf{a}_y\, 0.3 \cos 10^8 t$ (A/m).

a) Determine the skin depth and the intrinsic impedance of the ocean water.

b) Find the expressions of $\mathbf{E}(z, t)$ and $\mathbf{H}(z, t)$ in the ocean.

c) Find the average power loss per unit area in the ocean as a function of z.

P.7-16 For uniform plane waves in medium 1 incident normally on an plane interface with medium 2, obtain the ratios:

a) H_{r0}/H_{i0} and compare with the reflection coefficient in Eq. (7-94), and

b) H_{t0}/H_{i0} and compare with the transmission coefficient in Eq. (7-95).

P.7-17 A right-hand circularly polarized plane wave represented by the phasor

$$\mathbf{E}(z) = E_0(\mathbf{a}_x - j\mathbf{a}_y)e^{-j\beta z}$$

impinges normally on a perfectly conducting wall at $z = 0$.

a) Determine the polarization of the reflected wave.

b) Find the induced current on the conducting wall.

c) Obtain the instantaneous expression of the total electric intensity based on a cosine time reference.

P.7-18 Determine the condition under which the magnitude of the reflection coefficient equals that of the transmission coefficient for a uniform plane wave at normal incidence on an interface between two lossless dielectric media. What is the standing-wave ratio in dB under this condition?

P.7-19 A uniform plane wave in air with $\mathbf{E}_i(z) = \mathbf{a}_x 10e^{-j6z}$ (V/m) is incident normally on an interface at $z = 0$ with a lossy medium having a dielectric constant 2.25 and a loss tangent 0.3. Find the following:

a) The phasor expressions for $\mathbf{E}_r(z)$, $\mathbf{H}_r(z)$, $\mathbf{E}_t(z)$, and $\mathbf{H}_t(z)$.

b) The standing-wave ratio for the wave in air.

c) The expressions for time-average Poynting vectors in air and in the lossy medium.

P.7-20 A uniform sinusoidal plane wave in air with the following phasor expression for electric intensity

$$\mathbf{E}_i(x, z) = \mathbf{a}_y 10e^{-j(6x+8z)} \qquad \text{(V/m)}$$

is incident on a perfectly conducting plane at $z = 0$.

a) Find the frequency and wavelength of the wave.

b) Write the instantaneous expressions for $\mathbf{E}_i(x, z; t)$ and $\mathbf{H}_i(x, z; t)$.

c) Determine the angle of incidence.

d) Find $\mathbf{E}_r(x, z)$ and $\mathbf{H}_r(x, z)$ of the reflected wave.

e) Find $\mathbf{E}_1(x, z)$ and $\mathbf{H}_1(x, z)$ of the total field in air.

P.7-21 For the case of oblique incidence of a uniform plane wave with perpendicular polarization on a plane boundary with $\epsilon_1 = \epsilon_0$, $\epsilon_2 = 2.25\epsilon_0$, $\mu_1 = \mu_2 = \mu_0$, as shown in Fig. 7-14, assume $E_{i0} = 20$ (V/m), $f = 100$ (MHz), and $\theta_i = 30°$.

a) Find the reflection and transmission coefficients.

b) Write the instantaneous expression for $\mathbf{E}_t(x, z; t)$ and $\mathbf{H}_t(x, z; t)$.

P.7-22 For the case of oblique incidence of a uniform plane wave with parallel polarization on a plane boundary with $\epsilon_1 = \epsilon_0$, $\epsilon_2 = 2.25\epsilon_0$, $\mu_1 = \mu_2 = \mu_0$, as shown in Fig. 7-15, assume $H_{i0} = 0.053$ (A/m), $f = 100$ (MHz), and $\theta_i = 30°$.

a) Find the reflection and transmission coefficients.

b) Write the instantaneous expressions for $\mathbf{E}_t(x, z; t)$ and $\mathbf{H}_t(x, z; t)$.

P.7-23

a) Suggest a method for measuring the maximum electron density in the ionosphere.

b) Assuming $N_{max} = 8 \times 10^{11}$ per cubic meter, discuss the choice of minimum usable frequency for communicating with a space station beyond the ionosphere.

c) What happens at oblique incidence on the ionosphere if a lower frequency is used?

P.7-24 A uniform plane wave is incident on the ionosphere at an angle of incidence $\theta_i = 60°$. Assuming a constant electron density and a wave frequency equal to one-half of the plasma frequency of the ionosphere, determine

a) $\Gamma_\perp$ and $\tau_\perp$, and

b) $\Gamma_\parallel$ and $\tau_\parallel$.

Interpret the significance of these complex quantities.

P.7-25 A 10-(kHz) electromagnetic wave in air with parallel polarization is incident obliquely on an ocean surface at a near-grazing angle $\theta_i = 88°$. Using $\epsilon_r = 81$, $\mu_r = 1$, and $\sigma = 4\,(\text{S/m})$ for seawater, find (a) the angle of refraction θ_t, (b) the transmission coefficient $\tau_\parallel$, and (c) the distance below the ocean surface where the field intensity has been diminished by 30 (dB).

P.7-26 A light ray is incident from air obliquely on a transparent sheet of thickness d with an index of refraction n, as shown in Fig. 7-16. The angle of incidence is θ_i. Find (a) θ_t, (b) the distance ℓ_1 at the point of exit, and (c) the amount of the lateral displacement ℓ_2 of the emerging ray.

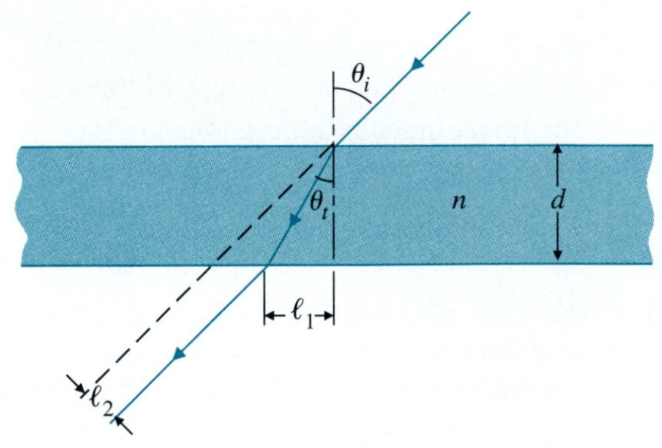

FIGURE 7-16 Light-ray impinging obliquely on a transparent sheet of refraction index n (Problem P.7-26).

P.7-27 A uniform plane wave with perpendicular polarization represented by Eqs. (7-135) and (7-136) is incident on a plane interface at $z = 0$, as shown in Fig. 7-14. Assuming $\epsilon_2 < \epsilon_1$ and $\theta_i > \theta_c$, (a) obtain the phasor expressions for the transmitted field $(\mathbf{E}_t, \mathbf{H}_t)$, and (b) verify that the average power transmitted into medium 2 is zero.

P.7-28 A uniform plane wave of angular frequency ω in medium 1 having a refractive index n_1 is incident on a plane interface at $z = 0$ with medium 2 having a refractive index $n_2\,(<n_1)$ at the critical angle. Let E_{io} and E_{to} denote the amplitudes of the incident and refracted electric field intensities, respectively.

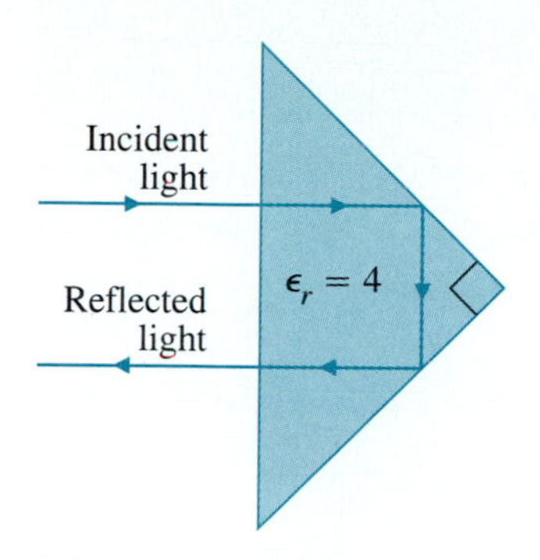

FIGURE 7-17 Light reflection by a right isosceles triangular prism (Problem P.7-30).

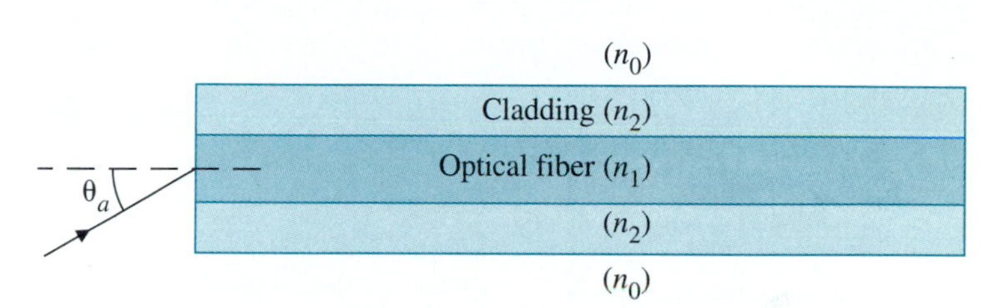

FIGURE 7-18 A cladded-core optical fiber (Problem P.7-31).

a) Find the ratio E_{t0}/E_{i0} for perpendicular polarization.

b) Find the ratio E_{t0}/E_{i0} for parallel polarization.

c) Write the instantaneous expressions of $\mathbf{E}_i(x, z; t)$ and $\mathbf{E}_t(x, z; t)$ for perpendicular polarization in terms of the parameters ω, n_1, n_2, θ_i, and E_{i0}.

P.7-29 An electromagnetic wave from an underwater source with perpendicular polarization is incident on a water–air interface at $\theta_i = 20°$. Using $\epsilon_r = 81$ and $\mu_r = 1$ for fresh water, find (a) critical angle θ_c, (b) reflection coefficient $\Gamma_\perp$, (c) transmission coefficient $\tau_\perp$, and (d) attenuation in dB for each wavelength into the air.

P.7-30 Glass isosceles triangular prisms shown in Fig. 7-17 are used in optical instruments. Assuming $\epsilon_r = 4$ for glass, calculate the percentage of the incident light power reflected back by the prism.

P.7-31 For preventing interference of waves in neighboring fibers and for mechanical protection, individual optical fibers are usually cladded by a material of a lower refractive index, as shown in Fig. 7-18, where $n_2 < n_1$.

a) Express the maximum angle of incidence θ_a in terms of n_0, n_1, and n_2 for meridional rays incident on the core's end face to be trapped inside the core by total internal reflection. (***Meridional rays*** are those that pass through the fiber axis. The angle θ_a is called the ***acceptance angle***, and $\sin\theta_a$ the ***numerical aperture*** (N.A) of the fiber.)

b) Find θ_a and N.A. if $n_1 = 2$, $n_2 = 1.74$, and $n_0 = 1$.

P.7-32 Prove that, under the condition of no reflection at an interface, the sum of the Brewster angle and the angle of refraction is $\pi/2$ for:

a) perpendicular polarization ($\mu_1 \neq \mu_2$), and

b) parallel polarization ($\epsilon_1 \neq \epsilon_2$).

P.7-33 For an incident wave with parallel polarization, find the relation between the critical angle θ_c and the Brewster angle $\theta_{B\parallel}$ for two contiguous media of equal permeability.

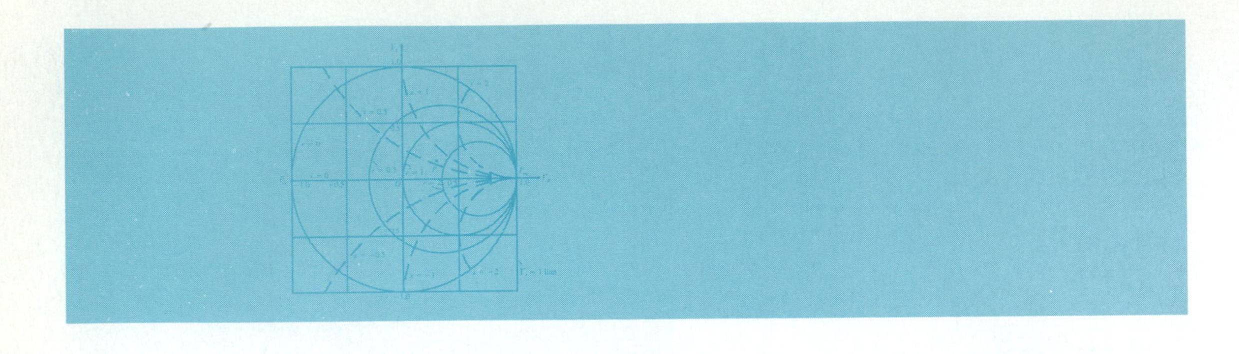

CHAPTER 8

8-1 OVERVIEW

According to our electromagnetic model we understand that time-varying charges and currents are sources of electromagnetic fields and waves. The waves carry electromagnetic power and propagate in the surrounding medium with the velocity of light. The effect of the waves on a receiver depends, among other things, on the average power density of the waves at the receiver site. At large distances this power density (power per unit area) is very low because of the very large total area of the great sphere with its center at the sources. Thus the transmission of power from an omnidirectional electromagnetic source to a receiver is very inefficient. Even when a source radiates with the aid of a highly directive antenna, its power still spreads over a very wide area at large distances.

Power transmission through radiation is very inefficient.

For efficient *point-to-point* transmission of power and information, the source energy must be guided. In this chapter we study transverse electromagnetic (TEM) waves guided by transmission lines. We will show that many of the characteristics of TEM waves guided by transmission lines are the same as those of a uniform plane wave propagating in an unbounded dielectric medium discussed in Chapter 7.

Three most common types of two-conductor transmission lines

The three most common types of two-conductor transmission lines that support TEM waves are:

a) *Parallel-plate transmission line.* This type of transmission line consists of two parallel conducting plates separated by a dielectric slab of a uniform

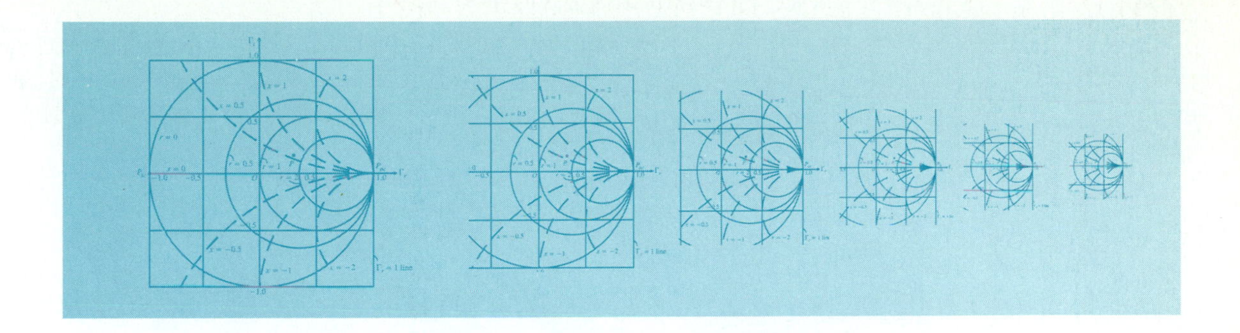

Transmission Lines

thickness. [See Fig. 8-1(a).] At microwave frequencies, parallel-plate transmission lines can be fabricated inexpensively on a dielectric substrate using printed-circuit technology. They are often called ***striplines***. A diagram showing two types of striplines is given in Fig. 8-3.

b) *Two-wire transmission line.* This transmission line consists of a pair of parallel conducting wires separated by a uniform distance. [See Fig. 8-1(b).] Examples are the ubiquitous overhead power and telephone lines seen in rural areas and the flat lead-in lines from a rooftop antenna to a television receiver.

c) *Coaxial transmission line.* This consists of an inner conductor and a coaxial outer conducting sheath separated by a dielectric medium. [See Fig. 8-1(c).] This structure has the important advantage of confining the electric and magnetic fields entirely within the dielectric region, and little external interference is coupled into the line. Examples are telephone and TV cables and the input cables to high-frequency precision measuring instruments.

The general transmission-line equations can be derived from a circuit model in terms of the resistance, inductance, conductance, and capacitance per unit length of a line. From the transmission-line equations, all the characteristics of wave propagation along a given line can be derived and studied.

The study of time-harmonic steady-state properties of transmission lines is greatly facilitated by the use of graphical charts, which avert the necessity of

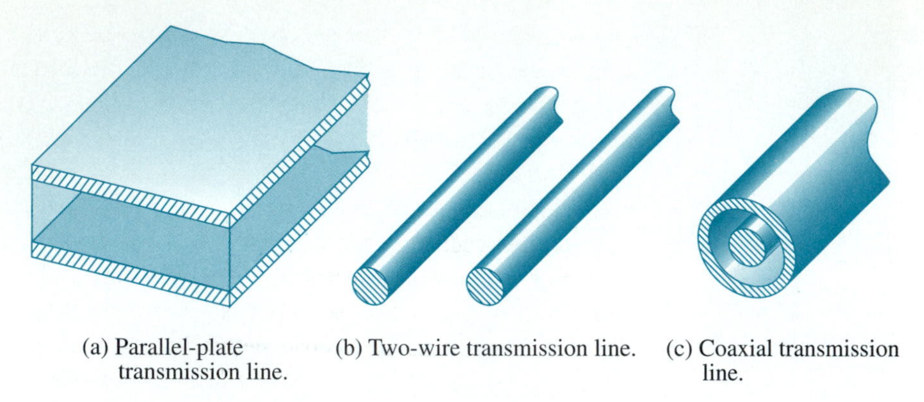

(a) Parallel-plate transmission line. (b) Two-wire transmission line. (c) Coaxial transmission line.

FIGURE 8-1 Common types of transmission lines.

repeated calculations with complex numbers. The best known and most widely used graphical chart is the ***Smith chart***. The use of the Smith chart for determining wave characteristics on a transmission line and for impedance matching will be discussed.

EXERCISE 8.1 Half-wavelength antennas are used to transmit

(a) UHF-TV signals at 800 (MHz),
(b) VHF-TV signals at 60 (MHz),
(c) FM signals at 95 (MHz), and
(d) radio navigation signals at 100 (kHz).

What are the respective antenna lengths?

ANS. (a) 18.75 (cm), (d) 1,500 (m).

REVIEW QUESTIONS

Q.8-1 What are the three most common types of guiding structures that support TEM waves?

Q.8-2 Compare the advantages and disadvantages of coaxial cables and two-wire transmission lines.

8-2 GENERAL TRANSMISSION-LINE EQUATIONS

We consider a uniform transmission line consisting of two parallel perfect conductors. The distance of separation between the conductors is small in comparison with the operating wavelength. We could begin the analysis by writing down Maxwell's equations for the transverse components of **E** and **H**. We will find that the dependence of these field components on the transverse coordinates are the same as those in static and source-free situations. Thus the parameters of a transmission line can be determined by methods used under static conditions. The voltage between the conductors and the current along the line are closely related to the transverse components of **E** and **H**, respectively, through Eqs. (3-28) and (5-63). Consequently, we can discuss the operating characteristics of a two-conductor uniform transmission line carrying a TEM wave in terms of voltage and current waves.

Ordinary electric networks have physical dimensions much smaller than operating wavelength and can be represented by discrete lumped parameters.

In this section we use a circuit model to derive the equations that govern general two-conductor uniform transmission lines. Transmission lines differ from ordinary electric networks in one essential feature. Whereas the physical dimensions of electric networks are very much smaller than the operating wavelength, transmission lines are usually a considerable fraction of a wavelength and may even be many wavelengths long. The circuit elements in an ordinary electric network can be considered discrete and as such may be described by lumped parameters. It is assumed that currents flowing in lumped-circuit elements do not vary spatially over the elements, and that no standing waves exist. A transmission line, on the other hand, is a distributed-parameter network and must be described by circuit parameters that are distributed throughout its length. Except under matched conditions (to be explained later in this chapter), standing waves exist in a transmission line.

Transmission lines are usually long in terms of wavelength and are represented by distributed parameters.

Consider a differential length Δz of a transmission line that is described by the following four parameters:

R, resistance per unit length (both conductors), in Ω/m.

L, inductance per unit length (both conductors), in H/m.

G, conductance per unit length, in S/m.

C, capacitance per unit length, in F/m.

Note that R and L are series elements and G and C are shunt elements. Figure 8-2 shows the equivalent electric circuit of such a line segment. The quantities $v(z, t)$ and $v(z + \Delta z, t)$ denote the instantaneous voltages at z and $z + \Delta z$,

FIGURE 8-2 Equivalent circuit of a differential length Δz of a two-conductor transmission line.

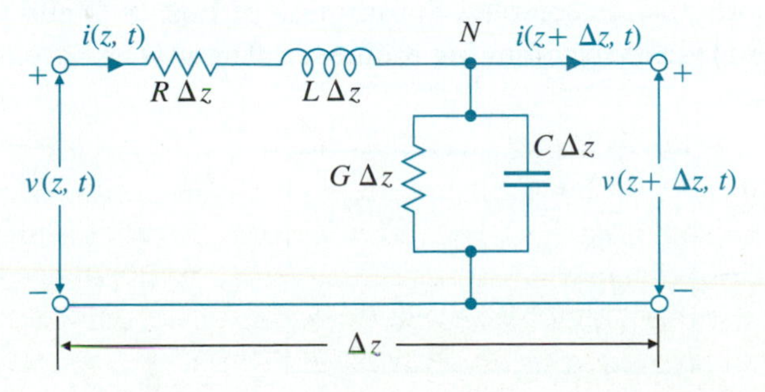

respectively. Similarly, $i(z, t)$ and $i(z + \Delta z, t)$ denote the instantaneous currents at z and $z + \Delta z$, respectively. Applying Kirchhoff's voltage law, we obtain

$$v(z, t) - R\,\Delta z i(z, t) - L\,\Delta z \frac{\partial i(z, t)}{\partial t} - v(z + \Delta z, t) = 0, \tag{(8-1)}$$

which leads to

$$-\frac{v(z + \Delta z, t) - v(z, t)}{\Delta z} = Ri(z, t) + L\frac{\partial i(z, t)}{\partial t}. \tag{8-2}$$

On the limit as $\Delta z \to 0$, Eq. (8-2) becomes

Equations (8-3) and (8-5): A pair of coupled partial differential transmission-line equations, in instantaneous voltage and current functions

$$-\frac{\partial v(z, t)}{\partial z} = Ri(z, t) + L\frac{\partial i(z, t)}{\partial t}. \tag{8-3}$$

Similarly, applying Kirchhoff's current law to the node N in Fig. 8-2, we have

$$i(z, t) - G\,\Delta z v(z + \Delta z, t) - C\,\Delta z \frac{\partial v(z + \Delta z, t)}{\partial t} - i(z + \Delta z, t) = 0. \tag{8-4}$$

On dividing by Δz and letting Δz approach zero, Eq. (8-4) becomes

$$-\frac{\partial i(z, t)}{\partial z} = Gv(z, t) + C\frac{\partial v(z, t)}{\partial t}. \tag{8-5}$$

Equations (8-3) and (8-5) are a pair of first-order partial differential equations in $v(z, t)$ and $i(z, t)$. They are the ***general transmission-line equations.***

For harmonic time dependence the use of phasors simplifies the transmission-line equations to ordinary differential equations. For a cosine reference we write

$$v(z, t) = \mathcal{R}e[V(z)e^{j\omega t}], \tag{8-6}$$

$$i(z, t) = \mathcal{R}e[I(z)e^{j\omega t}], \tag{8-7}$$

where phasors $V(z)$ and $I(z)$ are functions of the space coordinate z only and both may be complex. Substitution of Eqs. (8-6) and (8-7) in Eqs. (8-3) and (8-5) yields the following ordinary differential equations for phasors $V(z)$ and $I(z)$:

A pair of coupled ordinary differential transmission-line equations, in phasor voltage and current functions

$$-\frac{dV(z)}{dz} = (R + j\omega L)I(z), \tag{8-8}$$

$$-\frac{dI(z)}{dz} = (G + j\omega C)V(z). \tag{8-9}$$

Equations (8-8) and (8-9) are coupled ***time-harmonic transmission-line equations***. They can be combined to solve for $V(z)$ and $I(z)$. We obtain the following one-dimensional second-order ordinary differential equations:

Second-order ordinary differential transmission-line equation for phasor voltage

$$\frac{d^2V(z)}{dz^2} = \gamma^2 V(z) \tag{8-10}$$

and

Second-order ordinary differential transmission-line equation for phasor current

$$\frac{d^2I(z)}{dz^2} = \gamma^2 I(z), \tag{8-11}$$

where

$$\gamma = \alpha + j\beta = \sqrt{(R + j\omega L)(G + j\omega C)} \qquad (m^{-1}) \tag{8-12}$$

Propagation constant, attenuation constant, and phase constant of transmission line

is the ***propagation constant*** whose real and imaginary parts, α and β, are the ***attenuation constant*** (Np/m) and ***phase constant*** (rad/m) of the line, respectively. The nomenclature here is similar to that for plane-wave propagation in lossy media as defined in Section 7-3. These quantities are not really constants because, in general, they depend on ω in a complicated way. Equations (8-10) and (8-11) are basic equations, from which we derive all time-harmonic properties of both infinitely long and finite transmission lines. Noting the similarities between Eqs. (8-10) and (8-11) and Eq. (7-45b), we may conclude that many of the relations and conclusions we obtained on **E** and **H** in Chapter 7 for plane-wave propagation (not including those for oblique incidence) would also apply to voltage and current waves, $V(z)$ and $I(z)$, on transmission lines.

Close similarities between E and H for plane-wave propagation and *V* and *I* on a transmission line

8-3 TRANSMISSION-LINE PARAMETERS

The electrical properties of a transmission line at a given frequency are completely characterized by its four distributed parameters R, L, G, and C. In this section we list the formulas for these parameters in terms of the physical dimensions and the medium constitutive parameters for the three basic types of transmission lines.

Our basic premise is that the conductivity of the conductors in a transmission line is usually so high that the effect of the series resistance on the computation of the propagation constant is negligible, the implication being that the waves on the line are approximately TEM. We may write, in dropping R from Eq. (8-12),

$$\gamma = j\omega\sqrt{LC}\left(1 + \frac{G}{j\omega C}\right)^{1/2}. \tag{8-13}$$

From Eq. (7-43) we know that the propagation constant for a TEM wave in a medium with constitutive parameters (μ, ϵ, σ) is

$$\gamma = j\omega\sqrt{\mu\epsilon}\left(1 + \frac{\sigma}{j\omega\epsilon}\right)^{1/2}. \tag{8-14}$$

But

$$\frac{G}{C} = \frac{\sigma}{\epsilon} \tag{8-15}$$

in accordance with Eq. (4-38); hence comparison of Eqs. (8-13) and (8-14) yields

Relation between L and C of a transmission line and medium parameters

$$\boxed{LC = \mu\epsilon.} \tag{8-16}$$

Equation (8-16) is a very useful relation because if L is known for a line with a given medium, C can be determined, and vice versa. Knowing C, we can find G from Eq. (8-15). The series resistance R is to be determined from the power loss in the conductors.

A. *Parallel-plate transmission line*

Two parallel conducting plates each of width w and separated by a dielectric medium (ϵ, μ) of thickness d.

From Eq. (3-87) we have, neglecting fringing effects,

$$\boxed{C = \epsilon\frac{w}{d} \quad \text{(F/m).}} \quad \text{(parallel-plate line)} \tag{8-17}$$

Using Eq. (8-17) in Eqs. (8-16) and (8-15), we obtain

$$\boxed{L = \mu\frac{d}{w} \quad \text{(H/m)}} \quad \text{(parallel-plate line)} \tag{8-18}$$

and

$$\boxed{G = \sigma\frac{w}{d} \quad \text{(S/m).}} \quad \text{(parallel-plate line)} \tag{8-19}$$

The determination of a series resistance per unit length, R, implies that the metal plates are not infinitely conducting. The tangential component of the electric field does not vanish, and waves along lossy transmission lines are therefore strictly not TEM. For good conductors at sufficiently high frequencies the skin depth is very small. We can obtain an approximate expression for R by considering the total current in the conductors to be uniformly distributed over the skin depth δ. Let σ_c and μ_c denote, respectively, the conductivity and permeability of the conductors. The resistance of a unit length of material having a width w and depth δ is, from Eq. (4-16),

$$R = \frac{1}{\sigma_c S}, \tag{8-20}$$

where $S = w\delta$. For two plate conductors, we have

$$R = \frac{2}{\sigma_c w\delta}, \tag{8-21}$$

which becomes, in view of Eq. (7-56),

$$R = \frac{2}{w}\sqrt{\frac{\pi f \mu_c}{\sigma_c}} \quad (\Omega/\text{m}). \qquad \text{(parallel-plate line)} \tag{8-22}$$

B. *Two-wire transmission line*

Two conducting wires, each of radius a and separated by a distance D in a dielectric medium (ϵ, μ).

From Eq. (3-165) we have

$$C = \frac{\pi\epsilon}{\cosh^{-1}(D/2a)} \quad (\text{F/m}). \qquad \text{(two-wire line)} \tag{8-23}^{\dagger}$$

Using Eq. (8-23) in Eqs. (8-16) and (8-15), we obtain

$$L = \frac{\mu}{\pi}\cosh^{-1}\left(\frac{D}{2a}\right) \quad (\text{H/m}) \qquad \text{(two-wire line)} \tag{8-24}^{\dagger}$$

and

$$G = \frac{\pi\sigma}{\cosh^{-1}(D/2a)} \quad (\text{S/m}). \qquad \text{(two-wire line)} \tag{8-25}^{\dagger}$$

† $\cosh^{-1}(D/2a) \simeq \ln(D/a)$ if $(D/2a)^2 \gg 1$.

To find R, we use Eq. (8-20). The cross-sectional area of current flow in each wire with a skin depth δ is approximately $2\pi a\delta$. For two wires we have

$$R = \frac{1}{\sigma_c(\pi a\delta)}, \tag{8-26}$$

or

$$\boxed{R = \frac{1}{\pi a}\sqrt{\frac{\pi f \mu_c}{\sigma_c}} \quad (\Omega/\text{m}).} \quad \text{(two-wire line)} \tag{8-27}$$

C. *Coaxial transmission line*

An inner conductor of radius a separated from a concentric outer conducting tube of inner radius b by a dielectric medium (ϵ, μ).

From Eq. (3-90) we have

$$\boxed{C = \frac{2\pi\epsilon}{\ln(b/a)} \quad (\text{F/m}),} \quad \text{(coaxial line)} \tag{8-28}$$

which yields from Eqs. (8-16) and (8-15),

$$\boxed{L = \frac{\mu}{2\pi}\ln\frac{b}{a} \quad (\text{H/m})} \quad \text{(coaxial line)} \tag{8-29}$$

and

$$\boxed{G = \frac{2\pi\sigma}{\ln(b/a)} \quad (\text{S/m}).} \quad \text{(coaxial line)} \tag{8-30}$$

At high frequencies we assume that current flows uniformly in both the inner and outer conductors with a skin depth δ. The cross-sectional area for current flow in the inner conductor is then $S_i = 2\pi a\delta$, and in the outer conductor, $S_o = 2\pi b\delta$. Equation (8-20) then gives

$$R = \frac{1}{\sigma_c S_i} + \frac{1}{\sigma_c S_o} = \frac{1}{2\pi\sigma_c\delta}\left(\frac{1}{a} + \frac{1}{b}\right), \tag{8-31}$$

or

$$\boxed{R = \frac{1}{2\pi}\left(\frac{1}{a} + \frac{1}{b}\right)\sqrt{\frac{\pi f \mu_c}{\sigma_c}} \quad (\Omega/\text{m}).} \quad \text{(coaxial line)} \tag{8-32}$$

The R, L, G, and C parameters for the three types of transmission lines are listed in Table 8-1, where R_s is the real part of the intrinsic impedance of the conductor: $\eta_c = R_s + jX_s = (1 + j)\sqrt{\pi f \mu_c/\sigma_c}$, as given in Eq. (7-53). The internal self-inductance of the conductors that we discussed in Examples 5-10 and 5-11 is not included in the parameter L. Calculations in those examples were based on a direct current distributed uniformly over the entire cross section of the conductors. At high frequencies the current shifts to the surface due to skin effect, and the internal inductance is reduced to a negligible value.

EXERCISE 8.2 Obtain the low-frequency formula for the resistance per unit length of

a) a two-wire line, each having a radius a, and

b) a coaxial cable with an inner conductor of radius a and an outer conductor having an inner radius b and an outer radius d.

HINT: Use the resistance formula for direct currents.

ANS. (a) $\dfrac{2}{\sigma_c \pi a^2}$, (b) $\dfrac{1}{\sigma_c \pi}\left(\dfrac{1}{a^2} + \dfrac{1}{d^2 - b^2}\right)$.

REVIEW QUESTIONS

Q.8-3 What is the essential difference between a transmission line and an ordinary electric network?

Q.8-4 Explain why waves along a lossy transmission line cannot be purely TEM.

Q.8-5 Do the transmission-line equations (8-8) through (8-11) hold for pulse-type voltages and currents? Explain.

Distributed parameters of transmission lines

TABLE 8-1 DISTRIBUTED PARAMETERS OF PARALLEL-PLATE, TWO-WIRE, AND COAXIAL TRANSMISSION LINES

Parameter	Parallel-Plate Line	Two-Wire Line	Coaxial Line	Unit
R	$\dfrac{2}{w}R_s$	$\dfrac{R_s}{\pi a}$	$\dfrac{R_s}{2\pi}\left(\dfrac{1}{a}+\dfrac{1}{b}\right)$	Ω/m
L	$\mu\dfrac{d}{w}$	$\dfrac{\mu}{\pi}\cosh^{-1}\left(\dfrac{D}{2a}\right)$	$\dfrac{\mu}{2\pi}\ln\dfrac{b}{a}$	H/m
G	$\sigma\dfrac{w}{d}$	$\dfrac{\pi\sigma}{\cosh^{-1}(D/2a)}$	$\dfrac{2\pi\sigma}{\ln(b/a)}$	S/m
C	$\epsilon\dfrac{w}{d}$	$\dfrac{\pi\epsilon}{\cosh^{-1}(D/2a)}$	$\dfrac{2\pi\epsilon}{\ln(b/a)}$	F/m

Note: $R_s = \sqrt{\pi f \mu_c/\sigma_c}$; $\cosh^{-1}(D/2a) \cong \ln(D/a)$ if $(D/2a)^2 \gg 1$. Internal inductance is not included.

REMARKS

1. There is a close analogy between the characteristics of **E** and **H** for plane-wave propagation and those of $V(z)$ and $I(z)$ on transmission lines.
2. Propagation "constant" $\gamma = \alpha + j\beta$. Both attenuation "constant" α and phase "constant" β along a line may depend on frequency in a complicated way.
3. It is important to distinguish μ_c and σ_c for conductors from μ and σ for a dielectric medium.
4. The series resistance, R, of a transmission line is *not* the reciprocal of the parallel conductance, G.

8-3.1 MICROSTRIP LINES

Striplines are modified forms of parallel-plate transmission lines.

The development of solid-state microwave devices and systems has led to the widespread use of a form of parallel-plate transmission lines called microstrip lines or simply ***striplines***. A stripline usually consists of a dielectric substrate sitting on a grounded conducting plane with a thin narrow metal strip on top of the substrate, as shown in Fig. 8-3(a). Since the advent of printed-circuit techniques, striplines can be easily fabricated and integrated with other circuit components. However, because the results that we have derived for the parallel-plate transmission line were based on the assumption of two wide conducting plates (with negligible fringing effect) of equal width, they are not expected to apply here exactly. The approximation is closer if the width of the metal strip is much greater than the substrate thickness.

When the substrate has a high dielectric constant, a TEM approximation is found to be reasonably satisfactory. An exact analytical solution of the stripline in Fig. 8-3(a) satisfying all the boundary conditions is a difficult problem. Not all the fields will be confined in the dielectric substrate; some will stray from the top strip into the region outside of the strip, thus causing interference in the neighboring circuits. Semiempirical modifications to the formulas for the distributed parameters and the characteristic impedance are necessary for more accurate calculations.† All of these quantities tend to be frequency-dependent, and striplines are dispersive.

One method for reducing the stray fields of striplines is to have a grounded conducting plane on both sides of the dielectric substrate and to

†See, for instance, K. F. Sander and G. A. L. Reed, *Transmission and Propagation of Electromagnetic Waves*, 2nd edition, Sec. 6.5.6, Cambridge University Press, New York, 1986. Also see D. M. Pozar, *Microwave Engineering*, Sec. 4.7 and 4.8, Addison-Wesley Publishing Co., Reading, Mass., 1990.

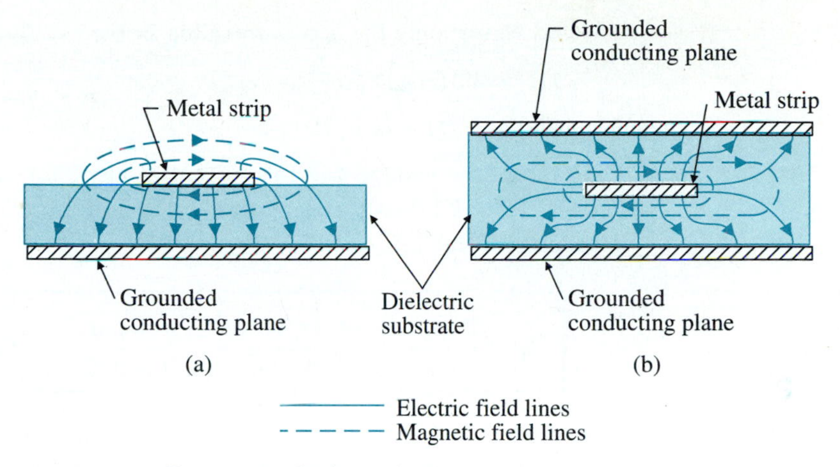

FIGURE 8-3 Two types of microstrip lines.

Triplate lines minimize stray fields.

put the thin metal strip in the middle as in Fig. 8-3(b). This arrangement is known as a ***triplate line***. We can appreciate that triplate lines are more difficult and costly to fabricate and that the characteristic impedance of a triplate line is one-half that of a corresponding stripline.

8-4 WAVE CHARACTERISTICS ON AN INFINITE TRANSMISSION LINE

In Section 8-2 we derived the governing equations (8-10) and (8-11) for time-harmonic $V(z)$ and $I(z)$ on a transmission line. Let us now examine their characteristics on an infinite line. The solutions of Eqs. (8-10) and (8-11) are

$$\begin{aligned} V(z) &= V^+(z) + V^-(z) \\ &= V_0^+ e^{-\gamma z} + V_0^- e^{\gamma z}, \end{aligned} \tag{8-33}$$

$$\begin{aligned} I(z) &= I^+(z) + I^-(z) \\ &= I_0^+ e^{-\gamma z} + I_0^- e^{\gamma z}, \end{aligned} \tag{8-34}$$

where the plus and minus superscripts denote waves traveling in the $+z$- and $-z$-directions, respectively. Wave amplitudes (V_0^+, I_0^+) and (V_0^-, I_0^-) are related by Eqs. (8-8) and (8-9), and we can verify (Exercise 8.3) that

$$\frac{V_0^+}{I_0^+} = -\frac{V_0^-}{I_0^-} = \frac{R + j\omega L}{\gamma}. \tag{8-35}$$

For an infinite line (actually a semi-infinite line with the source at the left end) the terms containing the $e^{\gamma z}$ factor must vanish. If not, these terms would increase indefinitely with z, a physical impossibility. There are no reflected waves; only the waves traveling in the $+z$-direction exist. Thus

$$V(z) = V^+(z) = V_0^+ e^{-\gamma z}, \qquad (8\text{-}36)$$

$$I(z) = I^+(z) = I_0^+ e^{-\gamma z}. \qquad (8\text{-}37)$$

The ratio of the voltage and the current at any z for an infinitely long line, $V^+(z)/I^+(z) = V_0^+/I_0^+$, is independent of z and is called the ***characteristic impedance*** of the line.

Characteristic impedance of transmission line

$$Z_0 = \frac{R + j\omega L}{\gamma} = \frac{\gamma}{G + j\omega C} = \sqrt{\frac{R + j\omega L}{G + j\omega C}} \quad (\Omega). \qquad (8\text{-}38)$$

Note that γ and Z_0 are characteristic properties of a transmission line whether or not the line is infinitely long. They depend on R, L, G, C, and ω—not on the length of the line. An infinite line simply implies that there are no reflected waves.

EXERCISE 8.3 Verify the relation in Eq. (8-35) by substituting Eqs. (8-33) and (8-34) in Eq. (8-8).

Two special cases for γ in Eq. (8-12) and Z_0 in Eq. (8-38) are of particular interest.

1. *Lossless Line* ($R = 0$, $G = 0$).

A lossless line has zero α and X_0, and constant u_p and R_0.

a) Propagation constant:

$$\gamma = \alpha + j\beta = j\omega\sqrt{LC}; \qquad (8\text{-}39)$$

$$\alpha = 0, \qquad (8\text{-}40)$$

$$\beta = \omega\sqrt{LC} \quad \text{(a linear function of } \omega\text{)}. \qquad (8\text{-}41)$$

b) Phase velocity:

$$u_p = \frac{\omega}{\beta} = \frac{1}{\sqrt{LC}} \quad \text{(constant)}. \qquad (8\text{-}42)$$

In view of Eq. (8-16), we see that the phase velocity of waves on a lossless transmission is equal to the velocity of propagation, $1/\sqrt{\mu\epsilon}$, of unguided plane waves in the medium of the line.

c) Characteristic impedance:

$$Z_0 = R_0 + jX_0 = \sqrt{\frac{L}{C}}; \qquad (8\text{-}43)$$

$$R_0 = \sqrt{\frac{L}{C}} \quad \text{(constant)}, \qquad (8\text{-}44)$$

$$X_0 = 0. \qquad (8\text{-}45)$$

2. *Distortionless Line* ($R/L = G/C$). If the condition

Condition for a distortionless transmission line

$$\boxed{\frac{R}{L} = \frac{G}{C}} \tag{8-46}$$

is satisfied, the expressions for both γ and Z_0 simplify.

a) Propagation constant:

$$\gamma = \alpha + j\beta = \sqrt{(R + j\omega L)\left(\frac{RC}{L} + j\omega C\right)}$$

$$= \sqrt{\frac{C}{L}}(R + j\omega L); \tag{8-47}$$

$$\alpha = R\sqrt{\frac{C}{L}}, \tag{8-48}$$

$$\beta = \omega\sqrt{LC} \qquad \text{(a linear function of } \omega\text{)}. \tag{8-49}$$

b) Phase velocity:

A lossy distortionless line has the same characteristics as those of a lossless line, except for a nonzero attenuation constant.

$$u_p = \frac{\omega}{\beta} = \frac{1}{\sqrt{LC}} \qquad \text{(constant)}. \tag{8-50}$$

c) Characteristic impedance:

$$Z_0 = R_0 + jX_0 = \sqrt{\frac{R + j\omega L}{(RC/L) + j\omega C}} = \sqrt{\frac{L}{C}}; \tag{8-51}$$

$$R_0 = \sqrt{\frac{L}{C}} \qquad \text{(constant)}, \tag{8-52}$$

$$X_0 = 0, \tag{8-53}$$

Thus, except for a nonvanishing attenuation constant, the characteristics of a distortionless line are the same as those of a lossless line—namely, a constant phase velocity ($u_p = 1/\sqrt{LC}$) and a constant real characteristic impedance ($Z_0 = R_0 = \sqrt{L/C}$).

A constant phase velocity is a direct consequence of the linear dependence of the phase constant β on ω. Since a signal usually consists of a band of frequencies, it is essential that the different frequency components travel along a transmission line at the same velocity in order to avoid distortion. This condition is satisfied by a lossless line and is approximated by a line with very low losses. For a lossy line, wave amplitudes will be attenuated, and distortion will result when different frequency components attenuate differently, even when they travel with the same velocity. The condition specified in Eq. (8-46) leads to both a constant α and a constant u_p—thus the name ***distortionless line***.

The phase constant of a lossy transmission line is determined by expanding the expression for γ in Eq. (8-12). In general, the phase constant is

Signal distortion (dispersion) occurs when phase constant is not proportional to frequency.

not a linear function of ω; thus it will lead to a frequency-dependent phase velocity u_p. As the different frequency components of a signal propagate along the line with different velocities, the signal suffers distortion or ***dispersion***. A general, lossy, transmission line is therefore *dispersive*, as is a lossy dielectric.

EXAMPLE 8-1

Neglecting losses and fringing effects and assuming the substrate of a stripline to have a thickness 0.4 (mm) and a dielectric constant 2.25, (a) determine the required width w of the metal strip in order for the stripline to have a characteristic resistance of 50 (Ω); (b) determine L and C of the line; and (c) determine u_p along the line. (d) Repeat parts (a), (b), and (c) for a characteristic resistance of 75 (Ω).

SOLUTION

a) We use Eqs. (8-17) and (8-18) in Eq. (8-43) to find w:

$$w = \frac{d}{Z_0}\sqrt{\frac{\mu}{\epsilon}} = \frac{0.4 \times 10^{-3}}{50}\frac{\eta_0}{\sqrt{\epsilon_r}}$$

$$= \frac{0.4 \times 10^{-3} \times 377}{50\sqrt{2.25}} = 2 \times 10^{-3} \text{ (m)}, \quad \text{or } 2 \text{ (mm)}.$$

b) $$L = \mu\frac{d}{w} = 4\pi 10^{-7} \times \frac{0.4}{2} = 2.51 \times 10^{-7} \quad \text{(H/m)}, \quad \text{or } 0.251 \quad (\mu\text{H/m}).$$

$$C = \epsilon_0\epsilon_r\frac{w}{d} = \frac{10^{-9}}{36\pi} \times 2.25 \times \frac{2}{0.4} = 99.5 \times 10^{-12} \quad \text{(F/m)},$$

99.5 (pF/m). or,

c) $$u_p = \frac{1}{\sqrt{LC}} = \frac{1}{\sqrt{\mu\epsilon}} = \frac{c}{\sqrt{\epsilon_r}} = \frac{c}{\sqrt{2.25}} = \frac{c}{1.5} = 2 \times 10^8 \quad \text{(m/s)}.$$

d) Since w is inversely proportional to Z_0, we have, for $Z_0' = 75$ (Ω),

$$w' = \left(\frac{Z_0}{Z_0'}\right)w = \frac{50}{75} \times 2 = 1.33 \quad \text{(mm)}.$$

$$L' = \left(\frac{w}{w'}\right)L = \left(\frac{2}{1.33}\right) \times 0.251 = 0.377 \quad (\mu\text{H/m}).$$

$$C' = \left(\frac{w'}{w}\right)C = \left(\frac{1.33}{2}\right) \times 99.5 = 66.2 \quad \text{(pF/m)}.$$

$$u_p' = u_p = 2 \times 10^8 \quad \text{(m/s)}.$$

8-4.1 ATTENUATION CONSTANT FROM POWER RELATIONS

The attenuation constant of a traveling wave on a transmission line is the real part of the propagation constant; it can be determined from the basic definition in Eq. (8-12). The attenuation constant can also be found from a power relationship. Using (Eqs. (8-36) and (8-37), we write (dropping the plus superscript for simplicity):

$$V(z) = V_0 e^{-(\alpha + j\beta)z}, \tag{8-54}$$

$$I(z) = \frac{V_0}{Z_0} e^{-(\alpha + j\beta)z}. \tag{8-55}$$

The time-average power propagated along the line at any z is

$$P(z) = \tfrac{1}{2}\mathcal{Re}[V(z)I^*(z)]$$
$$= \frac{V_0^2}{2|Z_0|^2} R_0 e^{-2\alpha z}. \tag{8-56}$$

The law of conservation of energy requires that the rate of decrease of $P(z)$ with distance along the line equals the time-average power loss P_L per unit length. Thus,

$$-\frac{\partial P(z)}{\partial z} = P_L(z)$$
$$= 2\alpha P(z),$$

from which we obtain the following formula:

Determining attenuation constant from power relation

$$\boxed{\alpha = \frac{P_L(z)}{2P(z)} \quad \text{(Np/m)}.} \tag{8-57}$$

EXAMPLE 8-2

a) Use Eq. (8-57) to find the attenuation constant of a lossy transmission line with distributed parameters R, L, G, and C.

b) Specialize the result in part (a) to obtain the attenuation constants of a low-loss line and of a distortionless line.

SOLUTION

a) For a lossy transmission line the time-average power loss per unit length is

$$P_L(z) = \tfrac{1}{2}[|I(z)|^2 R + |V(z)|^2 G]$$
$$= \frac{V_0^2}{2|Z_0|^2}(R + G|Z_0|^2)e^{-2\alpha z} \tag{8-58}$$

Substitution of Eqs. (8-56) and (8-58) in Eq. (8-57) gives

Attenuation constant of lossy transmission line

$$\alpha = \frac{1}{2R_0}(R + G|Z_0|^2) \qquad \text{(Np/m)}. \tag{8-59}$$

b) For a low-loss line, $Z_0 \cong R_0 = \sqrt{L/C}$, Eq. (8-59) becomes

$$\begin{aligned}\alpha &\cong \frac{1}{2}\left(\frac{R}{R_0} + GR_0\right) \\ &= \frac{1}{2}\left(R\sqrt{\frac{C}{L}} + G\sqrt{\frac{L}{C}}\right).\end{aligned} \tag{8-60}$$

For a distortionless line, (Eq. (8-52) gives $Z_0 = R_0 = \sqrt{L/C}$. Thus Eq. (8-60) applies and yields

$$\alpha = \frac{1}{2}R\sqrt{\frac{C}{L}}\left(1 + \frac{G}{R}\frac{L}{C}\right),$$

which, in view of the condition in Eq. (8-46), reduces to

$$\alpha = R\sqrt{\frac{C}{L}}. \tag{8-61}$$

EXERCISE 8.4 A parallel-plate transmission line is constructed of highly conducting plates of width w that are separated by a lossless dielectric of dielectric constant ϵ_r and thickness d. Neglecting fringing effect, explain how the following should be changed if it is desired to double its characteristic impedance by changing only one parameter: (a) w, (b) d, and (c) ϵ_r.

EXERCISE 8.5 The attenuation constant at 10 (MHz) of a distortionless coaxial transmission line is found to be 0.1 (dB/km). Determine its value

a) at 50 (MHz), and

b) at 10 (MHz) if the dielectric constant of the insulating material is doubled.

ANS. (a) 0.224 (dB/km), (b) 0.141 (dB/km).

REVIEW QUESTIONS

Q.8-6 Define *propagation constant* and *characteristic impedance* of a transmission line. Write their general expressions in terms of R, L, G, and C for sinusoidal excitation.

Q.8-7 What are *striplines*?

Q.8-8 What is a *triplate line*? How does the characteristic impedance of a triplate line compare with that of a corresponding stripline? Explain.

Q.8-9 What is meant by a "distortionless line"? What relation must the distributed parameters of a line satisfy in order for the line to be distortionless?

Q.8-10 What is the relation between the attenuation constant and the power loss along a transmission line?

REMARKS

1. The velocity of wave propagation on a lossless transmission line and on a distortionless line is the same as the velocity of propagation, $1/\sqrt{\mu\epsilon}$, of unguided plane wave in the dielectric medium of the line.
2. Exact determination of the distributed parameters and characteristic impedance of microstrip lines is very difficult. Usually semiempirical formulas are used.
3. A lossless line is distortionless; but a distortionless line can be lossy.
4. Lossy lines are, in general, dispersive, unless Eq. (8-46) is satisfied.

8-5 WAVE CHARACTERISTICS ON FINITE TRANSMISSION LINES

In Section 8-2 we indicated that the general solutions for the time-harmonic one-dimensional Helmholtz equations, Eqs. (8-10) and (8-11), for transmission lines are

$$V(z) = V_0^+ e^{-\gamma z} + V_0^- e^{\gamma z} \tag{8-62}$$

and

$$I(z) = I_0^+ e^{-\gamma z} + I_0^- e^{\gamma z}, \tag{8-63}$$

where

$$\frac{V_0^+}{I_0^+} = -\frac{V_0^-}{I_0^-} = Z_0, \tag{8-64}$$

as given in Section 8-4. For waves launched on an infinitely long line at $z = 0$ there can be only forward waves traveling in the $+z$-direction, and the second terms on the right side of Eqs. (8-62) and (8-63), representing reflected waves, vanish.

Let us now consider the general case of a finite transmission line having a characteristic impedance Z_0 terminated in an arbitrary load impedance Z_L, as depicted in Fig. 8-4. The length of the line is ℓ. A *sinusoidal* voltage source $V_g\underline{/0^\circ}$ with an internal impedance Z_g is connected to the line at $z = 0$. In such a case,

$$\left(\frac{V}{I}\right)_{z=\ell} = \frac{V_L}{I_L} = Z_L, \tag{8-65}$$

which obviously cannot be satisfied without the second terms on the right side of Eqs. (8-62) and (8-63) unless $Z_L = Z_0$. Given the characteristic γ and Z_0 of the line and its length ℓ, there are four unknowns V_0^+, V_0^-, I_0^+, and I_0^- in Eqs. (8-62) and (8-63). These four unknowns are not all independent because they are constrained by the relations at $z = 0$ and at $z = \ell$.

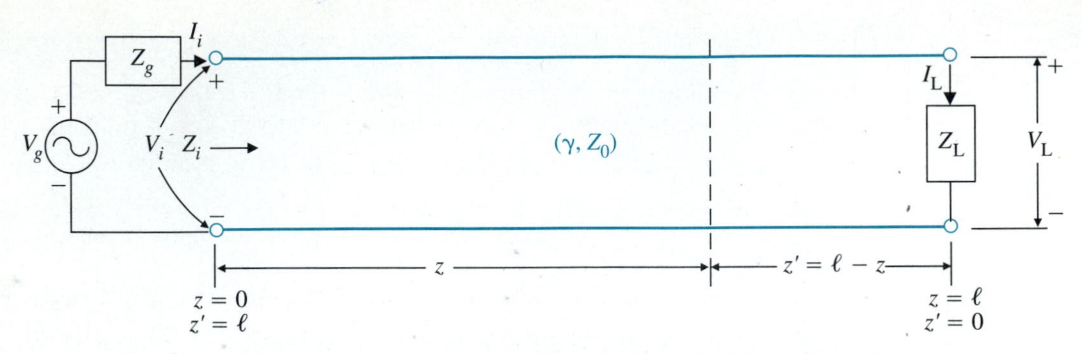

FIGURE 8-4 Finite transmission line terminated with load impedance Z_L.

Letting $z = \ell$ in Eqs. (8-62) and (8-63), and using Eq. (8-64), we have

$$V_L = V_0^+ e^{-\gamma\ell} + V_0^- e^{\gamma\ell}, \tag{8-66}$$

$$I_L = \frac{V_0^+}{Z_0} e^{-\gamma\ell} - \frac{V_0^-}{Z_0} e^{\gamma\ell}. \tag{8-67}$$

Solving Eqs. (8-66) and (8-67) for V_0^+ and V_0^-, we have

$$V_0^+ = \tfrac{1}{2}(V_L + I_L Z_0)e^{\gamma\ell} = \frac{I_L}{2}(Z_L + Z_0)e^{\gamma\ell}, \tag{8-68}$$

$$V_0^- = \tfrac{1}{2}(V_L - I_L Z_0)e^{-\gamma\ell} = \frac{I_L}{2}(Z_L - Z_0)e^{-\gamma\ell}. \tag{8-69}$$

Using Eqs. (8-68) and (8-69) in Eqs. (8-62) and (8-63), we obtain

$$V(z) = \frac{I_L}{2}[(Z_L + Z_0)e^{\gamma(\ell - z)} + (Z_L - Z_0)e^{-\gamma(\ell - z)}], \tag{8-70}$$

$$I(z) = \frac{I_L}{2Z_0}[(Z_L + Z_0)e^{\gamma(\ell - z)} - (Z_L - Z_0)e^{-\gamma(\ell - z)}]. \tag{8-71}$$

Since ℓ and z appear together in the combination $(\ell - z)$, it is expedient to introduce a new variable $z' = \ell - z$, which is the distance measured backward from the load. Equations (8-70) and (8-71) then become

$$V(z') = \frac{I_L}{2}[(Z_L + Z_0)e^{\gamma z'} + (Z_L - Z_0)e^{-\gamma z'}], \tag{8-72}$$

$$I(z') = \frac{I_L}{2Z_0}[(Z_L + Z_0)e^{\gamma z'} - (Z_L - Z_0)e^{-\gamma z'}]. \tag{8-73}$$

The use of hyperbolic functions simplifies the equations above. Recalling the relations

$$e^{\gamma z'} + e^{-\gamma z'} = 2\cosh \gamma z' \qquad \text{and} \qquad e^{\gamma z'} - e^{-\gamma z'} = 2\sinh \gamma z',$$

we may write Eqs. (8-72) and (8-73) as

$V(z')$ and $I(z')$ of a finite line in terms of γ, Z_0, Z_L, I_L, and z'

$$V(z') = I_L(Z_L \cosh \gamma z' + Z_0 \sinh \gamma z'), \tag{8-74}$$

$$I(z') = \frac{I_L}{Z_0}(Z_L \sinh \gamma z' + Z_0 \cosh \gamma z'), \tag{8-75}$$

which can be used to find the voltage and current at any point along a transmission line in terms of I_L, Z_L, γ, and Z_0.

The ratio $V(z')/I(z')$ is the impedance when we look toward the load end of the line *at a distance z′ from the load.*

$$Z(z') = \frac{V(z')}{I(z')} = Z_0 \frac{Z_L \cosh \gamma z' + Z_0 \sinh \gamma z'}{Z_L \sinh \gamma z' + Z_0 \cosh \gamma z'}, \tag{8-76}$$

or

Impedance $Z(z')$ in terms of γ, Z_0, Z_L, and z'

$$Z(z') = Z_0 \frac{Z_L + Z_0 \tanh \gamma z'}{Z_0 + Z_L \tanh \gamma z'} \quad (\Omega). \tag{8-77}$$

At the source end of the line, $z' = \ell$, the generator looking into the line sees an ***input impedance*** Z_i.

General formula for input impedance Z_i of a line of length ℓ

$$Z_i = (Z)_{\substack{z=0 \\ z'=\ell}} = Z_0 \frac{Z_L + Z_0 \tanh \gamma \ell}{Z_0 + Z_L \tanh \gamma \ell} \quad (\Omega). \tag{8-78}$$

As far as the conditions at the generator are concerned, the terminated finite transmission line can be replaced by Z_i, as shown in Fig. 8-5. The input voltage V_i and input current I_i in Fig. 8-4 are found easily from the equivalent circuit in Fig. 8-5.

A transmission line is matched when $Z_L = Z_0$.

It is obvious from Eq. (8-78) that when $Z_L = Z_0$, $Z_i = Z_0$ irrespective of the length, ℓ, of the line. Under this condition, the line is said to be *matched.* We will say more about this condition later.

In most cases, transmission-line segments can be considered lossless: $\gamma = j\beta$, $Z_0 = R_0$, and $\tanh \gamma \ell = \tanh(j\beta\ell) = j \tan \beta\ell$. The formula in Eq.

FIGURE 8-5 Equivalent circuit for finite transmission line in Figure 8-4 at generator end.

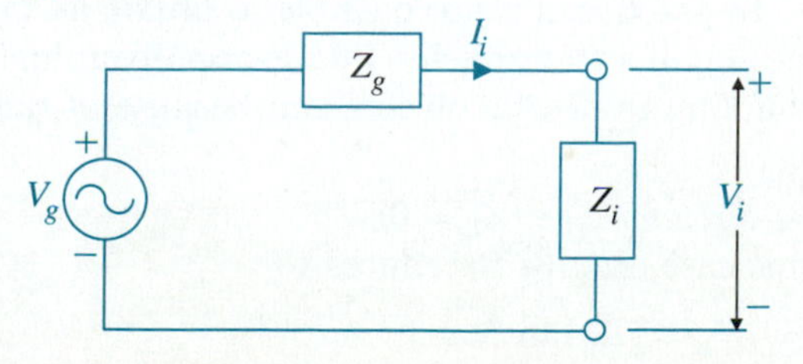

(8-78) for the input impedance Z_i of a lossless line of length ℓ terminated in Z_L becomes

Formula for input impedance Z_i of a lossless line of length ℓ

$$Z_i = R_0 \frac{Z_L + jR_0 \tan \beta\ell}{R_0 + jZ_L \tan \beta\ell} \quad (\Omega). \quad \text{(Lossless line)} \tag{8-79}$$

8-5.1 OPEN-CIRCUITED AND SHORT-CIRCUITED LINES

Transmission lines can be used not only as wave-guiding structures for transmitting power and information from one place to another, but at ultra high frequencies (UHF)—from 300 (MHz) to 3 (GHz)—they may serve as circuit elements. At these frequencies, lumped circuit elements such as inductances and capacitances are difficult to make, and stray fields become important. Open- and short-circuited elements may act as inductances and/or capacitances. For frequencies above the UHF range the physical dimensions of transmission-line circuit lines become inconveniently small, and it would be advantageous to use waveguide components to be discussed in the next chapter.

A. *Open-circuited line* ($Z_L \to \infty$).
From Eq. (8-79) we have

Input impedance of a lossless open-circuited line

$$Z_{io} = jX_{io} = -\frac{jR_0}{\tan \beta\ell} = -jR_0 \cot \beta\ell. \tag{8-80}$$

Equation (8-80) shows that the input impedance of an open-circuited lossless line is purely reactive. The line can, however, be either capacitive or inductive because the function $\cot \beta\ell$ can be either positive or negative, depending on the value of $\beta\ell\ (=2\pi\ell/\lambda)$.

If the length of an open-circuited line is very short in comparison with a wavelength, $\beta\ell \ll 1$, we can obtain a very simple formula for its capacitive reactance by noting that $\tan \beta\ell \cong \beta\ell$. From Eq. (8-80) we have

Input impedance of a very short open-circuited line is purely capacitive.

$$Z_i = jX_{io} \cong -j\frac{R_0}{\beta\ell} = -j\frac{\sqrt{L/C}}{\omega\sqrt{LC}\ell} = -j\frac{1}{\omega C\ell}, \tag{8-81}$$

which is the impedance of a capacitance of $C\ell$ farads.

Open-circuited lines are seldom used as circuit elements.

In practice, it is not possible to obtain an infinite load impedance at the end of a transmission line, especially at high frequencies, because of coupling to nearby objects and because of radiation from the open end.

B. *Short-circuited line* ($Z_L = 0$).
In this case, Eq. (8-79) reduces to

Input impedance of a lossless short-circuited line

$$Z_{is} = jX_{is} = jR_0 \tan \beta\ell. \tag{8-82}$$

A short-circuited quarter-wavelength line is effectively an open circuit.

Since $\tan \beta\ell$ can range from $-\infty$ to $+\infty$, the input impedance of a short-circuited lossless line can also be either purely inductive or purely capacitive, depending on the value of $\beta\ell$. In particular, when $\beta\ell = \pi/2$, or $\ell = \lambda/4$, Z_i in Eq. (8-82) becomes infinite. Thus, a short-circuited quarter-wavelength transmission line is effectively an open-circuit.

If the length of a short-circuited line is very short in comparison with a wavelength, $\beta\ell \ll 1$, Eq. (8-82) becomes approximately

Input impedance of a very short short-circuited line is purely inductive.

$$Z_{is} = jX_{is} \cong jR_0\beta\ell = j\sqrt{\frac{L}{C}}\,\omega\sqrt{LC}\,\ell = j\omega L\ell, \tag{8-83}$$

which is the impedance of an inductance of $L\ell$ henries.

■ **EXERCISE 8.6** Prove that a quarter-wavelength lossless line transforms a load impedance to the input terminals as its inverse multiplied by the square of the characteristic resistance; that is, $Z_i = R_0^2/Z_L$.

■ **EXERCISE 8.7** Prove that a half-wavelength line transfers a load impedance to the input terminals without change.

8-5.2 CHARACTERISTIC IMPEDANCE AND PROPAGATION CONSTANT FROM INPUT MEASUREMENTS

We can determine the characteristic impedance and the propagation constant of a transmission line by measuring the input impedance of a line section under open- and short-circuit conditions. The following expressions follow directly from Eq. (8-78):

Open-circuited line, $Z_L \to \infty$: $\quad Z_{io} = Z_0 \coth \gamma\ell$. (8-84a)

Short-circuited line, $Z_L = 0$: $\quad Z_{is} = Z_0 \tanh \gamma\ell$. (8-84b)

From Eqs. (8-84a) and (8-84b) we have

Calculating Z_0 and γ of a line from input impedances measured under open-circuit and short-circuit conditions

$$\boxed{Z_0 = \sqrt{Z_{io}Z_{is}} \quad (\Omega)} \tag{8-85}$$

and

$$\boxed{\gamma = \frac{1}{\ell}\tanh^{-1}\sqrt{\frac{Z_{is}}{Z_{io}}} \quad (\text{m}^{-1}).} \tag{8-86}$$

Equations (8-85) and (8-86) are easily evaluated.

EXAMPLE 8-3

A signal generator having an internal resistance 1 (Ω) and an open-circuit voltage $v_g(t) = 0.3\cos 2\pi 10^8 t$ (V) is connected to a 50-(Ω) lossless transmission line. The line is 4 (m) long, and the velocity of wave propagation on the line is 2.5×10^8 (m/s). For a matched load, find (a) the instantaneous expressions for the voltage and current at an arbitrary location on the line, (b) the instantaneous expressions for the voltage and current at the load, and (c) the average power transmitted to the load.

SOLUTION

a) In order to find the voltage and current at an arbitrary location on the line, it is first necessary to obtain those at the input end ($z = 0$, $z' = \ell$). The given quantities are as follows:

$V_g = 0.3\underline{/0^\circ}$ (V), a phasor with a cosine reference,

$Z_g = R_g = 1$ (Ω),

$Z_0 = R_0 = 50$ (Ω),

$\omega = 2\pi \times 10^8$ (rad/s),

$u_p = 2.5 \times 10^8$ (m/s),

$\ell = 4$ (m).

Since the line is terminated with a matched load, $Z_i = Z_0 = 50\,(\Omega)$. The voltage and current at the input terminals can be evaluated from the equivalent circuit in Fig. 8-5. We have

$$V_i = \frac{Z_i}{Z_g + Z_i} V_g = \frac{50}{1 + 50} \times 0.3\underline{/0^\circ} = 0.294\underline{/0^\circ} \quad \text{(V)},$$

$$I_i = \frac{V_g}{Z_g + Z_i} = \frac{0.3\underline{/0^\circ}}{1 + 50} = 0.0059\underline{/0^\circ} \quad \text{(A)}.$$

Since only forward-traveling waves exist on a matched line, we use Eqs. (8-54) and (8-55) for the voltage and current, respectively, at an arbitrary location. For the given line, $\alpha = 0$ and

$$\beta = \frac{\omega}{u_p} = \frac{2\pi \times 10^8}{2.5 \times 10^8} = 0.8\pi \quad \text{(rad/m)}.$$

Thus,

$$V(z) = 0.294e^{-j0.8\pi z} \quad \text{(V)},$$
$$I(z) = 0.0059e^{-j0.8\pi z} \quad \text{(A)}.$$

These are phasors. The corresponding instantaneous expressions are, from Eqs. (8-6) and (8-7),

$$v(z, t) = \mathcal{R}e[0.294e^{j(2\pi 10^8 t - 0.8\pi z)}]$$
$$= 0.294 \cos(2\pi 10^8 t - 0.8\pi z) \quad \text{(V)},$$
$$i(z, t) = \mathcal{R}e[0.0059e^{j(2\pi 10^8 t - 0.8\pi z)}]$$
$$= 0.0059 \cos(2\pi 10^8 t - 0.8\pi z) \quad \text{(A)}.$$

b) At the load, $z = \ell = 4$ (m),

$$v(4, t) = 0.294 \cos(2\pi 10^8 t - 3.2\pi) \quad \text{(V)},$$
$$i(4, t) = 0.0059 \cos(2\pi 10^8 t - 3.2\pi) \quad \text{(A)}.$$

c) The average power transmitted to the load on a lossless line is equal to that at the input terminals.

$$(P_{av})_L = (P_{av})_i = \tfrac{1}{2}\mathscr{R}e[V(z)I^*(z)]$$
$$= \tfrac{1}{2}(0.294 \times 0.0059) = 8.7 \times 10^{-4} \text{ (W)} = 0.87 \quad \text{(mW)}.$$

EXAMPLE 8-4

The open-circuit and short-circuit impedances measured at the input terminals of a lossless transmission line of length 1.5 (m), which is less than a quarter wavelength, are $-j54.6$ (Ω) and $j103$ (Ω), respectively.

a) Find Z_0 and γ of the line.

b) Without changing the operating frequency, find the input impedance of a short-circuited line that is twice the given length.

c) How long should the short-circuited line be in order for it to appear as an open circuit at the input terminals?

SOLUTION

The given quantities are

$$Z_{io} = -j54.6, \qquad Z_{is} = j103, \qquad \ell = 1.5.$$

a) Using Eqs. (8-85) and (8-86), we find

$$Z_0 = \sqrt{-j54.6(j103)} = 75 \quad (\Omega),$$

$$\gamma = \frac{1}{1.5}\tanh^{-1}\sqrt{\frac{j103}{-j54.6}} = \frac{j}{1.5}\tan^{-1} 1.373 = j0.628 \quad \text{(rad/m)}.$$

b) For a short-circuited line twice as long, $\ell = 3.0$ (m),

$$\gamma\ell = j0.628 \times 3.0 = j1.884 \quad \text{(rad)}.$$

The input impedance is, from Eq. (8-84b),

$$Z_{is} = 75 \tanh(j1.884) = j75 \tan 108^\circ$$
$$= j75(-3.08) = -j231 \quad (\Omega).$$

Note that Z_{is} for the 3-(m) shorted line is now a capacitive reactance, whereas that for the given 1.5-(m) shorted line is an inductive reactance.

c) In order for a short-circuited line to appear as an open circuit at the input terminals, it should be an odd multiple of a quarter-wavelength long, which will make $\tan \beta\ell \to \infty$ in Eq. (8-82).

$$\lambda = \frac{2\pi}{\beta} = \frac{2\pi}{0.628} = 10 \quad \text{(m)}.$$

Hence the required line length is

$$\ell = \frac{\lambda}{4} + (n - 1)\frac{\lambda}{2}$$
$$= 2.5 + 5(n - 1) \quad \text{(m)}, \qquad n = 1, 2, 3, \ldots.$$

REVIEW QUESTIONS

Q.8-11 What does "matched transmission line" mean?

Q.8-12 On what factors does the input impedance of a transmission line depend?

Q.8-13 What is the input impedance of an open-circuited lossless transmission line if the length of the line is (a) $\lambda/4$, (b) $\lambda/2$, and (c) $3\lambda/4$?

Q.8-14 What is the input impedance of a short-circuited lossless transmission line if the length of the line is (a) $\lambda/4$, (b) $\lambda/2$, and (c) $3\lambda/4$?

Q.8-15 Is the input reactance of a transmission line $\lambda/8$ long inductive or capacitive if it is (a) open-circuited, and (b) short-circuited?

Q.8-16 What is the input impedance of a lossless transmission line of length ℓ that is terminated in a load impedance Z_L if (a) $\ell = \lambda/2$, and (b) $\ell = \lambda$?

Q.8-17 Describe a method for determining the characteristic impedance and the propagation constant of a transmission line.

REMARKS

1. The input impedance of a transmission line terminated in its characteristic impedance is equal to the characteristic impedance, irrespective of the length of the line.
2. As in the case of an infinite transmission line, only forward waves exist on a line terminated in its characteristic impedance. At the input (sending) end, the circuit conditions for an infinite line and for a matched line are the same.
3. The input impedance of a short-circuited line of length ℓ is $jR_0 \tan \beta\ell$; it can be either inductive or capacitive.
4. The characteristic impedance and propagation constant of a line can be determined by measuring the input impedances with the line open-circuited and with the line short-circuited.

8-5.3 REFLECTION COEFFICIENT AND STANDING-WAVE RATIO

On a finite transmission line such as that shown in Fig. 8-4, the voltage and current waves traveling from the generator end toward the load Z_L will give rise to reflected voltage and current waves if $Z_L \neq Z_0$. The voltage at any distance $z' = \ell - z$ from the load has been obtained in Eq. (8-72), in which the term with $e^{\gamma z'}$ represents the incident voltage wave and the term with $e^{-\gamma z'}$ represents the reflected voltage wave. We may write

$$V(z') = \frac{I_L}{2}(Z_L + Z_0)e^{\gamma z'}\left[1 + \frac{Z_L - Z_0}{Z_L + Z_0}e^{-2\gamma z'}\right]$$

$$= \frac{I_L}{2}(Z_L + Z_0)e^{\gamma z'}[1 + \Gamma e^{-2\gamma z'}], \tag{8-87}$$

where

Definition of voltage reflection coefficient of load impedance Z_L

$$\boxed{\Gamma = \frac{Z_L - Z_0}{Z_L + Z_0} = |\Gamma|e^{j\theta_\Gamma} \qquad \text{(Dimensionless)}} \tag{8-88}$$

is the ratio of the complex amplitudes of the reflected and incident voltage waves at the load ($z' = 0$), and is called the ***voltage reflection coefficient of the load impedance*** Z_L. It is of the same form as the definition of the reflection coefficient in Eq. (7-94) for a plane wave incident normally on a plane interface between two dielectric media. It is, in general, a complex quantity with a magnitude $|\Gamma| \leq 1$.

The current equation corresponding to $V(z')$ in Eq. (8-87) is, from Eq. (8-73),

$$I(z') = \frac{I_L}{2Z_0}(Z_L + Z_0)e^{\gamma z'}[1 - \Gamma e^{-2\gamma z'}]. \tag{8-89}$$

Current reflection coefficient equals the negative of voltage reflection coefficient.

The current reflection coefficient, defined as the ratio of the complex amplitudes of the reflected and incident current waves at the load ($z' = 0$), is the negative of the voltage reflection coefficient, as is evident from Eq. (8-64). In what follows we shall refer only to the voltage reflection coefficient. If a transmission line is matched—that is, if $Z_L = Z_0$—$\Gamma = 0$ and there is no reflection at the load. When $Z_L \neq Z_0$, standing voltage and current waves exist along the line in accordance with Eqs. (8-87) and (8-89), and exhibit maxima and minima.

Analogous to the plane-wave case in Eq. (7-98), we define the ratio of the maximum to the minimum voltages along a finite, terminated line as the ***standing-wave ratio* (SWR),** S:

Definition of standing-wave ratio (SWR)

$$\boxed{S = \frac{|V_{max}|}{|V_{min}|} = \frac{1 + |\Gamma|}{1 - |\Gamma|} \qquad \text{(Dimensionless).}} \tag{8-90}$$

Depending on the value of Z_L, SWR can range from 1 ($|\Gamma| = 0$, matched load) to ∞ ($|\Gamma| = 1$, open circuit or short circuit). Because of the wide range of S, it is customary to express it on a logarithmic scale: $20\log_{10} S$ in (dB). A high standing-wave ratio on a line is undesirable because it results in a large power loss. The inverse relation of Eq. (8-90) is

Calculating $|\Gamma|$ from SWR

$$|\Gamma| = \frac{S-1}{S+1} \quad \text{(Dimensionless).} \tag{8-91}$$

The SWR on a transmission line can be easily measured by taking the ratio of the maximum and the minimum detected electric field intensities picked up by a small probe that protrudes into the line through a narrow slot cut along a section of the line. From Eq. (8-90), $S = |V_{max}|/|V_{min}|$, and $|\Gamma|$ is obtained by using Eq. (8-91). The angle θ_Γ can be determined from the location of V_{max} or V_{min}. (The distance between successive voltage maxima, or successive voltage minima, is a half-wavelength.) Having found $|\Gamma|$ and θ_Γ, Z_L can be calculated from Eq. (8-88), as will be shown in Example 8-5.

For a *lossless* transmission line, $\gamma = j\beta$, Eqs. (8-87) and (8-89) become

$$V(z') = \frac{I_L}{2}(Z_L + R_0)e^{j\beta z'}[1 + |\Gamma|e^{j(\theta_\Gamma - 2\beta z')}] \tag{8-92}$$

and

$$I(z') = \frac{I_L}{2R_0}(Z_L + R_0)e^{j\beta z'}[1 - |\Gamma|e^{j(\theta_\Gamma - 2\beta z')}]. \tag{8-93}$$

EXAMPLE 8-5

The standing-wave ratio on a lossless 50-(Ω) transmission line terminated in an unknown load impedance is found to be 3.0. The distance between successive voltage minima is 20 (cm), and the first minimum is located at 5 (cm) from the load. Determine (a) the reflection coefficient Γ, and (b) the load impedance Z_L.

SOLUTION

a) The distance between successive voltage minima is a half-wavelength.

$$\lambda = 2 \times 0.2 = 0.4 \quad \text{(m)}, \qquad \beta = \frac{2\pi}{\lambda} = \frac{2\pi}{0.4} = 5\pi \quad \text{(rad/m)}.$$

We find the magnitude of the reflection coefficient, $|\Gamma|$, from the given SWR $S = 3$ and Eq. (8-91)

$$|\Gamma| = \frac{S-1}{S+1} = \frac{3-1}{3+1} = 0.5.$$

In order to determine the angle θ_Γ, we observe from Eq. (8-92) that the first voltage minimum occurs when

$$\theta_\Gamma - 2\beta z'_m = -\pi,$$

where z'_m denotes the location of the first voltage minimum. We have

$$\theta_\Gamma = 2\beta z'_m - \pi$$
$$= 2 \times 5\pi \times 0.05 - \pi = -0.5\pi = -\pi/2 \quad \text{(rad)}.$$

Thus,

$$\Gamma = |\Gamma|e^{j\theta_\Gamma} = 0.5e^{-j\pi/2} = -j0.5.$$

b) The load impedance Z_L is determined from Eq. (8-88).

$$\frac{Z_L - 50}{Z_L + 50} = -j0.5,$$

$$Z_L = \frac{50 - j25}{1 + j0.5} = 30 - j40 \quad (\Omega).$$

EXERCISE 8.8 Repeat Example 8-5 for the same lossless 50-(Ω) transmission line operating at the same frequency but with a different unknown load resistance. If the SWR is 2.5 and a voltage maximum appears at the load, find (a) the reflection coefficient Γ_L, and (b) the load impedance Z_L.

ANS. (a) 0.43, (b) 125.4 (Ω).

In Eqs. (8-87) and (8-89) we expressed the voltage $V(z')$ and current $I(z')$ on a finite transmission line terminated in Z_L in terms of the load current I_L and the voltage reflection coefficient Γ of the load. No mention was made of the conditions at the input or generator end; but, of course, I_L depends on the conditions at the input end. If a generator of voltage V_g and internal impedance Z_g is connected to the input end of the line as in Fig. 8-4, we have the following condition:

$$V_i = V_g - I_i Z_g, \tag{8-94}$$

where V_i and I_i are obtained by setting $z' = \ell$ in Eqs. (8-87) and (8-89) respectively. In terms of traveling voltage waves, an input voltage $V_i = V_g Z_0/(Z_0 + Z_g)$ travels toward the load with a velocity $u_p = \omega/\beta$ as soon as the generator is connected to the input terminals of the line. At the load, if $Z_L \neq Z_0$, the incident wave is reflected with a reflection coefficient Γ. This reflected wave travels back to the input end with the same u_p. If $Z_g \neq Z_0$, another reflection takes place with a reflection coefficient $\Gamma_g = (Z_g - Z_0)/(Z_g + Z_0)$. The above process will repeat indefinitely with reflections at both ends, and the standing wave $V(z')$ is the sum of all the waves traveling in both directions.

Reflection also takes place at the generator (input) end if $Z_g \neq Z_0$.

If $Z_L = Z_0$ (matched load), $\Gamma = 0$, there will be only one wave traveling from the generator and ending at the load. If $Z_L \neq Z_0$ but $Z_g = Z_0$, there will be one initial wave traveling from the generator to the load (incident wave) and one reflected wave going from the load and ending at the generator.

EXAMPLE 8-6

A 100-(MHz) generator with $V_g = 10\underline{/0^\circ}$ (V) and internal resistance 50 (Ω) is connected to a lossless 50-(Ω) air line that is 3.6 (m) long and terminated in a $25 + j25$ (Ω) load. Find (a) $V(z)$ at a location z from the generator, (b) V_i at the input terminals and V_L at the load, (c) the standing-wave ratio on the line, and (d) the average power delivered to the load.

SOLUTION

Referring to Fig. 8-4, the given quantities are

$$V_g = 10\underline{/0^\circ} \quad \text{(V)}, \qquad Z_g = 50 \quad (\Omega), \qquad f = 10^8 \quad \text{(Hz)},$$
$$R_0 = 50 \quad (\Omega), \qquad Z_L = 25 + j25 = 35.36\underline{/45^\circ} \quad (\Omega), \qquad \ell = 3.6 \quad \text{(m)}.$$

Thus,

$$\beta = \frac{\omega}{c} = \frac{2\pi 10^8}{3 \times 10^8} = \frac{2\pi}{3} \quad \text{(rad/m)}, \qquad \beta\ell = 2.4\pi \quad \text{(rad)},$$

$$\Gamma = \frac{Z_L - Z_0}{Z_L + Z_0} = \frac{(25 + j25) - 50}{(25 + j25) + 50} = \frac{-25 + j25}{75 + j25} = \frac{35.36\underline{/135^\circ}}{79.1\underline{/18.4^\circ}}$$
$$= 0.447\underline{/116.6^\circ} = 0.447\underline{/0.648\pi},$$

$\Gamma_g = 0$. (Only one incident wave and one reflected wave.)

a) To find $V(z)$, we substitute $V(z' = \ell) = V_i$ from Eq. (8-87) and $I(z' = \ell) = I_i$ from Eq. (8-89) into Eq. (8-94) to determine

$$I_L(Z_L + Z_0)e^{\gamma\ell} = V_g. \tag{8-95}$$

Using Eq. (8-95) in Eq. (8-87), we obtain

$$V(z) = \frac{V_g}{2} e^{-j\beta z}[1 + \Gamma e^{-j2\beta(\ell - z)}]. \tag{8-96}$$

For this problem,

$$V(z) = \frac{10}{2} e^{-j2\pi z/3}[1 + 0.447\, e^{j(0.648 - 4.8)\pi}\, e^{j4\pi z/3}]$$
$$= 5[e^{-j2\pi z/3} + 0.447 e^{j(2z/3 - 0.152)\pi}] \quad \text{(V)}.$$

b) At the input terminals,

$$V_i = V(0) = 5(1 + 0.447 e^{-j0.152\pi})$$
$$= 5(1.396 - j0.207)$$
$$= 7.06\underline{/-8.43^\circ} \quad \text{(V)}.$$

At the load,

$$V_L = V(3.6) = 5[e^{-j0.4\pi} + 0.447 e^{j0.248\pi}]$$
$$= 5(0.627 - j0.637) = 4.47\underline{/-45.5^\circ} \quad \text{(V)}.$$

c) The standing-wave ratio (SWR) is

$$S = \frac{1 + |\Gamma|}{1 - |\Gamma|} = \frac{1 + 0.447}{1 - 0.447} = 2.62.$$

d) The average power delivered to the load is

$$P_{av} = \frac{1}{2}\left|\frac{V_L}{Z_L}\right|^2 R_L = \frac{1}{2}\left(\frac{4.47}{35.36}\right)^2 \times 25 = 0.20 \quad \text{(W)}.$$

EXERCISE 8.9 For the transmission-line circuit in Example 8-6, find the average power delivered to a matched load $Z_L = Z_0 = 50 + j0\ (\Omega)$. Explain the difference between your answer and the result obtained in part (d) of the Example.

ANS. 0.25 (W).

REVIEW QUESTIONS

Q.8-18 Define *voltage reflection coefficient.* Is it the same as "current reflection coefficient"? Explain.

Q.8-19 Define *standing-wave ratio.* How is it related to voltage and current reflection coefficients?

Q.8-20 Why is a high standing-wave ratio on a transmission line undesirable?

Q.8-21 What are Γ and S for a line with an open-circuit termination? A short-circuit termination?

Q.8-22 Explain how the value of a terminating resistance can be determined by measuring the standing-wave ratio on a lossless transmission line.

REMARKS

1. When a transmission line is not matched ($Z_L \neq Z_0, \Gamma \neq 0$), a standing wave will exist on the line. The distance between successive voltage maxima (or successive voltage minima) is $\lambda/2$, and the distance between neighboring V_{max} and V_{min} is $\lambda/4$.
2. For a lossless line with characteristic resistance R_0 terminated in a resistance R_L, a V_{max} appears at the load if $R_L > R_0$, and a V_{min} appears at the load if $R_L < R_0$.—This can be seen from Eq. (8-92) by setting $z' = 0$ and noting from Eq. (8-88) that $\theta_\Gamma = 0$ for $R_L > R_0$ and $\theta_\Gamma = \pi$ for $R_L < R_0$.
3. Refer to the finite transmission-line circuit in Fig. 8-4, and assume the generator to be switched on at $t = 0$. A steady state on the line will be reached: (a) at $t_1 = \ell/u_p = \beta\ell/\omega$ if $Z_L = Z_0$ (whether $Z_g = Z_0$ or $Z_g \neq Z_0$) and (b) at $2t_1$ if $Z_L \neq Z_0$ and $Z_g = Z_0$.

8-6 THE SMITH CHART

Graphical charts simplify transmission-line calculations.

Transmission-line calculations often involve tedious manipulations of complex numbers. This tedium can be alleviated by using a graphical method of solution. The best known and most widely used graphical chart is the ***Smith chart*** devised by P. H. Smith.† A Smith chart is a graphical plot of normalized resistance and reactance functions in the reflection-coefficient plane.

A Smith chart

To understand how the Smith chart for a *lossless* transmission line is constructed let us examine the voltage reflection coefficient of the load impedance defined in Eq. (8-88):

$$\Gamma = \frac{Z_L - R_0}{Z_L + R_0} = |\Gamma| e^{j\theta_\Gamma}. \tag{8-97}$$

Let the load impedance Z_L be normalized with respect to the characteristic impedance $R_0 = \sqrt{L/C}$ of the line.

$$z_L = \frac{Z_L}{R_0} = \frac{R_L}{R_0} + j\frac{X_L}{R_0} = r + jx \qquad \text{(Dimensionless)}, \tag{8-98}$$

where r and x are the normalized resistance and normalized reactance, respectively. Equation (8-97) can be rewritten as

$$\Gamma = \Gamma_r + j\Gamma_i = \frac{z_L - 1}{z_L + 1}, \tag{8-99}$$

where Γ_r and Γ_i are the real and imaginary parts, respectively, of the voltage reflection coefficient Γ. The inverse relation of Eq. (8-99) is

$$z_L = \frac{1 + \Gamma}{1 - \Gamma} = \frac{1 + |\Gamma| e^{j\theta_\Gamma}}{1 - |\Gamma| e^{j\theta_\Gamma}}. \tag{8-100}$$

Expressing both z_L and Γ in their real and imaginary components, we have

$$r + jx = \frac{(1 + \Gamma_r) + j\Gamma_i}{(1 - \Gamma_r) - j\Gamma_i}. \tag{8-101}$$

Multiplying both the numerator and the denominator of Eq. (8-101) by the complex conjugate of the denominator and separating the real and imaginary parts, we obtain

$$r = \frac{1 - \Gamma_r^2 - \Gamma_i^2}{(1 - \Gamma_r)^2 + \Gamma_i^2} \tag{8-102}$$

and

$$x = \frac{2\Gamma_i}{(1 - \Gamma_r)^2 + \Gamma_i^2}. \tag{8-103}$$

† P. H. Smith, "Transmission-line calculator," *Electronics*, vol. 12, p. 29, January 1939; and "An improved transmission-line calculator," *Electronics*, vol. 17, p. 130, January 1944.

If Eq. (8-102) is plotted in the $\Gamma_r - \Gamma_i$ plane for a given value of r, the resulting graph is the locus for this r. The locus can be recognized when the equation is rearranged as

Equation of constant-r circles on a Smith chart

$$\left(\Gamma_r - \frac{r}{1+r}\right)^2 + \Gamma_i^2 = \left(\frac{1}{1+r}\right)^2. \tag{8-104}$$

It is the equation for a circle having a radius $1/(1+r)$ and centered at $\Gamma_r = r/(1+r)$ and $\Gamma_i = 0$. Different values of r yield circles of different radii with centers at different positions on the Γ_r-axis. A family of r-circles are shown in solid lines in Fig. 8-6. Since $|\Gamma| \leq 1$ for a lossless line, only that part of the graph lying within the unit circle on the $\Gamma_r - \Gamma_i$ plane is meaningful; everything outside can be disregarded. Note that all circles pass through the (1, 0) point. The $r = 0$ circle, having a unity radius and centered at the origin, is the largest.

Similarly, Eq. (8-103) may be rearranged as

Equation of constant-x circles on a Smith chart

$$(\Gamma_r - 1)^2 + \left(\Gamma_i - \frac{1}{x}\right)^2 = \left(\frac{1}{x}\right)^2. \tag{8-105}$$

FIGURE 8-6 Smith chart with rectangular coordinates.

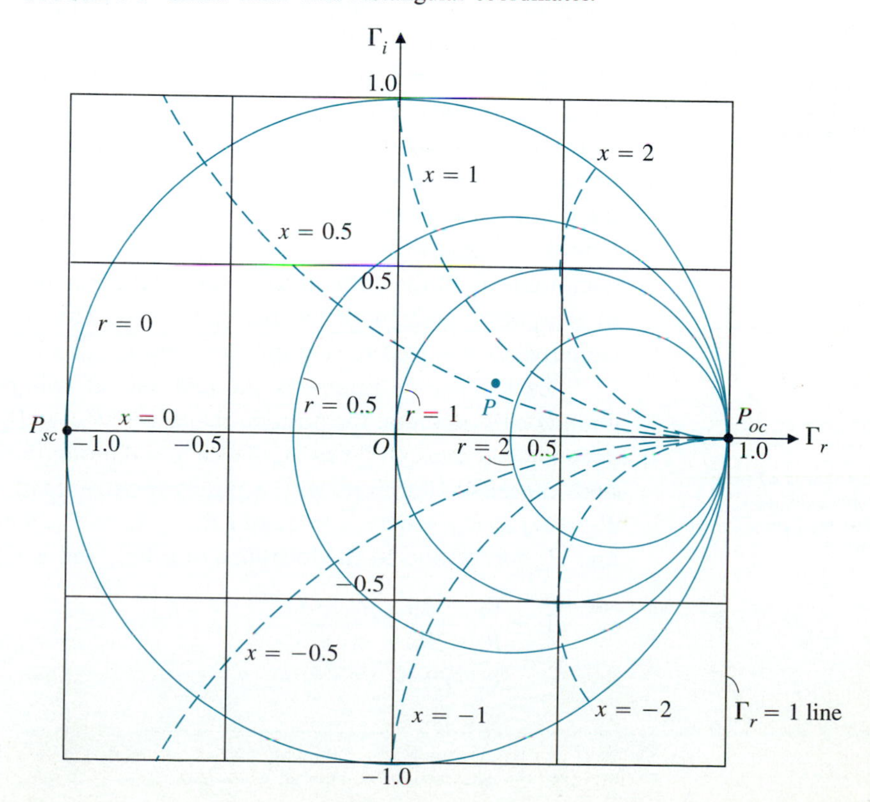

This is the equation for a circle having a radius $1/|x|$ and centered at $\Gamma_r = 1$ and $\Gamma_i = 1/x$. Different values of x yield circles of different radii with centers at different positions on the $\Gamma_r = 1$ line. A family of the portions of x-circles lying inside the $|\Gamma| = 1$ boundary are shown in dashed lines in Fig. 8-6. All x-circles also pass through the (1, 0) point. Their centers lie above the Γ_r-axis for $x > 0$ (inductive reactances) and below the Γ_r−axis for $x < 0$ (capacitive reactances). The radius of the x-circle is larger as $|x|$ decreases, and the locus for $x = 0$ degenerates into the Γ_r-axis.

It can be proved that the r- and x-circles are everywhere orthogonal to one another. The intersection of an r-circle and an x-circle defines a point that represents a normalized load impedance $z_L = r + jx$. The actual load impedance is $Z_L = R_0(r + jx)$.

Locations of short-circuit point P_{sc} and open-circuit point P_{oc} on a Smith chart

As an illustration, point P in Fig. 8-6 is the intersection of the $r = 1.7$ circle and the $x = 0.6$ circle. (The same point P is more exactly located on the detailed chart in Fig. 8-8.) Hence it represents $z_L = 1.7 + j0.6$. The point P_{sc} at ($\Gamma_r = -1, \Gamma_i = 0$) corresponds to $r = 0$ and $x = 0$ and therefore represents a short circuit. The point P_{oc} at ($\Gamma_r = 1$, $\Gamma_i = 0$) corresponds to an infinite impedance and represents an open circuit.

Smith chart on reflection-coefficient plane can be marked with $\Gamma_r - \Gamma_i$ rectangular coordinates or with $|\Gamma| - \theta_\Gamma$ polar coordinates.

The Smith chart in Fig. 8-6 is marked with Γ_r and Γ_i rectangular coordinates. The same chart can be marked with polar coordinates, such that every point in the Γ-plane is specified by a magnitude $|\Gamma|$ and a phase angle θ_Γ. This is illustrated in Fig. 8-7, where several $|\Gamma|$-circles are shown in dashed black lines and some θ_Γ-angles are marked around the $|\Gamma| = 1$ circle which is the same as the $r = 0$ circle. The $|\Gamma|$-circles are normally not shown on commercially available Smith charts; but once the point representing a certain $z_L = r + jx$ is located, it is a simple matter to draw a circle centered at the origin through the point. The fractional distance from the center to the point (compared with the unity radius to the edge of the chart) is equal to the magnitude $|\Gamma|$ of the voltage reflection coefficient; and the angle that the line to the point makes with the real axis is θ_Γ. This graphical determination circumvents the need for computing Γ by Eq. (8-99).

Locations of points representing V_{max} and V_{min} on a Smith chart

Each $|\Gamma|$-circle intersects the real axis at two points. In Fig. 8-7 we designate the point on the positive-real axis (OP_{oc}) as P_M and the point on the negative-real axis (OP_{sc}) as P_m. Since $x = 0$ along the real axis, P_M and P_m both represent situations with a purely resistive load, $Z_L = R_L$. Obviously, $R_L > R_0$ at P_M, where $r_L > 1$; and $R_L < R_0$ at P_m, where $r_L < 1$. Points P_M and P_m correspond to the locations of a V_{max} and a V_{min}, respectively. Now,

$$\Gamma = \frac{R_L - R_0}{R_L + R_0} = \frac{r_L - 1}{r_L + 1}$$

$$= \frac{S - 1}{S + 1}. \tag{8-106}$$

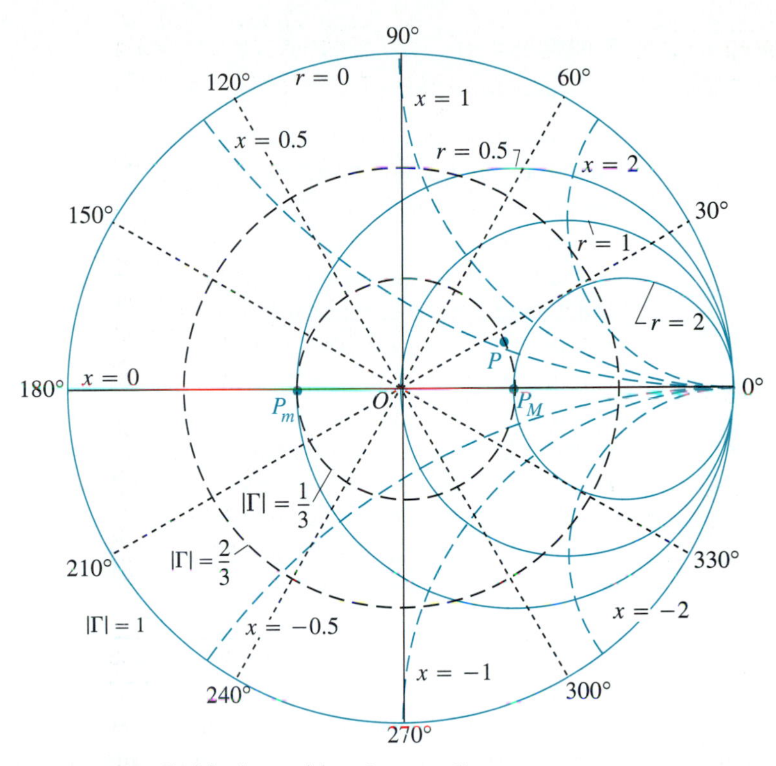

FIGURE 8-7 Smith chart with polar coordinates.

Finding SWR, S, on a Smith chart

Thus, $S = r_L = R_L/R_0$ for $R_L > R_0$. It follows that *the value of the r-circle passing through the point P_M is numerically equal to the standing-wave ratio.* For the $z_L = 1.7 + j0.6$ point, marked P in Fig. 8-7, we find $|\Gamma| = 1/3$ and $\theta_\Gamma = 28°$. At P_M, $r = S = 2.0$. These results can be verified analytically.

So far we have based the construction of the Smith chart on the definition of the voltage reflection coefficient of the load impedance, as given in Eq. (8-88). The input impedance looking toward the load at a distance z' from the load is the ratio of $V(z')$ and $I(z')$. From Eqs. (8-87) and (8-89) we have, by writing $j\beta$ for γ for a lossless line,

$$Z_i(z') = \frac{V(z')}{I(z')} = R_0\left[\frac{1 + \Gamma e^{-j2\beta z'}}{1 - \Gamma e^{-j2\beta z'}}\right]. \tag{8-107}$$

The normalized input impedance is

$$z_i = \frac{Z_i}{R_0} = \frac{1 + \Gamma e^{-j2\beta z'}}{1 - \Gamma e^{-j2\beta z'}}$$

$$= \frac{1 + |\Gamma| e^{j\phi}}{1 - |\Gamma| e^{j\phi}}, \tag{8-108}$$

where

$$\phi = \theta_\Gamma - 2\beta z'. \tag{8-109}$$

We note that Eq. (8-108) relating z_i and $\Gamma e^{-j2\beta z'} = |\Gamma| e^{j\phi}$ is of exactly the same form as Eq. (8-100) relating z_L and $\Gamma = |\Gamma| e^{j\theta_\Gamma}$. In fact, the latter is a special case of the former for $z' = 0$ ($\phi = \theta_\Gamma$). The magnitude, $|\Gamma|$, of the reflection coefficient and therefore the standing-wave ratio S, are not changed by the additional line length z'. Thus, just as we can use the Smith chart to find $|\Gamma|$ and θ_Γ for a given z_L at the load, we can keep $|\Gamma|$ constant and *subtract* (rotate in the *clockwise* direction) from θ_Γ an angle equal to $2\beta z' = 4\pi z'/\lambda$. This will locate the point for $|\Gamma| e^{j\phi}$, which determines z_i, the normalized input impedance looking into a lossless line of characteristic impedance R_0, length z', and a normalized load impedance z_L. Two additional scales in $\Delta z'/\lambda$ are usually provided along the perimeter of the $|\Gamma| = 1$ circle for easy reading of the phase change $2\beta(\Delta z')$ due to a change in line length $\Delta z'$: The outer scale is marked "wavelengths toward generator" in the clockwise direction (increasing z'); and the inner scale is marked "wavelengths toward load" in the counterclockwise direction (decreasing z'). Figure 8-8 is a typical Smith chart, which is commercially available.† It has a complicated appearance, but it actually consists merely of constant-r and constant-x circles. We note that a change of half a wavelength in line length ($\Delta z' = \lambda/2$) corresponds to a $2\beta(\Delta z') = 2\pi$ change in ϕ. A complete revolution around a $|\Gamma|$-circle returns to the same point and results in no change in impedance.

Finding normalized input impedance, z_i, on a Smith chart

We shall illustrate the use of the Smith chart for solving some typical transmission-line problems by several examples.

EXAMPLE 8-7

Use the Smith chart to find the input impedance of a section of a 50-(Ω) lossless transmission line that is 0.1 wavelength long and is terminated in a short-circuit.

SOLUTION

Given

$$z_L = 0,$$

$$R_0 = 50 \quad (\Omega),$$

$$z' = 0.1\lambda.$$

1. Enter the Smith chart at the intersection of $r = 0$ and $x = 0$ (point P_{sc} on the extreme left of the chart; see Fig. 8-9).
2. Move along the perimeter of the chart ($|\Gamma| = 1$) by 0.1 "wavelengths toward generator" in a clockwise direction to P_1.

† All of the Smith charts used in this book are reprinted with permission of Emeloid Industries, Inc., New Jersey.

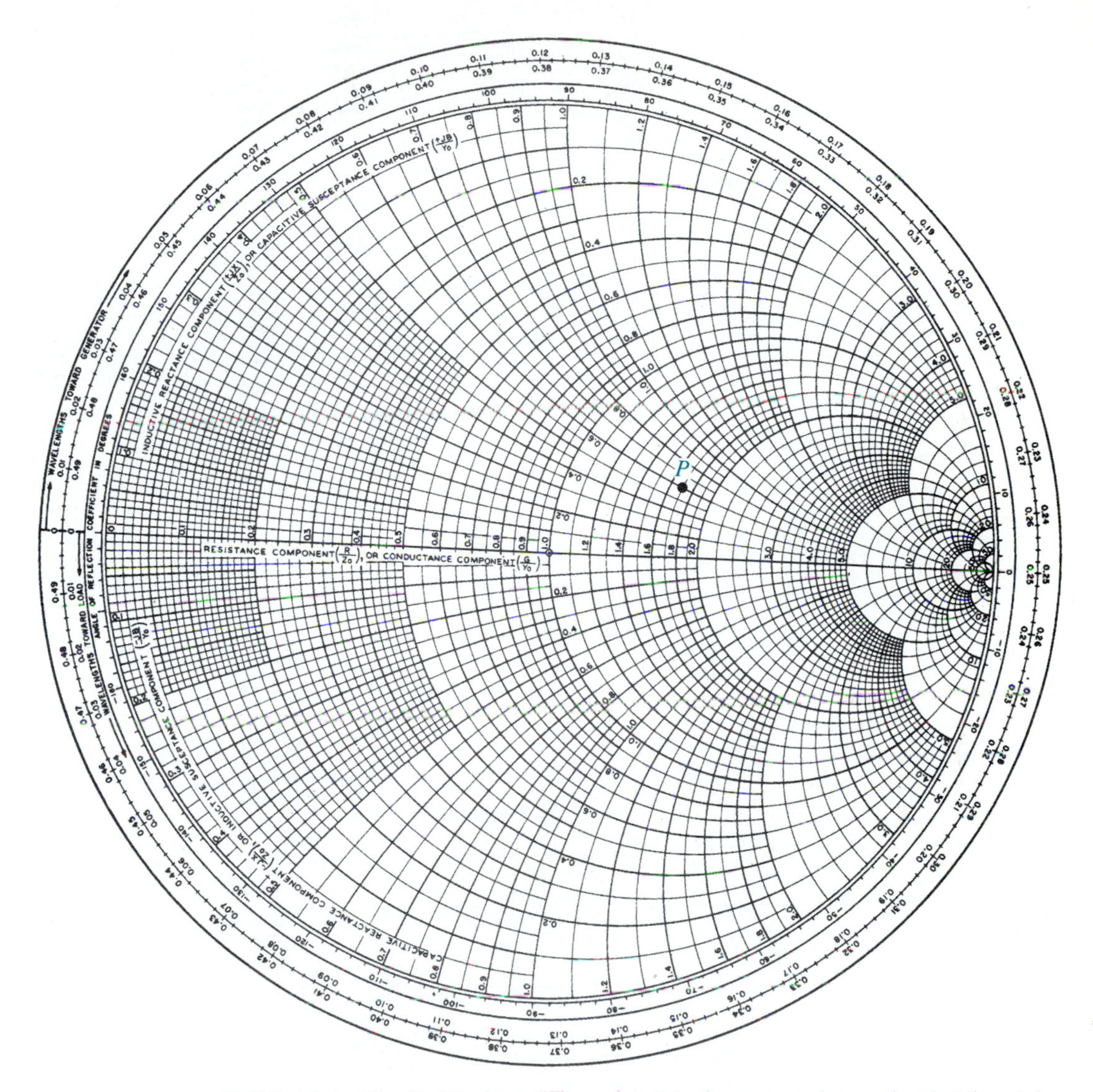

FIGURE 8-8 The Smith chart. (The point P is the same point as that in Figs. 8-6 and 8-7.)

3. At P_1, read $r = 0$ and $x \cong 0.725$, or $z_i = j0.725$. Thus, $Z_i = R_0 z_i = 50(j0.725) = j36.3$ (Ω). (The input impedance is purely inductive.)

This result can be checked readily by using Eq. (8-82):

$$Z_i = jR_0 \tan \beta\ell = j50 \tan\left(\frac{2\pi}{\lambda}\right)0.1\lambda$$

$$= j50 \tan 36^\circ = j36.3 \quad (\Omega).$$

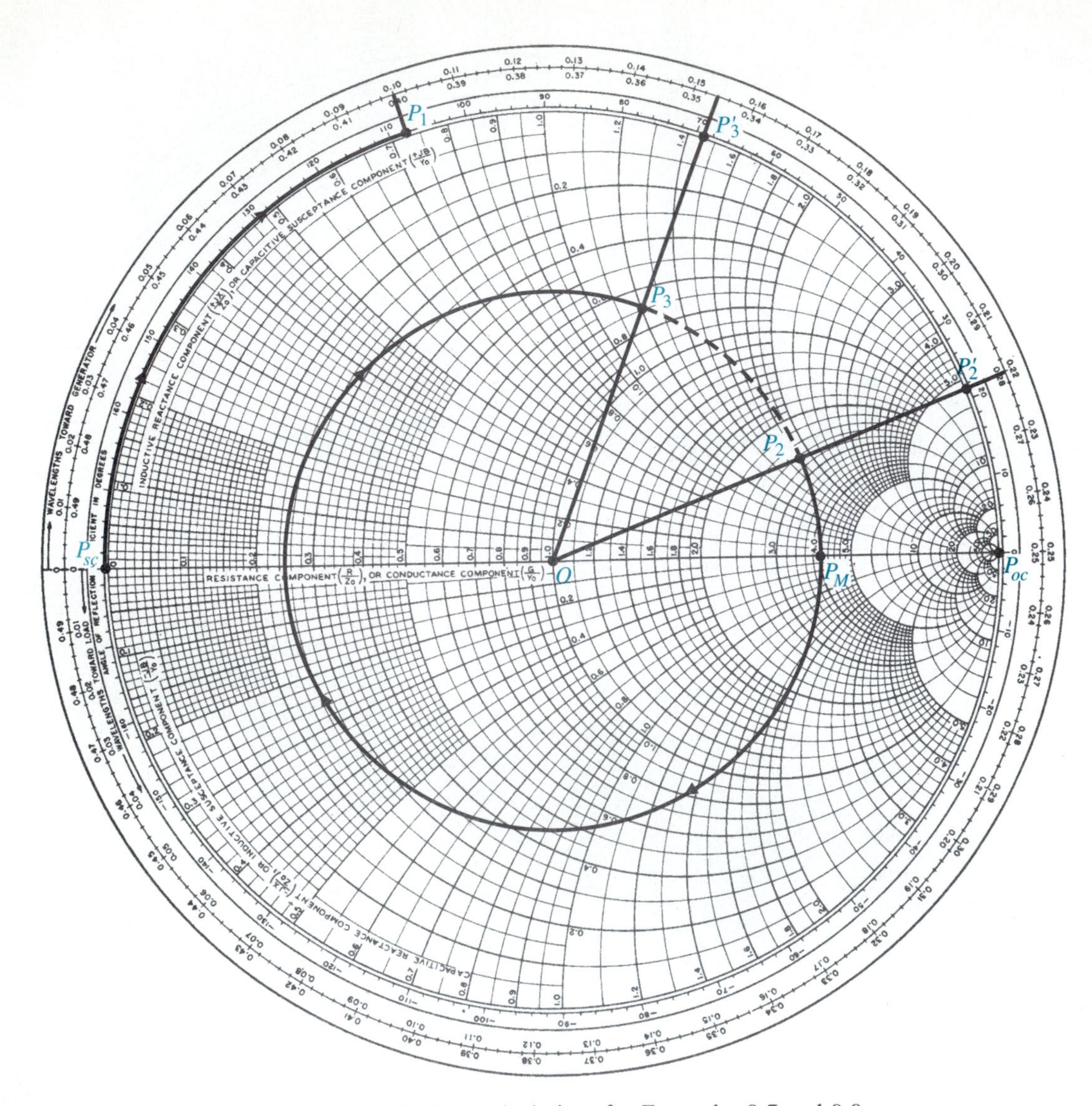

FIGURE 8-9 Smith-chart calculations for Examples 8-7 and 8-8.

EXERCISE 8.10 The input impedance of an open-circuited 75-(Ω) lossless transmission line is a capacitive reactance of 90 (Ω). Use the Smith chart to determine the length of the line in terms of wavelength.

ANS. 0.11λ.

EXAMPLE 8-8

A lossless transmission line of length 0.434λ and characteristic impedance 100 (Ω) is terminated in an impedance $260 + j180$ (Ω). Find (a) the voltage reflection coefficient, (b) the standing-wave ratio, (c) the input impedance, and (d) the location of a voltage maximum closest to the load.

SOLUTION

Given

$$z' = 0.434\lambda,$$
$$R_0 = 100 \quad (\Omega),$$
$$Z_L = 260 + j180 \quad (\Omega).$$

a) We find the voltage reflection coefficient in several steps:

1. Enter the Smith chart at $z_L = Z_L/R_0 = 2.6 + j1.8$ (point P_2 in Fig. 8-9).

2. With the center at the origin, draw a circle of radius $\overline{OP}_2 = |\Gamma| = 0.60$. (The radius of the chart $\overline{OP}_{sc}$ equals unity.)

3. Draw the straight line OP_2 and extend it to P_2' on the periphery. Read 0.220 on "wavelengths toward generator" scale. The phase angle θ_Γ of the reflection coefficient is $(0.250 - 0.220) \times 4\pi = 0.12\pi$ (rad) or 21.6°. (We multiply the change in wavelengths by 4π because angles on the Smith chart are measured in $2\beta z'$ or $4\pi z'/\lambda$.) This angle can also be read from the markings on the periphery. The answer to part (a) is then

$$\Gamma = |\Gamma| e^{j\theta_\Gamma} = 0.60\underline{/21.6°}.$$

b) The $|\Gamma| = 0.60$ circle intersects with the positive-real axis OP_{oc} at $r = S = 4$. Thus the voltage standing-wave ratio is 4.

c) To find the input impedance, we proceed as follows:

1. Move P_2' at 0.220 clockwise by a total of 0.434 "wavelengths toward generator," first to 0.500 (same as 0.000) and then further to 0.154[(0.500 − 0.220) + 0.154 = 0.434] to P_3'.

2. Join O and P_3' by a straight line, which intersects the $|\Gamma| = 0.60$ circle at P_3.

3. Read $r = 0.69$ and $x = 1.2$ at P_3. Hence,

$$Z_i = R_0 z_i = 100(0.69 + j1.2) = 69 + j120 \quad (\Omega).$$

d) In going from P_2 to P_3, the $|\Gamma| = 0.60$ circle intersects the positive-real axis OP_{oc} at P_M where the voltage is a maximum. Thus a voltage maximum appears at $(0.250 - 0.220)\lambda$ or 0.030λ from the load.

EXERCISE 8.11 It is decided to reduce the standing-wave ratio on the line in Example 8-8 from 4 to 2 by changing the terminating impedance to a resistive load R_L. (a) What should R_L be? (b) What will be the input impedance?

ANS. (a) 200 (Ω), (b) $13.5 + j76$ (Ω).

REVIEW QUESTIONS

Q.8-23 What is a Smith chart and why is it useful in making transmission-line calculations?

Q.8-24 What are the rectangular coordinates of a Smith chart?

Q.8-25 What are the polar coordinates of a Smith chart?

Q.8-26 Where is the point representing a matched load on a Smith chart?

Q.8-27 For a given load impedance Z_L on a lossless line of characteristic impedance Z_0, how do we use a Smith chart to determine (a) the reflection coefficient, and (b) the standing-wave ratio?

REMARKS

1. The r- and x-circles on a Smith chart are everywhere orthogonal to one another, and they all pass through the (1, 0) point.
2. Smith charts are applicable to transmission lines having any characteristic resistance.
3. The value of the r-circle passing through the intersection of a $|\Gamma|$-circle and the positive real axis equals the standing-wave ratio S.
4. A change of half a wavelength corresponds to a complete revolution on a Smith chart.

8-6.1 ADMITTANCES ON SMITH CHART

So far we have discussed the Smith chart in terms of normalized impedances $z = r + jx$; but the Smith chart can also be used to make admittance calculations. We recall from Eq. (8-79) the formula for the input impedance Z_i of a lossless line of length ℓ terminated in an impedance Z_L:

$$Z_i = R_0 \frac{Z_L + jR_0 \tan \beta\ell}{R_0 + jZ_L \tan \beta\ell}. \tag{8-79)(8-110}$$

When $\ell = \lambda/4$: $\beta\ell = \pi/2$, $\tan \beta\ell \to \infty$, and Eq. (8-110) becomes (see Exercise 8.6)

Property of quarter-wave transformers

$$\boxed{Z_i = \frac{R_0^2}{Z_L} \qquad \text{(Quarter-wave line).}} \tag{8-111}$$

Hence a quarter-wave lossless line acts as an impedance inverter and is often referred to as a ***quarter-wave transformer***.

Now let $Y_L = 1/Z_L$ denote the load admittance. The normalized load impedance is

$$z_L = \frac{Z_L}{R_0} = \frac{1}{R_0 Y_L} = \frac{1}{y_L}, \tag{8-112}$$

where

$$y_L = R_0 Y_L$$
$$= Y_L / Y_0 = g + jb \qquad \text{(Dimensionless)} \qquad (8\text{-}113)$$

is the normalized load admittance having normalized conductance g and normalized susceptance b as its real and imaginary parts, respectively. Equation (8-112) suggests that a quarter-wave line with a unity normalized characteristic impedance will transform z_L to y_L, and vice versa. On the Smith chart we need only move the point representing z_L along the $|\Gamma|$-circle by a quarter-wavelength to locate the point representing y_L. Since a $\lambda/4$-change in line length corresponds to a change of π radians on the Smith chart, ***the points representing z_L and y_L are*** then ***diametrically opposite to each other on the $|\Gamma|$-circle.*** This observation enables us to find y_L from z_L, and z_L from y_L, on the Smith chart in a very simple manner. (Remember that both z_L and y_L are dimensionless.)

Finding y_L from z_L, and vice versa, on a Smith chart

EXAMPLE 8-9

Given $Z = 95 + j20$ (Ω), use a Smith chart to find Y.

SOLUTION

This problem has nothing to do with any transmission line. In order to use the Smith chart we can choose an arbitrary normalizing constant; for instance, $R_0 = 50\,(\Omega)$. Thus,

$$z = \tfrac{1}{50}(95 + j20) = 1.9 + j0.4.$$

Enter z as point P_1 on the Smith chart in Fig. 8-10. The point P_2 on the other

FIGURE 8-10 Finding admittance from impedance (Example 8-9).

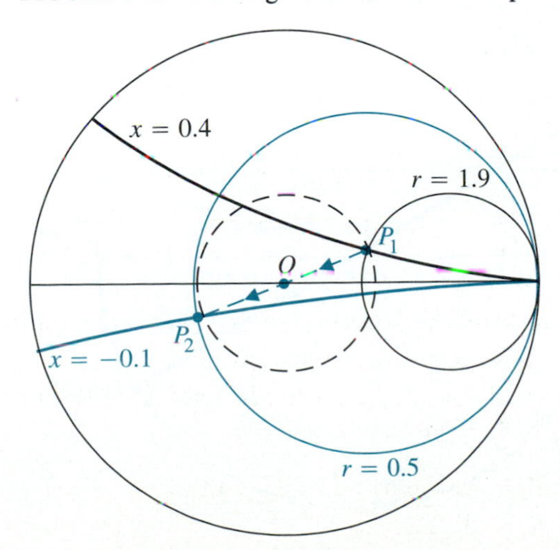

side of the line joining P_1 and O represents $y: \overline{OP_2} = \overline{OP}_1$.

$$Y = \frac{1}{R_0} y = \frac{1}{50}(0.5 - j0.1) = 10 - j2 \quad \text{(mS)}.$$

EXERCISE 8.12 Given $Y = 6 + j11$ (mS), use a Smith chart to find Z.

ANS. $38 - j70$ (Ω).

EXAMPLE 8-10

Use a Smith chart to find the input admittance of an open-circuited line of characteristic impedance 300 (Ω) and length 0.04λ.

SOLUTION

1. For an open-circuited line we start from the point P_{oc} on the extreme right of the impedance Smith chart, at 0.25 in Fig. 8-11.
2. Move along the perimeter of the chart by 0.04 "wavelengths toward generator" to P_3 (at 0.29).
3. Draw a straight line from P_3 through O, intersecting at P_3' on the opposite side.
4. Read at P_3'

   $y_i = 0 + j0.26$.

FIGURE 8-11 Finding input admittance of open-circuited line (Example 8-10).

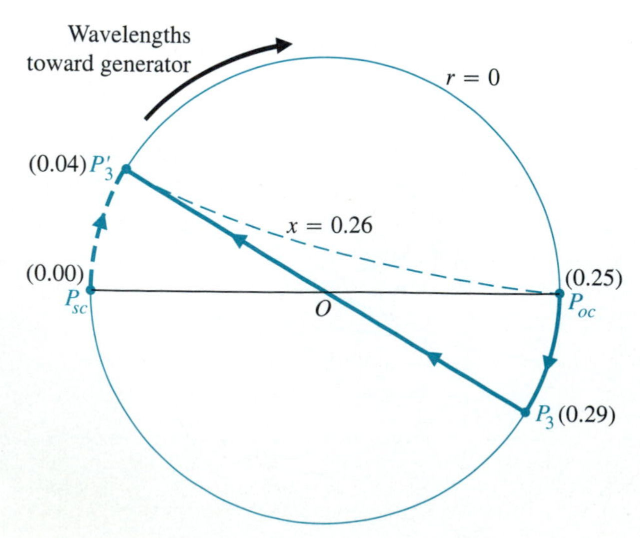

Thus,

$$Y_i = \frac{1}{300}(0 + j0.26) = j0.87 \quad \text{(mS)}.$$

A Smith chart can be used either as an impedance chart or as an admittance chart.

In the preceding two examples we have made admittance calculations by using the Smith chart as an impedance chart. The Smith chart can also be used as an admittance chart, in which case the r- and x-circles would be g- and b-circles. The points representing an open- and a short-circuit termination would be the points on the extreme left and the extreme right, respectively, on an admittance chart. For Example 8-10, we could then start from extreme left point on the chart, at 0.00 in Fig. 8-11, and move 0.04 "wavelengths toward generator" to P_3' directly.

The Smith chart as an admittance chart is actually more useful than as an impedance chart in solving problems involving parallel connections of lines because admittances add in parallel connections. This will become clear when we discuss impedance-matching in the following section.

8-7 TRANSMISSION-LINE IMPEDANCE MATCHING

The importance of impedance-matching on transmission lines

Transmission lines are used for the transmission of power and information. For radio-frequency power transmission it is highly desirable that as much power as possible is transmitted from the generator to the load and as little power as possible is lost on the line itself. This will require that the load be matched to the characteristic impedance of the line so that the standing-wave ratio on the line is as close to unity as possible. For information transmission it is essential that the lines be matched because reflections from mismatched loads and junctions will result in echoes and will distort the information-carrying signal. In this section we discuss a simple single-stub method for impedance-matching on lossless transmission lines.

Short-circuited (instead of open-circuited) stubs are used for impedance-matching on transmission lines.

An arbitrary load impedance can be matched to a transmission line by placing a single short-circuited stub *in parallel* with the line at a suitable location, as shown in Fig. 8-12. This is the *single-stub method* for impedance matching.† Since we deal with a parallel connection, it is more convenient to explain the method in terms of admittances. Short-circuited stubs are usually used in preference to open-circuited stubs because an infinite terminating impedance is more difficult to realize than a zero terminating impedance. Radiation from an open end and coupling with neighboring objects make the

† An alternative method, using two stubs spaced a fixed distance apart, is also used for impedance matching—the double-stub method. (See D. K. Cheng, *Field and Wave Electromagnetics*, 2/E, pp 504–509, Addison-Wesley Publishing Co., Reading, MA, 1989.)

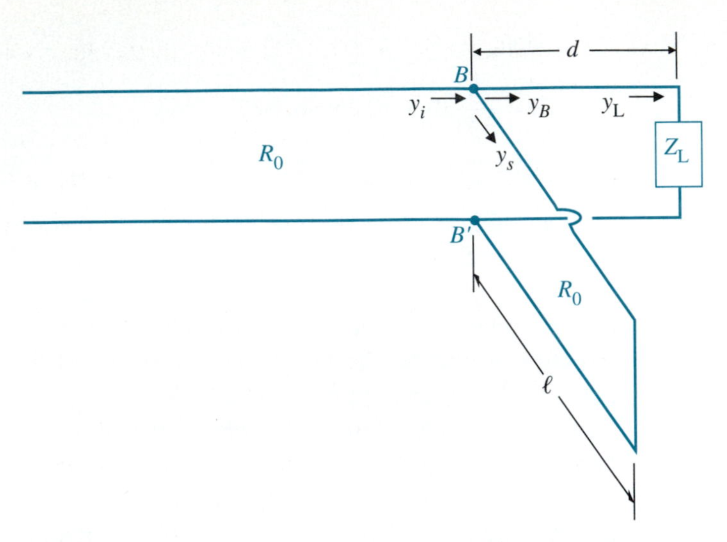

FIGURE 8-12 Impedance matching by single-stub method.

impedance non-infinite. Moreover, a short-circuited stub of an adjustable length and a constant characteristic resistance is much easier to construct (by simply changing the location of the short-circuit) than an open-circuited one whose length has to be cut precisely. Of course, the difference in the required length for an open-circuited stub and that for a short-circuited stub is an odd multiple of a quarter-wavelength.

Assuming Y_B to be the input admittance at B–B' looking toward the load in Fig. 8-12 without the stub, our impedance- (or admittance-) matching problem is to determine the location d and the length ℓ of the stub such that

$$Y_i = Y_0 = Y_B + Y_s, \tag{8-114}$$

where $Y_0 = 1/R_0$. In terms of normalized admittances, Eq. (8-114) becomes

$$1 = y_B + y_s, \tag{8-115}$$

where $y_B = R_0 Y_B$ is for the load section and $y_s = R_0 Y_s$ is for the short-circuited stub. However, since the input admittance of a short-circuited stub is purely susceptive, y_s is purely imaginary. As a consequence, Eq. (8-115) can be satisfied only if

$$y_B = 1 + jb_B \tag{8-116}$$

and

$$y_s = -jb_B, \tag{8-117}$$

where b_B can be either positive or negative. Our objectives, then, are (1) to find the length d such that the admittance, y_B, of the load section looking to the right of terminals B–B' has a *unity real part*, and (2) to find the length ℓ of the stub required to *cancel the imaginary part*.

Using the Smith chart as an admittance chart, we proceed as follows for single-stub matching:

Procedure for single-stub impedance-matching

1. Enter the point representing the normalized load admittance y_L.
2. Draw the $|\Gamma|$-circle for y_L, which will intersect the $g = 1$ circle at two points. At these points, $y_{B1} = 1 + jb_{B1}$ and $y_{B2} = 1 + jb_{B2}$. Both are possible solutions.
3. Determine load-section lengths d_1 and d_2 from the angles between the point representing y_L and the points representing y_{B1} and y_{B2}.
4. Determine stub lengths ℓ_1 and ℓ_2 from the angles between the short-circuit point P_{sc} on the *extreme right* of the chart to the points representing $-jb_{B1}$ and $-jb_{B2}$, respectively.

The following example will illustrate the necessary steps.

EXAMPLE 8-11

A 50-(Ω) transmission line is connected to a load impedance $Z_L = 35 - j47.5$ (Ω). Find the position and length of a short-circuited stub required to match the line.

SOLUTION

Given

$$R_0 = 50\ (\Omega)$$

$$Z_L = 35 - j47.5\ (\Omega)$$

$$z_L = Z_L/R_0 = 0.70 - j0.95.$$

1. Enter z_L on the Smith chart as P_1 (Fig. 8-13).
2. Draw a $|\Gamma|$-circle centered at O with radius $\overline{OP}_1$.
3. Draw a straight line from P_1 through O to point P_2' on the perimeter, intersecting the $|\Gamma|$-circle at P_2, which represents y_L. Note 0.109 at P_2' on the "wavelengths toward generator" scale.
4. Note the two points of intersection of the $|\Gamma|$-circle with the $g = 1$ circle.

   At P_3: $y_{B1} = 1 + j1.2 = 1 + jb_{B1}$;

   At P_4: $y_{B2} = 1 - j1.2 = 1 + jb_{B2}$.
5. Solutions for the position of the stub:

   For P_3 (from P_2' to P_3'): $d_1 = (0.168 - 0.109)\lambda = 0.059\lambda$;

   For P_4 (from P_2' to P_4'): $d_2 = (0.332 - 0.109)\lambda = 0.223\lambda$.
6. Solutions for the length of the short-circuited stub to provide $y_s = -jb_B$:

   For P_3 (from P_{sc} on the extreme right of chart to P_3'', which represents $-jb_{B1} = -j1.2$):

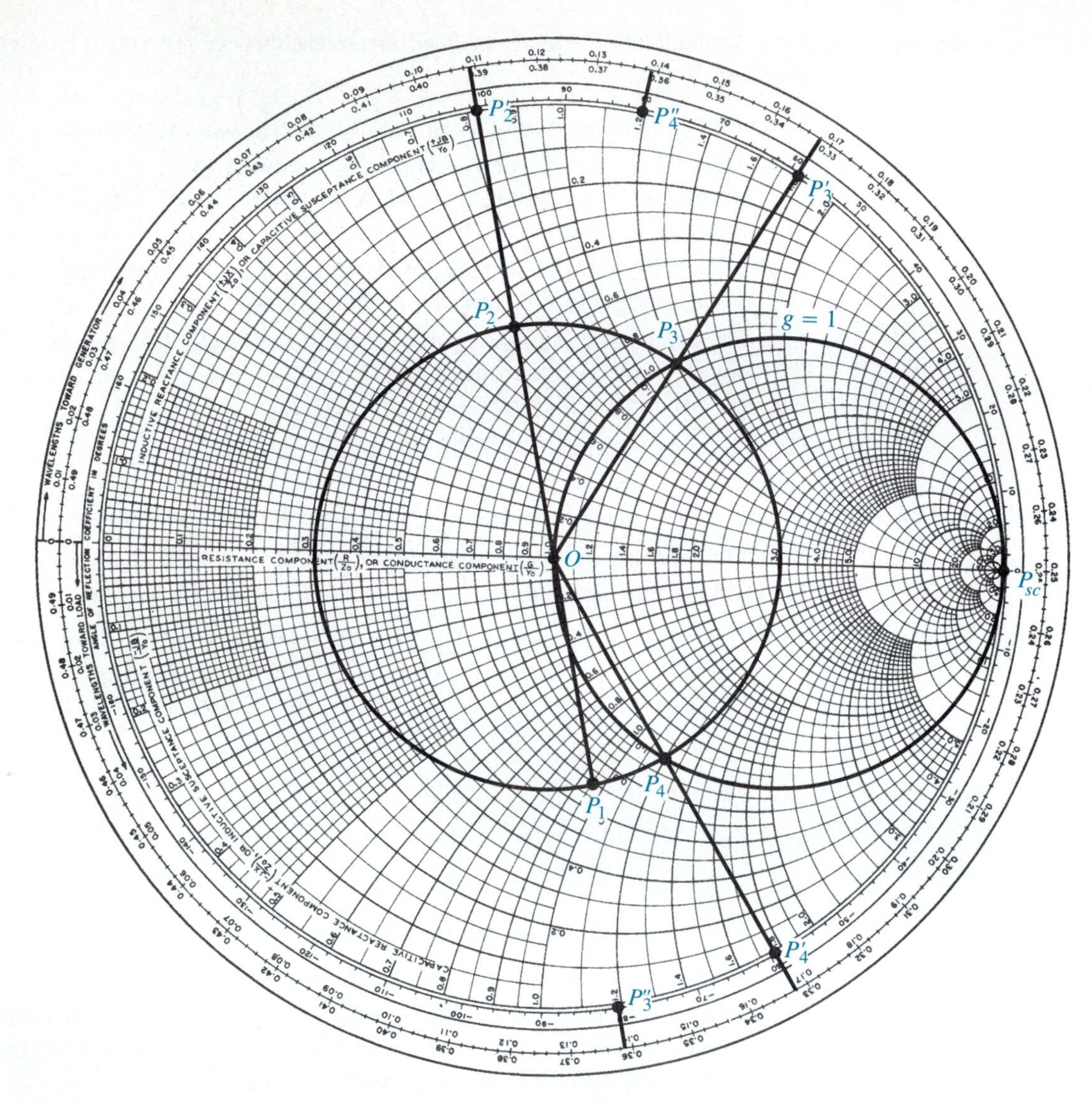

FIGURE 8-13 Construction for single-stub matching on Smith admittance chart (Example 8-11).

$\ell_1 = (0.361 - 0.250)\lambda = 0.111\lambda$;

For P_4 (from P_{sc} to P_4'', which represents $-jb_2 = j1.2$):

$\ell_2 = (0.139 + 0.250)\lambda = 0.389\lambda$.

In general, the solution with the shorter lengths is preferred unless there are other practical constraints. The exact length, ℓ, of the short-circuited stub may require fine adjustments in the actual matching procedure; hence the shorted matching sections are sometimes called ***stub tuners***.

■ **EXERCISE 8.13** Measurements on a 50-(Ω) lossless transmission line show that the consecutive voltage minima are 6 (cm) apart. It is desired to match a load impedance $Z_L = 75 + j100$ (Ω) to the line by a single short-circuited stub. Find (a) the location of the stub closest to the load, (b) the shortest required stub length, (c) the SWR on the line between the stub and the load, and (d) the SWR on the line between the stub and the source.

ANS. (a) 2.78 (cm), (b) 1.02 (cm), (c) 4.62, (d) 1.00.

REVIEW QUESTIONS

Q.8-28 Why does a change of half a wavelength in line length correspond to a complete revolution on a Smith chart?

Q.8-29 Given an impedance $Z = R + jX$, what procedure do we follow to find the admittance $Y = 1/Z$ on a Smith chart?

Q.8-30 Given an admittance $Y = G + jB$, how do we use a Smith chart to find the impedance $Z = 1/Y$?

Q.8-31 Is the standing-wave ratio constant on a transmission line even when the line is lossy? Explain.

Q.8-32 Why is it more convenient to use a Smith chart as an admittance chart for solving impedance-matching problems than to use it as an impedance chart?

Q.8-33 Why is it desirable to achieve an impedance match on a transmission line?

Q.8-34 Why are the stubs used in impedance matching usually of the short-circuited type instead of the open-circuited type?

REMARKS

1. The Smith chart can be used either as an impedance chart (for dimensionless impedances, $z = Z/R_0$) or as an admittance chart (for dimensionless admittances, $y = R_0 Y$). The point representing a short-circuit, P_{sc}, is at (1, 0) on a Smith admittance chart.
2. The principle of single-stub impedance matching is to connect a short-circuited stub of a proper length in parallel with the main line at an appropriate distance from the load such that the input admittance at the junctions of the parallel combination is $1 + j0$.
3. Transmission-line impedance-matching schemes are frequency-sensitive: The required stub location and stub length depend on the operating frequency.

SUMMARY

Transmission lines are used for efficient point-to-point transmission of energy and information. We have devoted this chapter to studying the method of analysis and the behavior of the transverse electromagnetic (TEM) waves guided by transmission lines. In particular, we

- discussed the characteristics of the three most common types of transmission lines (the parallel-plate line, the two-wire line, and the coaxial line),
- derived the general transmission-line equations, which combined to yield one-dimensional, second-order, ordinary differential equations under time-harmonic conditions,
- examined the wave characteristics on infinite transmission lines,
- determined the propagation constant, phase velocity, and characteristic impedances of lossless lines and distortionless lines,
- expressed the attenuation constant of a traveling wave on a lossy line in terms of power relations,
- analyzed the wave characteristics on finite transmission lines in terms of propagation constant, input impedance, reflection coefficient, and SWR,
- examined the properties of open-circuited and short-circuited lines,
- solved transmission-line circuit problems,
- studied the principle of construction and the applications of the Smith chart, and
- explained the single-stub method for impedance matching.

PROBLEMS

P.8-1 Consider lossless stripline designs for a given characteristic impedance.

a) How should the dielectric thickness, d, be changed for a given plate width, w, if the dielectric constant, ϵ_r, is doubled?

b) How should w be changed for a given d if ϵ_r is doubled?

c) How should w be changed for a given ϵ_r if d is doubled?

d) Will the velocity of propagation remain the same as that for the original line after the changes specified in parts (a), (b), and (c)? Explain.

P.8-2 Consider a transmission line made of two parallel brass strips—$\sigma_c = 1.6 \times 10^7$ (S/m)—of width 20 (mm) and separated by a lossy dielectric slab—$\mu = \mu_0$, $\epsilon_r = 3$, $\sigma = 10^{-3}$ (S/m)—of thickness 2.5 (mm). The operating frequency is 500 (MHz).

a) Calculate the R, L, G, and C per unit length.

b) Compare the magnitudes of the axial and transverse components of the electric field.

c) Find γ and Z_0.

P.8-3 It is desired to construct uniform transmission lines using polyethylene ($\epsilon_r = 2.25$) as the dielectric medium. Assuming negligible losses, (a) find the distance of separation for a 300-(Ω) two-wire line, where the radius of the conducting wires is 0.6 (mm); and (b) find the inner radius of the outer conductor for a 75-(Ω) coaxial line, where the radius of the center conductor is 0.6 (mm).

P.8-4 Calculate the attenuation constant at 1 (MHz) of a distortionless copper coaxial transmission line whose inner conductor has a radius 0.6 (mm) and outer conductor has an inner radius 3.91 (mm). The dielectric constant of the medium in-between is 2.25.

P.8-5 The following characteristics have been measured on a lossy transmission line at 100 (MHz):

$$Z_0 = 50 + j0 \quad (\Omega),$$
$$\alpha = 0.01 \quad (\text{dB/m}),$$
$$\beta = 0.8\pi \quad (\text{rad/m}).$$

Determine R, L, G, and C for the line.

P.8-6 Prove that a maximum power is transferred from a voltage source with an internal impedance Z_g to a load impedance Z_L over a lossless transmission line when $Z_i = Z_g^*$, where Z_i is the impedance looking into the loaded line. What is the maximum power-transfer efficiency?

P.8-7 Express $V(z)$ and $I(z)$ in terms of the voltage V_i and current I_i at the input end and γ and Z_0 of a transmission line (a) in exponential form, and (b) in hyperbolic form.

P.8-8 A d-c generator of voltage V_g and internal resistance R_g is connected to a lossy transmission line characterized by a resistance per unit length R and a conductance per unit length G.

a) Write the governing voltage and current transmission-line equations.

b) Find the general solutions for $V(z)$ and $I(z)$.

c) Specialize the solutions in part (b) to those for an infinite line.

d) Specialize the solutions in part (b) to those for a finite line of length ℓ that is terminated in a load resistance R_L.

P.8-9 A generator with an open-circuit voltage $v_g(t) = 10 \sin 8000\pi t$ (V) and internal impedance $Z_g = 40 + j30$ (Ω) is connected to a 50-(Ω) distortionless line. The line has a resistance of 0.5 (Ω/m), and its lossy dielectric medium has a loss tangent of 0.18%. The line is 50 (m) long and is terminated in a matched load. Find (a) the instantaneous expressions for the voltage and current at an arbitrary location on the line, (b) the instantaneous expressions for the voltage and current at the load, and (c) the average power transmitted to the load.

P.8-10 Find the input impedance of a low-loss quarter-wavelength line ($\alpha\lambda \ll 1$):

a) terminated in a short circuit.

b) terminated in an open circuit.

P.8-11 A 2-(m) lossless air-spaced transmission line having a characteristic impedance 50 (Ω) is terminated with an impedance $40 + j30$ (Ω) at an operating frequency of 200 (MHz). Find the input impedance.

P.8-12 The open-circuit and short-circuit impedances measured at the input terminals of an air-spaced transmission line 4 (m) long are $250\underline{/-50°}$ (Ω) and $360\underline{/20°}$ (Ω), respectively.

a) Determine Z_0, α, and β of the line.

b) Determine R, L, G, and C.

P.8-13 Measurements on a 0.6-(m) lossless coaxial cable at 100 (kHz) show a capacitance of 54 (pF) when the cable is open-circuited and an inductance of 0.30 (μH) when it is short-circuited.

a) Determine Z_0 and the dielectric constant of its insulating medium.

b) Calculate the X_{io} and X_{is} at 10 (MHz).

P.8-14 A 75-(Ω) lossless line is terminated in a load impedance $Z_L = R_L + jX_L$.

a) What must be the relation between R_L and X_L in order that the standing-wave ratio on the line be 3?

b) Find X_L, if $R_L = 150$ (Ω).

P.8-15 Consider a lossless transmission line.

a) Determine the line's characteristic resistance so that it will have a minimum possible standing-wave ratio for a load impedance $40 + j30$ (Ω).

b) Find this minimum standing-wave ratio and the corresponding voltage reflection coefficient.

P.8-16 A transmission line of characteristic impedance $R_0 = 50$ (Ω) is to be matched to a load impedance $Z_L = 40 + j10$ (Ω) through a length ℓ' of another transmission line of characteristic impedance R_0'. Find the required ℓ' and R_0' for matching.

P.8-17 Obtain the formulas for finding the length ℓ and the terminating resistance R_L of a lossless line having a characteristic impedance R_0 such that the input impedance equals $Z_i = R_i + jX_i$.

P.8-18 The standing-wave ratio on a lossless 300-(Ω) transmission line terminated in an unknown load impedance is 2.0, and the nearest voltage minimum is at a distance 0.3λ from the load. Determine (a) the reflection coefficient Γ of the load, and (b) the unknown load impedance Z_L.

P.8-19 A sinusoidal voltage generator with $V_g = 0.1\underline{/0°}$ (V) and internal impedance 50 (Ω) is connected to a lossless transmission line having a characteristic impedance $R_0 = 50$ (Ω). The line is $\lambda/8$ long and is terminated in a load resistance $R_L = 25$ (Ω). Find (a) V_i, I_i, V_L, and I_L; (b) the standing-wave ratio on the line; and (c) the average power delivered to the load. Compare the result in part (c) with the case where $R_L = 50$ (Ω).

P.8-20 The characteristic impedance of a given lossless transmission line is 75 (Ω). Use a Smith chart to find the input impedance at 200 (MHz) of such a line that is: (a) 1 (m) long and open-circuited, and (b) 0.8 (m) long and short-circuited. Then (c) determine the corresponding input admittances for the lines in parts (a) and (b).

P.8-21 A load impedance $30 + j10$ (Ω) is connected to a lossless transmission line of length 0.101λ and characteristic impedance 50 (Ω). Use a Smith chart to find (a) the standing-wave ratio, (b) the voltage reflection coefficient, (c) the input impedance, (d) the input admittance, and (e) the location of the voltage minimum on the line.

P.8-22 Repeat Problem P.8-21 for a load impedance $30 - j10$ (Ω).

P.8-23 In a laboratory experiment conducted on a 50-(Ω) lossless transmission line terminated in an unknown load impedance, it is found that the standing-wave ratio is 2.0. The successive voltage minima are 25 (cm) apart, and the first minimum occurs at 5 (cm) from the load. Find (a) the load impedance, and (b) the reflection coefficient of the load. (c) Where would the first voltage minimum be located if the load were replaced by a short-circuit?

P.8-24 A dipole antenna having an input impedance of 73 (Ω) is fed by a 200-(MHz) source through a 300-(Ω) two-wire transmission line. Design a quarter-wave two-wire air line with a 2-(cm) spacing to match the antenna to the 300-(Ω) line.

P.8-25 The single-stub method is used to match a load impedance $25 + j25$ (Ω) to a 50-(Ω) transmission line. Use a Smith chart to find the required length and position (in terms of wavelength) of a short-circuited stub made of a section of the same 50-(Ω) line.

P.8-26 Repeat Problem P.8-25 using a short-circuited stub made of a section of a line that has a characteristic impedance of 75 (Ω).

P.8-27 Measurements on a lossless transmission line of characteristic resistance 75 (Ω) show a standing-wave ratio of 2.4 and the first two voltage minima nearest to the load at 0.335(m) and 1.235(m). Use a Smith chart to: (a) determine the load impedance Z_L, and (b) find the location nearest to the load and the length of a short-circuited stub required to match Z_L to the line.

P.8-28 A load impedance can be matched to a transmission line also by using a single stub placed in series with the load at an appropriate location, as shown in Fig. 8-14. Assuming that $Z_L = 25 + j25$ (Ω), $R_0 = 50$ (Ω), and $R'_0 = 35$ (Ω), find d and ℓ required for matching.

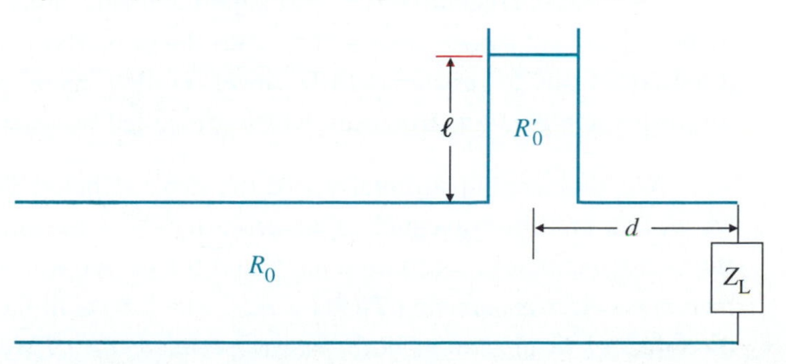

FIGURE 8-14 Impedance matching by a series stub (Problem P.8-28).

CHAPTER 9

9-1 OVERVIEW

In the preceding chapter we studied the characteristic properties of transverse electromagnetic (TEM) waves guided by transmission lines. TEM waves, however, are not the only mode of guided waves that can propagate on transmission lines; nor are the three types of transmission lines (parallel-plate, two-wire, and coaxial) mentioned in Section 8-1 the only possible wave-guiding structures. For applications above the SHF and EHF range ($f > 3\,\text{GHz}$, $\lambda < 10\,\text{cm}$) the use of the two-conductor transmission lines listed in Section 8-1 becomes impractical because the attenuation constant of TEM waves along a line resulting from a finite conductivity of the conductors increases with the resistance per unit length, R, of the line, which is directly proportional to the square-root of the frequency (see Table 8-1). At microwave frequencies the attenuation would be prohibitively high. In this chapter we study the characteristics of electromagnetic waves propagating inside hollow metal pipes. Since metal pipes are single conductors with large surface areas, we expect the attenuation due to resistance to be lower. Hollow metal pipes are one form of uniform wave-guiding structures, which are called ***waveguides***.

Hollow metal pipes are one form of wavelength guide at microwave frequencies.

We first present an analysis of the general behaviors of electromagnetic waves in a uniform waveguide of an arbitrary cross section. The starting point is the vector Helmholtz's equation for **E** and for **H**. We will see that, in addition to ***transverse electromagnetic (TEM) waves***, which have no field components in the direction of propagation, both ***transverse magnetic (TM) waves*** with a longitudinal electric-field component and ***transverse electric (TE) waves*** with a

Three types of propagating electromagnetic waves: TEM, TM, and TE

Waveguides and Cavity Resonators

longitudinal magnetic-field component can also exist. However, it will be clear that TEM waves cannot exist in a single-conductor hollow (or dielectric-filled) waveguide. The characteristics of both TM and TE modes in a rectangular waveguide will be studied in detail.

At microwave frequencies, ordinary lumped-parameter elements (such as inductances and capacitances) connected by wires are no longer practical as circuit elements or as resonant circuits because the dimensions of the elements would have to be extremely small, because the resistance of the wire circuits becomes very high as a result of the skin effect, and because of radiation. A hollow conducting box with proper dimensions can be used as a resonant device with a very high Q. Such a box, which is essentially a segment of a waveguide with closed end faces, is called a ***cavity resonator***. We will discuss the different mode patterns of the fields inside simple rectangular cavity resonators.

Cavity resonators are waveguide sections with endfaces.

9-2 General Wave Behaviors Along Uniform Guiding Structures

In this section we examine some general characteristics for waves propagating along straight guiding structures with a uniform cross section. We will assume that the waves propagate in the $+z$-direction with a propagation constant $\gamma = \alpha + j\beta$ that is yet to be determined. For harmonic time dependence with an angular frequency ω, the dependence on z and t for all field components can be described by the exponential factor

$$e^{-\gamma z}e^{j\omega t} = e^{(j\omega t - \gamma z)} = e^{-\alpha z}e^{j(\omega t - \beta z)}. \tag{9-1}$$

As an example, for a cosine reference we may write the instantaneous expression for the **E** field in Cartesian coordinates as

$$\mathbf{E}(x, y, z; t) = \mathcal{R}e[\mathbf{E}^0(x, y)e^{(j\omega t - \gamma z)}], \tag{9-2}$$

where $\mathbf{E}^0(x, y)$ is a two-dimensional vector phasor that depends only on the cross-sectional coordinates. Hence, in using a phasor representation in equations relating field quantities we may replace partial derivatives with respect to t and z simply by products with $(j\omega)$ and $(-\gamma)$, respectively; the common factor $e^{(j\omega t - \gamma z)}$ can be dropped.

We consider a straight waveguide in the form of a dielectric-filled metal tube having an arbitrary cross section and lying along the z-axis, as shown in Fig. 9-1. According to Eqs. (6-98) and (6-99), the electric and magnetic field intensities in the charge-free dielectric region inside satisfy the following homogeneous vector Helmholtz's equations:

$$\nabla^2\mathbf{E} + k^2\mathbf{E} = 0 \tag{9-3}$$

and

$$\nabla^2\mathbf{H} + k^2\mathbf{H} = 0, \tag{9-4}$$

where **E** and **H** are three-dimensional vector phasors, and k is the wavenumber:

$$k = \omega\sqrt{\mu\epsilon}. \tag{9-5}$$

The three-dimensional Laplacian operator ∇^2 may be broken into two parts: $\nabla^2_{u_1u_2}$ for the cross-sectional coordinates and ∇^2_z for the longitudinal coordinate. For waveguides with a rectangular cross section we use Cartesian coordinates:

$$\begin{aligned}\nabla^2\mathbf{E} = (\nabla^2_{xy} + \nabla^2_z)\mathbf{E} &= \left(\nabla^2_{xy} + \frac{\partial^2}{\partial z^2}\right)\mathbf{E} \\ &= \nabla^2_{xy}\mathbf{E} + \gamma^2\mathbf{E}.\end{aligned} \tag{9-6}$$

Combination of Eqs. (9-3) and (9-6) gives

$$\nabla^2_{xy}\mathbf{E} + (\gamma^2 + k^2)\mathbf{E} = 0. \tag{9-7}$$

Similarly, from Eq. (9-4) we have

$$\nabla^2_{xy}\mathbf{H} + (\gamma^2 + k^2)\mathbf{H} = 0. \tag{9-8}$$

We note that each of Eqs. (9-7) and (9-8) is really three second-order partial differential equations, one for each component of **E** and **H**. The exact solution of these component equations depends on the cross-sectional geometry and the boundary conditions. (See Section 9-3.)

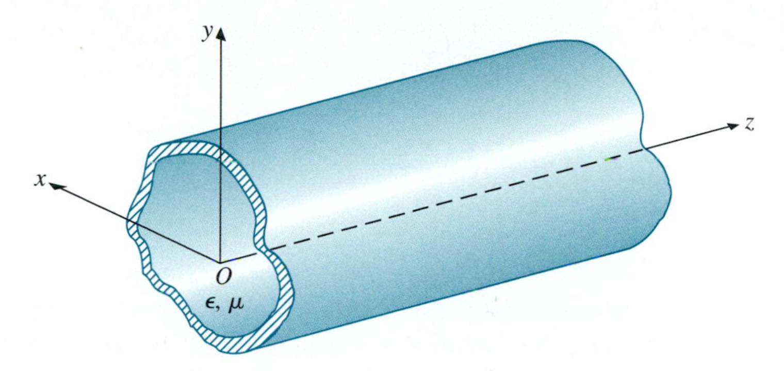

FIGURE 9-1 A uniform waveguide with an arbitrary cross section.

Of course, the various components of **E** and **H** are not all independent, and it is not necessary to solve all six second-order partial differential equations for the six components of **E** and **H**. Let us examine the interrelationships among the six components in Cartesian coordinates by expanding the two source-free curl equations, Eqs. (6-80a) and (6-80b) with $\mathbf{J} = 0$:

From $\nabla \times \mathbf{E} = -j\omega\mu\mathbf{H}$:	From $\nabla \times \mathbf{H} = j\omega\epsilon\mathbf{E}$:
$\frac{\partial E_z^0}{\partial y} + \gamma E_y^0 = -j\omega\mu H_x^0$ (9-9a)	$\frac{\partial H_z^0}{\partial y} + \gamma H_y^0 = j\omega\epsilon E_x^0$ (9-10a)
$-\gamma E_x^0 - \frac{\partial E_z^0}{\partial x} = -j\omega\mu H_y^0$ (9-9b)	$-\gamma H_x^0 - \frac{\partial H_z^0}{\partial x} = j\omega\epsilon E_y^0$ (9-10b)
$\frac{\partial E_y^0}{\partial x} - \frac{\partial E_x^0}{\partial y} = -j\omega\mu H_z^0$ (9-9c)	$\frac{\partial H_y^0}{\partial x} - \frac{\partial H_x^0}{\partial y} = j\omega\epsilon E_z^0$ (9-10c)

Note that partial derivatives with respect to z have been replaced by multiplications by $(-\gamma)$. All the component field quantities in the equations above are phasors that depend only on x and y, the common $e^{-\gamma z}$ factor for z-dependence having been omitted. By manipulating these equations we can express the transverse field components H_x^0, H_y^0, and E_x^0, and E_y^0 in terms of the two longitudinal components E_z^0 and H_z^0. For instance, Eqs. (9-9a) and (9-10b) can be combined to eliminate E_y^0 and obtain H_x^0 in terms of E_z^0 and H_z^0. We have

Expressing transverse field components in terms of E_z and H_z

$$H_x^0 = -\frac{1}{h^2}\left(\gamma\frac{\partial H_z^0}{\partial x} - j\omega\epsilon\frac{\partial E_z^0}{\partial y}\right), \tag{9-11}$$

$$H_y^0 = -\frac{1}{h^2}\left(\gamma\frac{\partial H_z^0}{\partial y} + j\omega\epsilon\frac{\partial E_z^0}{\partial x}\right), \tag{9-12}$$

$$E_x^0 = -\frac{1}{h^2}\left(\gamma\frac{\partial E_z^0}{\partial x} + j\omega\mu\frac{\partial H_z^0}{\partial y}\right), \tag{9-13}$$

$$E_y^0 = -\frac{1}{h^2}\left(\gamma\frac{\partial E_z^0}{\partial y} - j\omega\mu\frac{\partial H_z^0}{\partial x}\right), \tag{9-14}$$

where

$$h^2 = \gamma^2 + k^2. \tag{9-15}$$

Procedure for determining wave behavior in a waveguide

The wave behavior in a waveguide can be analyzed by solving Eqs. (9-7) and (9-8) for the longitudinal components, E_z^0 and H_z^0, respectively, subject to the required boundary conditions, and then by using Eqs. (9-11) through (9-14) to determine the other components.

It is convenient to classify the propagating waves in a uniform waveguide into three types according to whether E_z or H_z exists.

Definition of three types of propagating waves in a uniform waveguide

1. *Transverse electromagnetic (TEM) waves.* These are waves that contain neither E_z nor H_z. We encountered TEM waves in Chapter 7 when we discussed plane waves and in Chapter 8 on waves along transmission lines.
2. *Transverse magnetic (TM) waves.* These are waves that contain a nonzero E_z but $H_z = 0$.
3. *Transverse electric (TE) waves.* These are waves that contain a nonzero H_z but $E_z = 0$.

The propagation characteristics of the various types of waves are different; they will be discussed in subsequent subsections.

9-2.1 TRANSVERSE ELECTROMAGNETIC WAVES

Since $E_z = 0$ and $H_z = 0$ for TEM waves within a guide, we see that Eqs. (9-11) through (9-14) constitute a set of trivial solutions (all field components vanish) unless the denominator h^2 also equals zero. In other words, TEM waves exist only when

$$\gamma_{\text{TEM}}^2 + k^2 = 0, \tag{9-16}$$

or

$$\gamma_{\text{TEM}} = jk = j\omega\sqrt{\mu\epsilon}, \tag{9-17}$$

which is exactly the same expression for the propagation constant of a uniform plane wave in an unbounded medium characterized by constitutive parameters ϵ and μ. We recall that Eq. (9-17) also holds for a TEM wave on a lossless transmission line. It follows that the velocity of propagation (phase velocity) for TEM waves is

Phase velocity for TEM waves

$$u_{p(\text{TEM})} = \frac{\omega}{k} = \frac{1}{\sqrt{\mu\epsilon}} \qquad \text{(m/s)}. \tag{9-18}$$

We can obtain the ratio between E_x^0 and H_y^0 from Eqs. (9-9b) and (9-10a) by setting E_z and H_z to zero. This ratio is called the ***wave impedance***. We have

$$Z_{\text{TEM}} = \frac{E_x^0}{H_y^0} = \frac{j\omega\mu}{\gamma_{\text{TEM}}} = \frac{\gamma_{\text{TEM}}}{j\omega\epsilon}, \tag{9-19}$$

which becomes, in view of Eq. (9-17),

Wave impedance for TEM waves

$$Z_{\text{TEM}} = \sqrt{\frac{\mu}{\epsilon}} = \eta \qquad (\Omega). \tag{9-20}$$

We note that Z_{TEM} is the same as the intrinsic impedance of the dielectric medium, as given in Eq. (7-14). Equations (9-18) and (9-20) assert that ***the phase velocity and the wave impedance for TEM waves are independent of the frequency of the waves.***

Single-conductor waveguides cannot support TEM waves. In Section 5-2 we pointed out that magnetic flux lines always close upon themselves. Hence if a TEM wave were to exist in a waveguide, the field lines of **B** and **H** would form closed loops in a transverse plane. However, the generalized Ampère's circuital law, Eq. (6-46b), requires that the line integral of the magnetic field around any closed loop in a transverse plane equals the sum of the longitudinal conduction and displacement currents through the loop. Without an inner conductor there is no longitudinal conduction current inside the waveguide. By definition, a TEM wave does not have an E_z-component; consequently, there is no longitudinal displacement current. The total absence of a longitudinal current inside a waveguide leads to the conclusion that there can be no closed loops of magnetic field lines in any transverse plane. Therefore, we conclude that ***TEM waves cannot exist in a single-conductor hollow (or dielectric-filled) waveguide of any shape.***

Single-conductor waveguides cannot support TEM waves.

9-2.2 Transverse Magnetic Waves

Transverse magnetic (TM) waves do not have a component of the magnetic field in the direction of propagation, $H_z = 0$. The behavior of TM waves can be analyzed by solving Eq. (9-7) for E_z subject to the boundary conditions of the guide and using Eqs. (9-11) through (9-14) to determine the other components. Writing Eq. (9-7) for E_z, we have

$$\nabla_{xy}^2 E_z^0 + (\gamma^2 + k^2)E_z^0 = 0, \tag{9-21}$$

or

$$\nabla_{xy}^2 E_z^0 + h^2 E_z^0 = 0. \tag{9-22}$$

Equation (9-22) is a second-order partial differential equation, which can be solved for E_z^0. In this section we wish to discuss only the general properties of the various wave types. The actual solution of Eq. (9-22) will wait until the next section when we discuss rectangular waveguides. Once E_z^0 is determined, all other field components can be found from Eqs. (9-11) through (9-14) with $H_z^0 = 0$. We can express the relation between the transverse components of magnetic field intensity, H_x^0 and H_y^0, and those of electric field intensity, E_x^0 and E_y^0, in terms of a wave impedance, Z_{TM}, for the TM mode.

Wave impedance for TM waves

$$Z_{TM} = \frac{E_x^0}{H_y^0} = -\frac{E_y^0}{H_x^0} = \frac{\gamma}{j\omega\epsilon} \quad (\Omega). \tag{9-23}$$

It is important to note that Z_{TM} is *not* equal to $j\omega\mu/\gamma$, because γ for TM waves, unlike γ_{TEM}, is *not* equal to $j\omega\sqrt{\mu\epsilon}$.

When we undertake to solve the two-dimensional homogeneous Helmholtz equation, Eq. (9-22), subject to the boundary conditions of a given waveguide, we will discover that solutions are possible only for *discrete values of h*. There may be an infinity of these discrete values, but solutions are not possible for all values of h. The values of h for which a solution of Eq. (9-22) exists are called the ***characteristic values***, or ***eigenvalues***, of the boundary-value problem. Each of the eigenvalues determines the characteristic properties of a particular TM mode of the given waveguide.

Definition of characteristic values, or eigenvalues

In the following sections we will also discover that the eigenvalues of the various waveguide problems are real numbers. From Eq. (9-15) we have

$$\begin{aligned}\gamma &= \sqrt{h^2 - k^2} \\ &= \sqrt{h^2 - \omega^2\mu\epsilon}.\end{aligned} \tag{9-24}$$

Two distinct ranges of the values for the propagation constant are noted, the dividing point being $\gamma = 0$, where

$$\omega_c^2 \mu\epsilon = h^2, \tag{9-25}$$

or

Relation between eigenvalue h and cutoff frequency f_c

$$f_c = \frac{h}{2\pi\sqrt{\mu\epsilon}} \quad \text{(Hz)}. \tag{9-26}$$

Definition of cutoff frequency f_c

The frequency, f_c, at which $\gamma = 0$ is called a ***cutoff frequency***. *The value of f_c for a particular mode in a waveguide depends on the eigenvalue, h, of this mode.* Using Eq. (9-26), we can write Eq. (9-24) as

$$\gamma = h\sqrt{1 - \left(\frac{f}{f_c}\right)^2}. \tag{9-27}$$

The two distinct ranges of γ can be defined in terms of the ratio $(f/f_c)^2$ as compared to unity.

a) $\left(\frac{f}{f_c}\right)^2 > 1$, or $f > f_c$. In this range, $\omega^2\mu\epsilon > h^2$ and γ is imaginary. We have, from Eqs. (9-24) and (9-26),

$$\gamma = j\beta = jk\sqrt{1 - \left(\frac{h}{k}\right)^2} = jk\sqrt{1 - \left(\frac{f_c}{f}\right)^2}. \tag{9-28}$$

It is a propagating mode with a phase constant β:

Waves with $f > f_c$ are propagating modes.

$$\boxed{\beta = k\sqrt{1 - \left(\frac{f_c}{f}\right)^2} \qquad \text{(rad/m).}} \tag{9-29}$$

The corresponding wavelength in the guide is

Guide wavelengths are longer than the corresponding wavelengths in unbounded media.

$$\boxed{\lambda_g = \frac{2\pi}{\beta} = \frac{\lambda}{\sqrt{1 - (f_c/f)^2}},} \tag{9-30}$$

where

$$\lambda = \frac{2\pi}{k} = \frac{1}{f\sqrt{\mu\epsilon}} = \frac{u}{f} \tag{9-31}$$

is the wavelength of a plane wave with a frequency f in an unbounded dielectric medium characterized by μ and ϵ, and $u = 1/\sqrt{\mu\epsilon}$ is the velocity of light in the medium. Equation (9-30) can be rearranged to give a simple relation connecting λ, the guide wavelength λ_g, and the cutoff wavelength $\lambda_c = u/f_c$:

$$\boxed{\frac{1}{\lambda^2} = \frac{1}{\lambda_g^2} + \frac{1}{\lambda_c^2}.} \tag{9-32}$$

The phase velocity of the propagating wave in the guide is

$$u_p = \frac{\omega}{\beta} = \frac{u}{\sqrt{1 - (f_c/f)^2}} = \frac{\lambda_g}{\lambda}u > u. \tag{9-33}$$

We see from Eq. (9-33) that the phase velocity within a waveguide is always higher than that in an unbounded medium and is frequency-dependent. Hence ***single-conductor waveguides are dispersive transmission systems***, although an unbounded lossless dielectric medium is nondispersive. Substitution of Eq. (9-28) in Eq. (9-23) yields

Single-conductor waveguides are dispersive.

$$Z_{TM} = \eta\sqrt{1-\left(\frac{f_c}{f}\right)^2} \quad (\Omega). \tag{9-34}$$

The wave impedance for propagating TM modes is less than the intrinsic impedance of the guide medium.

Thus, the wave impedance of propagating TM modes in a waveguide with a lossless dielectric is purely resistive and is always less than the intrinsic impedance of the dielectric medium.

b) $\left(\frac{f}{f_c}\right)^2 < 1$, or $f < f_c$. When the operating frequency is lower than the cutoff frequency, γ is real and Eq. (9-27) can be written as

$$\gamma = \alpha = h\sqrt{1-\left(\frac{f}{f_c}\right)^2}, \qquad f < f_c, \tag{9-35}$$

which is, in fact, an attenuation constant. Since all field components contain the propagation factor $e^{-\gamma z} = e^{-\alpha z}$, the wave diminishes rapidly with z and is said to be ***evanescent***. Therefore, ***a waveguide exhibits the property of a high-pass filter***. *For a given mode, only waves with a frequency higher than the cutoff frequency of the mode can propagate in the guide.* Substitution of Eq. (9-35) in Eq. (9-23) gives an imaginary wave impedance of TM modes for $f < f_c$. Thus, the wave impedance of evanescent TM modes at frequencies below cutoff is purely reactive, indicating that there is no power flow associated with evanescent waves.

Waves with $f < f_c$ are evanescent, or nonpropagating.

Waveguides are high-pass devices.

9-2.3 TRANSVERSE ELECTRIC WAVES

Transverse electric (TE) waves do not have a component of the electric field in the direction of propagation, $E_z = 0$. The behavior of TE waves can be analyzed by first solving Eq. (9-8) for H_z:

$$\nabla_{xy}^2 H_z + h^2 H_z = 0. \tag{9-36}$$

Proper boundary conditions at the guide walls must be satisfied. The transverse field components can then be found by substituting H_z into the reduced Eqs. (9-11) through (9-14) with E_z set to zero.

The transverse components of electric field intensity, E_x^0 and E_y^0, are related to those of magnetic field intensity, H_x^0 and H_y^0, through the wave impedance. We have,

Wave impedance for TE waves

$$Z_{TE} = \frac{E_x^0}{H_y^0} = -\frac{E_y^0}{H_x^0} = \frac{j\omega\mu}{\gamma} \quad (\Omega). \tag{9-37}$$

Note that Z_{TE} in Eq. (9-37) is quite different from Z_{TM} in Eq. (9-23) because γ for TE waves, unlike γ_{TEM}, is *not* equal to $j\omega\sqrt{\mu\epsilon}$.

Since we have not changed the relation between γ and h, Eqs. (9-24) through (9-33) pertaining to TM waves also apply to TE waves. There are also two distinct ranges of γ, depending on whether the operating frequency is higher or lower than the cutoff frequency, f_c, given in Eq. (9-26).

a) $\left(\frac{f}{f_c}\right)^2 > 1$, or $f > f_c$. In this range, γ is imaginary, and we have a propagating mode. The expression for γ is the same as that given in Eq. (9-28):

$$\gamma = j\beta = jk\sqrt{1 - \left(\frac{f_c}{f}\right)^2}. \tag{9-38}$$

Consequently, the formulas for β, λ_g, and u_p in Eqs. (9-29), (9-30) and (9-33), respectively, also hold for TE waves. Using Eq. (9-38) in Eq. (9-37), we obtain

The wave impedance for propagating TE modes is larger than the intrinsic impedance of the guide medium.

$$Z_{TE} = \frac{\eta}{\sqrt{1 - (f_c/f)^2}} \quad (\Omega), \tag{9-39}$$

which is obviously different from the expression for Z_{TM} in Eq. (9-34). Equation (9-39) indicates that ***the wave impedance of propagating TE modes in a waveguide with a lossless dielectric is purely resistive and is always larger than the intrinsic impedance of the dielectric medium.***

b) $\left(\frac{f}{f_c}\right)^2 < 1$, or $f < f_c$. In this case, γ is real and we have an evanescent or non-propagating mode:

$$\gamma = \alpha = h\sqrt{1 - \left(\frac{f}{f_c}\right)^2}, \qquad f < f_c. \tag{9-40}$$

Since γ in Eq. (9-40) is purely real, the wave impedance of TE modes in Eq. (9-37) for $f < f_c$,

Waves with $f < f_c$ are evanescent, or nonpropagating.

$$Z_{TE} = j\frac{\omega\mu}{h\sqrt{1 - (f/f_c)^2}}, \qquad f < f_c, \tag{9-41}$$

is purely reactive, indicating again that there is no power flow for evanescent waves at $f < f_c$.

■ **EXERCISE 9.1**

(a) Determine the wave impedance and guide wavelength (in terms of their values for the TEM mode) at a frequency equal to twice the cutoff frequency in a waveguide for TM and TE modes.

(b) Repeat part (a) for a frequency equal to one-half of the cutoff frequency.

ANS. (a) 0.866η, 1.155λ; 1.155η, 1.155λ. (b) $-j0.276h/f_c\epsilon$; $j3.63f_c\mu/h$.

■ **EXERCISE 9.2**

Obtain from Eq. (9-33) an expression for the group velocity, u_g, in a waveguide in terms of f_c and f, and prove that

$$u_g u_p = u^2. \tag{9-43}$$

EXAMPLE 9-1

Equation (9-29) gives the relation between the phase constant β and the frequency f of propagating modes in a waveguide. Plot the $\omega - \beta$ graph for TM and TE modes, and discuss how the phase and group velocities of a propagating wave in the guide can be determined from the graph.

SOLUTION

Equation (9-29) holds for both TM and TE propagating modes. Since $k = \omega/u$, where $u = 1/\sqrt{\mu\epsilon}$ is the velocity of wave propagation in the unbounded medium, we can rewrite Eq. (9-29) as

$$\omega = \frac{\beta u}{\sqrt{1 - (\omega_c/\omega)^2}}. \tag{9-43}$$

Dispersion diagram is an ω-β graph.

The $\omega - \beta$ graph, called a ***dispersion diagram***, is plotted in Fig. 9-2 as the solid curve. It intersects the ω-axis ($\beta = 0$) at $\omega = \omega_c$. The slope of the line joining

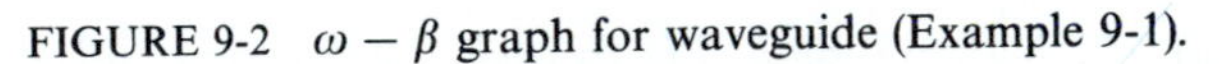

FIGURE 9-2 $\omega - \beta$ graph for waveguide (Example 9-1).

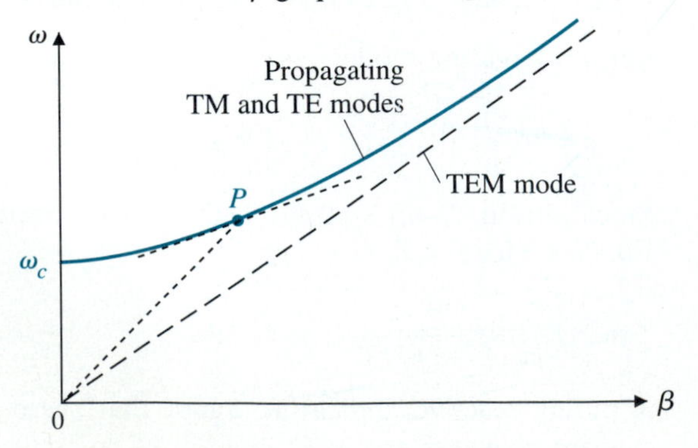

the origin and any point, such as P, on the curve is equal to the phase velocity, u_p, for a particular mode having a cutoff frequency f_c and operating at a particular frequency. The local slope of the ω–β curve at P is the group velocity, u_g. We note that, for propagating TM and TE waves in a waveguide, $u_p > u$, $u_g < u$, and Eq. (9-43) holds. As the operating frequency increases much above the cutoff frequency, both u_p and u_g approach u asymptotically. The exact value of ω_c depends on the eigenvalue h in Eq. (9-26)—that is, on the particular TM or TE mode.

EXAMPLE 9-2

Consider a parallel-plate waveguide of two perfectly conducting plates separated by a distance b and filled with a dielectric medium having constitutive parameters (ϵ, μ), as shown in Fig. 9-3. The plates are assumed to be infinite in extent in the x-direction. (Fields do not vary in the x-direction.)

a) Obtain the time-harmonic field expressions for TM modes in the guide.

b) Determine the cutoff frequency.

SOLUTION

a) Let the waves propagate in the $+z$-direction. For TM modes, $H_z = 0$. With harmonic time dependence it is convenient to work with field intensity phasors and write $E_z(y, z)$ as $E_z^0(y)e^{-\gamma z}$. Since there is no variation in the x-direction, Eq. (9-22) becomes

$$\frac{d^2E_z^0(y)}{dy^2} + h^2E_z^0(y) = 0. \qquad (9\text{-}44)$$

The general solution for Eq. (9-44) is

$$E_z^0(y) = A_n \sin hy + B_n \cos hy. \qquad (9\text{-}45)$$

FIGURE 9-3 An infinite parallel-plate waveguide.

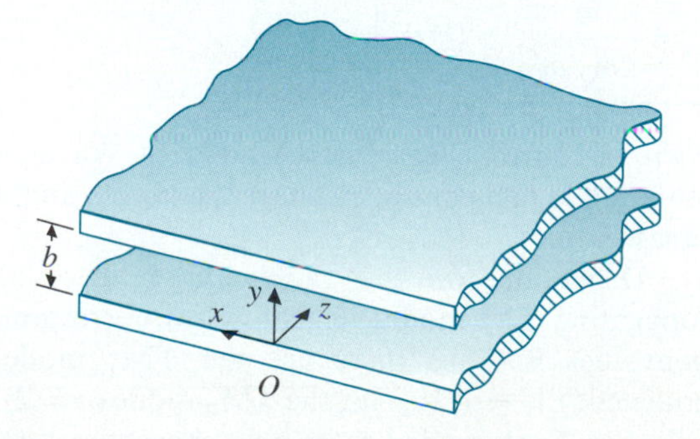

Since the tangential component of the electric field must vanish on the surface of the perfectly conducting plates, the following boundary conditions should be satisfied:

(i) At $y = 0$, $E_z^0(0) = 0$,

and

(ii) At $y = b$, $E_z^0(b) = 0$.

Boundary condition (i) requires $B_n = 0$; and boundary condition (ii) requires $\sin hb = 0$, or $hb = n\pi$, which determines the value of the eigenvalue h:

$$h = \frac{n\pi}{b}, \qquad n = 1, 2, 3, \ldots. \tag{9-46}$$

Hence, $E_z^0(y)$ in Eq. (9-45) must be of the following form:

$$E_z^0(y) = A_n \sin\left(\frac{n\pi y}{b}\right), \tag{9-47}$$

where the amplitude A_n depends on the strength of excitation of the particular TM wave. The only other nonzero field components are obtained from Eqs. (9-11) and (9-14), keeping in mind that $H_z^0 = 0$ and $\partial E_z^0/\partial x = 0$.

$$H_x^0(y) = \frac{j\omega\epsilon}{h} A_n \cos\left(\frac{n\pi y}{b}\right), \tag{9-48}$$

$$E_y^0(y) = -\frac{\gamma}{h} A_n \cos\left(\frac{n\pi y}{b}\right). \tag{9-49}$$

The γ in Eq. (9-49) is the propagation constant that can be determined from Eq. (9-24):

$$\gamma = \sqrt{\left(\frac{n\pi}{b}\right)^2 - \omega^2\mu\epsilon}. \tag{9-50}$$

b) The cutoff frequency is the frequency that makes $\gamma = 0$. We have

Cutoff frequency of a parallel-plate waveguide

$$\boxed{f_c = \frac{n}{2b\sqrt{\mu\epsilon}} \quad \text{(Hz)},} \tag{9-51}$$

which, of course, checks with Eq. (9-26). Waves with $f > f_c$ propagate with a phase constant β, given in Eq. (9-29); and waves with $f \leq f_c$ are evanescent.

Depending on the values of n, there are different possible propagating TM modes (eigenmodes) corresponding to the different eigenvalues h. Thus there are the TM_1 mode ($n = 1$) with cutoff frequency $(f_c)_1 = 1/2b\sqrt{\mu\epsilon}$, the TM_2 mode ($n = 2$) with $(f_c)_2 = 1/b\sqrt{\mu\epsilon}$, and so on. Each mode has its own characteristics. When $n = 0$, $E_z = 0$

and only the transverse components H_x and E_y exist. Hence TM_0 mode is the TEM mode, a special case, for which $f_c = 0$.

REVIEW QUESTIONS

Q.9-1 Why are the common types of transmission lines not useful for long-distance signal transmission at microwave frequencies in the TEM mode?

Q.9-2 Why are lumped-parameter elements connected by wires not useful as resonant circuits at microwave frequencies?

Q.9-3 What are the three basic types of propagating waves in a uniform waveguide?

Q.9-4 Explain why single-conductor hollow or dielectric-filled waveguides cannot support TEM waves.

Q.9-5 Define *wave impedance.*

Q.9-6 In what way does the wave impedance in a waveguide depend on frequency:

a) For a propagating TEM wave?

b) For a propagating TM wave?

c) For a propagating TE wave?

Q.9-7 What are *eigenvalues* of a boundary-value problem?

Q.9-8 What is meant by a *cutoff frequency* of a waveguide?

Q.9-9 Can a waveguide have more than one cutoff frequency? On what factors does the cutoff frequency of a waveguide depend?

Q.9-10 Is the guide wavelength of a propagating wave in a waveguide longer or shorter than the wavelength in the corresponding unbounded dielectric medium?

Q.9-11 What is an *evanescent mode*?

REMARKS

1. A waveguide exhibits the properties of a high-pass filter, letting only those frequencies that are higher than a cutoff frequency propagate.
2. The wave impedances for both TM and TE propagating modes in lossless waveguides are purely real and change with the ratio (f_c/f); those for evanescent modes are purely imaginary.
3. The wavelength (λ_g) and phase velocity (u_p) for both TM and TE modes in a waveguide are greater than the wavelength (λ) and velocity (u), respectively, of wave propagation in the corresponding unbounded medium.
4. The group velocity (u_g) of wave propagation in a waveguide is *not* equal to the product $f\lambda_g$, but the phase velocity (u_p) is.
5. Single-conductor waveguides are dispersive transmission systems. (u_p is not proportional to f.)
6. The cutoff frequency of a parallel-plate waveguide is inversely proportional to the plate separation.

9-3 RECTANGULAR WAVEGUIDES

The parallel-plate waveguide analyzed in Example 9-2 assumed the plates to be of an infinite extent in the transverse x-direction; that is, the fields do not vary with x. In practice, these plates are always finite in width, with fringing fields at the edges. Electromagnetic energy will leak through the sides of the guide and create undesirable stray couplings to other circuits and systems. Thus practical waveguides are usually uniform structures of a cross section of the enclosed variety. In this section we will analyze the wave behavior in hollow rectangular waveguides.

In the following discussion we draw on the material in Section 9-2 concerning general wave behaviors along uniform guiding structures. Propagation of time-harmonic waves in the $+z$-direction with a propagation constant γ is considered. TM and TE modes will be discussed separately. As we have noted previously, TEM waves cannot exist in a single-conductor hollow or dielectric-filled waveguide.

9-3.1 TM WAVES IN RECTANGULAR WAVEGUIDES

Consider the waveguide sketched in Fig. 9-4, with its rectangular cross section of sides a and b. The enclosed dielectric medium is assumed to have constitutive parameters ϵ and μ. For TM waves, $H_z = 0$ and E_z is to be solved from Eq. (9-22). Writing $E_z(x, y, z)$ as

$$E_z(x, y, z) = E_z^0(x, y)e^{-\gamma z}, \tag{9-52}$$

we solve the following second-order partial differential equation:

$$\left(\frac{\partial^2}{\partial x^2} + \frac{\partial^2}{\partial y^2} + h^2\right) E_z^0(x, y) = 0. \tag{9-53}$$

FIGURE 9-4 A rectangular waveguide.

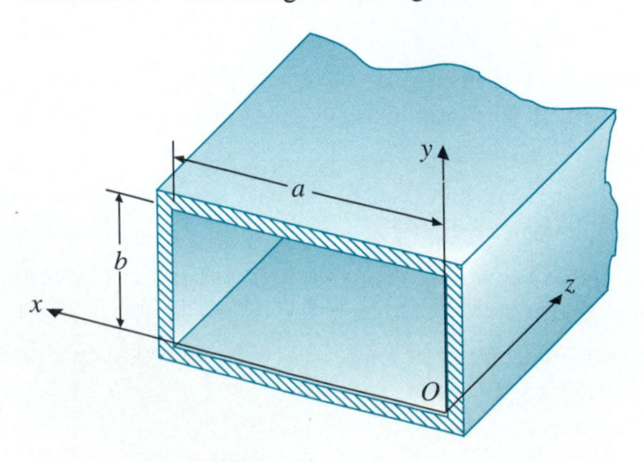

We assume that the solution $E_z^0(x, y)$ can be expressed as the product of a function $X(x)$ depending only on x and a function $Y(y)$ depending only on y:

$$E_z^0(x, y) = X(x)Y(y). \tag{9-54}$$

In Subsection 3-11.5 we mentioned the uniqueness theorem, which guarantees that a solution of Eq. (9-54), however obtained, is the only possible solution if it satisfies the problem's boundary conditions. By assuming a product solution such as that in Eq. (9-54), we follow the *method of separation of variables*. Substituting Eq. (9-54) in Eq. (9-53) and dividing the resulting equation by $X(x)Y(y)$, we have

The method of separation of variables

$$-\frac{1}{X(x)}\frac{d^2X(x)}{dx^2} = \frac{1}{Y(y)}\frac{d^2Y(y)}{dy^2} + h^2. \tag{9-55}$$

Since the left side of Eq. (9-55) is a function of x only and the right side is a function of y only, both sides must equal a constant in order for the equation to hold for all values of x and y. Calling this constant (separation constant) k_x^2, we obtain two separate ordinary differential equations:

$$\frac{d^2X(x)}{dx^2} + k_x^2X(x) = 0, \quad \text{and} \tag{9-56}$$

$$\frac{d^2Y(y)}{dy^2} + k_y^2Y(y) = 0, \tag{9-57}$$

where

$$k_y^2 = h^2 - k_x^2. \tag{9-58}$$

The general solutions of Eqs. (9-56) and (9.57) are

$$X(x) = A_1 \sin k_x x + A_2 \cos k_x x \quad \text{and} \tag{9-59}$$

$$Y(y) = B_1 \sin k_y y + B_2 \cos k_y y. \tag{9-60}$$

The appropriate forms to be chosen for $X(x)$ and $Y(y)$ must be such that their product in Eq. (9-54) satifies the following boundary conditions:

1. In the x-direction:

$$E_z^0(0, y) = 0, \quad \text{and} \tag{9-61}$$

$$E_z^0(a, y) = 0. \tag{9-62}$$

2. In the y-direction:

$$E_z^0(x, 0) = 0, \quad \text{and} \tag{9-63}$$

$$E_z^0(x, b) = 0. \tag{9-64}$$

Obviously, then, we must choose:

$X(x)$ in the form of $\sin k_x x$,

$$k_x = \frac{m\pi}{a}, \qquad m = 1, 2, 3, \ldots;$$

$Y(y)$ in the form of $\sin k_y y$,

$$k_y = \frac{n\pi}{b}, \qquad n = 1, 2, 3, \ldots;$$

and the proper solution for $E_z^0(x, y)$ is

$$E_z^0(x, y) = E_0 \sin\left(\frac{m\pi}{a}x\right)\sin\left(\frac{n\pi}{b}y\right) \qquad \text{(V/m)}, \tag{9-65}$$

where E_0 has been written for the product $A_1 B_1$, which is to be determined from the excitation conditions of the waveguide. The other field components can be obtained from Eqs. (9-11) through (9-14) by setting $H_z^0 = 0$.

The eigenvalue h and propagation constant γ are related to k_x and k_y by Eqs. (9-58) and (9-24), respectively. We have

$$h^2 = \left(\frac{m\pi}{a}\right)^2 + \left(\frac{n\pi}{b}\right)^2, \tag{9-66}$$

$$\gamma = j\beta = j\sqrt{k^2 - h^2} = j\sqrt{\omega^2\mu\epsilon - \left(\frac{m\pi}{a}\right)^2 - \left(\frac{n\pi}{b}\right)^2}. \tag{9-67}$$

Every combination of the integers m and n defines a possible mode that may be designated as the TM_{mn} mode; thus there are a double infinite number of TM modes. The first subscript denotes the number of half-cycle variations of the fields in the x-direction, and the second subscript denotes the number of half-cycle variations of the fields in the y-direction. The cutoff of a particular mode is the condition that makes γ vanish. For the TM_{mn} mode, the cutoff frequency is, from Eq. (9-26),

Cutoff frequency of TM_{mn} mode

$$(f_c)_{mn} = \frac{h}{2\pi\sqrt{\mu\epsilon}} = \frac{1}{2\sqrt{\mu\epsilon}}\sqrt{\left(\frac{m}{a}\right)^2 + \left(\frac{n}{b}\right)^2} \qquad \text{(Hz)}. \tag{9-68}$$

Alternatively, we may write $\lambda_c = u/f_c = 2\pi/h$, or

Cutoff wavelength of TM_{mn} mode

$$(\lambda_c)_{mn} = \frac{2}{\sqrt{\left(\frac{m}{a}\right)^2 + \left(\frac{n}{b}\right)^2}} \qquad \text{(m)}, \tag{9-69}$$

where λ_c is the ***cutoff wavelength***.

For TM modes in rectangular waveguides, neither m nor n can be zero. Otherwise, $E_z^0(x, y)$ in Eq. (9-65) and all other field components would vanish. Hence, the TM_{11} mode has the lowest cutoff frequency of all TM modes in a rectangular waveguide. The expressions for the phase constant β and the wave impedance Z_{TM} for propagating modes in Eqs. (9-29) and (9-34), respectively, apply here directly.

EXAMPLE 9-3

A rectangular waveguide having interior dimensions $a = 2.3$ (cm) and $b = 1.0$ (cm) is filled with a medium characterized by $\epsilon_r = 2.25$, $\mu_r = 1$.

a) Find h, f_c, and λ_c for TM_{11} mode.

b) If the operating frequency is 15% higher than the cutoff frequency, find $(Z)_{\text{TM}_{11}}$, $(\beta)_{\text{TM}_{11}}$, and $(\lambda_g)_{\text{TM}_{11}}$. Assume the waveguide to be lossless for propagating modes.

SOLUTION

a) For the TM_{11} mode, we use $m = n = 1$ in Eqs. (9-66), (9-68), and (9-69).

$$(h)_{\text{TM}_{11}} = \sqrt{\left(\frac{\pi}{2.3 \times 10^{-2}}\right)^2 + \left(\frac{\pi}{1.0 \times 10^{-2}}\right)^2} = 342.6 \quad (\text{m}^{-1}).$$

$$(f_c)_{\text{TM}_{11}} = \frac{h}{2\pi\sqrt{\mu\epsilon}} = \frac{hc}{2\pi\sqrt{\epsilon_r}} = \frac{342.6 \times (3 \times 10^8)}{2\pi\sqrt{2.25}} = 10.9 \quad (\text{GHz}).$$

$$(\lambda_c)_{\text{TM}_{11}} = \frac{2\pi}{h} = \frac{2\pi}{342.6} = 1.83 \quad (\text{cm}).$$

b) At $f = 1.15f_c = 1.15 \times 10.9 = 12.54$ (GHz), TM_{11} is a propagating mode. The expressions for Z_{TM}, β, and λ_g in Eqs. (9-34), (9-29), and (9-30) all contain the factor $\sqrt{1 - (f_c/f)^2} = \sqrt{1 - (1/1.15)^2} = 0.494$. We have

$$(Z)_{\text{TM}_{11}} = \eta\sqrt{1 - \left(\frac{f_c}{f}\right)^2} = \frac{377}{\sqrt{2.25}} \times 0.494 = 124.2 \quad (\Omega),$$

$$(\beta)_{\text{TM}_{11}} = \omega\sqrt{\mu\epsilon}\sqrt{1 - \left(\frac{f_c}{f}\right)^2} = \frac{2\pi \times (12.54 \times 10^9)\sqrt{2.25}}{3 \times 10^8} \times 0.494$$

$$= 194.5 \quad (\text{rad/m}),$$

and

$$(\lambda_g)_{\text{TM}_{11}} = \frac{2\pi}{(\beta)_{\text{TM}_{11}}} = \frac{2\pi}{194.5} = 0.0323 \text{ (m)} = 3.23 \quad (\text{cm}).$$

EXERCISE 9.3 For $f < (f_c)_{TM_{11}}$, TM_{11} is an evanescent mode. Find the attenuation constant for TM_{11} mode in the waveguide in Example 9-3 at $f = 0.85 f_c$. What is the distance in the waveguide over which field amplitudes will be reduced by a factor of e^{-1} or 36.8%?

ANS. 180.5 (Np/m), 5.5 (mm).

EXERCISE 9.4 For the waveguide in Example 9-3 find the cutoff frequencies for TM_{12} and TM_{21} modes.

ANS. 20.5 (GHz), 13.3 (GHz).

9-3.2 TE WAVES IN RECTANGULAR WAVEGUIDES

For transverse electric waves, $E_z = 0$, we solve Eq. (9-36) for H_z. We write

$$H_z(x, y, z) = H_z^0(x, y)e^{-\gamma z}, \tag{9-70}$$

where $H_z^0(x, z)$ satisfies the following second-order partial-differential equation:

$$\left(\frac{\partial^2}{\partial x^2} + \frac{\partial^2}{\partial y^2} + h^2\right) H_z^0(x, y) = 0. \tag{9-71}$$

Equation (9-71) is seen to be of exactly the same form as Eq. (9-53). The solution for $H_z^0(x, y)$ must satisfy the following boundary conditions.

1. In the x-direction:

$$\frac{\partial H_z^0}{\partial x} = 0 \ (E_y = 0) \qquad \text{at } x = 0, \tag{9-72}$$

$$\frac{\partial H_z^0}{\partial x} = 0 \ (E_y = 0) \qquad \text{at } x = a. \tag{9-73}$$

2. In the y-direction:

$$\frac{\partial H_z^0}{\partial y} = 0 \ (E_x = 0) \qquad \text{at } y = 0, \tag{9-74}$$

$$\frac{\partial H_z^0}{\partial y} = 0 \ (E_x = 0) \qquad \text{at } y = b. \tag{9-75}$$

It is readily verified that the appropriate solution for $H_z^0(x, y)$ is

$$\boxed{H_z^0(x, y) = H_0 \cos\left(\frac{m\pi}{a} x\right) \cos\left(\frac{n\pi}{b} y\right) \qquad \text{(A/m).}} \tag{9-76}$$

The other field components are obtained from Eqs. (9-11) through (9-14):

$$E_x^0(x, y) = \frac{j\omega\mu}{h^2}\left(\frac{n\pi}{b}\right) H_0 \cos\left(\frac{m\pi}{a} x\right) \sin\left(\frac{n\pi}{b} y\right), \tag{9-77}$$

$$E_y^0(x, y) = -\frac{j\omega\mu}{h^2}\left(\frac{m\pi}{a}\right)H_0 \sin\left(\frac{m\pi}{a}x\right)\cos\left(\frac{n\pi}{b}y\right), \tag{9-78}$$

$$H_x^0(x, y) = \frac{\gamma}{h^2}\left(\frac{m\pi}{a}\right)H_0 \sin\left(\frac{m\pi}{a}x\right)\cos\left(\frac{n\pi}{b}y\right), \tag{9-79}$$

$$H_y^0(x, y) = \frac{\gamma}{h^2}\left(\frac{n\pi}{b}\right)H_0 \cos\left(\frac{m\pi}{a}x\right)\sin\left(\frac{n\pi}{b}y\right), \tag{9-80}$$

where h and γ have the same expressions as those in Eqs. (9-66) and (9-67), respectively, for TM modes.

Equation (9-68) for cutoff frequency also applies here. For TE modes, either m or n (but not both) can be zero. If $a > b$, $h = \pi/a$ is the smallest eigenvalue and the cutoff frequency is the *lowest* when $m = 1$ and $n = 0$:

TE_{10} mode has the lowest cutoff frequency of all TE modes in a rectangular waveguide with $a > b$.

$$(f_c)_{TE_{10}} = \frac{1}{2a\sqrt{\mu\epsilon}} = \frac{u}{2a} \quad \text{(Hz)}. \tag{9-81}$$

The corresponding cutoff wavelength is

Cutoff wavelength of TE_{10} mode

$$(\lambda_c)_{TE_{10}} = 2a \quad \text{(m)}. \tag{9-82}$$

Dominant mode

The importance of the TE_{10} mode in a rectangular waveguide

The mode having the lowest cutoff frequency (longest cutoff wavelength) is called the ***dominant mode***. Hence *the* TE_{10} *mode is the dominant mode of a rectangular waveguide with* $a > b$. Because the TE_{10} mode has the lowest attenuation of all modes in a rectangular waveguide and its electric field is definitely polarized in one direction everywhere, it is of particular importance. We will examine its field structure and other characteristics more in detail later in this chapter.

■ EXERCISE 9.5

(a) What is the dominant mode of an $a \times b$ rectangular waveguide if $a < b$? What is its cutoff wavelength? (b) What are the cutoff frequencies in a square waveguide ($a = b$) for TM_{11}, TE_{20}, and TE_{02} modes?

ANS. (a) $2b$, (b) $1/a\sqrt{2\mu\epsilon}$.

EXAMPLE 9-4

A TE_{10} wave at 10 (GHz) propagates in a rectangular waveguide with inner dimensions $a = 1.5$ (cm) and $b = 0.6$ (cm), which is filled with polyethylene ($\epsilon_r = 2.25$, $\mu_r = 1$). Determine (a) the phase constant, (b) the guide wavelength, (c) the phase velocity, and (d) the wave impedance.

SOLUTION

At $f = 10^{10}$ (Hz) the wavelength in *unbounded* polyethylene is

$$\lambda = \frac{u}{f} = \frac{3 \times 10^8}{\sqrt{2.25} \times 10^{10}} = \frac{2 \times 10^8}{10^{10}} = 0.02 \quad \text{(m)}.$$

The cutoff frequency for the TE_{10} mode is, from Eq. (9-81),

$$f_c = \frac{u}{2a} = \frac{2 \times 10^8}{2 \times (1.5 \times 10^{-2})} = 0.667 \times 10^{10} \quad \text{(Hz)}.$$

a) The phase constant is, from Eq. (9-38),

$$\beta = \frac{\omega}{u}\sqrt{1 - \left(\frac{f_c}{f}\right)^2} = \frac{2\pi 10^{10}}{2 \times 10^8}\sqrt{1 - 0.667^2}$$

$$= 74.5\pi = 234 \quad \text{(rad/m)}.$$

b) The guide wavelength is, from Eq. (9-30).

$$\lambda_g = \frac{\lambda}{\sqrt{1 - (f_c/f)^2}} = \frac{0.02}{0.745} = 0.0268 \quad \text{(m)}.$$

c) The phase velocity is, from Eq. (9-33),

$$u_p = \frac{u}{\sqrt{1 - (f_c/f)^2}} = \frac{2 \times 10^8}{0.745} = 2.68 \times 10^8 \quad \text{(m/s)}.$$

d) The wave impedance is, from Eq. (9-39),

$$(Z_{TE})_{10} = \frac{\sqrt{\mu/\epsilon}}{\sqrt{1 - (f_c/f)^2}} = \frac{377/\sqrt{2.25}}{0.745} = 337.4 \quad (\Omega).$$

EXERCISE 9.6 Which TM and TE modes can propagate in the polyethylene-filled rectangular waveguide in Example 9-4 if the operating frequency is 19 (GHz)? What are their cutoff frequencies?

ANS. TE_{10}, TE_{20}, TE_{01}, TE_{11}, and TM_{11}. Cutoff frequencies in (GHz): 6.67, 13.3, 16.7, 17.9, and 17.9.

EXAMPLE 9-5

(a) Write the instantaneous field expressions for the TE_{10} mode in a rectangular waveguide having sides a and b. (b) Sketch the electric and magnetic field lines in typical xy-, yz-, and xz-planes. (c) Sketch the surface currents on the guide walls.

SOLUTION

a) The instantaneous field expressions for the dominant TE_{10} mode are obtained by multiplying the phasor expressions in Eqs. (9-76) through

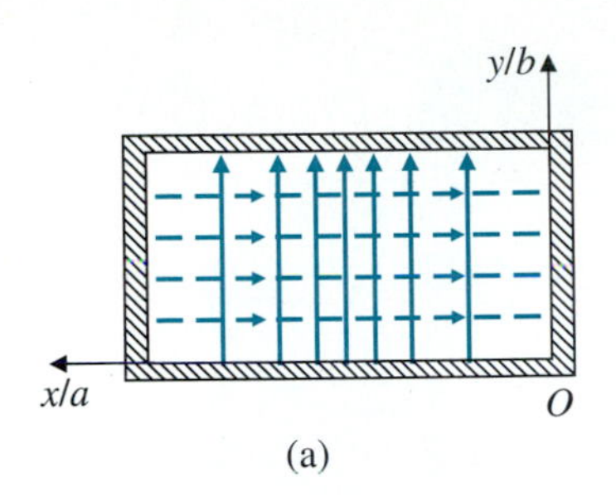

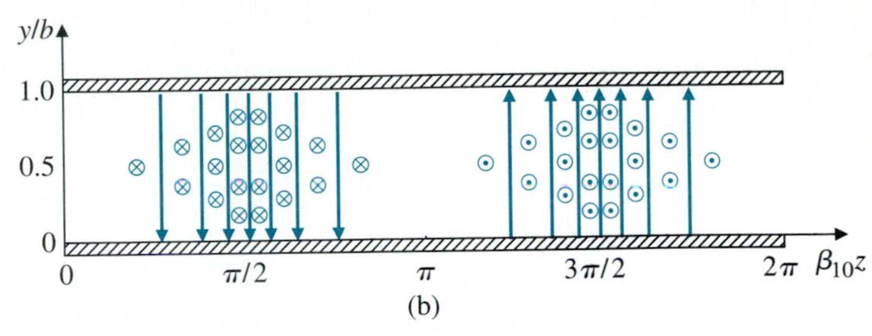

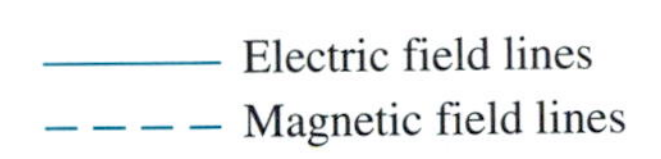

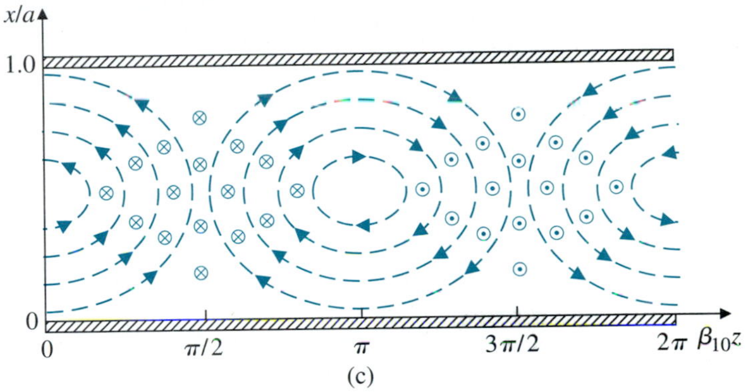

FIGURE 9-5 Field lines for TE_{10} mode in rectangular waveguide.

(9-80) with $e^{j(\omega t - \beta z)}$ and then taking the real part of the product. We have, for $m = 1$ and $n = 0$ ($h_{10} = \pi/a$),

$$E_x(x, y, z; t) = 0, \tag{9-83}$$

$$E_y(x, y, z; t) = \frac{\omega\mu a}{\pi} H_0 \sin\left(\frac{\pi}{a}x\right) \sin(\omega t - \beta_{10}z), \tag{9-84}$$

$$E_z(x, y, z; t) = 0, \tag{9-85}$$

$$H_x(x, y, z; t) = -\frac{\beta_{10}a}{\pi} H_0 \sin\left(\frac{\pi}{a}x\right) \sin(\omega t - \beta_{10}z), \tag{9-86}$$

$$H_y(x, y, z; t) = 0, \tag{9-87}$$

$$H_z(x, y, z; t) = H_0 \cos\left(\frac{\pi}{a}x\right) \cos(\omega t - \beta_{10}z), \tag{9-88}$$

where

$$\beta = \sqrt{k^2 - h^2} = \sqrt{\omega^2\mu\epsilon - \left(\frac{\pi}{a}\right)^2}. \tag{9-89}$$

b) We see from Eqs. (9-83) through (9-88) that the TE_{10} mode has only three nonzero field components—namely, E_y, H_x, and H_z. In a typical xy-plane, say, when $\sin(\omega t - \beta z) = 1$, both E_y and H_x vary as $\sin(\pi x/a)$ and are independent of y, as shown in Fig. 9-5(a). In a typical yz-plane, for example at $x = a/2$ or $\sin(\pi x/a) = 1$ and $\cos(\pi x/a) = 0$, we only

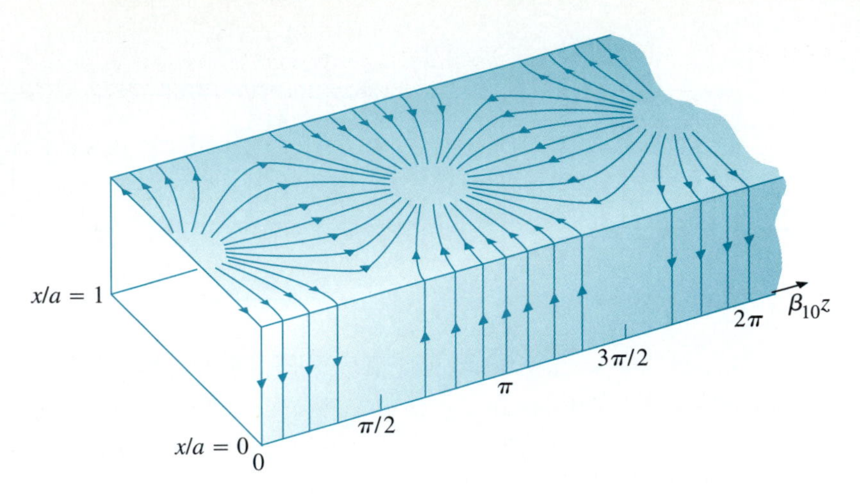

FIGURE 9-6 Surface currents on guide walls for TE_{10} mode in rectangular waveguide.

have E_y and H_x, both of which vary sinusoidally with βz. A sketch of E_y and H_x at $t = 0$ is given in Fig. 9-5(b).

The sketch in an xz-plane will show all three nonzero field components—E_y, H_x, and H_z. The slope of the **H** lines at $t = 0$ is governed by the following equation:

$$\left(\frac{dx}{dz}\right)_{\mathbf{H}} = \frac{\beta}{h^2}\left(\frac{\pi}{a}\right)\tan\left(\frac{\pi}{a}x\right)\tan\beta z, \tag{9-90}$$

which can be used to draw the **H** lines in Fig. 9-5(c). These lines are independent of y.

c) The surface current density on guide walls, $\mathbf{J}_s$, is related to the magnetic field intensity by Eq. (6-47b):

$$\mathbf{J}_s = \mathbf{a}_n \times \mathbf{H}, \tag{9-91}$$

where $\mathbf{a}_n$ is the *outward* normal from the wall surface and **H** is the magnetic field intensity at the wall. We have, at $t = 0$,

$$\mathbf{J}_s(x = 0) = -\mathbf{a}_y H_z(0, y, z; 0) = -\mathbf{a}_y H_0 \cos\beta z, \tag{9-92}$$

$$\mathbf{J}_s(x = a) = \mathbf{a}_y H_z(a, y, z; 0) = \mathbf{J}_s(x = 0), \tag{9-93}$$

$$\begin{aligned}\mathbf{J}_s(y = 0) &= \mathbf{a}_x H_z(x, 0, z; 0) - \mathbf{a}_z H_x(x, 0, z; 0)\\ &= \mathbf{a}_x H_0 \cos\left(\frac{\pi}{a}x\right)\cos\beta z - \mathbf{a}_z \frac{\beta}{h^2}\left(\frac{\pi}{a}\right) H_0 \sin\left(\frac{\pi}{a}x\right)\sin\beta z,\end{aligned} \tag{9-94}$$

$$\mathbf{J}_s(y = b) = -\mathbf{J}_s(y = 0). \tag{9-95}$$

The surface currents on the inside walls at $x = 0$ and at $y = b$ are sketched in Fig. 9-6.

■ **EXERCISE 9.7**

As waves propagate along a waveguide, charges are deposited on the inner surface of its walls. Find the expressions for the surface charge distributions

a) along the center lines of the top and bottom walls, and

b) along the side walls for the TE_{10} mode at $t = 0$

ANS. (a) $\rho_s = (\omega\mu\epsilon a/\pi)H_0 \sin\beta_{10}z$ (C/m²) on top wall, (b) 0.

EXAMPLE 9-6

Standard air-filled rectangular waveguides have been designed for the radar bands listed in Table 6-4. The inner dimensions of a waveguide suitable for X-band applications are: $a = 2.29$ cm (0.90 in.) and $b = 1.02$ cm (0.40 in.). If it is desired that the waveguide operates only in the dominant TE_{10} mode and that the operating frequency be at least 25% above the cutoff frequency of the TE_{10} mode but no higher than 95% of the next higher cutoff frequency, what is the allowable operating-frequency range?

SOLUTION

For $a = 2.29 \times 10^{-2}$ (m) and $b = 1.02 \times 10^{-2}$ (m), the two modes having the lowest cutoff frequencies are TE_{10} and TE_{20}. Using Eq. (9-68), we find

$$(f_c)_{10} = \frac{c}{2a} = \frac{3 \times 10^8}{2 \times 2.29 \times 10^{-2}} = 6.55 \times 10^9 \quad \text{(Hz)},$$

$$(f_c)_{20} = \frac{c}{a} = 13.10 \times 10^9 \quad \text{(Hz)}.$$

Thus the allowable operating-frequency range under the specified conditions is

$$1.25(f_c)_{TE_{10}} \leq f \leq 0.95(f_c)_{TE_{10}}$$

or

$$8.19 \quad \text{(GHz)} \leq f \leq 12.45 \quad \text{(GHz)}.$$

9-3.3 ATTENUATION IN RECTANGULAR WAVEGUIDES

Causes for attenuation in waveguides

Attenuation in a waveguide arises from two sources: lossy dielectric and imperfectly conducting walls. Losses modify the electric and magnetic fields within the guide, making exact solutions difficult to obtain. However, in practical waveguides the losses are usually very small, and we can assume that the transverse field patterns of the propagating modes are not appreciably affected by them. A real part of the propagation constant will now appear as the attenuation constant, which accounts for power losses. The attenuation constant consists of two parts:

$$\alpha = \alpha_d + \alpha_c, \tag{9-96}$$

where α_d is the attenuation constant due to losses in the dielectric and α_c is that due to ohmic power loss in the imperfectly conducting walls. The analytical determination of α_d and α_c is somewhat tedious. Here we will merely outline the general procedure for their determination.

To find α_d we note from Section 7-3 that the effects of a lossy dielectric on wave propagation can be studied by considering a complex permittivity

$$\begin{aligned} \epsilon_d &= \epsilon' - j\epsilon'' \\ &= \epsilon' - j\frac{\sigma_d}{\omega}, \end{aligned} \tag{9-97}$$

Attenuation due to a lossy dielectric increases with dielectric conductivity and with frequency.

where $\epsilon''/\epsilon' = \sigma_d/\omega\epsilon'$ is the loss tangent and σ_d is the equivalent conductivity of the dielectric. Use of ϵ_d in place of ϵ in Eq. (9-24) will yield a complex propagation constant γ, the real part of which is α_d. It will be found that α_d is directly proportional to σ_d and decreases with the ratio (f_c/f).

To find α_c we use Eq. (8-57), which was derived from the law of conservation of energy,

$$\alpha_c = \frac{P_{\rm L}(z)}{2P(z)}, \tag{9-98}$$

Attenuation due to finite wall conductivity is inversely proportional to the square root of wall conductivity, but depends on the mode and the frequency in a complicated way.

where $P(z)$ is the time-average power propagated along the guide at z, and $P_{\rm L}(z)$ is the time-average power loss per unit length at z. $P(z)$ can be found from the transverse electric and magnetic field intensities using Eq. (7-79). In order to calculate $P_{\rm L}(z)$, it is necessary to consider the power loss in all four guide walls due to a finite conductivity σ_c. This requires the integration of $|J_s|^2R_s$ over the inner surfaces per unit length of the walls, where J_s and R_s denote, respectively, the surface current density on, and the intrinsic resistance (see Eq. 7-53) of the wall conductors. The expressions of α_c for TM_{mn} and TE_{mn} are different because of different current distributions. They depend on guide dimensions and the ratio (f_c/f) in a complicated way. For all propagating modes, α_c is directly proportional to R_s, which, in turn, is inversely proportional to the square root of the wall conductivity σ_c. Naturally α_c is zero (no power loss) if the guide walls are infinitely conducting $(\sigma_c \to \infty)$.

TE_{10} mode has the lowest attenuation in a rectangular waveguide.

In Fig. 9-7 are plotted typical α_c curves as functions of the operating frequency f for TE_{10} and TM_{11} modes in a 2.29 (cm) × 1.02 (cm) rectangular copper waveguide. From Eq. (9-68) we find $(f_c)_{10} = 6.55$ (GHz) and $(f_c)_{11} = 16.10$ (GHz). The curves show that the attenuation constant increases rapidly toward infinity as the operating frequency approaches the cutoff frequency. In the operating range $(f > f_c)$, both curves possess a broad minimum. The attenuation constant of the TE_{10} mode is everywhere lower than that of the TM_{11} mode. These facts have direct relevance in the choice of operating modes and frequencies.

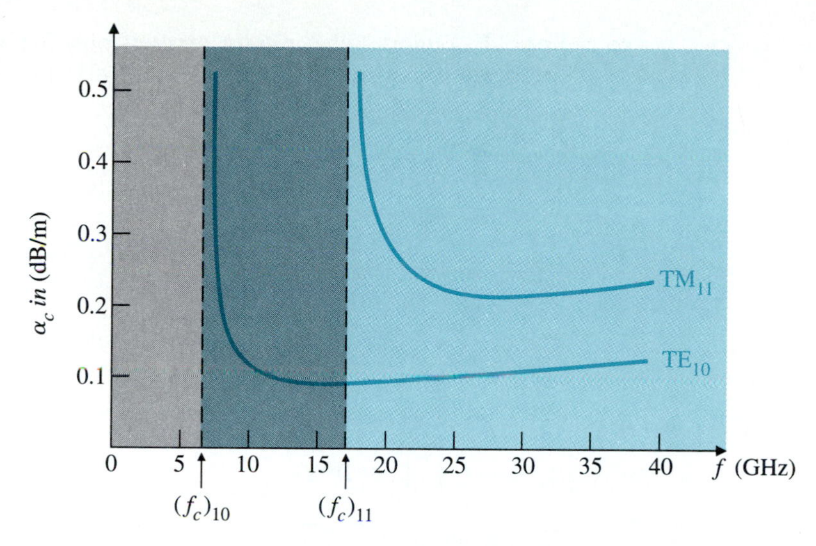

FIGURE 9-7 Attenuation due to wall losses in rectangular copper waveguide for TE_{10} and TM_{11} modes. $a = 2.29$ (cm), $b = 1.02$ (cm).

EXAMPLE 9-7

An air-filled 5.0 (cm) × 2.5 (cm) rectangular waveguide, 0.8 (m) long, is to deliver 1.2 (kW) to a matched load at 4.5 (GHz). Assuming an attenuation constant of 0.05 (dB/m), find (a) the required average input power to the waveguide, (b) the total amount of power dissipated in the walls of the waveguide, and (c) the maximum value of the electric intensity within the guide.

SOLUTION

Given: $a = 5.0 \times 10^{-2}$ (m), $b = 2.5 \times 10^{-2}$ (m).

$$\text{Dominant-mode } (f_c)_{10} = \frac{c}{2a} = \frac{3 \times 10^8}{2 \times 5.0 \times 10^{-2}}$$

$$= 3 \times 10^9 \text{ (Hz)} = 3 \text{ (GHz)}.$$

The next higher modes are TE_{20} and TE_{01}, both of which have a cutoff frequency equal to 6 (GHz) > 4.5 (GHz). Thus, the only propagating mode at 4.5 (GHz) is the TE_{10} mode.

a) $\alpha = 5 \times 10^{-2}$ (dB/m) $= 5.75 \times 10^{-3}$ (Np/m).

$$P_{in} = P_{load} e^{2\alpha\ell} = 1.2 \times 10^3 e^{2 \times (5.75 \times 10^{-3}) \times 0.8}$$

$$= 1.2 \times 10^3 e^{0.0092} = 1{,}211 \quad \text{(W)}.$$

b) Power dissipated $= P_{in} - P_{load}$

$$= 1{,}211 - 1{,}200 = 11 \quad \text{(W)}.$$

c) For TE_{10} mode, the phasor expressions for the transverse field components can be written from Eqs. (9-84) and (9-86) as

$$E_y^0 = E_0 \sin\left(\frac{\pi}{a}\right) x \tag{9-99}$$

and

$$H_x^0 = -\frac{E_0}{\eta_0}\sqrt{1-\left(\frac{f_c}{f}\right)^2}\sin\left(\frac{\pi}{a}x\right), \tag{9-100}$$

where $\eta_0 = \sqrt{\mu_0/\epsilon_0}$, and β_{10} in Eq. (9-86) has been written as $\omega\sqrt{\mu_0\epsilon_0}\sqrt{1-(f_c/f)^2}$ (see Eq. 9-38). Since the maximum fields occur at the input end, we have, from Eq. (7-79),

$$\begin{aligned} P_{in} &= -\frac{1}{2}\int_0^b \int_0^a E_y^0 H_x^0 \, dx\, dy \\ &= \frac{E_0^2 ab}{4\eta_0}\sqrt{1-\left(\frac{f_c}{f}\right)^2}. \end{aligned} \tag{9-101†}$$

Substituting the numbers in Eq. (9-101), we obtain the following:

$$1{,}211 = \frac{E_0^2(5.0\times 10^{-2})(2.5\times 10^{-2})}{4\times 377}\sqrt{1-\left(\frac{3}{4.5}\right)^2},$$

from which we find

$$E_0 = 44{,}283 \quad \text{(V/m)}.$$

REVIEW QUESTIONS

Q.9-12 Explain the method of separation of variables for solving partial differential equations.

Q.9-13 What is meant by the *dominant mode* of a waveguide? What is the dominant mode of a parallel-plate waveguide?

Q.9-14 State the boundary conditions to be satisfied by E_z for TM waves in a rectangular waveguide.

Q.9-15 Which TM mode has the lowest cutoff frequency of all the TM modes in a rectangular waveguide?

Q.9-16 State the boundary conditions to be satisfied by H_z for TE waves in a rectangular waveguide.

---

† From Eq. (7-79),

$$P_{in} = \frac{1}{2}\int_0^b \int_0^a \mathcal{R}e(\mathbf{E}\times\mathbf{H}^*)\cdot\mathbf{a}_z \, dx\, dy,$$

where $\mathbf{E} = \mathbf{a}_y E_y^0$ and $\mathbf{H} = \mathbf{a}_x H_x^0 + \mathbf{a}_z H_z^0$. Hence, $\mathcal{R}e(\mathbf{E}\times\mathbf{H}^*) = -\mathbf{a}_z E_y^0 H_x^0$, since H_z^0 is 90° out-of-phase with E_y^0 and their product has no real part.

Q.9-17 Which mode is the dominant mode in a rectangular waveguide if (a) $a > b$, (b) $a < b$, and (c) $a = b$?

Q.9-18 What is the cutoff wavelength of the TE_{10} mode in a rectangular waveguide if $a > b$?

Q.9-19 Which are the nonzero field components for the TE_{10} mode in a rectangular waveguide?

Q.9-20 Why is the TE_{10} mode in a rectangular waveguide of particular practical importance?

Q.9-21 Why are waveguides not used for VHF and lower frequency bands?

EXERCISE 9.8 Find the maximum amount of 10-(GHz) average power that can be transmitted through an air-filled $a = 2.25$ (cm), $b = 1.00$ (cm) rectangular waveguide at the TE_{10} mode without a breakdown.

REMARKS

1. There are a double infinite number of TM and TE modes in a waveguide.
2. The eigenvalues of TM and TE modes in a rectangular waveguide are discrete real numbers.
3. The cutoff frequencies for TM_{mn} and TE_{mn} modes are the same.
4. Cutoff frequencies are inversely proportional to $\sqrt{\epsilon_r}$ of the dielectric medium.
5. The lowest-order TM mode in a rectangular waveguide is TM_{11}.
6. The dominant mode in a rectangular waveguide with $a > b$ is TE_{10}, whose cutoff wavelength is $2a$.
7. The TE_{10} mode in a rectangular waveguide has the lowest attenuation constant due to wall losses.
8. The attenuation constant due to wall losses in a waveguide is inversely proportional to the square-root of wall conductivity.

9-4 OTHER WAVEGUIDE TYPES

We recall from Sec. 9-2 that the procedure used for analyzing the wave behaviors along uniform waveguides started with the solution of a homogeneous vector Helmholtz's equation in the plane of the guide cross section. A complete solution depends on the shape and dimensions of the cross section. Following this procedure, we obtained the operating characteristics of rectangular waveguides in Section 9-3. In this section we briefly discuss some other waveguide types that also find practical applications.

First, we mention hollow circular waveguides, which are, in fact, round metal pipes. A complete analysis of the wave behaviors along a circular waveguide involves the solution of a two-dimensional Helmholtz's equation in the circular cross section of the guide in polar coordinates (r, ϕ). This, in turn, requires a knowledge of Bessel's differential equation and Bessel functions. We shall not attempt such a solution in this book, except to point out that, as in rectangular waveguides, both TM and TE modes with characteristic cutoff frequencies may exist.

Dielectric slabs and rods without conducting walls can also support TM and TE guided-wave modes that are confined essentially within the dielectric medium. Outside the guide medium the fields fall exponentially. These are stray fields and may cause interference problems between neighboring circuits.

Optical fibers are waveguides at optical frequencies. They are flexible, and have very low attenuation and very large bandwidth.

A type of waveguide of particular importance for optical frequencies consists of a very thin fiber of a dielectric material, typically glass, cladded with a sheath having a slightly lower index of refraction. Such optical waveguides are generally called ***optical fibers***. An illustration of a cladded optical fiber was shown in Fig. 7-18. The operation of an optical fiber can be explained in terms of total internal reflection, as was done in Example 7-10 for Fig. 7-13. The core diameter of optical fibers is usually in the range of 25 to 100 (μm), and an attenuation as low as 1/4 (dB/km) can be attained at infrared frequencies. Compared with an attenuation of about 30 (dB/km) for hollow metal waveguides and hundreds of dB/km for ordinary coaxial cables, this low-attenuation feature is a tremendous advantage for optical fibers. Moreover, at infrared frequencies the available bandwidth is such that a single fiber-optic circuit can carry about 20 million telephone channels or 20 thousand television channels.

Optical fibers are hair-thin and flexible, and thousands of them can be bundled together to form an important part of an endoscope, a medical instrument for examining the interior of a hollow human organ such as the bronchial tubes, the colon, the bladder, and so on. Images are effectively transmitted through the optical waveguides.

9-5 CAVITY RESONATORS

Cavity resonators are enclosed metal boxes.

We have previously pointed out that at microwave frequencies, ordinary lumped-circuit elements such as R, L, and C are difficult to make, and stray fields become important. Circuits with dimensions comparable to the operating wavelength become efficient radiators and will interfere with other circuits and systems. Furthermore, conventional wire circuits tend to have a high effective resistance both because of energy loss through radiation and as

Electromagnetic fields are confined inside the boxes. Radiation and high-resistance effects are eliminated, resulting in a very high Q.

a result of skin effect. To provide a resonant circuit at UHF and higher frequencies, we look to an enclosure (a cavity) completely surrounded by conducting walls. Such a shielded enclosure confines electromagnetic fields inside and furnishes large areas for current flow, thus eliminating radiation and high-resistance effects. These enclosures have natural resonant freq-uencies and a very high Q (quality factor), and are called ***cavity resonators***.

9-5.1 RECTANGULAR CAVITY RESONATORS

Consider a rectangular waveguide with both ends closed by a conducting wall. The interior dimensions of the cavity are a, b, and d, as shown in Fig. 9-8. Let us disregard for the moment the probe-excitation part of the figure. Since both TM and TE modes can exist in a rectangular guide, we expect TM and TE modes in a rectangular resonator too. However, the designation of TM and TE modes in a resonator is *not unique* because we are free to choose x or y or z as the "direction of propagation"; that is, there is no unique "longitudinal direction." For example, a TE mode with respect to the z-axis could be a TM mode with respect to the y-axis.

For our purposes *we choose the z-axis as the reference* "direction of propagation." In actuality, the existence of conducting end walls at $z = 0$ and $z = d$ gives rise to multiple reflections and sets up standing waves; no wave propagates in an enclosed cavity. A three-symbol (***mnp***) subscript is needed to designate a TM or TE standing wave pattern in a cavity resonator.

FIGURE 9-8 Excitation of cavity modes by a coaxial line.

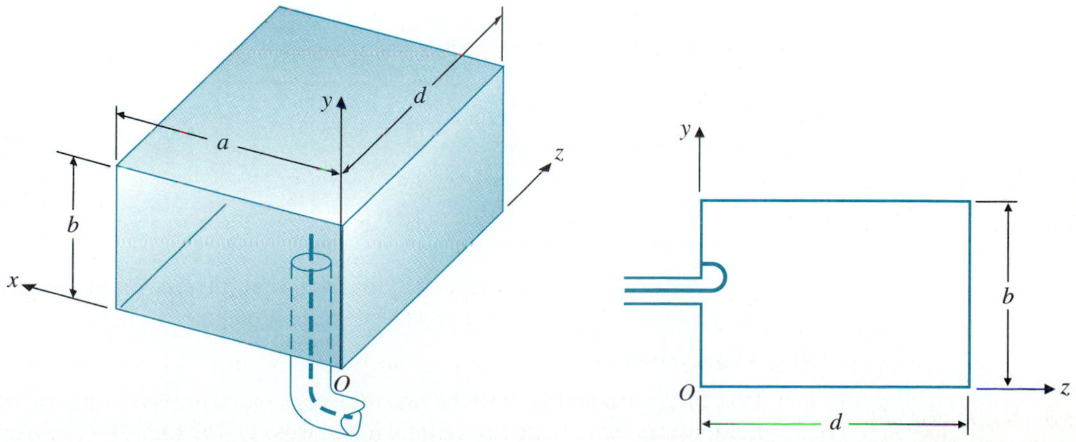

(a) Probe excitation.

(b) Loop excitation.

(A) TM_{mnp} Modes

The phasor expression for the sole longitudinal field component, $E_z(x, y, z) = E_z^0(z, y)e^{-\gamma z}$, for TM_{mn} modes in a waveguide has been given in Eqs. (9-52) and (9-65). Note that the longitudinal variation for a wave traveling in the $+z$-direction is described by the factor $e^{-\gamma z}$ or $e^{-j\beta z}$. This wave will be reflected by the end wall at $z = d$; and the reflected wave, going in the $-z$-direction, is described by a factor $e^{j\beta z}$. The superposition of a term with $e^{-j\beta z}$ and another of the same amplitude with $e^{j\beta z}$ results in a standing wave of the $\sin\beta z$ or $\cos\beta z$ type. Which should it be? The answer to this question depends on the particular field component.

Consider the transverse component $E_y(x, y, z)$. Boundary conditions at the conducting surfaces require that it be zero at $z = 0$ and $z = d$. This means that (1) its z-dependence must be of the $\sin\beta z$ type, and that (2) $\beta = p\pi/d$. The same argument applies to the other transverse electric field component $E_x(x, y, z)$. The relations between the transverse components, E_x^0 and E_y^0, and E_z^0 have been given in Eqs. (9-13) and (9-14), in which H_z^0 vanish for TM modes. We recall that the appearance of the factor $(-\gamma)$ in Eqs. (9-13) and (9-14) is the result of a differentiation with respect to z. Thus, if $E_y(x, y, z)$ depends on $\sin\beta z$, we can conclude from Eq. (9-14), which contains the factor $(-\gamma)$, that $E_z(x, y, z)$ must vary according to $\cos\beta z$. We have, for the TM_{mnp} mode,

For TM_{mnp} modes, $m\neq 0$ and $n\neq 0$

$$E_z(x, y, z) = E_z^0(x, y)\cos\left(\frac{p\pi}{d}z\right) = E_0 \sin\left(\frac{m\pi}{a}x\right)\sin\left(\frac{n\pi}{b}y\right)\cos\left(\frac{p\pi}{d}z\right). \tag{9-102}$$

All other field components can be written by using E_z in Eqs. (9-11) through (9-14) and noting that multiplication by $(-\gamma)$ signifies a partial differentiation with respect to z.

Substituting $\beta = p\pi/d$ in Eq. (9-67), we obtain the resonant frequency for TM_{mnp} modes ($u = 1/\sqrt{\mu\epsilon}$):

Resonant frequency of cavity resonator

$$f_{mnp} = \frac{u}{2}\sqrt{\left(\frac{m}{a}\right)^2 + \left(\frac{n}{b}\right)^2 + \left(\frac{p}{d}\right)^2} \quad \text{(Hz)}. \tag{9-103}$$

Equation (9-103) states the obvious fact that the resonant frequency increases as the order of a mode becomes higher.

(B) TE_{mnp} Modes

For TE_{mnp} modes ($E_z = 0$) the phasor expressions for the standing-wave field components can be written from Eqs. (9-76) and (9-77) through (9-80). We follow the same rules as those we used for TM_{mnp} modes; namely, (1) the transverse (tangential) electric field components must vanish at $z = 0$ and $z = d$, and (2) the factor γ indicates a negative partial differentiation with respect to z. The first rule requires a $\sin(p\pi z/d)$ factor in $E_x(x, y, z)$ and $E_y(x, y, z)$, as well as in $H_z(x, y, z)$; and the second rule indicates a $\cos(p\pi z/d)$ factor in $H_x(x, y, z)$ and $H_y(x, y, z)$. We have

For TE_{mnp} modes, $p \neq 0$; m and n are not both zero.

$$H_z(x, y, z) = H_z^0(x, y) \sin\left(\frac{p\pi}{d} z\right)$$

$$= H_0 \cos\left(\frac{m\pi}{a} x\right) \cos\left(\frac{n\pi}{b} y\right) \sin\left(\frac{p\pi}{d} z\right). \tag{9-104}$$

All other field components can be written by using H_z in Eqs. (9-11) through (9-14) and by noting that multiplication by $(-\gamma)$ signifies a partial differentiation with respect to z.

The expression for resonant frequency, f_{mnp}, remains the same as that obtained for TM_{mnp} modes in Eq. (9-103). Different modes having the same resonant frequency are called ***degenerate modes***. Thus TM_{mnp} and TE_{mnp} modes are always degenerate if none of the mode indices is zero. The mode with the lowest resonant frequency for a given cavity size is referred to as the ***dominant mode*** (see Example 9-8).

Degenerate modes

A cavity resonator may be excited by a small probe or loop.

A particular mode in a cavity resonator (or a waveguide) may be excited from a coaxial line by means of a small probe or loop antenna. In Fig. 9-8(a) a probe is shown that is the tip of the inner conductor of a coaxial cable; it protrudes into a cavity at a location where the electric field is a maximum for the desired mode. The probe is, in fact, an antenna that couples electromagnetic energy into the resonator. Alternatively, a cavity resonator may be excited through the introduction of a small loop at a place where the magnetic flux of the desired mode linking the loop is a maximum. Figure 9-8(b) illustrates such an arrangement. Of course, the source frequency from the coaxial line must be the same as the resonant frequency of the desired mode in the cavity.

As an example, for the TE_{101} mode in an $a \times b \times d$ rectangular cavity, there are only three nonzero field components:

$$E_y = -\frac{j\omega\mu a}{\pi} H_0 \sin\left(\frac{\pi}{a} x\right) \sin\left(\frac{\pi}{d} z\right), \tag{9-105}$$

$$H_x = -\frac{a}{d} H_0 \sin\left(\frac{\pi}{a} x\right) \cos\left(\frac{\pi}{d} z\right), \tag{9-106}$$

$$H_z = H_0 \cos\left(\frac{\pi}{a} x\right) \sin\left(\frac{\pi}{d} z\right). \tag{9-107}$$

This mode may be excited by a probe inserted in the center region of the top or bottom face where E_y is maximum, as shown in Fig. 9-8(a), or by a loop to couple a maximum H_x placed inside the front or back face, as shown in Fig. 9-8(b). The best location of a probe or a loop is affected by the impedance-matching requirements of the microwave circuit of which the resonator is a part.

A cavity resonator may also be excited through an iris.

A commonly used method for coupling energy from a waveguide to a cavity resonator is the introduction of a hole or an iris at an appropriate location in the cavity wall. The field in the waveguide at the hole must have a component that is favorable in exciting the desired mode in the resonator.

EXAMPLE 9-8

Determine the dominant modes and their frequencies in an air-filled rectangular cavity resonator for (a) $a > b > d$, (b) $a > d > b$, and (c) $a = b = d$, where a, b, and d are the dimensions in the x-, y-, and z-directions, respectively.

SOLUTION

As usual, we choose the z-axis as the reference "direction of propagation." For TM_{mnp} modes, Eq. (9-102) shows that neither m nor n can be zero, but that p can be zero. For TE_{mnp} modes, Eq. (9-104) indicates that H_z does not vanish even if both m and n are zero, provided p is not zero. However, if H_z is independent of both x and y, Eqs. (9-11) through (9-14) show that there will be no transverse field components at all. Thus, for TE_{mnp} modes, p cannot be zero, and either m or n (but not both) can be zero.

The modes of the lowest orders in a rectangular cavity resonator are

TM_{110}, TE_{011}, and TE_{101}.

The resonant frequency for both TM and TE modes is given by Eq. (9-103).

a) For $a > b > d$, the lowest resonant frequency is

$$f_{110} = \frac{c}{2}\sqrt{\frac{1}{a^2} + \frac{1}{b^2}}, \tag{9-108}$$

where c is the velocity of light in free space. Therefore TM_{110} is the dominant mode.

b) For $a > d > b$, the lowest resonant frequency is

$$f_{101} = \frac{c}{2}\sqrt{\frac{1}{a^2} + \frac{1}{d^2}}, \tag{9-109}$$

and TE_{101} is the dominant mode.

c) For $a = b = d$, all three of the lowest-order modes (namely, TM_{110}, TE_{011}, and TE_{101}) have the same field patterns. The resonant frequency of these degenerate modes is

$$f_{110} = \frac{c}{\sqrt{2}a}. \tag{9-110}$$

■ **EXERCISE 9.9** Determine the four lowest resonant frequencies of an air-filled 2.5 (cm) × 1.5 (cm) × 5.0 (cm) rectangular cavity resonator and identify their modes.

ANS. 6.71, 8.49, 10.44, 11.66 (GHz).

9-5.2 QUALITY FACTOR OF CAVITY RESONATOR

A cavity resonator stores energy in the electric and magnetic fields for any particular mode pattern. In any practical cavity the walls have a finite conductivity (a nonzero surface resistance), and the resulting power loss causes a decay of the stored energy. The ***quality factor***, or Q, of a resonator, like that of any resonant circuit, is a measure of the bandwidth of the resonator and is defined as

Definition of the Q of a resonator

$$Q = 2\pi \frac{\text{Time-average energy stored at a resonant frequency}}{\text{Energy dissipated in one period of this frequency}} \quad \text{(Dimensionless).} \tag{9-111}$$

Let W be the total time-average energy in a cavity resonator. We write

$$W = W_e + W_m, \tag{9-112}$$

where W_e and W_m denote the energies stored in the electric and magnetic fields, respectively. If P_L is the time-average power dissipated in the cavity, then the energy dissipated in one period is P_L divided by frequency, and Eq. (9-111) can be written as

Formula for computing the Q of a cavity resonator

$$Q = \frac{\omega W}{P_L} \quad \text{(Dimensionless).} \tag{9-113}$$

In determining the Q of a cavity at a resonant frequency, it is customary to assume that the loss is small enough to allow the use of the field patterns without loss.

We will now find the Q of an $a \times b \times d$ cavity for the TE_{101} mode that has three nonzero field components given in Eqs. (9-105), (9-106) and (9-107). The time-average stored electric energy is

$$\begin{aligned} W_e &= \frac{\epsilon_0}{4} \int |E_y|^2 \, dv \\ &= \frac{\epsilon_0 \omega_{101}^2 \mu_0^2 a^2}{4\pi^2} H_0^2 \int_0^d \int_0^b \int_0^a \sin^2\left(\frac{\pi}{a}x\right) \sin^2\left(\frac{\pi}{d}z\right) dx\,dy\,dz \\ &= \frac{\epsilon_0 \omega_{101}^2 \mu_0^2 a^2}{4\pi^2} H_0^2 \left(\frac{a}{2}\right) b \left(\frac{d}{2}\right) = \frac{1}{4} \epsilon_0 \mu_0^2 a^3 b d f_{101}^2 H_0^2. \end{aligned} \tag{9-114}$$

The total time-average stored magnetic energy is

$$
\begin{aligned}
W_m &= \frac{\mu_0}{4}\int \{|H_x|^2 + |H_z|^2\}\,dv \\
&= \frac{\mu_0}{4}H_0^2 \int_0^d\int_0^b\int_0^a \left\{\frac{a^2}{d^2}\sin^2\left(\frac{\pi}{a}x\right)\cos^2\left(\frac{\pi}{d}z\right)\right. \\
&\qquad \left. +\cos^2\left(\frac{\pi}{a}x\right)\sin^2\left(\frac{\pi}{d}z\right)\right\}dx\,dy\,dz \\
&= \frac{\mu_0}{4}H_0^2\left\{\frac{a^2}{d^2}\left(\frac{a}{2}\right)b\left(\frac{d}{2}\right) + \left(\frac{a}{2}\right)b\left(\frac{d}{2}\right)\right\} = \frac{\mu_0}{16}abd\left(\frac{a^2}{d^2}+1\right)H_0^2.
\end{aligned}
\tag{9-115}
$$

From Eq. (9-103) the resonant frequency for the TE_{101} mode is

$$
f_{101} = \frac{1}{2\sqrt{\mu_0\epsilon_0}}\sqrt{\frac{1}{a^2}+\frac{1}{d^2}}. \tag{9-116}
$$

Substitution of f_{101} from Eq. (9-116) in Eq. (9-114) proves that, *at the resonant frequency*, $W_e = W_m$. Thus,

$$
W = 2W_e = 2W_m = \frac{\mu_0 H_0^2}{8}abd\left(\frac{a^2}{d^2}+1\right). \tag{9-117}
$$

To find P_L, we note that the power loss per unit area is

$$
\mathscr{P}_{av} = \tfrac{1}{2}|J_s|^2 R_s = \tfrac{1}{2}|H_t|^2 R_s, \tag{9-118}
$$

where $|H_t|$ denotes the magnitude of the tangential component of the magnetic field at the cavity walls. The power loss in the $z = d$ (back) wall is the same as that in the $z = 0$ (front) wall. Similarly, the power loss in the $x = a$ (left) wall is the same as that in the $x = 0$ (right) wall; and the power loss in the $y = b$ (upper) wall is the same as that in the $y = 0$ (lower) wall. We have

$$
\begin{aligned}
P_L &= \oint \mathscr{P}_{av}\,ds = R_s\left\{\int_0^b\int_0^a |H_x(z=0)|^2\,dx\,dy + \int_0^d\int_0^b |H_z(x=0)|^2\,dy\,dz\right. \\
&\qquad \left. + \int_0^d\int_0^a |H_x|^2\,dx\,dz + \int_0^d\int_0^a |H_z|^2\,dx\,dz\right\} \\
&= \frac{R_s H_0^2}{2}\left\{\frac{a^2}{d}\left(\frac{b}{d}+\frac{1}{2}\right) + d\left(\frac{b}{a}+\frac{1}{2}\right)\right\}.
\end{aligned}
\tag{9-119}
$$

Using Eqs. (9-117) and (9-119) in Eq. (9-113), we obtain

$$
Q_{101} = \frac{\pi f_{101}\mu_0 abd(a^2+d^2)}{R_s[2b(a^3+d^3) + ad(a^2+d^2)]} \qquad (\text{TE}_{101}\text{ mode}), \tag{9-120}
$$

where f_{101} has been given in Eq. (9-116).

■ **EXERCISE 9.10** The total stored energy, W, in a lossy cavity decays exponentially according to $e^{-2\alpha t}$, and the rate of change of W with time is equal to the power, P_L, dissipated in the cavity walls. Show that the attenuation constant α is related to cavity Q by the formula $\alpha = \omega/2Q$.

EXAMPLE 9-9

a) What should be the size of a hollow cubic cavity made of copper in order for it to have a dominant resonant frequency of 10 (GHz)?

b) Find the Q at that frequency.

SOLUTION

a) For a cubic cavity, $a = b = d$. From Example 9-8, part (c), we know that TM_{110}, TE_{011}, and TE_{101} are degenerate dominant modes having the same field patterns and that

$$f_{101} = \frac{3 \times 10^8}{\sqrt{2}a} = 10^{10} \quad \text{(Hz)}.$$

Therefore,

$$a = \frac{3 \times 10^8}{\sqrt{2} \times 10^{10}} = 2.12 \times 10^{-2} \quad \text{(m)}$$

$$= 21.2 \quad \text{(mm)}.$$

b) The expression of Q in Eq. (9-120) for a cubic cavity reduces to

$$Q_{101} = \frac{\pi f_{101}\mu_0 a}{3R_s} = \frac{a}{3}\sqrt{\pi f_{101}\mu_0\sigma}. \tag{9-121}$$

For copper, $\sigma = 5.80 \times 10^7$ (S/m), we have

$$Q_{101} = \left(\frac{2.12}{3} \times 10^{-2}\right)\sqrt{\pi 10^{10}(4\pi 10^{-7})(5.80 \times 10^7)} = 10{,}693.$$

The Q of a cavity resonator is thus extremely high in comparison with that obtainable from lumped L–C resonant circuits. In practice, the preceding value is somewhat lower because of losses through feed connections and surface irregularities.

A high-Q cavity resonator has a very narrow bandwidth.

As we know from circuit theory, the response variable (voltage or current) in a resonant circuit is a maximum at the resonant frequency and falls off sharply in a high-Q circuit as the frequency deviates from the resonant frequency on either side. Thus, a high-Q cavity resonator is very selective and has a narrow bandwidth.

■ **EXERCISE 9.11** Assuming that the cubic cavity in Example 9-9 is made of brass and filled with a lossless dielectric material ($\epsilon_r = 3$, $\mu_r = 1$), determine

a) the lowest resonant frequency, and

b) the quality factor Q.

ANS. 5.78 (GHz), (b) 4,230.

REVIEW QUESTIONS

Q.9-22 What are cavity resonators? What are their most desirable properties?

Q.9-23 Are the field patterns in a cavity resonator traveling waves or standing waves? How do they differ from those in a waveguide?

Q.9-24 Referring to the z-axis, which of the following modes cannot exist in a rectangular cavity resonator: TM_{011}, TM_{101}, TM_{110}, TE_{011}, TE_{101}, TE_{110}? Explain.

Q.9-25 What is meant by *degenerate modes*?

Q.9-26 Define the quality factor, Q, of a resonator.

Q.9-27 What fundamental assumption is made in the derivation of the formulas for the Q of cavity resonators?

Q.9-28 Explain why the measured Q of a cavity resonator is lower than the calculated value.

REMARKS

1. TM_{mnp} and TE_{mnp} modes ($m, n, p \neq 0$) in a rectangular resonator have the same resonant frequencies.
2. The subscripts m, n, and p denote, respectively, the number of half-wavelengths of field variations in the x-, y-, and z-directions.
3. The Q of a cavity resonator is directly proportional to the square-root of wall conductivity.

SUMMARY

In this chapter on waveguides and cavity resonators we

- discussed the method for analyzing the wave behavior along uniform guiding structures by solving homogeneous vector Helmholtz's equations,
- examined the general characteristics of TM and TE waves,
- explained the cutoff and high-pass properties of waveguides,
- analyzed the field and current distributions of the dominant TE_{10} mode in a rectangular waveguide,

- discussed the method for determining the attenuation constant of propagating modes in a rectangular waveguide, and showed typical α_c curves as functions of frequency, and
- explained the wave modes, determined the resonant frequencies, and verified the high-Q property of rectangular cavity resonators.

PROBLEMS

P.9-1 (a) Plot the wave impedances for an air-filled waveguide versus the ratio (f/f_c) for TM and TE modes. (b) Compare the values of Z_{TM} and Z_{TE} at $f = 1.1f_c$ and $2.2f_c$.

P.9-2 For uniform waveguides, use appropriate relations in Section 9-2 to:

a) prove that the universal diagram relating u_g/u and f_c/f is a quarter-circle with a unity radius,

b) plot the universal graph of λ_g/λ versus f/f_c, and

c) find u_g/u, λ_g/λ and u_p/u at $f = 1.25f_c$.

P.9-3 Assume that a TE wave of a frequency f is launched along the z-direction in the parallel-plate waveguide in Fig. 9-3. The dielectric medium between the plates has constitutive parameters ϵ and μ. (a) Find the phasor expression for $H_z^0(y)$. (b) Find the cutoff frequency for the TE_1 mode. (c) Write the instantaneous expression for all the field components of the TE_1 mode.

P.9-4 For the air-filled parallel-plate waveguide in Fig. 9-3,

a) obtain the phasor expressions of all field components for TE modes,

b) determine the cutoff frequency for the TE_n mode, and

c) find the surface current densities on the conducting plates. Do the currents on the two plates flow in the same direction, or in opposite directions?

P.9-5 Guide wavelength and impedances can be measured by means of a detector attached to a probe moving along a slotted section of a waveguide. Assuming that when a shorting conducting plane is placed at the load end of a lossless hollow 2.50 (cm) × 1.25 (cm) rectangular waveguide operating at the TE_{10} mode, adjacent voltage minima are 2.65 (cm) apart. When the shorting plane is replaced by a load, it is found that the SWR is 2.0 and that the voltage minima have been shifted toward the load by 0.80 (cm). Find (a) the operating frequency, (b) the load impedance, and (c) the power delivered to the load for an input power of 10 (W).

P.9-6 For an air-filled $a \times b$ rectangular waveguide operating at frequency f in the TM_{11} mode, (a) write the phasor expressions for all the field components, and (b) find f_c, λ_c, and λ_g.

P.9-7 A standard air-filled S-band rectangular waveguide has dimensions

$a = 7.21$ (cm) and $b = 3.40$ (cm). What mode types can be used to transmit electromagnetic waves having the following wavelengths?

a) $\lambda = 10$ (cm), and **b)** $\lambda = 5$ (cm).

P.9-8 Calculate and list in ascending order the cutoff frequencies (in terms of the cutoff frequency of the dominant mode) of an $a \times b$ rectangular waveguide for the following modes: TE_{01}, TE_{10}, TE_{11}, TE_{02}, TE_{20}, TM_{11}, TM_{12}, and TM_{22} (a) if $a = 2b$, and (b) if $a = b$.

P.9-9 An air-filled $a \times b$ ($b < a < 2b$) rectangular waveguide is to be constructed to operate at 3 (GHz) in the dominant mode. We desire the operating frequency to be at least 20% higher than the cutoff frequency of the dominant mode and also at least 20% below the cutoff frequency of the next higher-order mode.

a) Give a typical design for the dimensions a and b.

b) Calculate for your design β, u_p, λ_g, and the wave impedance at the operating frequency.

P.9-10 Calculate and compare the values of β, u_p, u_g, λ_g, and $Z_{\text{TE}_{10}}$ for a 2.5 (cm) $\times$ 1.5 (cm) rectangular waveguide operating at 7.5 (GHz):

a) if the waveguide is hollow, and

b) if the waveguide is filled with a dielectric medium characterized by $\epsilon_r = 2$, $\mu_r = 1$ and $\sigma = 0$.

P.9-11 Starting from Eq. (9-65),

a) obtain the expressions of $E_x^0(x, y)$ $E_y^0(x, y)$, $H_x^0(x, y)$, and $H_y^0(x, y)$ for the TM_{11} mode, and

b) obtain a formula for the average power P_{av} transmitted along an $a \times b$ waveguide.

P.9-12 The instantaneous expression for E_z of a TM mode in an air-filled 5.0 (cm) $\times$ 2.5 (cm) rectangular waveguide is

$$E_z = E_0 \sin(100\pi x) \sin(100\pi y) \cos(2\pi 10^{10} t - \beta z) \quad \text{(V/m)}.$$

a) What is the mode of operation?

b) Calculate f_c, β, Z_{TM}, and λ_g.

P.9-13 The instantaneous expression for H_z of a TE mode in an air-filled 2.5 (cm) $\times$ 2.5 (cm) square waveguide is

$$H_z = 0.3 \cos(80\pi y) \cos(\omega t - 280z) \quad \text{(A/m)}.$$

a) What is the mode of operation?

b) Calculate f_c, f, Z_{TE}, and λ_g.

c) Assuming negligible losses, calculate the average power flow in the waveguide.

P.9-14 The attenuation of propagating modes in a waveguide due to a lossy dielectric can be studied in terms of a complex permittivity ϵ_d and an equivalent conductivity σ_d shown in Eq. (9-97). (a) Substitute Eq. (9-97) in Eq. (9-24) to obtain a formula for the attenuation constant α_d due to lossy dielectric in terms of the ratio f_c/f. (b) Calculate α_d in a 2.50 (cm) $\times$ 1.25 (cm) rectangular waveguide operating at 4.0 (GHz). The dielectric medium has a dielectric constant 4 and an equivalent conductivity 3×10^{-5} (S/m).

P.9-15 An electromagnetic wave is to propagate along an air-filled $a \times b$ rectangular waveguide at the dominant mode. Assume $a = 2.50$ (cm) and the usable bandwidth to be between $1.15(f_c)_{10}$ and 15% below the cutoff frequency of the next higher mode.

a) Calculate and compare the permissible bandwidth for $b = 0.25a$, $b = 0.50a$, and $b = 0.75a$.

b) Calculate and compare the average powers transmitted along the three guides in part (a) at 7 (GHz) if the maximum electric intensity is 10 (kV/m). Neglect the losses.

P.9-16 Given an air-filled lossless rectangular cavity resonator with dimensions 8 (cm) $\times$ 6 (cm) $\times$ 5 (cm), find the first eight lowest-order modes and their resonant frequencies.

P.9-17 An air-filled rectangular cavity with brass walls—ϵ_0, μ_0, $\sigma = 1.57 \times 10^7$ (S/m)—has the following dimensions: $a = 4$ (cm), $b = 3$ (cm), and $d = 5$ (cm).

a) Determine the dominant mode and its resonant frequency for this cavity.

b) Find the Q and the time-average stored electric and magnetic energies at the resonant frequency, assuming H_0 to be 0.1 (A/m).

P.9-18 If the rectangular cavity in Problem P.9-17 is filled with a lossless dielectric material having a dielectric constant 2.5, find

a) the resonant frequency of the dominant mode,

b) the Q, and

c) the time-average stored electric and magnetic energies at the resonant frequency, assuming H_0 to be 0.1 (A/m).

P.9-19 Equation (9-121) indicates that the quality factor Q_{101} for TE_{101} mode in a cubic resonant cavity ($a = b = d$) can be written as

$$Q_{101} = \frac{a}{3\delta}, \tag{9-122}$$

where δ is the skin depth of the cavity walls.

a) If the cavity is made of brass, determine a so that a quality factor of 6,500 is obtained.

b) Find the resonant frequency.

c) What will Q_{101} be if the cavity is made of copper?

P.9-20 For an air-filled rectangular copper cavity resonator,

a) calculate its Q for the TE_{101} mode if its dimensions are $a = d = 1.8b = 3.6$ (cm), and

b) determine how much b should be increased in order to make Q 20% higher.

CHAPTER 10

10-1 OVERVIEW

In Chapter 7 we studied the propagation characteristics of plane electromagnetic waves in source-free media without considering how the waves were generated. Of course, the waves must originate from sources, which in electromagnetic terms are time-varying charges and currents. In order to radiate electromagnetic energy efficiently in prescribed directions, the charges and currents must be distributed in specific ways. ***Antennas*** are structures designed for radiating and receiving electromagnetic energy effectively in a prescribed manner. Each antenna has a characteristic input impedance, and can be viewed as a transducer for matching the feeding transmission line or waveguide to the intrinsic impedance of the surrounding medium. Without an efficient antenna, electromagnetic energy would be localized, and wireless transmission of information over long distances would not be possible.

Functions of antennas

In this chapter we will first study the radiation fields and characteristic properties of an elemental electric dipole. We then consider finite-length thin linear antennas, of which the half-wavelength dipole is an important special case. The radiation characteristics of a linear antenna are largely determined by its length and the manner in which it is excited. To obtain more directivity and other desirable properties, a number of such antennas may be arranged together to form an ***antenna array***. Some basic properties of simple arrays will be considered. We will discuss the concepts of the effective area of receiving antennas and the backscattering cross section of scatterers. We will also examine the power

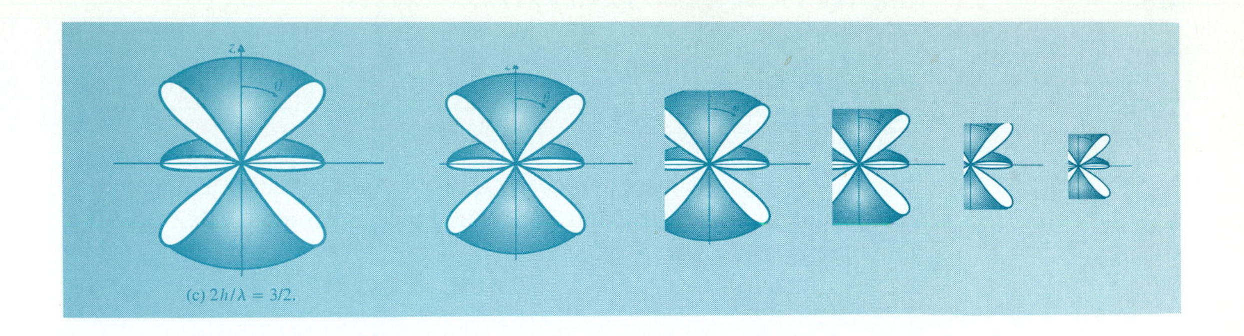

Antennas and Antenna Arrays

transmission relation between transmitting and receiving antennas, including the radar equation.

Procedure for determining radiation characteristics of an antenna

In general, the analysis of the radiation characteristics of an antenna follows the three steps below:

1. Determine the magnetic potential **A** from known or assumed current distribution **J** on the antenna. For harmonic time dependence, the phasor retarded vector potential is, from Eq. (6-85),

$$\mathbf{A} = \frac{\mu}{4\pi}\int_{V'} \frac{\mathbf{J}e^{-jkR}}{R}\,dv', \tag{10-1}$$

where $k = \omega\sqrt{\mu\epsilon} = 2\pi/\lambda$ is the wavenumber.

2. Find the magnetic field intensity **H** from **A**. See Eq. (6-50).

$$\mathbf{H} = \frac{1}{\mu}\nabla \times \mathbf{A}. \tag{10-2}$$

3. Find the electric field intensity **E** from **H**. See Eq. (6-80b) with $\mathbf{J} = 0$ in space.

$$\mathbf{E} = \frac{1}{j\omega\epsilon}\nabla \times \mathbf{H}. \tag{10-3}$$

After knowing **E** and **H**, all other radiation characteristics of the antenna can be determined.

10-2 THE ELEMENTAL ELECTRIC DIPOLE

We first consider the radiation characteristics of a very short (compared to the operating wavelength), thin, conducting wire of length $d\ell$ that carries a time-harmonic current

$$i(t) = I \cos \omega t = \mathscr{R}e[Ie^{j\omega t}], \tag{10-4}$$

as shown in Fig. 10-1. Such a current element is a building block of linear antennas, and is called a ***Hertzian dipole***.

Hertzian dipole is a very short radiating current element.

To determine the electromagnetic field of a Hertzian dipole, we follow the three steps outlined in Section 10-1.

STEP 1 Find the phasor representation of the retarded vector potential **A**. From Eq. (10-1) we have

$$\mathbf{A} = \mathbf{a}_z \frac{\mu_0 I\, d\ell}{4\pi} \left(\frac{e^{-j\beta R}}{R} \right), \tag{10-5}$$

where $\beta = k_0 = \omega/c = 2\pi/\lambda$. Since (see Eq. 2-47)

$$\mathbf{a}_z = \mathbf{a}_R \cos\theta - \mathbf{a}_\theta \sin\theta, \tag{10-6}$$

the spherical components of $\mathbf{A} = \mathbf{a}_R A_R + \mathbf{a}_\theta A_\theta + \mathbf{a}_\phi A_\phi$ are

$$A_R = A_z \cos\theta = \frac{\mu_0 I\, d\ell}{4\pi} \left(\frac{e^{-j\beta R}}{R} \right) \cos\theta, \tag{10-6a}$$

$$A_\theta = -A_z \sin\theta = -\frac{\mu_0 I\, d\ell}{4\pi} \left(\frac{e^{-j\beta R}}{R} \right) \sin\theta, \tag{10-6b}$$

$$A_\phi = 0. \tag{10-6c}$$

STEP 2 Find **H** from **A**.

From the geometry of Fig. 10-1 we expect no variation with respect to the coordinate ϕ. We have, from Eq. (2-99),

FIGURE 10-1 A Hertzian dipole.

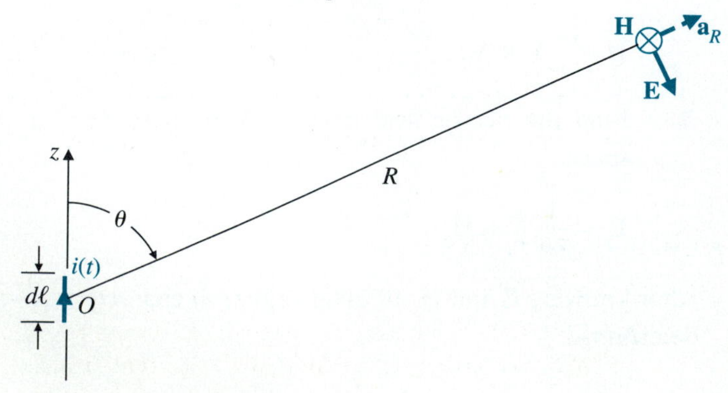

$$\mathbf{H} = \frac{1}{\mu_0}\nabla \times \mathbf{A} = \mathbf{a}_\phi \frac{1}{\mu_0 R}\left[\frac{\partial}{\partial R}(RA_\theta) - \frac{\partial A_R}{\partial \theta}\right]$$

$$= -\mathbf{a}_\phi \frac{I\,d\ell}{4\pi}\beta^2 \sin\theta \left[\frac{1}{j\beta R} + \frac{1}{(j\beta R)^2}\right]e^{-j\beta R}. \tag{10-7}$$

STEP 3 Find **E** from **H**.

$$\mathbf{E} = \frac{1}{j\omega\epsilon_0}\nabla \times \mathbf{H}$$

$$= \frac{1}{j\omega\epsilon_0}\left[\mathbf{a}_R \frac{1}{R\sin\theta}\frac{\partial}{\partial\theta}(H_\phi \sin\theta) - \mathbf{a}_\theta \frac{1}{R}(RH_\phi)\right], \tag{10-8}$$

which gives

$$E_R = -\frac{I\,d\ell}{4\pi}\eta_0\beta^2 2\cos\theta\left[\frac{1}{(j\beta R)^2} + \frac{1}{(j\beta R)^3}\right]e^{-j\beta R}, \tag{10-8a}$$

$$E_\theta = -\frac{I\,d\ell}{4\pi}\eta_0\beta^2\sin\theta\left[\frac{1}{j\beta R} + \frac{1}{(j\beta R)^2} + \frac{1}{(j\beta R)^3}\right]e^{-j\beta R}, \tag{10-8b}$$

$$E_\phi = 0, \tag{10-8c}$$

where $\eta_0 = \sqrt{\mu_0/\epsilon_0} \cong 120\pi\ (\Omega)$.

Equations (10-7) and (10-8) constitute the electromagnetic fields of a Hertzian dipole. These expressions are fairly complicated. However, in antenna problems we are primarily interested in the fields at distances very far from the antenna; that is, in regions where $R \gg \lambda/2\pi$, or $\beta R = 2\pi R/\lambda \gg 1$. Under these circumstances (in the *far zone*) we could neglect $1/(\beta R)^2$ and $1/(\beta R)^3$ terms and write the ***far field***, or ***radiation field***, of the elemental electric dipole as

Far zone

Radiation fields are far fields.

Far fields of a Hertzian dipole

$$H_\phi = j\frac{I\,d\ell}{4\pi}\left(\frac{e^{-j\beta R}}{R}\right)\beta\sin\theta \qquad \text{(A/m)}, \tag{10-9}$$

$$E_\theta = j\frac{I\,d\ell}{4\pi}\left(\frac{e^{-j\beta R}}{R}\right)\eta_0\beta\sin\theta = \eta_0 H_\phi \qquad \text{(V/m)}. \tag{10-10}$$

The other field components can be neglected.

REVIEW QUESTIONS

Q.10-1 What are the essential functions of antennas?

Q.10-2 State the procedure for finding the electromagnetic field due to an assumed time-harmonic current distribution on an antenna structure.

Q.10-3 What is a *Hertzian dipole*?

Q.10-4 Define the *far zone* of an antenna.

Q.10-5 What is meant by the *radiation fields* of an antenna?

REMARKS

1. The radiation field of a vertical Hertzian dipole consists of H_ϕ and $E_\theta = \eta_0 H_\phi$.
2. E_θ and H_ϕ are in space quadrature and in time phase, both varying inversely with the distance to the dipole.

10-3 ANTENNA PATTERNS AND DIRECTIVITY

Radiation pattern of an antenna, the antenna pattern

No physical antennas radiate uniformly in all directions in space. The graph that describes the relative far-zone field strength versus direction at a fixed distance from an antenna is called the ***radiation pattern*** of the antenna, or simply the ***antenna pattern***. In general, an antenna pattern is three-dimensional, varying with both θ and ϕ in a spherical coordinate system. The difficulties of making three-dimensional plots can be avoided by plotting separately the magnitude of the normalized field strength (with respect to the peak value) versus θ for a constant ϕ (an ***E-plane pattern***) and the magnitude of the normalized field strength versus ϕ for $\theta = \pi/2$ (the ***H-plane pattern***).

***E*-plane and *H*-plane radiation patterns**

EXAMPLE 10-1

Plot the E-plane and H-plane radiation patterns of a Hertzian dipole.

SOLUTION

Since E_θ and H_ϕ in the far zone are proportional to each other, we need only consider the normalized magnitude of E_θ.

a) *E-plane pattern.* At a given R, E_θ is independent of ϕ; and from Eq. (10-10) the normalized magnitude of E_θ is

$$\text{Normalized } |E_\theta| = |\sin\theta|. \tag{10-11}$$

This is the E-plane ***pattern function*** of a Hertzian dipole. For any given ϕ, Eq. (10-11) represents a pair of circles, as shown in Fig. 10-2(a).

Pattern function is the normalized electric intensity function describing an antenna pattern.

b) *H-plane pattern.* At a given R and for $\theta = \pi/2$ the normalized magnitude of E_θ is $|\sin\theta| = 1$. The H-plane pattern is then simply a circle of unity radius centered at the z-directed dipole, as shown in Fig. 10-2(b).

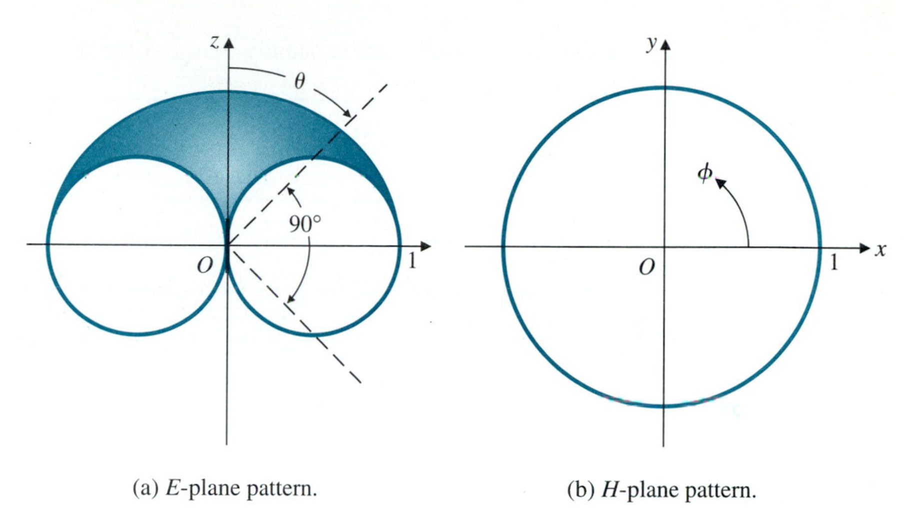

(a) E-plane pattern. (b) H-plane pattern.

FIGURE 10-2 Radiation patterns of a Hertzian dipole.

A commonly used parameter to measure the overall ability of an antenna to direct radiated power in a given direction is ***directive gain***, which may be defined in terms of radiation intensity. ***Radiation intensity*** is the time-average power per unit solid angle. The SI unit for radiation intensity is watt per steradian (W/sr). Since there are R^2 square meters of spherical surface area for each unit solid angle, radiation intensity, U, equals R^2 times the time-average power per unit area or R^2 times the magnitude of the time-average Poynting vector, $\mathscr{P}_{av}$:

Definition of and SI unit for radiation intensity

$$U = R^2 \mathscr{P}_{av} \qquad \text{(W/sr)}. \tag{10-12}$$

The total time-average power radiated is

$$P_r = \oint \boldsymbol{\mathscr{P}}_{av} \cdot d\mathbf{s} = \oint U \, d\Omega \qquad \text{(W)}, \tag{10-13}$$

where $d\Omega$ is the differential solid angle, $d\Omega = \sin\theta \, d\theta \, d\phi$.

The ***directive gain***, $G_D(\theta, \phi)$, of an antenna pattern is the ratio of the radiation intensity in the direction (θ, ϕ) to the average radiation intensity:

Directive gain

$$G_D(\theta, \phi) = \frac{U(\theta, \phi)}{P_r/4\pi} = \frac{4\pi U(\theta, \phi)}{\oint U \, d\Omega}. \tag{10-14}$$

Obviously, the directive gain of an isotropic or omnidirectional antenna (an antenna that radiates uniformly in all directions) is unity. However, an isotropic antenna does not exist in practice.

Isotropic or omnidirectional antennas do not exist in practice.

The maximum directive gain of an antenna is called the ***directivity*** of the antenna. It is the ratio of the maximum radiation intensity to the average

Antenna directivity

radiation intensity and is usually denoted by D:

$$D = \frac{U_{max}}{U_{av}} = \frac{4\pi U_{max}}{P_r} \quad \text{(Dimensionless).} \tag{10-15}$$

In terms of electric field intensity, D can be expressed as

Calculating directivity from far-zone electric field intensity

$$D = \frac{4\pi |E_{max}|^2}{\int_0^{2\pi}\int_0^{\pi} |E(\theta, \phi)|^2 \sin\theta \, d\theta \, d\phi} \quad \text{(Dimensionless).} \tag{10-16}$$

Directivity is frequently expressed in decibels, referring to unity.

EXAMPLE 10-2

Find the directive gain and the directivity of a Hertzian dipole.

SOLUTION

For a Hertzian dipole the magnitude of the time-average Poynting vector is

$$\mathscr{P}_{av} = \tfrac{1}{2}\mathscr{R}e|\mathbf{E} \times \mathbf{H}^*| = \tfrac{1}{2}|E_\theta|\,|H_\phi|. \tag{10-17}$$

Hence from Eqs. (10-9), (10-10), and (10-12),

$$U = \frac{(I\,d\ell)^2}{32\pi^2}\eta_0 \beta^2 \sin^2\theta. \tag{10-18}$$

The directive gain can be obtained from Eq. (10-14):

$$\begin{aligned} G_D(\theta, \phi) &= \frac{4\pi \sin^2\theta}{\int_0^{2\pi}\int_0^{\pi} (\sin^2\theta)\sin\theta \, d\theta \, d\phi} \\ &= \tfrac{3}{2}\sin^2\theta. \end{aligned} \tag{10-19}$$

The directivity is the maximum value of $G_D(\theta, \phi)$:

$$D = G_D\left(\frac{\pi}{2}, \phi\right) = 1.5,$$

which corresponds to $10\log_{10} 1.5$ or 1.76 (dB).

Definition of antenna gain

A measure of antenna efficiency is the power gain. The ***power gain***, or simply the ***gain***, G_P, of an antenna referred to an isotropic source is the ratio of its maximum radiation intensity to the radiation intensity of a lossless isotropic source with the same power input. The directive gain as defined in Eq. (10-14) is based on radiated power P_r. Because of ohmic power loss, P_ℓ, in the antenna itself as well as in nearby lossy structures including the ground, P_r is less than the total input power P_i. We have

$$P_i = P_r + P_\ell. \tag{10-20}$$

The power gain of an antenna is then

Definition of radiation efficiency

$$G_P = \frac{4\pi U_{\max}}{P_i} \quad \text{(Dimensionless)}. \tag{10-21}$$

The ratio of the gain to the directivity of an antenna is the ***radiation efficiency***, ζ_r:

$$\zeta_r = \frac{G_P}{D} = \frac{P_r}{P_i} \quad \text{(Dimensionless)}. \tag{10-22}$$

Normally, the efficiency of well-constructed antennas is very close to 100%.

Antenna radiation resistance

A useful measure of the amount of power radiated by an antenna is radiation resistance. The ***radiation resistance*** of an antenna is the value of a hypothetical resistance that would dissipate an amount of power equal to the radiated power P_r when the current in the resistance is equal to the maximum current along the antenna. A high radiation resistance is a desirable property for an antenna.

EXAMPLE 10-3

Find the radiation resistance of a Hertzian dipole.

SOLUTION

If we assume no ohmic losses, the time-average power radiated by a Hertzian dipole for an input time-harmonic current with an amplitude I is

$$P_r = \frac{1}{2}\int_0^{2\pi}\int_0^{\pi} E_\theta H_\phi^* R^2 \sin\theta\, d\theta\, d\phi. \tag{10-23}$$

Using the far-zone fields in Eqs. (10-9) and (10-10), we find

$$\begin{aligned} P_r &= \frac{I^2(d\ell)^2}{32\pi^2}\eta_0\beta^2 \int_0^{2\pi}\int_0^{\pi} \sin^3\theta\, d\theta\, d\phi \\ &= \frac{I^2(d\ell)^2}{12\pi}\eta_0\beta^2 = \frac{I^2}{2}\left[80\pi^2\left(\frac{d\ell}{\lambda}\right)^2\right]. \end{aligned} \tag{10-24}$$

In this last expression we have used 120π for the intrinsic impedance of free space, η_0, and substituted $2\pi/\lambda$ for β.

Since the current along the short Hertzian dipole is uniform, we refer the power dissipated in the radiation resistance R_r to I. Equating $I^2R_r/2$ to P_r, we obtain

Radiation resistance of a Hertzian dipole

$$R_r = 80\pi^2\left(\frac{d\ell}{\lambda}\right)^2 \quad (\Omega). \tag{10-25}$$

As an example, if $d\ell = 0.01\lambda$, R_r is only about 0.08 (Ω), an extremely small value. Hence a short dipole antenna is a poor radiator of electromagnetic power. However, it is erroneous to say without qualification that the radiation resistance of a dipole antenna increases as the square of its length because Eq. (10-24) holds only if $d\ell \ll \lambda$.

EXAMPLE 10-4

Find the radiation efficiency of an isolated Hertzian dipole made of a metal wire of radius a, length d, and conductivity σ.

SOLUTION

Let I be the amplitude of the current in the wire dipole having a loss resistance R_ℓ. Then the ohmic power loss is

$$P_\ell = \tfrac{1}{2}I^2R_\ell. \tag{10-26}$$

In terms of radiation resistance R_r the radiated power is

$$P_r = \tfrac{1}{2}I^2R_r. \tag{10-27}$$

From Eqs. (10-20) and (10-22) we have

$$\zeta_r = \frac{P_r}{P_r + P_\ell} = \frac{R_r}{R_r + R_\ell} = \frac{1}{1 + (R_\ell/R_r)}, \tag{10-28}$$

where R_r has been found in Eq. (10-25). The loss resistance R_ℓ of the metal wire can be expressed in terms of the surface resistance R_s:

$$R_\ell = R_s\left(\frac{d\ell}{2\pi a}\right), \tag{10-29}$$

where

$$R_s = \sqrt{\frac{\pi f \mu_0}{\sigma}} \tag{10-30}$$

as given in Eq. (7-53). Using Eqs. (10-25) and (10-29) in Eq. (10-28), we obtain the radiation efficiency of an isolated Hertzian dipole:

$$\zeta_r = \frac{1}{1 + \dfrac{R_s}{160\pi^3}\left(\dfrac{\lambda}{a}\right)\left(\dfrac{\lambda}{d\ell}\right)}. \tag{10-31}$$

Assume that $a = 1.8$ (mm), $d\ell = 2$ (m), operating frequency $f = 1.5$ (MHz), and σ (for copper) $= 5.80 \times 10^7$ (S/m). We find that

$$\lambda = \frac{c}{f} = \frac{3 \times 10^8}{1.5 \times 10^6} = 200 \quad \text{(m)},$$

$$R_s = \sqrt{\frac{\pi \times (1.50 \times 10^6) \times (4\pi 10^{-7})}{5.80 \times 10^7}} = 3.20 \times 10^{-4} \quad (\Omega),$$

$$R_\ell = 3.20 \times 10^{-4} \times \left(\frac{2}{2\pi 1.8 \times 10^{-3}}\right) = 0.057 \quad (\Omega),$$

$$R_r = 80\pi^2 \left(\frac{2}{200}\right)^2 = 0.079 \quad (\Omega),$$

and

$$\zeta_r = \frac{0.079}{0.079 + 0.057} = 58\%,$$

which is very low. Equation (10-31) shows that smaller values of (a/λ) and $(d\ell/\lambda)$ lower the radiation efficiency.

EXERCISE 10.1 The normalized pattern function in the E-plane of a vertical antenna over a ground plane is given as $\sqrt{\sin\theta}$, $(0 \le \theta \le \pi/2$, and $0 \le \phi \le 2\pi)$.

a) Obtain an expression for the directive gain.

b) Calculate its directivity.

ANS. (a) $(8/\pi)\sin\theta$, (b) 2.55, or 4.06 (dB).

REVIEW QUESTIONS

Q.10-6 Define *antenna pattern.*

Q.10-7 Describe the E-plane and H-plane patterns of a Hertzian dipole.

Q.10-8 Define *radiation intensity.*

Q.10-9 Define *directive gain* and *directivity* of an antenna.

Q.10-10 Define *power gain* and *radiation efficiency* of an antenna.

Q.10-11 Define *radiation resistance* of an antenna.

REMARKS

1. A radiation-intensity pattern (or power pattern) of an antenna plots the square of field intensity versus θ or ϕ at a fixed distance.
2. Directive gain is *not* the same as power gain.
3. Directivity is *not* the same as radiation efficiency.
4. Radiation resistance is *not* the same as the real part of input impedance.

10-4 THIN LINEAR ANTENNAS

Linear dipole antenna

We have just indicated that a short dipole antenna is not a good radiator of electromagnetic power because of its low radiation resistance and low radiation efficiency. We now examine the radiation characteristics of a center-fed thin straight antenna having a length comparable to a wavelength, as shown in Fig. 10-3. Such an antenna is a ***linear dipole antenna.*** If the current distribution along the antenna is known, we can find its radiation field by integrating the radiation field due to an elemental dipole over the entire

FIGURE 10-3 A center-fed linear dipole with sinusoidal current distribution.

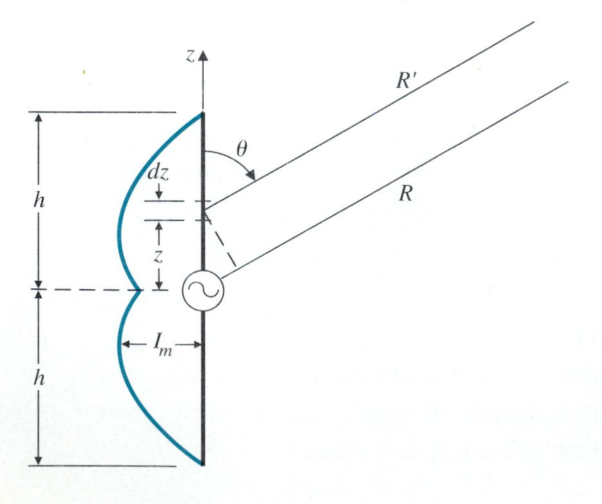

length of the antenna. The determination of the exact current distribution on such a seemingly simple geometrical configuration (a straight wire of a finite radius) is, however, a very difficult boundary-value problem. For our purposes we assume a sinusoidal space variation for the current on a very thin, straight dipole. Such a current distribution constitutes a kind of standing wave over the dipole, as sketched in Fig. 10-3; it represents a good approximation.

Since the dipole is center-driven, the currents on the two halves of the dipole are symmetrical and go to zero at the ends. We write the current phasor as

$$I(z) = I_m \sin \beta(h - |z|)$$
$$= \begin{cases} I_m \sin \beta(h - z), & z > 0, \\ I_m \sin \beta(h + z), & z < 0. \end{cases} \tag{10-32}$$

We are interested only in the far-zone fields. The far-field contribution from the differential current element $I\,dz$ is, from Eqs. (10-9) and (10-10),

$$dE_\theta = \eta_0\, dH_\phi = j\frac{I\,dz}{4\pi}\left(\frac{e^{-j\beta R'}}{R'}\right)\eta_0 \beta \sin\theta. \tag{10-33}$$

Now R' in Eq. (10-33) is slightly different from R measured to the origin of the spherical coordinates, which coincides with the center of the dipole. In the far zone, $R \gg h$,

$$R' \cong R - z\cos\theta. \tag{10-34}$$

The magnitude difference between $1/R'$ and $1/R$ is insignificant, but the approximate relation in Eq. (10-34) must be retained in the phase term. Using Eqs. (10-32) and (10-34) in Eq. (10-33) and integrating, we have

$$E_\theta = \eta_0 H_\phi$$
$$= j\frac{I_m \eta_0 \beta \sin\theta}{4\pi R} e^{-j\beta R} \int_{-h}^{h} \sin\beta(h - |z|) e^{j\beta z \cos\theta}\, dz. \tag{10-35}$$

The integrand in Eq. (10-35) is a product of an even function of z, $\sin\beta(h - |z|)$, and

$$e^{j\beta z\cos\theta} = \cos(\beta z \cos\theta) + j\sin(\beta z\cos\theta),$$

where $\sin(\beta z\cos\theta)$ is an odd function of z. Integrating between symmetrical limits $-h$ and h, we know that only the part of the integrand containing the product of two even functions of z, $\sin\beta(h - |z|)\cos(\beta z\cos\theta)$, yields a nonzero value. Equation (10-35) then reduces to

$$E_\theta = \eta_0 H_\phi = j\frac{I_m \eta_0 \beta \sin\theta}{2\pi R} e^{-j\beta R} \int_0^h \sin\beta(h - z)\cos(\beta z\cos\theta)\, dz$$

$$= \frac{j60I_m}{R} e^{-j\beta R} F(\theta), \tag{10-36}$$

where

Pattern function of a linear dipole antenna having half-length h

$$F(\theta) = \frac{\cos(\beta h \cos\theta) - \cos\beta h}{\sin\theta}. \tag{10-37}$$

The factor $|F(\theta)|$ is the E-plane ***pattern function*** of a linear dipole antenna. The exact shape of the radiation pattern represented by $|F(\theta)|$ in Eq. (10-37) depends on the value of $\beta h = 2\pi h/\lambda$ and can be quite different for different antenna lengths. The radiation pattern, however, is always symmetrical with respect to the $\theta = \pi/2$ plane. Figure 10-4 shows the E-plane patterns for four different dipole lengths measured in terms of wavelength: $2h/\lambda = \frac{1}{2}$, 1, $\frac{3}{2}$ and 2. The H-plane patterns are circles inasmuch as $F(\theta)$ is independent of ϕ. From the patterns in Fig. 10-4 we see that the direction of maximum radiation tends to shift away from the $\theta = 90°$ plane when the dipole length approaches $3\lambda/2$. For $2h = 2\lambda$ there is no radiation in the $\theta = 90°$ plane.

FIGURE 10-4 E-plane radiation patterns for center-fed dipole antennas.

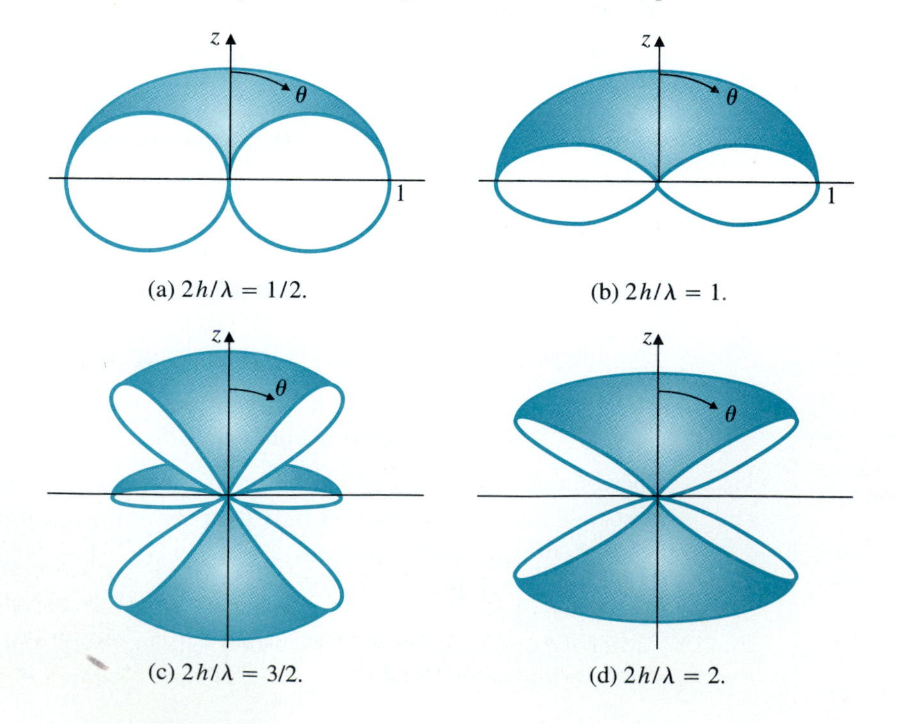

(a) $2h/\lambda = 1/2$.

(b) $2h/\lambda = 1$.

(c) $2h/\lambda = 3/2$.

(d) $2h/\lambda = 2$.

10-4.1 THE HALF-WAVE DIPOLE

The half-wave dipole having a length $2h = \lambda/2$ is of particular practical importance because of its desirable pattern and impedance characteristics. With $\beta h = 2\pi h/\lambda = \pi/2$, the pattern function in Eq. (10-37) becomes

Pattern function of a half-wave dipole

$$F(\theta) = \frac{\cos[(\pi/2)\cos\theta]}{\sin\theta}. \tag{10-38}$$

This function has a maximum equal to unity at $\theta = 90°$ and has nulls at $\theta = 0°$ and $180°$. The corresponding E-plane radiation pattern is sketched in Fig. 10-4(a). The far-zone field phasors are, from Eq. (10-36),

$$E_\theta = \eta_0 H_\phi = \frac{j60I_m}{R} e^{-j\beta R} \left\{\frac{\cos[(\pi/2)\cos\theta]}{\sin\theta}\right\}. \tag{10-39}$$

The magnitude of the time-average Poynting vector is

$$\mathscr{P}_{av}(\theta) = \frac{1}{2} E_\theta H_\phi^* = \frac{15I_m^2}{\pi R^2} \left\{\frac{\cos[(\pi/2)\cos\theta]}{\sin\theta}\right\}^2. \tag{10-40}$$

The total power radiated by a half-wave dipole is obtained by integrating $\mathscr{P}_{av}$ over the surface of a great sphere:

$$\begin{aligned} P_r &= \int_0^{2\pi}\int_0^{\pi} \mathscr{P}_{av}(\theta) R^2 \sin\theta \, d\theta \, d\phi \\ &= 30I_m^2 \int_0^{\pi} \frac{\cos^2[(\pi/2)\cos\theta]}{\sin\theta} d\theta. \end{aligned} \tag{10-41}$$

The integral in Eq. (10-41) can be evaluated numerically to give a value 1.218. Hence

$$P_r = 36.54 I_m^2 \quad \text{(W)}, \tag{10-42}$$

from which we obtain the radiation resistance of a free-standing half-wave dipole:

Radiation resistance of a half-wave dipole

$$R_r = \frac{2P_r}{I_m^2} = 73.1 \quad (\Omega). \tag{10-43}$$

Neglecting losses, we find that the input resistance of a thin half-wave dipole equals 73.1 (Ω) and that the input reactance is a small positive number that can be made to vanish when the dipole length is adjusted to be slightly shorter than $\lambda/2$. (As we have indicated before, the actual calculation of the input impedance is tedious and is beyond the scope of this book.)

The directivity of a half-wave dipole can be found by using Eq. (10-15). We have, from Eqs. (10-12) and (10-40),

$$U_{\max} = R^2 \mathscr{P}_{av}(90°) = \frac{15}{\pi} I_m^2 \tag{10-44}$$

and

Directivity of a half-wave dipole

$$D = \frac{4\pi U_{\max}}{P_r} = \frac{60}{36.54} = 1.64, \tag{10-45}$$

which corresponds to $10 \log_{10} 1.64$ or 2.15 (dB) referring to an omnidirectional radiator.

EXAMPLE 10-5

A thin quarter-wavelength vertical antenna over a perfectly conducting ground is excited by a time-harmonic source at its base. Find its radiation pattern, radiation resistance, and directivity.

SOLUTION

Since current is charge in motion, we can use the method of images discussed in Subsection 3-11.5 and replace the conducting ground by the image of the vertical antenna. A little thought will convince us that the image of a vertical antenna carrying a current I is another vertical antenna of the same length located below the ground. The image antenna carries the same current in the *same direction* as the original antenna. The electromagnetic field *in the upper half-space* due to the quarter-wave vertical antenna in Fig. 10-5(a) is, then, the same as that of the half-wave antenna in Fig. 10-5(b). The pattern function in Eq. (10-38) applies here for $0 \le \theta \le \pi/2$, and the radiation pattern drawn in dashed lines in Fig. 10-5(b) is the upper half of that in Fig. 10-4(a).

The magnitude of the time-average Poynting vector, $\mathscr{P}_{av}$, in Eq. (10-40), holds for $0 \le \theta \le \pi/2$. Inasmuch as the quarter-wave antenna (a ***monopole***) radiates only into the upper half-space, its total radiated power is only one-half that given in Eq. (10-42):

$$P_r = 18.27 I_m^2 \quad \text{(W)}.$$

Consequently, the radiation resistance is

$$R_r = \frac{2P_r}{I_m^2} = 36.54 \quad (\Omega), \tag{10-46}$$

which is one-half of the radiation resistance of a half-wave antenna in free-space.

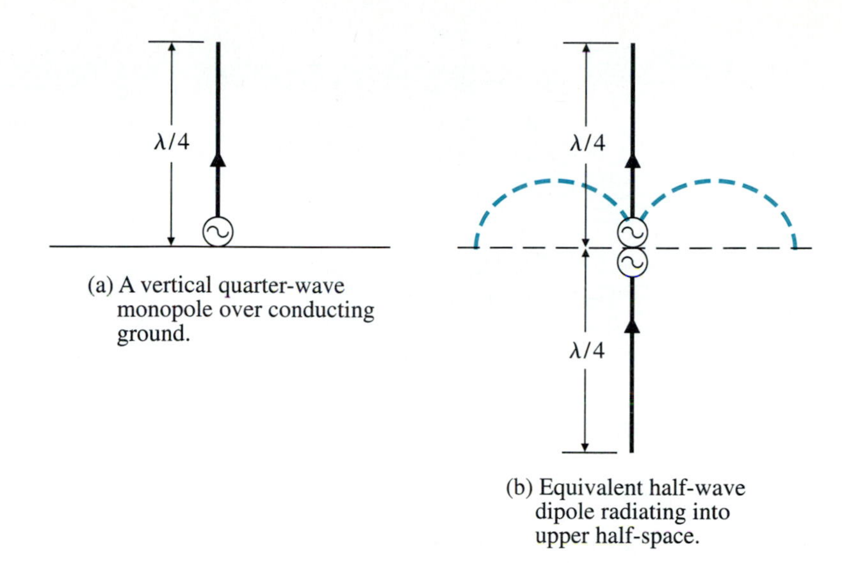

FIGURE 10-5 Quarter-wave monopole over a conducting ground and its equivalent half-wave dipole.

To calculate directivity, we note that both the maximum radiation intensity, $U_{\max}$, and the average radiation intensity, $P_r/2\pi$, remain the same as those for the half-wave dipole. Thus,

Directivity of a quarter-wave monopole

$$D = \frac{U_{\max}}{U_{av}} = \frac{U_{\max}}{P_r/2\pi} = 1.64, \tag{10-47}$$

which is the same as the directivity of a half-wave antenna. This antenna is known as a ***quarter-wave monopole***.

■ **EXERCISE 10.2** A center-fed dipole of length 25 (cm) operating at 600 (MHz) radiates a total average power of 475 (W). Find the magnitudes of the electric and magnetic field intensities at the point $P(100\,\text{m}, \pi/2, 0)$.

ANS. 2.17 (V/m), 5.75 (mA/m).

REVIEW QUESTIONS

Q.10-12 Describe qualitatively the E- and H-plane radiation patterns of a half-wave dipole antenna.

Q.10-13 What are the radiation resistance and directivity of a half-wave dipole antenna?

Q.10-14 What are the radiation resistance and directivity of a vertical quarter-wave monopole over a conducting ground?

Q.10-15 What is the image of a horizontal dipole over a conducting ground?

REMARKS

1. Time-harmonic current distributions on center-fed, thin, linear antennas are approximately sinusoidal standing waves vanishing at the ends.
2. The only far-zone field intensities of a vertical antenna are E_θ and H_ϕ.
3. The radiation resistance and directivity of a center-fed linear half-wave dipole antenna are, respectively, 73.1 (Ω) and 1.64.

10-5 ANTENNA ARRAYS

Antenna arrays

As we have seen, single-element linear antennas tend to spread radiated power over the broad beams in their radiation patterns. They have low directivity and their main beams point to fixed directions. These restrictions can be overcome by arranging a group of several antenna elements in various configurations (straight lines, circles, triangles, and so on) with proper amplitude and phase relations to give certain desired radiation characteristics. Such arrangements of antenna elements are called ***antenna arrays.*** In this section we examine the basic theories and characteristics of linear antenna arrays (radiating elements arranged along a straight line). We first consider the simplest case of two-element arrays. After some experience has been gained with them, we consider the basic properties of uniform linear arrays made up of many identical elements.

10-5.1 TWO-ELEMENT ARRAYS

The simplest array is one consisting of two identical radiating elements (antennas) spaced a distance apart. This arrangement is illustrated in Fig. 10-6. For simplicity, let us assume that the antennas are lined along the x-axis, and examine the far-zone electric field of the individual antennas in the θ-direction. The antennas are excited with a current of the same magnitude, but the phase in antenna 1 leads that in antenna 0 by an angle ξ. We have, at point $P(\theta, \phi)$,

$$E_0 = E_m F(\theta, \phi)\frac{e^{-j\beta R_0}}{R_0}, \qquad \text{and} \tag{10-48}$$

$$E_1 = E_m F(\theta, \phi)\frac{e^{j\xi}e^{-j\beta R_1}}{R_1}, \tag{10-49}$$

where $F(\theta, \phi)$ is the pattern function of the individual antennas, and E_m is an amplitude function. The electric field of the two-element array is the sum of

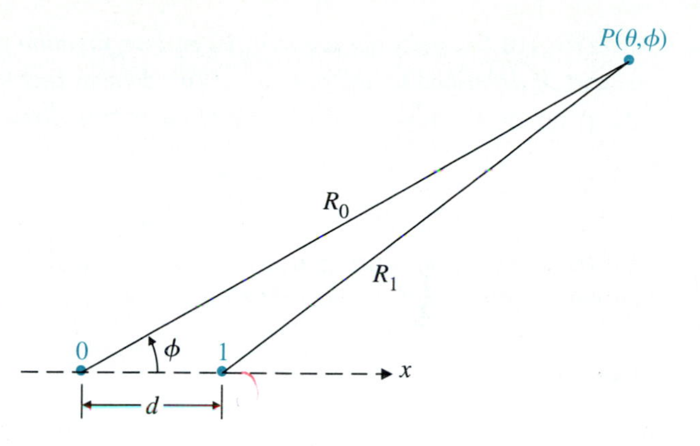

FIGURE 10-6 A two-element array.

E_0 and E_1. Hence,

$$E = E_0 + E_1 = E_m F(\theta, \phi)\left[\frac{e^{-j\beta R_0}}{R_0} + \frac{e^{j\xi}e^{-j\beta R_1}}{R_1}\right]. \tag{10-50}$$

In the far zone, $R_0 \gg d/2$, and the factor $1/R_1$ in the magnitude may be replaced approximately by $1/R_0$. However, a small difference between R_0 and R_1 in the exponents may lead to a significant phase difference, and a better approximation must be used. Because the lines joining the field point P and the two antennas are nearly parallel, we may write

$$R_1 \cong R_0 - d\sin\theta\cos\phi. \tag{10-51}$$

Substitution of Eq. (10-51) in Eq. (10-50) yields

$$\begin{aligned} E &= E_m \frac{F(\theta, \phi)}{R_0} e^{-j\beta R_0}[1 + e^{j\beta d \sin\theta\cos\phi} e^{j\xi}] \\ &= E_m \frac{F(\theta, \phi)}{R_0} e^{-j\beta R_0} e^{j\psi/2}\left(2\cos\frac{\psi}{2}\right), \end{aligned} \tag{10-52}$$

where

$$\psi = \beta d \sin\theta\cos\phi + \xi. \tag{10-53}$$

The magnitude of the electric field of the array is

$$|E| = \frac{2E_m}{R_0}|F(\theta, \phi)|\left|\cos\frac{\psi}{2}\right|, \tag{10-54}$$

where $|F(\theta, \phi)|$ may be called the ***element factor***, and $|\cos(\psi/2)|$ the normalized ***array factor***. The element factor is the magnitude of the pattern function of the individual radiating elements, and the array factor depends on array geometry as well as on the relative amplitudes and phases of the excitations in the elements. (In this particular case the excitation amplitudes are equal.)

Principle of pattern multiplication

From Eq. (10-54) we conclude that ***the pattern function of an array of identical elements is described by the product of the element factor and the array factor.*** This property is called the ***principle of pattern multiplication.***

EXAMPLE 10-6

Plot the H-plane radiation patterns of two parallel dipoles for the following two cases: (a) $d = \lambda/2$, $\xi = 0$, and (b) $d = \lambda/4$, $\xi = -\pi/2$.

SOLUTION

Let the dipoles be z-directed and placed along the x-axis, as shown in Fig. 10-6. In the H-plane ($\theta = \pi/2$), each dipole is omnidirectional, and the normalized pattern function is equal to the normalized array factor $|A(\phi)|$. Thus

$$|A(\phi)| = \left|\cos\frac{\psi}{2}\right| = \left|\cos\frac{1}{2}(\beta d\cos\phi + \xi)\right|.$$

a) $d = \lambda/2$ ($\beta d = \pi$), $\xi = 0$:

$$|A(\phi)| = \left|\cos\left(\frac{\pi}{2}\cos\phi\right)\right|. \tag{10-55}$$

Broadside array

The pattern has its maximum at $\phi_0 = \pm\pi/2$—that is, in the broadside direction. This is a type of ***broadside array***. Figure 10-7(a) shows this broadside pattern. Since the excitations in the two dipoles are in phase, their electric fields add in the broadside directions, $\phi = \pm\pi/2$. At $\phi = 0$ and π the electric fields cancel each other because the $\lambda/2$ separation leads to a phase difference of 180°.

FIGURE 10-7 H-plane radiation patterns of two-element parallel dipole array.

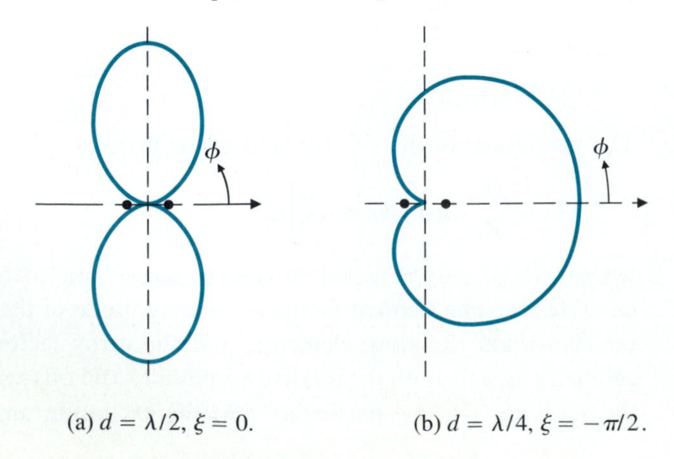

(a) $d = \lambda/2$, $\xi = 0$. (b) $d = \lambda/4$, $\xi = -\pi/2$.

b) $d = \lambda/4$ $(\beta d = \pi/2)$, $\xi = -\pi/2$:

$$|A(\phi)| = \left|\cos\frac{\pi}{4}(\cos\phi - 1)\right|, \tag{10-56}$$

which has a maximum at $\phi_0 = 0$ and vanishes at $\phi = \pi$. The pattern maximum is now in a direction *along* the line of the array, and the two dipoles constitute an ***endfire array***. Figure 10-7(b) shows this endfire pattern. In this case the phase in the right-hand dipole *lags* by $\pi/2$, which exactly compensates for the fact that its electric field arrives in the $\phi = 0$ direction a quarter of a cycle *earlier* than the electric field of the left-hand dipole. As a consequence, the electric fields add in the $\phi = 0$ direction. In the $\phi = \pi$ direction, the $\pi/2$ phase lag in the right-hand dipole plus the quarter-cycle delay results in a complete cancellation of the fields.

Endfire array

EXAMPLE 10-7

Discuss the radiation pattern of a linear array of the three isotropic sources spaces $\lambda/2$ apart. The excitations in the sources are in-phase and have amplitude ratios 1:2:1.

SOLUTION

This three-source array is equivalent to two two-element arrays displaced $\lambda/2$ from each other as depicted in Fig. 10-8. Each two-element array can be considered as a radiating source with an element factor as given by Eq. (10-55) and an array factor, which is also given by the same equation. By the principle of pattern multiplication we obtain

$$|E| = \frac{4E_m}{R_0}\left|\cos\left(\frac{\pi}{2}\cos\phi\right)\right|^2 = \frac{4E_m}{R_0}|A(\phi)|. \tag{10-57}$$

The radiation pattern represented by the normalized array function $|A(\phi)| = |\cos[(\pi/2)\cos\phi]|^2$ is sketched in Fig. 10-9. Compared to the pattern

FIGURE 10-8 A three-element array and its equivalent pair of displaced two-element arrays.

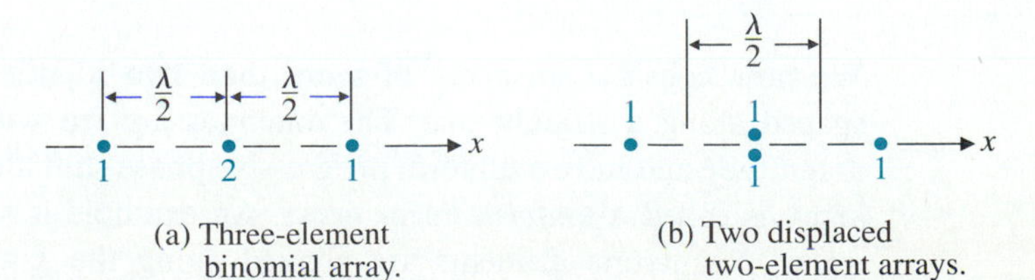

(a) Three-element binomial array.

(b) Two displaced two-element arrays.

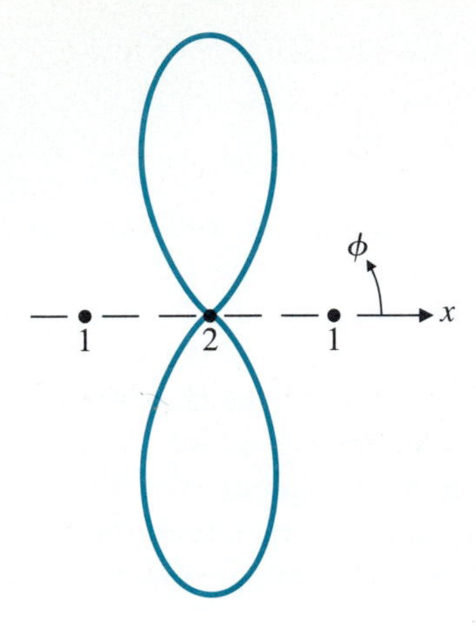

FIGURE 10-9 Radiation pattern of three-element broadside binomial array.

of the uniform two-element array in Fig. 10-7(a), this three-element broadside pattern is sharper (more directive). Both patterns have only main beams with no sidelobes.

Binomial arrays

The three-element broadside array is a special case of a class of *sidelobeless* arrays called ***binomial arrays***. In a binomial array of N elements, the array factor is a binomial function $(1 + e^{j\psi})^{N-1}$ and the excitation amplitudes vary according to the coefficients of a binomial expansion $\binom{N-1}{n}$, $n = 0, 1, 2, \ldots, N-1$. For $N = 3$ the relative excitation amplitudes are $\binom{2}{0} = 1$, $\binom{2}{1} = 2$ and $\binom{2}{2} = 1$, as in Example 10-7. To obtain a directive pattern without sidelobes, d in a binomial array is normally restricted to be $\lambda/2$.

EXERCISE 10.3

a) What are the relative excitation amplitudes of a binomial array of four isotropic equiphase sources having a uniform spacing of $\lambda/3$?

b) Assuming the sources to be located along the y-axis, obtain the normalized array factor in the $\theta = \pi/2$ plane.

ANS. (a) 1:3:3:1, (b) $\left|\cos\left(\frac{\pi}{3}\cos\phi\right)\right|^3$.

10-5.2 GENERAL UNIFORM LINEAR ARRAYS

We now consider an array of more than two identical antennas equally spaced along a straight line. The antennas are fed with currents of equal magnitude and have a uniform progressive phase shift along the line. Such an array is called a ***uniform linear array***. An example is shown in Fig. 10-10, where N antenna elements are aligned along the x-axis. Since the array elements are identical, the array pattern function is the product of the element factor and the array factor. The normalized array factor in the xy-plane is

$$|A(\psi)| = \frac{1}{N}|1 + e^{j\psi} + e^{j2\psi} + \cdots + e^{j(N-1)\psi}|, \tag{10-58}$$

where

$$\psi = \beta d \cos\phi + \xi. \tag{10-59}$$

The polynomial on the right side of Eq. (10-58) is a geometric progression and can be summed up in a closed form:

$$|A(\psi)| = \frac{1}{N}\left|\frac{1 - e^{jN\psi}}{1 - e^{j\psi}}\right|$$

or

Array factor of an *N*-element uniform linear array

$$|A(\psi)| = \frac{1}{N}\left|\frac{\sin(N\psi/2)}{\sin(\psi/2)}\right| \quad \text{(Dimensionless).} \tag{10-60}$$

Assuming omnidirectional H-plane patterns for the identical array elements, we may derive several significant properties from $|A(\psi)|$ given in Eq. (10-60).

Mainbeam occurs at $\Psi = 0$.

1. *Main-beam direction.* The maximum value of $|A(\psi)|$ occurs when $\psi = 0$, or when

   $$\beta d \cos\phi_0 + \xi = 0,$$

   which leads to

   $$\cos\phi_0 = -\frac{\xi}{\beta d}. \tag{10-61}$$

   Two special cases are of particular importance.

   a) *Broadside array.* For a broadside array, maximum radiation occurs at a direction perpendicular to the line of the array—that

FIGURE 10-10 A general uniform linear array.

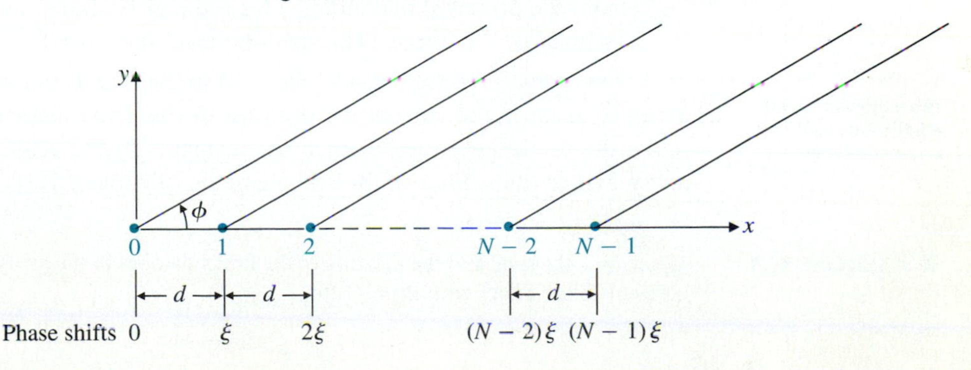

is, at $\phi_0 = \pm\pi/2$. This requires $\xi = 0$, which means that all the elements in a linear broadside array should be excited *in phase*, as was the case in Example 10-6(a).

b) *Endfire array.* For an endfire array, maximum radiation occurs at $\phi_0 = 0$. Equation (10-61) gives

$$\xi = -\beta d \cos\phi_0 = -\beta d.$$

We note that the direction of the main beam of a uniform linear array can be made to change (to scan) by changing the progressive phase shifts. Antenna arrays equipped with phase shifters to steer the main beam electronically are called *phased arrays.*

Phased arrays

Sidelobes

2. *Sidelobe locations.* Sidelobes are minor maxima that occur approximately when the numerator on the right side of Eq. (10-60) is a maximum—that is, when $|\sin(N\psi/2)| = 1$, or when

$$\frac{N\psi}{2} = \pm(2m+1)\frac{\pi}{2}, \qquad m = 1, 2, 3, \ldots.$$

The first sidelobes occur when

$$\frac{N\psi}{2} = \pm\frac{3}{2}\pi, \qquad (m = 1). \tag{10-62}$$

Note that $N\psi/2 = \pm\pi/2$ $(m = 0)$ does not represent locations of sidelobes because they are still within the main-beam region.

3. *First sidelobe level.* An important characteristic of the radiation pattern of an array is the level of the first sidelobes compared to that of the main beam, since the former is usually the highest of all sidelobes. All sidelobes should be kept as low as possible in order that most of the radiated power be concentrated in the main-beam direction and not be diverted to sidelobe regions. Substituting Eq. (10-62) in Eq. (10-59), we find the amplitude of the first sidelobes to be

$$\frac{1}{N}\left|\frac{1}{\sin(3\pi/2N)}\right| \cong \frac{1}{N}\left|\frac{1}{3\pi/2N}\right| = \frac{2}{3\pi} = 0.212$$

for large N. In logarithmic terms the first sidelobes of a uniform linear antenna array of many elements are $20\log_{10}(1/0.212)$ or 13.5 (dB) *down* from the principal maximum. This number is almost independent of N as long as N is large. (The sidelobe level is higher for small N.)

Tapering excitation amplitudes reduces array sidelobes.

One way to reduce the sidelobe level in the radiation pattern of a linear array is to taper the current distribution in the array elements—that is, to make the excitation amplitudes in the elements in the center portion of an array higher than those in the end elements. (Problem P.10-14.)

EXERCISE 10.4 Determine the level and the location of the first sidelobes in the array pattern of a five-element linear array with $d = \lambda/2$ for

a) broadside operation, and

b) endfire operation.

ANS. (a) -12.1 (dB) at $\phi = \pm 53.1°$ and $\pm 126.9°$. (b) -12.1 (dB) at $\phi = \pm 66.4°$.

EXAMPLE 10-8

For a five-element uniform linear array with $\lambda/2$ spacing, find the width of the main beam for (a) broadside operation, and (b) endfire operation.

SOLUTION

The width of the main beam is the region of the pattern between the first nulls on either side of the direction of maximum radiation. The first nulls of the array pattern occur at ψ_{01} that makes (see Eq. 10-60)

$$\frac{N\psi_{01}}{2} = \pm\pi. \tag{10-63}$$

For this example, $\psi_{01} = \pm 2\pi/5 = \pm 0.4\pi$. It is obvious that the corresponding null locations in ϕ are different for broadside and endfire arrays because of the different values of ξ implicit in ψ.

a) *Broadside operation.* $\xi = 0$, $\psi = \beta d \cos\phi = \pi \cos\phi$. At first nulls, $\pi \cos\phi_{01} = \pm 0.4\pi$, from which we obtain

$\phi_{01} = \cos^{-1}(\pm 0.4)$.

Taking the positive sign, we have $\phi_{01} = \pm 66.4°$. Taking the negative sign, we get $\phi_{01} = \pm 113.6°$. The main beams of a broadside array point in the broadside directions at $\phi_0 = \pm 90°$. Thus, the main-beam width is $113.6° - 66.4° = 47.2°$.

b) *Endfire operation.* $\xi = -\beta d$, $\psi = \beta d(\cos\phi - 1) = \pi(\cos\phi - 1)$. At first nulls, $\pi(\cos\phi_{01} - 1) = -0.4\pi$,† from which we obtain

$\phi_{01} = \cos^{-1} 0.6 = \pm 53.1°$.

Thus the width of the single main beam at $\phi_0 = 0°$ is $2 \times 53.1° = 106.2°$.

Width of main beam of endfire array is wider than that of the corresponding broadside array.

We see that the main beam of an endfire array is much broader than that of the corresponding broadside array.

A typical graph of the normalized array factor in Eq. (10-60) is shown in Fig. 10-11 for $N = 5$. It is a rectangular plot of $|A(\psi)|$ versus ψ. The actual normalized array pattern for the range of the azimuth angle $\phi = 0$ to 2π (the ***visible range***) depends on the relation between ψ and ϕ. As we have shown above:

†Note that the positive sign in Eq. (10-63) does not apply here because it would lead to the impossible situation of $\cos\phi_{01} = 1.4$.

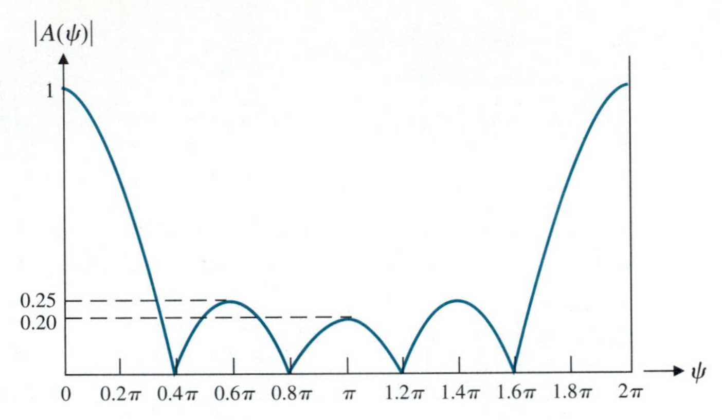

FIGURE 10-11 Normalized array factor of a five-element uniform linear array.

a) For broadside arrays, $\phi_0 = \pm\pi/2$, $\xi = \beta d \cos\phi$, (10-64a)

b) For endfire arrays, $\phi_0 = 0$, $\xi = \beta d(\cos\phi - 1)$. (10-64b)

The different transformations in Eqs. (10-64a) and (10-64b) lead to different array patterns versus ϕ for the same array factor.

EXERCISE 10.5 Use Fig. 10-11 to sketch the normalized array pattern $|A(\phi)|$ for ϕ from 0 to π for a five-element uniform broadside array with $d = \lambda/2$:

a) in rectangular coordinates, and

b) in polar coordinates.

EXERCISE 10.6 Use Fig. 10-11 to sketch the normalized array pattern $|A(\phi)|$ for ϕ from $-\pi/2$ to $+\pi/2$ for a five-element uniform endfire array with $d = \lambda/2$:

a) in rectangular coordinates, and

b) in polar coordinates.

REVIEW QUESTIONS

Q.10-16 What are the major advantages of antenna arrays compared to single-element antennas fed with the same input power?

Q.10-17 What is meant by the *normalized array factor* of an antenna array? How is it different from the pattern function of the individual antennas?

Q.10-18 State the *principle of pattern multiplication.*

Q.10-19 State the difference between a *broadside array* and an *endfire array.*

Q.10-20 What is a *binomial array*? What are the relative excitation amplitudes of a six-element binomial array?

Q.10-21 Is the radiation pattern of all linear binomial arrays sidelobeless? Explain.

Q.10-22 What is meant by a *uniform linear array*?

Q.10-23 What is a *phased array*?

Q.10-24 In the radiation pattern of a uniform linear array of many elements, how many decibels down from the principal maximum are the first sidelobes?

REMARKS

1. The principle of pattern multiplication applies only to arrays with identical elements.
2. The radiating elements in a broadside array are fed in phase.
3. The phase of a radiating element in an endfire array usually lags by an amount equal to $(2\pi/\lambda)$ multiplied by the distance that the element is displaced in the direction of maximum radiation.†
4. A broadside array has a narrower main beam and a higher directivity than those of a corresponding endfire array.
5. The sidelobe level in the radiation pattern of a linear array with a tapered amplitude distribution is lower than that of the corresponding array with a uniform amplitude distribution.

10-6 EFFECTIVE AREA AND BACKSCATTER CROSS SECTION

In the discussion of antennas and antenna arrays so far we have implied that they operate in a transmitting mode. In the transmitting mode a voltage source is applied to the input terminals of an antenna, setting up currents and charges on the antenna structure. The time-varying currents and charges, in turn, radiate electromagnetic waves, which carry energy and/or information. A transmitting antenna can then be regarded as a device that transforms energy from a source (a generator) to energy associated with an electromagnetic wave. A receiving antenna, on the other hand, extracts energy from an incident electromagnetic wave and delivers it to a load. By invoking reciprocity relations it is possible to justify the following conclusions:

Reciprocity relations for antennas in transmitting and receiving modes

1. The equivalent generator impedance of an antenna in the receiving mode is equal to the input impedance of the antenna in the transmitting mode, and
2. The directional pattern of an antenna for reception is identical with that for transmission.

We will accept these conclusions.

For a receiving antenna weakly coupled to a transmitting source, an approximate Thévenin's equivalent circuit at the receiving end is as shown in Fig. 10-12, where V_{oc} is the open-circuit voltage induced in the receiving

†The excitation phases in an endfire array may be adjusted in specific ways to improve array directivity. (See D. K. Cheng and P. D. Raymond, Jr., "Optimization of array directivity by phase adjustments," *Electronics Letters*, vol. 7, pp. 552–553, September 9, 1971.)

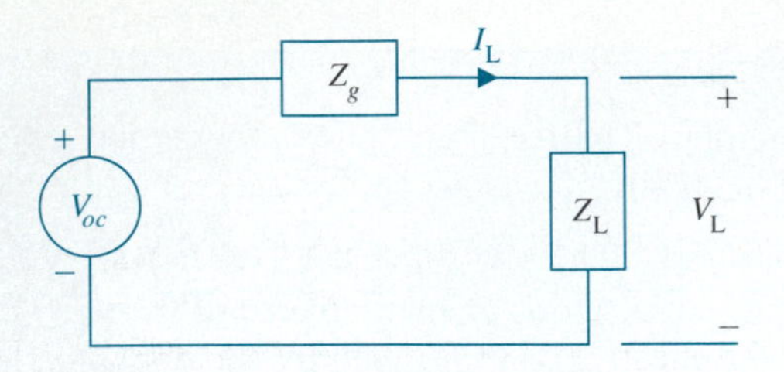

FIGURE 10-12 Thévenin's equivalent circuit for a receiving antenna with load.

antenna, Z_g is the equivalent generator internal impedance of the antenna in the receiving mode (equal to its input impedance in the transmitting mode), and Z_L is the load impedance. We will use this equivalent circuit to study the performance of receiving antennas.

10-6.1 EFFECTIVE AREA

Effective area of a receiving antenna

In discussing receiving antennas it is convenient to define a quantity called the ***effective area***.† The effective area, A_e, of a receiving antenna is the ratio of the average power, P_L, delivered to a *matched load* to the time-average power density $\mathscr{P}_{av}$, of the incident electromagnetic wave at the antenna. We write

$$A_e = \frac{P_L}{\mathscr{P}_{av}} \quad (\text{m}^2). \tag{10-65}$$

Under matched conditions,

$$Z_L = Z_g^* = Z_i^*. \tag{10-66}$$

Neglecting losses, the antenna input impedance Z_i in the transmitting mode may be written as

$$Z_i = R_r + jX_i, \tag{10-67}$$

where R_r denotes the radiation resistance. In view of Eqs. (10-66) and (10-67), the induced open-circuit voltage V_{oc} in Fig. 10-12 appears across a total resistance of $2R_r$, and the average power delivered to the matched load is

$$P_L = \frac{1}{2}\left\{\frac{|V_{oc}|}{2R_r}\right\}^2 R_r = \frac{|V_{oc}|^2}{8R_r}. \tag{10-68}$$

Let E_i denote the amplitude of the electric field intensity at the receiving antenna. Then the time-average power density at the receiving site is

† Also called ***effective aperture*** or ***receiving cross section***.

$$\mathscr{P}_{av} = \frac{E_i^2}{2\eta_0} = \frac{E_i^2}{240\pi}. \tag{10-69}$$

The ratio of P_L and $\mathscr{P}_{av}$ gives the effective area.

EXAMPLE 10-9

Determine the effective area, $A_e(\theta)$, of an elemental electric dipole of a length $d\ell$ ($\ll \lambda$) used to receive an incident plane electromagnetic wave of wavelength λ. Assume that the dipole axis makes an angle θ with the direction of the incident electromagnetic wave.

SOLUTION

Let E_i be the amplitude of the electric field intensity at the dipole. Then the induced open-circuit voltage is

$$V_{oc} = E_i\, d\ell \sin\theta. \tag{10-70}$$

The radiation resistance of the elemental electric dipole is, from Eq. (10-25),

$$R_r = 80\pi^2 \left(\frac{d\ell}{\lambda}\right)^2. \tag{10-71}$$

Use of V_{oc} from Eq. (10-70) and R_r from Eq. (10-71) in Eq. (10-68) gives

$$P_L = \frac{E_i^2}{640\pi^2}(\lambda \sin\theta)^2. \tag{10-72}$$

Substituting Eqs. (10-69) and (10-72) in Eq. (10-65), we obtain the effective area of the elemental electric dipole (Hertzian dipole):

Effective area of Hertzian dipole

$$A_e = \frac{3}{8\pi}(\lambda \sin\theta)^2. \tag{10-73}$$

Recalling from Eq. (10-19) in Example 10-2 that the directive gain of a Hertzian dipole is

$$G_D(\theta, \phi) = \tfrac{3}{2}\sin^2\theta, \tag{10-74}$$

we write the following relation for an antenna under matched-impedance conditions:

Relation between effective area and directive gain of an antenna

$$\boxed{A_e(\theta, \phi) = \frac{\lambda^2}{4\pi} G_D(\theta, \phi) \quad (\text{m}^2).} \tag{10-75}$$

It can be proved in general that the relation between A_e and G_D in Eq. (10-75) holds for any antenna.

EXERCISE 10.7 Calculate the maximum effective area of a Hertzian dipole at 3 (GHz).

ANS. 11.9 (cm^2).

10-6.2 BACKSCATTER CROSS SECTION

As we saw in the preceding subsection, the concept of effective area pertains to the power available to the matched load of a receiving antenna for a given incident power density. In cases in which the incident wave impinges on a passive object whose purpose is not to extract energy from the incident wave but whose presence creates a scattered field, it is appropriate to define a quantity called the ***backscatter cross section***, or ***radar cross section***. The backscatter cross section of an object is the equivalent area that would intercept that amount of incident power in order to produce the same scattered power density at the receiver site if the object scattered uniformly (isotropically) in all directions. Let

Backscatter cross section, (radar cross section)

$\mathscr{P}_i$ = Time-average incident power density at the object (W/m^2),

$\mathscr{P}_s$ = Time-average scattered power density at the receiver site (W/m^2),

σ_{bs} = Backscatter cross section (m^2),

r = Distance between scatterer and receiver (m).

Then,

$$\frac{\sigma_{bs}\mathscr{P}_i}{4\pi r^2} = \mathscr{P}_s,$$

or

$$\boxed{\sigma_{bs} = 4\pi r^2 \frac{\mathscr{P}_s}{\mathscr{P}_i} \quad (m^2).} \tag{10-76}$$

Note that $\mathscr{P}_s$ is inversely proportional to r^2 for large r and that σ_{bs} does not change with r.

Radar

The backscatter cross section is a measure of the detectability of the object (target) by ***radar*** (***ra***dio ***d***etection ***a***nd ***r***anging); hence the term radar cross section. It is a composite measure, depending on the geometry, orientation, constitutive parameters and surface conditions of the object, and on the frequency and polarization of the incident wave in a complicated way. The design of a stealth aircraft must be such that its backscatter or radar cross section is exceptionally small.

10-7 FRIIS TRANSMISSION FORMULA AND RADAR EQUATION

We now consider the power transmission relation between transmitting and receiving antennas. Assume that a communication link is established between stations 1 and 2 with antennas having effective areas A_{e1} and A_{e2}, respectively. The antennas are separated by a distance r. We wish to find a relation between the transmitted and received powers.

Let P_t be the total power radiated by antenna 1 having a directive gain G_{D1}. The average power density at antenna 2, $\mathscr{P}_{av}$, at a distance r away is

$$\mathscr{P}_{av} = \frac{P_t}{4\pi r^2} G_{D1}. \tag{10-77}$$

If antenna 2 has an effective area A_{e2}, it will receive a power P_L in a matched load (see Eq. 10-65):

$$P_L = A_{e2}\mathscr{P}_{av}. \tag{10-78}$$

Combining Eqs. (10-77) and (10-78) and using Eq. (10-75), we obtain

$$\frac{P_L}{P_t} = \left(\frac{A_{e2}}{4\pi r^2}\right) G_{D1} = \left(\frac{A_{e2}}{4\pi r^2}\right)\left(\frac{4\pi A_{e1}}{\lambda^2}\right),$$

or

Friis transmission formula

$$\boxed{\frac{P_L}{P_t} = \frac{A_{e1}A_{e2}}{r^2\lambda^2}.} \tag{10-79}$$

The relation in Eq. (10-79) is referred to as the ***Friis transmission formula***. For a given transmitted power, the received power is directly proportional to the product of the effective areas of the transmitting and receiving antennas and is inversely proportional to the square of the product of the distance of separation and wavelength.

Noting Eq. (10-75), we may write the Friis transmission formula in the following alternative form:

Alternative form of Friis transmission formula

$$\boxed{\frac{P_L}{P_t} = \frac{G_{D1}G_{D2}\lambda^2}{(4\pi r)^2}.} \tag{10-80}$$

The received power P_L in Eqs. (10-79) and (10-80) assumes a matched condition and disregards the power dissipated in the antenna itself. It also assumes that the transmitting and receiving antennas are in the far zone of each other.

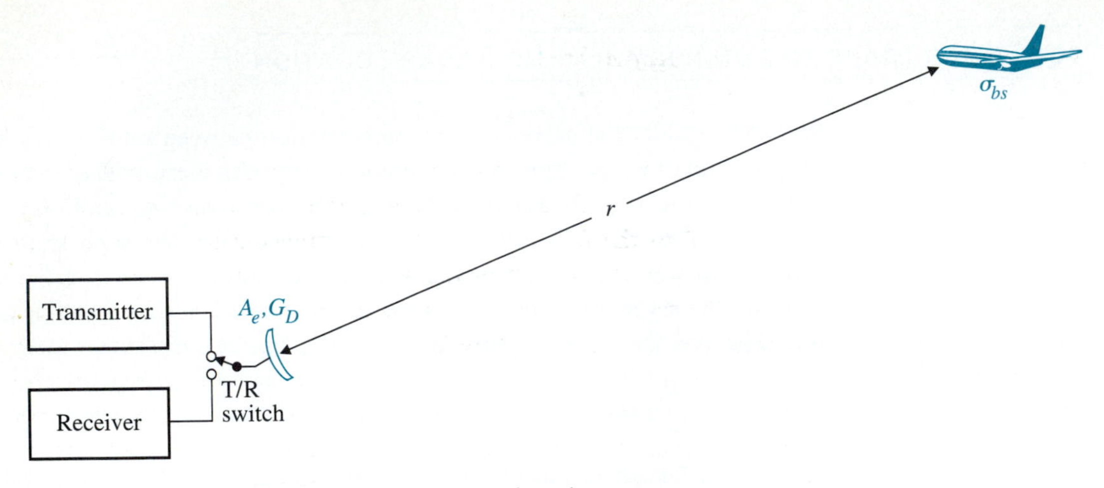

FIGURE 10-13 A monostatic radar system.

Now consider a radar system that uses the same antenna for transmitting short pulses of time-harmonic radiation and for receiving the energy scattered back from a target, as depicted in Fig. 10-13.† For a transmitted power P_t the power density at a target at a distance r away is (see Eq. 10-77)

$$\mathscr{P}_{av} = \frac{P_t}{4\pi r^2} G_D(\theta, \phi), \tag{10-81}$$

where $G_D(\theta, \phi)$ is the directive gain of the antenna in the direction of the target. If σ_{bs} denotes the backscatter or radar cross section of the target, then the equivalent power that is scattered isotropically is $\sigma_{bs}\mathscr{P}_{av}$, which results in a power density at the antenna $\sigma_{bs}\mathscr{P}_{av}/4\pi r^2$. Let A_e be the effective area of the antenna. We have the following expression for the received power:

$$\begin{aligned} P_{\mathrm{L}} &= A_e \sigma_{bs} \frac{\mathscr{P}_{av}}{4\pi r^2} \\ &= A_e \sigma_{bs} \frac{P_t}{(4\pi r^2)^2} G_D(\theta, \phi). \end{aligned} \tag{10-82}$$

By using Eq. (10-75), Eq. (10-82) becomes

Radar equation

$$\boxed{\frac{P_{\mathrm{L}}}{P_t} = \frac{\sigma_{bs}\lambda^2}{(4\pi)^3 r^4} G_D^2(\theta, \phi),} \tag{10-83}$$

which is called the ***radar equation***. In terms of the antenna effective area A_e instead of the directive gain $G_D(\theta, \phi)$, the radar equation can be written as

† A radar system employing a common antenna for transmitting and receiving at the same site and using a T/R (xmt/rcv) switch is called a *monostatic radar*.

Alternative form of radar equation

$$\frac{P_L}{P_t} = \frac{\sigma_{bs}}{4\pi}\left(\frac{A_e}{\lambda r^2}\right)^2. \tag{10-84}$$

Because radar signals have to make round trips from the antenna to the target and then back to the antenna, the received power is inversely proportional to the fourth power of the distance r of the target from the antenna. In practice, a part of the scattered power from the target, upon reaching the receiving antenna, is reflected or reradiated. Hence P_L will be somewhat lower than that given in Eq. (10-84).

A satellite communication system makes use of satellites traveling in orbits in the earth's equatorial plane. The speed of the satellites and the radius of their orbits are such that the period of rotation of the satellites around the earth is the same as that of the earth. Thus the satellites appear to be stationary with respect to the earth's surface, and they are said to be geostationary. The radius of the geosynchronous orbit is 42,300 (km). With an earth radius of 6380 (km) the satellites are about 36,000 (km) from the earth's surface.

Geosynchronous satellites

Signals are transmitted from a high-gain antenna at an earth station toward a satellite, which receives the signals, amplifies them, and retransmits them back toward the earth station at a different frequency. Three satellites equally spaced around the geosynchronous orbit would cover almost the entire earth's surface except the polar regions (see Problem P.10-21). A quantitative analysis of the power and antenna gain relations for a satellite communication circuit requires the application of the Friis transmission formula twice, once for the uplink (earth station to satellite) and once for the downlink (satellite to earth station).

EXAMPLE 10-10

A microwave link is to be established over a distance of 10 miles at 300 (MHz) by using two identical parabolic reflectors, each having a directive gain of 30 (dB). The transmitting antenna radiates a power of 500 (W). Neglecting losses, find (a) the power received, and (b) the magnitude of the electric field intensity at the receiving antenna.

SOLUTION

a) Let us first convert the 30-(dB) logarithmic directive gain into a number.

$$10 \log_{10}(G_D) = 30 \quad \text{(dB)},$$

$$G_D = 10^3 = 1{,}000.$$

$$r = 10 \times 1609 = 1.609 \times 10^4 \text{(m)}, \qquad \lambda = \frac{3 \times 10^8}{300 \times 10^6} = 1 \text{ (m)}.$$

Using Eq. (10-80), we have

$$P_L = P_t \left(\frac{G_D \lambda}{4\pi r}\right)^2$$

$$= 500 \left(\frac{1000 \times 1}{4\pi \times 1.609 \times 10^4}\right)^2$$

$$= 12.23 \times 10^{-3} \text{(W)} = 12.23 \text{ (mW)}.$$

b) From Eqs. (10-77) and (10-69),

$$\mathscr{P}_{av} = \frac{P_t G_D}{4\pi r^2} = \frac{E_i^2}{240\pi}.$$

Thus,

$$E_i = \frac{1}{r}\sqrt{60 P_t G_D}$$

$$= \frac{1}{1.609 \times 10^4}\sqrt{60 \times 500 \times 1000} = 0.341 \text{ (V/m)}.$$

EXAMPLE 10-11

Assume that 50 (kW) is fed into the antenna of a radar system operating at 3 (GHz). The antenna has an effective area of 4 (m^2) and a radiation efficiency of 90%. The minimum detectable signal power (over noise inherent in the receiving system and from the environment) is 1.5 (pW), and the power reflection coefficient for the antenna on receiving is 0.05. Determine the maximum usable range of the radar for detecting a target with a backscatter cross section of 1 (m^2).

SOLUTION

At $f = 3 \times 10^9$ (Hz), $\lambda = 0.1$ (m):

$$A_e = 4 \quad (\text{m}^2),$$

$$P_t = 0.90 \times 5 \times 10^4 = 4.5 \times 10^4 \quad \text{(W)},$$

$$P_L = 1.5 \times 10^{-12}\left(\frac{1}{1 - 0.05}\right) = 1.58 \times 10^{-12} \quad \text{(W)},$$

$$\sigma_{bs} = 1 \quad (\text{m}^2).$$

From Eq. (10-84),

$$r^4 = \frac{\sigma_{bs} A_e^2}{4\pi\lambda^2}\left(\frac{P_t}{P_L}\right),$$

and

$$r = 4.20 \times 10^4 \text{ (m)} = 42 \quad \text{(km)}.$$

EXERCISE 10.8 For the radar system in Example 10-11, calculate

a) the maximum range for detecting a target that has a backscatter cross section of 0.2 (m^2),

b) the directivity in (dB) of a new antenna necessary for detecting the new target at 42 (km), and

c) the total transmission loss in (dB) in the original case.

ANS. (a) 28.1 (km), (b) 40.5 (dB), (c) 155.2 (dB).

REVIEW QUESTIONS

Q.10-25 What are the important consequences of reciprocity relations concerning antennas that operate in the transmitting and receiving modes?

Q.10-26 Define *effective area* of an antenna.

Q.10-27 Define *backscatter cross section* of an object.

Q.10-28 Explain the principle of *radar*.

Q.10-29 What does the *Friis transmission formula* say in terms of effective areas of the antennas?

REMARKS

1. The ratio between the effective area and the directive gain of an antenna is a universal constant equal to $\lambda^2/4\pi$.
2. For a given incident power density, the power delivered to a matched load is proportional to the effective area (and therefore also to the directive gain) of an antenna.
3. Effective area is a property of antennas, and backscatter (radar) cross section is a property of passive objects.
4. For a given transmitted power, the power received in a monostatic radar system is proportional to the backscatter cross section of the target and to the square of the product of the antenna directive gain and the operating wavelength; it is also inversely proportional to the fourth power of the distance to the target.

SUMMARY

Antennas and antenna arrays are used to radiate, and/or receive, electromagnetic energy effectively in prescribed manners. In this chapter, we

- discussed the general procedure for determining the electromagnetic fields radiated by an antenna with an assumed current distribution,
- found the far-zone electric and magnetic field intensities of a radiating elemental electric (Hertzian) dipole,
- defined the essential radiation characteristics (directive gain, directivity, power gain, radiation resistance, radiation efficiency) of an antenna,
- examined the pattern functions of a general linear antenna, a half-wave dipole, and a quarter-wave monopole,
- explained the principle of pattern multiplication for antenna arrays of identical elements,
- pointed out the special feature of binomial arrays,
- discussed the general characteristics of array factors with special emphasis on broadside and endfire arrays,
- explained the concepts of effective area and backscatter cross section, and
- derived the Friis transmission formula and the radar equation.

PROBLEMS

P.10-1 Determine the maximum electric and magnetic field intensities at a distance of 10 (km) from a Hertzian dipole that has an input power of 15 (kW) and radiates at 70% efficiency.

P.10-2 The radiation intensity of an antenna is given as

$$U(\theta, \phi) = \begin{cases} 50 \sin^2 \theta \cos \phi; & 0 \le \theta \le \pi, \ -\pi/2 \le \phi \le \pi/2, \\ 0; & \text{elsewhere.} \end{cases}$$

Find (a) the directivity, and (b) the radiation resistance of the antenna if the magnitude of the input current is 2(A) and losses are negligible.

P.10-3 (a) Assume the spatial distribution of the current on a very thin center-fed half-wave dipole lying along the z-axis to be $I_0 \cos 2\pi z$. Find the charge distribution on the dipole. What is the wavelength? (b) Repeat part (a), assuming the current distribution along the dipole to be a triangular function described by

$$I(z) = I_0(1 - 4|z|).$$

P.10-4 A 1-(MHz) uniform current flows in a vertical antenna of length 15 (m). The antenna is a center-fed copper rod having a radius of 2 (cm). Find:

a) the radiation resistance,

b) the radiation efficiency, and

c) the maximum electric field intensity at a distance of 20 (km) if the radiated power of the antenna is 1.6 (kW).

P.10-5 Determine the radiation efficiency of a center-fed dipole of length 1.5 (m) operating at 100 (MHz). The dipole is made of brass and has a radius of 1 (mm).

P.10-6 The amplitude of the time-harmonic current distribution on a center-fed short dipole antenna of length $2h(h \ll \lambda)$ can be approximated by a triangular function

$$I(z) = I_0\left(1 - \frac{|z|}{h}\right).$$

Find (a) the far-zone electric and magnetic field intensities, (b) the radiation resistance, and (c) the directivity.

P.10-7 The transmitting antenna of a radio navigation system is a vertical metal mast 40 (m) in height insulated from the earth. A 180-(kHz) source sends a current having an amplitude of 100 (A) into the base of the mast. Assuming the current amplitude in the antenna to decrease linearly toward zero at the top of the mast and the earth to be a perfectly conducting plane, determine:

a) the maximum field intensity at a distance 160 (km) from the antenna,

b) the time-average radiated power, and

c) the radiation resistance.

P.10-8 (a) Verify the E-plane polar radiation patterns in Figs. 10-4(c) and 10-4(d) for center-fed dipole antennas with $2h/\lambda = 3/2$ and $2h/\lambda = 2$, respectively. (b) Plot these patterns in rectangular form with $F(\theta)$ versus θ. (c) Estimate the angles, θ_0, at which the patterns show a maximum.

P.10-9 The angle between the half-power points of the main beam of the radiation pattern of an antenna is often called the *beamwidth* of the pattern. (Half-power points are points at which the field strength is $1/\sqrt{2}$ of that in the direction of maximum radiation.) Find the beamwidth of the E-plane pattern of (a) a Hertzian dipole, and (b) a half-wave dipole.

P.10-10 Sketch the polar radiation pattern versus θ for a thin dipole antenna of total length $2h = 1.25\lambda$. Determine the width of the main beam between the first nulls.

P.10-11 Two elemental dipole antennas, each of length $2h$ ($h \ll \lambda$), are aligned colinearly along the z-axis with their centers spaced a distance d ($d > 2h$) apart. The excitations in the two antennas are of equal amplitude and equal phase.

a) Write the general expression for the far-zone electric field of this two-element colinear array.

b) Plot the normalized E-plane pattern for $d = \lambda/2$.

c) Repeat part (b) for $d = \lambda$.

P.10-12 Plot the H-plane polar radiation pattern of two parallel dipoles for

a) $d = \lambda/4$, $\xi = \pi/2$; **b)** $d = 3\lambda/4$, $\xi = \pi/2$.

P.10-13 For a five-element broadside binomial array:

a) Determine the relative excitation amplitudes in the array elements.

b) Plot the array factor for $d = \lambda/2$.

c) Determine the half-power beamwidth and compare it with that of a five-element uniform array having the same element spacings.

P.10-14 Find the array factor and plot the normalized radiation pattern of a broadside array of five isotropic elements spaced $\lambda/2$ apart and having excitation amplitude ratios 1:2:3:2:1. Compare the first sidelobe level with that of a five-element uniform array.

P.10-15 Obtain the pattern function of a uniformly excited rectangular array of $N_1 \times N_2$ parallel half-wave dipoles. Assume that the dipoles are parallel to the z-axis and their centers are spaced d_1 and d_2 apart in the x- and y-directions, respectively.

P.10-16 In dealing with thin linear antennas, it is sometimes convenient to define an effective length, ℓ_e, of the antenna, which is the current moment normalized with respect to the current at the feed point. For a center-fed dipole of half-length h, the maximum effective length (at $\theta = \pi/2$) is

$$\ell_e(\pi/2) = \frac{1}{I(0)} \int_{-h}^{+h} I(z)\,dz. \tag{10-85}$$

Determine the effective length of

a) a Hertzian dipole of length $d\ell$,

b) a half-wave dipole with sinusoidal current distribution $I_0 \cos \beta z$, and

c) a half-wave dipole with triangular current distribution $I_0(1 - 4|z|/\lambda)$.

P.10-17 When an antenna having an effective length ℓ_e as defined in Eq. (10-85) is used for receiving an incident electric field E_i parallel to the dipole, the product $|E_i \ell_e|$ equals the induced open-circuit voltage $|V_{oc}|$ in the receiving circuit. Assume that a half-wave dipole radiates 2 (kW) at 300 (MHz) and that a second half-wave dipole parallel to and 150 (m) away from the first is used as a receiving antenna. Neglecting losses, find (a) $|V_{oc}|$ in the equivalent receiving circuit, and (b) the power received in a matched load.

P.10-18 (a) Two parallel half-wave dipoles are 150 (m) apart. The transmitting dipole radiates 2 (kW) at 300 (MHz). Use Eq. (10-80) to find the power received at the receiving dipole. (b) Repeat part (a) assuming that both antennas are Hertzian dipoles.

P.10-19 Given a symmetrical dipole antenna of a half-length $\lambda/4$:

a) obtain an expression for effective area, $A_e(\theta)$,

b) calculate the maximum value of A_e for 100 (MHz), and

c) calculate the maximum value of A_e for 200 (MHz). Why is this answer smaller than that obtained in part (b)?

P.10-20 The antenna of a 120-(kW) monostatic radar operating at 3 (GHz) has a directive gain of 20 (dB). Suppose that it tracks a target 8 (km) away and that the backscatter cross section of the target is 15 (m^2). Determine

a) the magnitude of the electric intensity at the target,

b) the amount of power intercepted by the target, and

c) the amount of the reflected power absorbed by the antenna at the radar.

P.10-21 (a) Show that three satellites equally spaced around the geosynchronous orbit in the equatorial plane would cover almost the entire earth's surface. Explain why the polar regions are not covered. (b) Assuming the main beam of the radiation pattern of the satellite antenna to have the shape of a circular cone that just covers the earth with no spillover, find a relation between the main-lobe beamwidth and the directive gain of the antenna.

P.10-22 The antenna at the earth station of a satellite communication link having a gain of 55 (dB) at 14 (GHz) is aimed at a geostationary satellite 36,500 (km) away. Assume that the antenna on the satellite has a gain of 35 (dB) in transmitting the signal back toward the earth station at 12 (GHz). The minimum usable signal is 8 (pW).

a) Neglecting antenna ohmic and mismatch losses, find the minimum satellite transmitting power required.

b) Find the peak transmitting pulse power needed at the earth station in order to detect the satellite as a passive object, assuming the backscatter cross section of the satellite including its solar panels as 25 (m^2) and the minimum detectable return pulse power to be 0.5 (pW).

Appendix A

SYMBOLS AND UNITS

Quantity	Symbol	Unit	Abbreviation
Length	ℓ	meter	m
Mass	m	kilogram	kg
Time	t	second	s
Current	I, i	ampere	A

†Besides the MKSA system for the units of length, mass, time, and current, the SI adopted by the International Committee on Weights and Measures consists of two other fundamental units. They are Kelvin degree (K) for thermodynamic temperature and candela (cd) for luminous intensity.

A-2 DERIVED QUANTITIES

Quantity	Symbol	Unit	Abbreviation
Admittance	Y	siemens	S
Angular frequency	ω	radian/second	rad/s
Attenuation constant	α	neper/meter	Np/m
Capacitance	C	farad	F
Charge	Q, q	coulomb	C
Charge density (linear)	ρ_ℓ	coulomb/meter	C/m
Charge density (surface)	ρ_s	coulomb/meter^2	C/m^2
Charge density (volume)	ρ_v	coulomb/meter^3	C/m^3
Conductance	G	siemens	S
Conductivity	σ	siemens/meter	S/m
Current density (surface)	$\mathbf{J}_s$	ampere/meter	A/m
Current density (volume)	$\mathbf{J}$	ampere/meter^2	A/m^2
Dielectric constant (relative permittivity)	ϵ_r	(dimensionless)	—
Directivity	D	(dimensionless)	—
Electric dipole moment	$\mathbf{p}$	coulomb-meter	C · m
Electric displacement (Electric flux density)	$\mathbf{D}$	coulomb/meter^2	C/m^2
Electric field intensity	$\mathbf{E}$	volt/meter	V/m
Electric potential	V	volt	V
Electric susceptibility	χ_e	(dimensionless)	—
Electromotive force	$\mathcal{V}$	volt	V
Energy (work)	W	joule	J
Energy density	w	joule/meter^3	J/m^3
Force	$\mathbf{F}$	newton	N
Frequency	f	hertz	Hz
Impedance	Z, η	ohm	Ω
Inductance	L	henry	H
Magnetic dipole moment	$\mathbf{m}$	ampere-meter^2	A · m^2
Magnetic field intensity	$\mathbf{H}$	ampere/meter	A/m

Quantity	Symbol	Unit	Abbreviation
Magnetic flux	Φ	weber	Wb
Magnetic flux density	$\mathbf{B}$	tesla	T
Magnetic potential (vector)	$\mathbf{A}$	weber/meter	Wb/m
Magnetic susceptibility	χ_m	(dimensionless)	—
Magnetization	$\mathbf{M}$	ampere/meter	A/m
Magnetomotive force	$\mathscr{V}_m$	ampere	A
Permeability	μ, μ_0	henry/meter	H/m
Permittivity	ϵ, ϵ_0	farad/meter	F/m
Phase	ϕ	radian	rad
Phase constant	β	radian/meter	rad/m
Polarization vector	$\mathbf{P}$	coulomb/meter^2	C/m^2
Power	P	watt	W
Poynting vector (power density)	$\mathscr{P}$	watt/meter^2	W/m^2
Propagation constant	γ	meter^{-1}	m^{-1}
Radiation intensity	U	watt/steradian	W/sr
Reactance	X	ohm	Ω
Relative permeability	μ_r	(dimensionless)	—
Relative permittivity (dielectric constant)	ϵ_r	(dimensionless)	—
Reluctance	$\mathscr{R}$	henry^{-1}	H^{-1}
Resistance	R	ohm	Ω
Susceptance	B	siemens	S
Torque	T	newton-meter	N·m
Velocity	u	meter/second	m/s
Voltage	V	volt	V
Wavelength	λ	meter	m
Wavenumber	k	radian/meter	rad/m
Work (energy)	W	joule	J

A-3 Multiples and Submultiples of Units

Factor by Which Unit Is Multiplied	Prefix	Symbol
1 000 000 000 000 000 000 = 10^{18}	exa	E
1 000 000 000 000 000 = 10^{15}	peta	P
1 000 000 000 000 = 10^{12}	tera	T
1 000 000 000 = 10^{9}	giga	G
1 000 000 = 10^{6}	mega	M
1 000 = 10^{3}	kilo	k
100 = 10^{2}	hecto†	h
10 = 10^{1}	deka†	da
0.1 = 10^{-1}	deci†	d
0.01 = 10^{-2}	centi†	c
0.001 = 10^{-3}	milli	m
0.000 001 = 10^{-6}	micro	μ
0.000 000 001 = 10^{-9}	nano	n
0.000 000 000 001 = 10^{-12}	pico	p
0.000 000 000 000 001 = 10^{-15}	femto	f
0.000 000 000 000 000 001 = 10^{-18}	atto	a

†These prefixes are generally not used except for measurements of length, area, and volume.

Appendix B

SOME USEFUL MATERIAL CONSTANTS

B-1 CONSTANTS OF FREE SPACE

Constant	Symbol	Value
Velocity of light	c	$\sim 3 \times 10^8$ (m/s)
Permittivity	ϵ_0	$\sim \frac{1}{36\pi} \times 10^{-9}$ (F/m)
Permeability	μ_0	$4\pi \times 10^{-7}$ (H/m)
Intrinsic impedance	η_0	$\sim 120\pi$ or 377 (Ω)

B-2 PHYSICAL CONSTANTS OF ELECTRON AND PROTON

Constant	Symbol	Value
Rest mass of electron	m_e	9.107×10^{-31} (kg)
Charge of electron	$-e$	-1.602×10^{-19} (C)
Charge-to-mass ratio of electron	$-e/m_e$	-1.759×10^{11} (C/kg)
Radius of electron	R_e	2.81×10^{-15} (m)
Rest mass of proton	m_p	1.673×10^{-27} (kg)

B-3 RELATIVE PERMITTIVITIES (DIELECTRIC CONSTANTS)†

Material	Relative Permittivity, ϵ_r
Air	1.0
Bakelite	5.0
Glass	4–10
Mica	6.0
Oil	2.3
Paper	2–4
Paraffin wax	2.2
Plexiglass	3.4
Polyethylene	2.3
Polystyrene	2.6
Porcelain	5.7
Rubber	2.3–4.0
Soil (dry)	3–4
Teflon	2.1
Water (distilled)	80
Seawater	72

B-4 CONDUCTIVITIES†

Material	Conductivity, σ(S/m)	Material	Conductivity, σ(S/m)
Silver	6.17×10^7	Fresh water	10^{-3}
Copper	5.80×10^7	Distilled water	2×10^{-4}
Gold	4.10×10^7	Dry soil	10^{-5}
Aluminum	3.54×10^7	Transformer oil	10^{-11}
Brass	1.57×10^7	Glass	10^{-12}
Bronze	10^7	Porcelain	2×10^{-13}
Iron	10^7	Rubber	10^{-15}
Seawater	4	Fused quartz	10^{-17}

† Note that the constitutive parameters of some of the materials are frequency and temperature dependent. The listed constants are average low-frequency values at room temperature.

B-5 RELATIVE PERMEABILITIES†

Material	Relative Permeability, μ_r
Ferromagnetic (nonlinear)	
Nickel	250
Cobalt	600
Iron (pure)	4,000
Mumetal	100,000
Paramagnetic	
Aluminum	1.000021
Magnesium	1.000012
Palladium	1.00082
Titanium	1.00018
Diamagnetic	
Bismuth	0.99983
Gold	0.99996
Silver	0.99998
Copper	0.99999

†Note that the constitutive parameters of some of the materials are frequency and temperature dependent. The listed constants are average low-frequency values at room temperature.

Bibliography

The following books on electromagnetic fields and waves at a comparable level have been found useful as references. They are listed alphabetically by the names of the first authors.

Bewley, L. V., *Two Dimensional Fields in Electrical Engineering*, Dover Publications, New York, 1963.

Cheng, D. K., *Field and Wave Electromagnetics*, 2nd ed., Addison-Wesley, Reading, Mass., 1989.

Collin, R. E., *Antennas and Radiowave Propagation*, McGraw-Hill, New York, 1985.

Crowley, J. M., *Fundamentals of Applied Electrostatics*, Wiley, New York, 1986.

Feynman, R. P.; Leighton, R. O.; and Sands, M., *Lectures on Physics*, vol. 2, Addison-Wesley, Reading, Mass., 1964.

Javid, M., and Brown, P. M., *Field Analysis and Electromagnetics*, McGraw-Hill, New York, 1963.

Jordan, E. C., and Balmain, K. G., *Electromagnetic Waves and Radiating Systems*, 2nd ed., Prentice-Hall, Englewood Cliffs, N.J., 1968.

Kraus, J. D., *Electromagnetics*, 4th ed., McGraw-Hill, New York, 1992.

Lorrain, P., and Corson, D., *Electromagnetic Fields and Waves*, 2nd ed., Freeman, San Franscisco, Calif., 1970.

Neff, H. P., Jr., *Introductory Electromagnetics*, Wiley, New York, 1991.

Paris, D. T., and Hurd, F. K., *Basic Electromagnetic Theory*, McGraw-Hill, New York, 1969.

Parton, J. E.; Owen, S. J. T.; and Raven, M. S., *Applied Electromagnetics*, 2nd ed., Macmillan, London, 1986.

Paul, C. R., and Nasar, S. A., *Introduction to Electromagnetic Fields*, McGraw-Hill, New York, 1987.

Plonsey, R., and Collin, R. E., *Principles and Applications of Electromagnetic Fields*, 2nd ed., McGraw-Hill, New York, 1982.

Plonus, M. A., *Applied Electromagnetics*, McGraw-Hill, New York, 1978.

Popović, B. D., *Introductory Engineering Electromagnetics*, Addison-Wesley, Reading, Mass., 1971.

Pozar, D. M., *Microwave Engineering*, Addison-Wesley, Reading, Mass., 1990.

Ramo, S.; Whinnery, J. R.; and Van Duzer, T., *Fields and Waves in Communication Electronics*, 2nd ed., Wiley, New York, 1984.

Sander, K. F., and Reed, G. A. L., *Transmission and Propagation of Electromagnetic Waves*, 2nd ed., Cambridge University Press, Cambridge, England, 1986.

Seshadri, S. R., *Fundamentals of Transmission Lines and Electromagnetic Fields*, Addison-Wesley, Reading, Mass., 1971.

Shen, L. C., and Kong, J. A., *Applied Electromagnetism*, 2nd ed., PWS Engineering, Boston, Mass., 1987.

Zahn, M., *Electromagnetic Field Theory*, Wiley, New York, 1979.

Answers to Odd-Numbered Problems

CHAPTER 2

P.2-3 **a)** $(\mathbf{a}_x 2 - \mathbf{a}_y 3 + \mathbf{a}_z 6)/7$. **b)** 17.1. **c)** -1.71. **d)** -24. **e)** -3.43. **f)** $104.2°$. **g)** $-\mathbf{a}_x 4 - \mathbf{a}_y 3 - \mathbf{a}_z 10$. **h)** -118.

P.2-5 **a)** Right angle at P_1. **b)** 15.3.

P.2-7 **a)** $(\mathbf{a}_x 5 - \mathbf{a}_y 2 + \mathbf{a}_z)/\sqrt{30}$. **b)** $(\mathbf{a}_x 2 + \mathbf{a}_y 5)/\sqrt{29}$.

P.2-11 **a)** $(-3/2, -3\sqrt{3}/2, -4)$. **b)** $(5, 143.1°, 240°)$.

P.2-13 **a)** $A_x \cos\phi_1 + A_y \sin\phi_1$. **b)** $A_R(r_1/\sqrt{r_1^2 + z_1^2}) + A_\theta(z_1/\sqrt{r_1^2 + z_1^2})$.

P.2-15 **a)** $\mathbf{a}_R 2$; $-4/3$. **b)** $112.4°$.

P.2-17 **a)** $-(\mathbf{a}_x x + \mathbf{a}_y y + \mathbf{a}_z z)/R^3$. **b)** $-\mathbf{a}_R(1/R^2)$.

P.2-19 **a)** $\mathbf{a}_\phi$; $-\mathbf{a}_r$.

P.2-21 **a)** 3/2. **b)** $y + z + x$.

P.2-23 **a)** $2\pi R^3/3$. **b)** 1.

P.2-27 **a)** 1/2. **b)** $\mathbf{a}_z(3r - 5)\cos\phi$. **c)** 1/2.

CHAPTER 3

P.3-1 **a)** $\dfrac{m}{e}\left(\dfrac{u_0 h}{w}\right)^2$. **b)** $\dfrac{1}{2}\left(w + \dfrac{m u_0^2 D h}{e w V_{\max}}\right)$.

P.3-3 **a)** $Q_1/Q_2 = 4/3$. **b)** $Q_1/Q_2 = 3/4$.

P.3-5 $z = 8.66b$.

P.3-7 Assuming the semicircular line charge around the origin to lie in the upper half of xy-plane, $\mathbf{E} = -\mathbf{a}_y \rho_l/2\pi\epsilon_0 b$.

P.3-9 **a)** $E_r = 0$, $r < a$; $E_r = a\rho_{sa}/\epsilon_0 r$, $a < r < b$; $E_r = (a\rho_{sa} + b\rho_{sb})/\epsilon_0 r$, $r > b$.
b) $b/a = -\rho_{sa}/\rho_{sb}$.

P.3-11 **a)** $-30\,(\mu\text{J})$. **b)** $-60\,(\mu\text{J})$.

P.3-13 **a)** $\rho_{ps} = P_0 L/2$ on all six faces; $\rho_{pv} = -3P_0$.

P.3-15 **a)** $\rho_{ps} = P_0 r_o(3 + \sin^2\phi)$, $r = r_o$; $\rho_{ps} = -P_0 r_i(3 + \sin^2\phi)$, $r = r_i$; $\rho_{pv} = -7P_0$.

P.3-17 **a)** $150\,(k\text{V})$. **b)** $1{,}000\,(k\text{V})$. **c)** $130\,(k\text{V})$.

P.3-19 $\epsilon_{r2} = 1.667$.

P.3-21 **a)** $\mathbf{a}_r V_0/a \ln(b/a)$.
b) $a = b/e = b/2.718$.
c) eV_0/b.
d) $2\pi\epsilon(\text{F/m})$.

P.3-23 $4\pi\epsilon \bigg/ \left(\dfrac{1}{R_i} - \dfrac{1}{R_o}\right)$.

P.3-25 **a)** $27\,(\text{nJ})$. **b)** $27\,(\text{nJ})$.

P.3-27 $\mathbf{a}_x(\epsilon - \epsilon_0)V_0^2 w/2d$.

P.3-29 $V = c_1 \ln r + c_2 - Ar/\epsilon$; $c_1 = [A(b-a)/\epsilon - V_0]/\ln(b/a)$,
$c_2 = [V_0 \ln b + A(a \ln b - b \ln a)/\epsilon]/\ln(b/a)$.

P.3-31 **a)** $V(\theta) = V_0 \dfrac{\ln\left(\tan\dfrac{\theta}{2}\right)}{\ln\left(\tan\dfrac{\alpha}{2}\right)}$.

b) $\mathbf{E}(\theta) = -\mathbf{a}_\theta \dfrac{V_0}{R \ln[\tan(\alpha/2)] \sin\theta}$.

P.3-35 **a)** $d_i = 0.46\,(\text{mm})$. **b)** $2.96\,(\text{nF/m})$. **c)** $|\mathbf{E}| = 111.9\,(\text{V/m})$.

CHAPTER 4

P.4-1 **a)** $3.54 \times 10^7(\text{S/m})$.
b) $6 \times 10^{-3}(\text{V/m})$.
c) $1(\text{W})$.
d) $8.4 \times 10^{-6}(\text{m/s})$.

P.4-3 **a)** $\mathbf{a}_R\ 7.5 \times 10^9 Re^{-9.42 \times 10^{11}t}(\text{V/m})$, $R < b$; $\mathbf{a}_R(9/R^2) \times 10^6(\text{V/m})$, $R > b$.
b) $\mathbf{a}_R\ 7.5 \times 10^{10} Re^{-9.42 \times 10^{11}t}(\text{A/m}^2)$, $R < b$; 0, $R > b$.

P.4-5 $P_{R_1} = 3.33(\text{mW})$, $P_{R_2} = 8.00(\text{mW})$, $P_{R_3} = 5.31(\text{mW})$,
$P_{R_4} = 8.87(\text{mW})$, $P_{R_5} = 44.5(\text{mW})$. Total resistance $= 7(\Omega)$.

P.4-7 **a)** $\dfrac{d}{(\sigma_2 - \sigma_1)S} \ln\left(\dfrac{\sigma_2}{\sigma_1}\right)$.

b) $\dfrac{\epsilon_0(\sigma_2 - \sigma_1)V_0}{\sigma_2 d \ln(\sigma_2/\sigma_1)}$, $y = d$.

P.4-9 **a)** $C_i = \dfrac{2\pi\epsilon_1 L}{\ln(c/a)}$, $G_i = \dfrac{2\pi\sigma_1 L}{\ln(c/a)}$,

$C_o = \dfrac{2\pi\epsilon_2 L}{\ln(b/c)}$, $G_o = \dfrac{2\pi\sigma_2 L}{\ln(b/c)}$.

b) $J_i = J_o = \dfrac{\sigma_1\sigma_2 V_0}{r[\sigma_1 \ln(b/c) + \sigma_2 \ln(c/a)]}$

P.4-11 $R = \dfrac{2}{\pi\sigma h}\ln(b/a)$.

CHAPTER 5

P.5-1 $\mathbf{E} = u_0(\mathbf{a}_y B_z - \mathbf{a}_z B_y)$.

P.5-5 $\mathbf{B}_{P_2} = -\mathbf{a}_z \dfrac{\mu_0 I}{2\pi w}\ln\left(1 + \dfrac{w}{d_2}\right)$.

P.5-7 Assuming the current I to flow in the counterclockwise direction in a triangle lying in the xy-plane, $\mathbf{B} = \mathbf{a}_z \dfrac{9\mu_0 I}{2\pi w}$.

P.5-9 **a)** $\mathbf{A} = \mathbf{a}_z \dfrac{\mu_0 I}{2\pi}\ln\left(\dfrac{r_o}{r}\right)$.

b) $2.34(\mu Wb)$.

P.5-11 **a)** $\mathbf{a}_z \mu_0 H_0/\mu$.

b) $\mathbf{a}_z(H_0 - M_i)$.

P.5-13 **a)** $\mathbf{J}_{mv} = 0$, $\mathbf{J}_{ms} = \mathbf{a}_\phi M_0 \sin\theta$.

b) $(2/3)\mu_0 \mathbf{M}_0$.

P.5-15 $L = \mu_0 N^2(r_o - \sqrt{r_o^2 - b^2})$, $L \cong \mu_0 N^2 b^2/2r_o$.

P.5-17 $L_{12} = \dfrac{\mu_0 h_2}{2\pi}\ln\dfrac{(w_1 + d)(w_2 + d)}{d(w_1 + w_2 + d)}$.

P.5-19 $\mathbf{f} = \mathbf{a}_x \dfrac{\mu_0 I^2}{\pi w}\tan^{-1}\left(\dfrac{w}{2D}\right)$.

P.5-21 $\mathbf{T} = -\mathbf{a}_x 0.1(\mathrm{N\cdot m})$.

CHAPTER 6

P.6-1 $\mathscr{V} = -\oint_C \dfrac{\partial \mathbf{A}}{\partial t}\cdot d\boldsymbol{\ell}$.

P.6-3 $i_2(t) = -\dfrac{\omega\mu_0 I_1 h}{2\pi\sqrt{R^2 + \omega^2 L^2}}\ln\left(1 + \dfrac{w}{d}\right)\sin\left(\omega t + \tan^{-1}\dfrac{R}{\omega L}\right)$.

P.6-5 **a)** $i = 0.251 \sin 100\pi t\,(\mathrm{A})$.

b) $i = 0.104 \sin(100\pi t - 65.6°)\,(\mathrm{A})$.

P.6-7 **a)** 1 (GHz). **b)** 7.2 (MHz).

P.6-9 **a)** $\mathbf{H}_2 = \mathbf{a}_x 30 + \mathbf{a}_y 45 + \mathbf{a}_z 10\,(\mathrm{A/m})$.

b) $\mathbf{B}_2 = 2\mu_0 \mathbf{H}_2$. **c)** $\alpha_1 = 68.2°$. **d)** $\alpha_2 = 79.5°$.

P.6-15 $\alpha = \pi/6$, $H_0 = 1.73 \times 10^{-4}\,(\mathrm{A/m})$.

P.6-17 **a)** $\mathbf{A} = \mathbf{a}_R A_R + \mathbf{a}_\theta A_\theta + \mathbf{a}_\phi A_\phi$, where

$$A_R = A_z \cos\theta = \frac{\mu_0 I\, d\ell}{4\pi}\left(\frac{e^{-j\beta R}}{R}\right)\cos\theta,$$

$$A_\theta = -A_z \sin\theta = -\frac{\mu_0 I\, d\ell}{4\pi}\left(\frac{e^{-j\beta R}}{R}\right)\sin\theta,$$

$$A_\phi = 0.$$

b) $\mathbf{H} = -\mathbf{a}_\phi \dfrac{I\, d\ell}{4\pi}\beta^2 \sin\theta\left[\dfrac{1}{j\beta R} + \dfrac{1}{(j\beta R)^2}\right]e^{-j\beta R}.$

P.6-19 $k = 20\pi/3$ (rad/m).

$$\mathbf{H}(R, \theta; t) = \mathbf{a}_\phi \frac{10^{-3}}{120\pi R}\sin\theta\cos(2\pi 10^9 t - 20\pi R/3)\ \text{(A/m)}.$$

P.6-21 $\beta = 13.2\pi = 41.6$ (rad/m).

$$\mathbf{E}(x, z; t) = \mathbf{a}_x 496\cos(15\pi x)\sin(6\pi 10^9 t - 41.6z) + \mathbf{a}_z 565\sin(15\pi x)\cos(6\pi 10^9 t - 41.6z).$$

CHAPTER 7

P.7-1 **a)** $\nabla^2\mathbf{E} - \mu\sigma\dfrac{\partial E}{\partial t} - \mu\epsilon\dfrac{\partial^2 E}{\partial t^2} = 0.$

b) $\nabla^2\mathbf{E} - j\omega\mu\sigma\mathbf{E} + k^2\mathbf{E} = 0.$

P.7-3 $\mathbf{E}(\mathbf{R}) = -\eta\mathbf{a}_k \times \mathbf{H}(\mathbf{R}).$

P.7-5 **a)** $f = 1.59 \times 10^7$(Hz), $\lambda = 10.88$ (m).

b) $\epsilon_r = 3.$

c) Left-hand elliptically polarized.

d) $\mathbf{H}(z, t) = \dfrac{\sqrt{3}}{120\pi}\left[\mathbf{a}_z \sin(10^8 t - z/\sqrt{3}) + \mathbf{a}_y 2\cos(10^8 t - z/\sqrt{3})\right]$ (A/m).

P.7-7 **a)** 0.279 (m).

b) $\eta_c = 238\underline{/1.43^\circ}$, $\lambda = 0.063$ (m), $u_p = 1.897 \times 10^8$(m/s), $u_g = 1.898 \times 10^8$ (m/s).

c) $\mathbf{H}(x, t) = \mathbf{a}_z 0.21 e^{-2.48x}\sin(6\pi \times 10^9 t - 31.6\pi x + 0.325\pi)$ (A/m).

P.7-9 **a)** $\sigma = 9.9 \times 10^4$ (S/m). **b)** 0.175 (mm).

P.7-11 For $\mathbf{E}(z, t) = \mathbf{a}_x E_0\cos(\omega t - kz + \phi) + \mathbf{a}_y E_0 \sin(\omega t - kz + \phi)$, $\mathscr{P} = \mathbf{a}_z E_0^2/\eta$, which is independent of t and z.

P.7-15 **a)** $\delta = 6.3$ (cm), $\eta_c = 3.96 + j3.96\ (\Omega)$.

b) $\mathbf{E}(z, t) = \mathbf{a}_x 1.68 e^{-15.85z}\cos(10^8 t - 15.85z + 0.25\pi)$ (V/m), $\mathbf{H}(z, t) = \mathbf{a}_y 0.3 e^{-15.85z}\cos(10^8 t - 15.85z)$ (A/m).

c) $\mathscr{P}_{av} = \mathbf{a}_z 0.178 e^{-31.9z}$ (W/m^2).

P.7-17 **a)** $\mathbf{E}_r(z) = E_0(-\mathbf{a}_x + j\mathbf{a}_y)e^{j\beta z}$, a left-hand circularly polarized wave in $-z$ direction.

b) $\mathbf{J}_s = \dfrac{2E_0}{\eta_0}(\mathbf{a}_x - j\mathbf{a}_y).$

c) $\mathbf{E}_1(z, t) = 2E_0 \sin\beta z\,(\mathbf{a}_x \sin\omega t - \mathbf{a}_y \cos\omega t).$

P.7-19 **a)** $\mathbf{E}_r(z) = \mathbf{a}_x 2.08 e^{j(6z+159.7°)}$ (V/m),

$\mathbf{H}_r(z) = -\mathbf{a}_y 0.0055 e^{j(6z+159.7°)}$ (A/m),

$\mathbf{E}_t(z) = \mathbf{a}_x 8.08 e^{-1.35z} e^{-j(9.10x-5.1°)}$ (V/m),

$\mathbf{H}_t(z) = \mathbf{a}_y 0.032 e^{-1.35z} e^{-j(9.10x+3.4°)}$ (A/m).

b) $S = 1.53$.

c) $(\mathscr{P}_{av})_1 = \mathbf{a}_z 0.127 (\text{W/m}^2)$, $(\mathscr{P}_{av})_2 = \mathbf{a}_z 0.127 e^{-2.70z} (\text{W/m}^2)$.

P.7-21 **a)** $\Gamma = -0.241$, $\tau = 0.759$.

b) $\mathbf{E}_t(x, z; t) = \mathbf{a}_y 15.2 \cos(2\pi 10^8 t - 1.05x - 2.96z)$ (V/m),

$\mathbf{H}_t(x, z; t) = 0.06(-\mathbf{a}_x 0.943 + \mathbf{a}_z 0.333) \cos(2\pi 10^8 t - 1.05x - 2.96z)$ (A/m).

P.7-25 **a)** $\theta_t = 0.03°$. **b)** $\Gamma_{\|} \cong 0.0151(1+j)$. **c)** 8.69 (m).

P.7-27 **a)** $\mathbf{E}_t(x, z) = \mathbf{a}_y E_{t0} e^{-\alpha_2 z} e^{-j\beta_{2x} x}$,

$$\mathbf{H}_t(x, z) = \frac{E_{t0}}{\eta_2}\left(\mathbf{a}_x j\alpha_2 + \mathbf{a}_z \sqrt{\frac{\epsilon_1}{\epsilon_2}} \sin\theta_i\right) e^{-\alpha_2 z} e^{-j\beta_{2x} x},$$

where $\beta_{2x} = \beta_2 \sqrt{\dfrac{\epsilon_1}{\epsilon_2}} \sin\theta_i$, $\alpha_2 = \beta_2 \sqrt{\left(\dfrac{\epsilon_1}{\epsilon_2}\right)\sin^2\theta_i - 1}$,

and $$E_{t0} = \frac{2\eta_2 \cos\theta_i E_{i0}}{\eta_2 \cos\theta_i - j\eta_1 \sqrt{\left(\dfrac{\epsilon_1}{\epsilon_2}\right)\sin^2\theta_i - 1}}.$$

P.7-29 **a)** 6.38°. **b)** $e^{j0.66}$. **c)** $1.89 e^{j0.33}$. **d)** 159 (dB).

P.7-31 **a)** $\theta_a = \sin^{-1}\left(\dfrac{1}{n_0}\sqrt{n_1^2 - n_2^2}\right)$. **b)** 80.4°.

P.7-33 $\sin\theta_c = \tan\theta_{B\|}$.

CHAPTER 8

P.8-1 **a)** $d' = \sqrt{2}d$. **b)** $w' = w/\sqrt{2}$. **c)** $w' = 2w$.

d) $u_p' = u_p/\sqrt{2}$ for case **a** and **b**; $u_p' = u_p$ for case **c**.

P.8-3 **a)** 2.55 (cm). **b)** 3.91 (mm).

P.8-5 $R = 0.058\,(\Omega/\text{m})$, $L = 0.20\,(\mu\text{H/m})$, $C = 80\,(\text{pF/m})$, $G = 23\,(\mu\text{S/m})$.

P.8-7 **b)** $V(z) = V_i \cosh\gamma z - I_i Z_0 \sinh\gamma z$,

$$I(z) = I_i \cosh\gamma z - \frac{V_i}{Z_0}\sinh\gamma z.$$

P.8-9 **a)** $V(z, t) = 5.27 e^{-0.01z} \sin(8000\pi t - 5.55z - 0.322)$ (V),

$I(z, t) = 0.105 e^{-0.01z} \sin(8000\pi t - 5.55z - 0.322)$ (A).

b) $V(50, t) = 3.20 \sin(8000\pi t - 0.432\pi)$ (V),

$I(50, t) = 0.064 \sin(8000\pi t - 0.432\pi)$ (A).

c) 0.102 (W).

P.8-11 $Z_i = 26.3 - j9.87\,(\Omega)$.

P.8-13 **a)** $R_0 = 74.5\,(\Omega)$, $\epsilon_r = 4.05$.

b) $X_{io} = -290\,(\Omega)$, $X_{is} = 19.2\,(\Omega)$.

P.8-15 **a)** $Z_0 = 50\,(\Omega)$. **b)** Min. $S = 2$.

P.8-17 $\dfrac{R_L}{R_0} = \dfrac{1}{2r_i}\left[(1+r_i^2+x_i^2) \pm \sqrt{(1+r_i^2+x_i^2)^2 - 4r_i^2}\right]$; $r_i = \dfrac{R_i}{R_0}$, $x_i = \dfrac{X_i}{R_0}$,

$l = \dfrac{\lambda}{2\pi}\tan^{-1} t$; $t = \dfrac{1}{2x_i}\left\{-[1-(r_i^2+x_i^2)] \pm \sqrt{(1-r_i^2-x_i^2)^2 + 4x_i^2}\right\}$.

P.8-19 **a)** $V_i = 0.527\underline{/18.4^\circ}$ (V), $I_i = 1.05\underline{/-18.4^\circ}$ (mA),

$V_L = 0.033\underline{/-45^\circ}$ (V), $I_L = 1.33\underline{/-45^\circ}$ (mA).

b) $S = 2$.

c) 0.022 (mW); 0.025 (mW), if $R_L = 50(\Omega)$.

P.8-21 **a)** $S = 1.77$. **b)** $\Gamma = 0.28e^{j146^\circ} = 0.28e^{j2.55}$.

c) $Z_i = 50 + j29.5\,(\Omega)$. **d)** $Y_i = 0.015 - j0.009$ (S).

e) No voltage minimum on line, but $V_L < V_i$.

P.8-23 **a)** $Z_L = 33.75 - j23.75\,(\Omega)$. **b)** $\Gamma = \frac{1}{3}e^{j252.5^\circ} = \frac{1}{3}e^{j4.41}$.

c) At 25 (cm) from the short-circuit.

P.8-25 $d_1 = 0$ and $l_1 = 0.375\lambda$; or $d_2 = 0.324\lambda$ and $l_2 = 0.125\lambda$.

P.8-27 **a)** $Z_L = 104.3 - j73.5\,(\Omega)$.

b) $d = 0.173$ (m), $l = 0.238$ (m).

CHAPTER 9

P.9-1 **b)** At $f = 1.1f_c$: $Z_{TM} = 157\,(\Omega)$, $Z_{TE} = 904\,(\Omega)$.

At $f = 2.2f_c$: $Z_{TM} = 336\,(\Omega)$, $Z_{TE} = 423\,(\Omega)$.

P.9-3 **a)** $H_z^0(y) = B_n \cos\left(\dfrac{n\pi y}{b}\right)$.

b) $(f_c)_{TE_1} = \dfrac{1}{2b\sqrt{\mu\epsilon}}$.

c) $H_z(y, z; t) = B_1 \cos\left(\dfrac{\pi y}{b}\right)\cos(\omega t - \beta_1 z)$,

$H_y(y, z; t) = -\dfrac{\beta_1 b}{\pi} B_1 \sin\left(\dfrac{\pi y}{b}\right)\sin(\omega t - \beta_1 z)$,

$E_x(y, z; t) = -\dfrac{\omega\mu b}{\pi} B_n \sin\left(\dfrac{\pi y}{b}\right)\sin(\omega t - \beta_1 z)$,

$\beta_1 = \omega\sqrt{\mu\epsilon}\sqrt{1-\left(\dfrac{f_c}{f}\right)^2}$.

P.9-5 **a)** 8.25 (GHz). **b)** $544 + j390$. **c)** 8.89 (W).

P.9-7 **a)** TE_{10}. **b)** TE_{10}, TE_{20}, TE_{01}, TE_{11}, and TM_{11}.

P.9-9 **a)** A typical design: $a = 6.5$ (cm), $b = 3.5$ (cm).

b) $u_p = 4.70 \times 10^8$ (m/s), $\lambda_g = 15.7$ (cm),

$\beta = 40.1$ (rad/m), $(Z_{TE})_{10} = 590\,(\Omega)$.

P.9-11 **a)** $E_x^0(x, y) = -\frac{j\beta_{11}}{h^2}\left(\frac{\pi}{a}\right) E_0 \cos\left(\frac{\pi x}{a}\right) \sin\left(\frac{\pi y}{b}\right),$

$$E_y^0(x, y) = -\frac{j\beta_{11}}{h^1}\left(\frac{\pi}{b}\right) E_0 \sin\left(\frac{\pi x}{a}\right) \cos\left(\frac{\pi y}{b}\right),$$

$$H_x^0(x, y) = \frac{j\omega\epsilon}{h^2}\left(\frac{\pi}{b}\right) E_0 \sin\left(\frac{\pi x}{a}\right) \cos\left(\frac{\pi y}{b}\right),$$

$$H_y^0(x, y) = -\frac{j\omega\epsilon}{h^2}\left(\frac{\pi}{a}\right) E_0 \cos\left(\frac{\pi x}{a}\right) \sin\left(\frac{\pi y}{b}\right).$$

$$\beta_{11} = \omega\sqrt{\mu\epsilon}\sqrt{1-\left(\frac{f_c}{f}\right)^2},\ h^2 = \omega_c^2\mu\epsilon,$$

$$f_c = \frac{1}{2\sqrt{\mu\epsilon}}\sqrt{\frac{1}{a^2}+\frac{1}{b^2}}.$$

b) $P_{av}(z) = \dfrac{\omega\epsilon\beta_{11}\epsilon_0^2 ab}{8\left[\left(\frac{\pi}{a}\right)^2+\left(\frac{\pi}{b}\right)^2\right]}.$

P.9-13 **a)** TE_{02} mode.

b) $(f_c)_{02} = 12\,(\text{GHz})$, $f = 18\,(\text{GHz})$, $Z_{TE} = 506\,(\Omega)$, $\lambda_g = 2.24\,(\text{cm})$.

c) $P_{av} = 280\,(\text{W})$.

P.9-15 **a)** 3.3 (GHz); 3.3 (GHz); none.

b) 5.3 (W); 10.7 (W).

P.9-17 **a)** TE_{101} mode, $f_{101} = 4.802$ (GHz).

b) $Q_{101} = 6{,}869$. $W_e = W_m = 0.0773$ (pJ).

P.9-19 **a)** 2.89 (cm). **b)** 7.34 (GHz). **c)** 12,493.

CHAPTER 10

P10-1 $E_0 = 97.2$ (mV/m), $H_0 = 0.258$ (mA/m).

P.10-3 **a)** $\rho_l = -j(I_0/c)\sin 2\pi z$, $\lambda = 1$ (m).

b) $\rho_l = \begin{cases} -j2I_0/\pi c \text{ for } z > 0, \\ +j2I_0/\pi c \text{ for } z < 0. \end{cases}$

P.10-5 $\zeta_r = 99.2\%$.

P.10-7 **a)** $|E_\theta|_{max} = 2.82$ (mV/m).

b) $P_r = 1.14$ (kW).

c) $R_r = 0.227\,(\Omega)$.

P.10-9 **a)** 90°. **b)** 78°.

P.10-11 **a)** $E_\theta = \dfrac{j120Ih}{R}\beta e^{-j\beta(R - d/2\cos\theta)}F(\theta),$

where $F(\theta) = \sin\theta\cos\left(\frac{\beta d}{2}\cos\theta\right).$

P.10-11 **a)** $E_\theta = \dfrac{j120Ih}{R}\beta e^{-j\beta(R-d/2\cos\theta)}F(\theta)$,

where $F(\theta) = \sin\theta\cos\left(\dfrac{\beta d}{2}\cos\theta\right)$.

P.10-13 **a)** 1:4:6:4:1.

b) $|A(\phi)| = \left|\cos\left(\dfrac{\pi}{2}\cos\phi\right)\right|^4$.

c) 30.28° versus 20.78°.

P.10-15 $|F(\theta, \phi)| = \dfrac{1}{N_1N_2}\left|\dfrac{\cos\left(\dfrac{\pi}{2}\cos\theta\right)}{\sin\theta}\;\dfrac{\sin\left(\dfrac{N_1\psi_x}{2}\right)\sin\left(\dfrac{N_2\psi_y}{2}\right)}{\sin\left(\dfrac{\psi_x}{2}\right)\sin\left(\dfrac{\psi_y}{2}\right)}\right|$,

where $\psi_x = \dfrac{\beta d_1}{2}\sin\theta\cos\phi$, and

$\psi_y = \dfrac{\beta d_2}{2}\sin\theta\cos\phi$.

P.10-17 **a)** $|V_{oc}| = 0.942$ (V). **b)** $P_L = 1.52$ (mW).

P.10-19 **a)** $A_e(\theta) = 0.13\lambda^2\left[\dfrac{\cos\left(\dfrac{\pi}{2}\cos\theta\right)}{\sin\theta}\right]^2$.

b) 1.17 (m^2). **c)** 0.29 (m^2).

P.10-21 **b)** Main-lobe beamwidth $= 4/\sqrt{G_D}$.

Index

Some Useful Vector Identities

$$\mathbf{A}\cdot\mathbf{B}\times\mathbf{C} = \mathbf{B}\cdot\mathbf{C}\times\mathbf{A} = \mathbf{C}\cdot\mathbf{A}\times\mathbf{B}$$

$$\mathbf{A}\times(\mathbf{B}\times\mathbf{C}) = \mathbf{B}(\mathbf{A}\cdot\mathbf{C}) - \mathbf{C}(\mathbf{A}\cdot\mathbf{B})$$

$$\nabla(\psi V) = \psi\nabla V + V\nabla\psi$$

$$\nabla\cdot(\psi\mathbf{A}) = \psi\nabla\cdot\mathbf{A} + \mathbf{A}\cdot\nabla\psi$$

$$\nabla\times(\psi\mathbf{A}) = \psi\nabla\times\mathbf{A} + \nabla\psi\times\mathbf{A}$$

$$\nabla\cdot(\mathbf{A}\times\mathbf{B}) = \mathbf{B}\cdot(\nabla\times\mathbf{A}) - \mathbf{A}\cdot(\nabla\times\mathbf{B})$$

$$\nabla\cdot\nabla V = \nabla^2 V$$

$$\nabla\times\nabla\times\mathbf{A} = \nabla(\nabla\cdot\mathbf{A}) - \nabla^2\mathbf{A}$$

$$\nabla\times\nabla V = 0$$

$$\nabla\cdot(\nabla\times\mathbf{A}) = 0$$

$$\int_V \nabla\cdot\mathbf{A}\,dv = \oint_S \mathbf{A}\cdot d\mathbf{s} \qquad \text{(Divergence theorem)}$$

$$\int_S \nabla\times\mathbf{A}\cdot d\mathbf{s} = \oint_C \mathbf{A}\cdot d\ell \qquad \text{(Stokes's theorem)}$$

Gradient, Divergence, Curl, and Laplacian Operations

Cartesian Coordinates (x, y, z)

$$\nabla V = \mathbf{a}_x\frac{\partial V}{\partial x} + \mathbf{a}_y\frac{\partial V}{\partial y} + \mathbf{a}_z\frac{\partial V}{\partial z}$$

$$\nabla\cdot\mathbf{A} = \frac{\partial A_x}{\partial x} + \frac{\partial A_y}{\partial y} + \frac{\partial A_z}{\partial z}$$

$$\nabla\times\mathbf{A} = \begin{vmatrix} \mathbf{a}_x & \mathbf{a}_y & \mathbf{a}_z \\ \dfrac{\partial}{\partial x} & \dfrac{\partial}{\partial y} & \dfrac{\partial}{\partial z} \\ A_x & A_y & A_z \end{vmatrix} = \mathbf{a}_x\left(\frac{\partial A_z}{\partial y} - \frac{\partial A_y}{\partial z}\right) + \mathbf{a}_y\left(\frac{\partial A_x}{\partial z} - \frac{\partial A_z}{\partial x}\right) + \mathbf{a}_z\left(\frac{\partial A_y}{\partial x} - \frac{\partial A_x}{\partial y}\right)$$

$$\nabla^2 V = \frac{\partial^2 V}{\partial x^2} + \frac{\partial^2 V}{\partial y^2} + \frac{\partial^2 V}{\partial z^2}$$